104.00
68 I

Mechanics of Geomaterials
Rocks, Concretes, Soils

WILEY SERIES IN
NUMERICAL METHODS IN ENGINEERING

Consulting Editors
R. H. Gallagher, *College of Engineering,*
University of Arizona
and
O. C. Zienkiewicz, *Department of Civil Engineering,*
University of College of Swansea

Rock Mechanics in Engineering Practice
Edited by K. G. Stagg and O. C. Zienkiewicz

Optimum Structural Design: Theory and Applications
Edited by R. H. Gallagher and O. C. Zienkiewicz

Finite Elements in Fluids
Vol. 1 Viscous Flow and Hydrodynamics
Vol. 2 Mathematical Foundations, Aerodynamics and Lubrication
Edited by R. H. Gallagher, J. T. Oden, C. Taylor, and O. C. Zienkiewicz

Finite Elements for Thin Shells and Curved Members
Edited by D. G. Ashwell and R. H. Gallagher

Finite Elements in Geomechanics
Edited by G. Gudehus

Numerical Methods in Offshore Engineering
Edited by O. C. Zienkiewicz, R. W. Lewis, and K. G. Stagg

Finite Elements in Fluids
Vol. 3
Edited by R. H. Gallagher, O. C. Zienkiewicz, J. T. Oden, M. Morandi Cecchi, and C. Taylor

Energy Methods in Finite Element Analysis
Edited by R. Glowinski, E. Rodin, and O. C. Zienkiewicz

Finite Elements in Electrical and Magnetic Field Problems
Edited by M. V. K. Chari and P. Silvester

Numerical Methods in Heat Transfer
Edited by R. W. Lewis, K. Morgan, and O. C. Zienkiewicz

Finite Elements in Biomechanics
Edited by R. H. Gallagher, B. R. Simon, P. C. Johnson, and J. F. Gross

Soil Mechanics—Transient and Cyclic Loads
Edited by G. N. Pande and O. C. Zienkiewicz

Finite Elements in Fluids
Vol. 4
Edited by R. H. Gallagher, D. Norrie, J. T. Oden, and O. C. Zienkiewicz

Foundations of Structural Optimization: A Unified Approach
Edited by A. J. Morris

Creep and Shrinkage in Concrete Structures
Edited by Z. Bažant and F. Wittmann

Hybrid and Mixed Finite Element Methods
Edited by S. N. Atluri, R. H. Gallagher and O. C. Zienkiewicz

Numerical Methods in Heat Transfer Vol. II
Edited by R. W. Lewis, K. Morgan, and B. A. Shrefler

Numerical Methods in Coupled Systems
Edited by R. W. Lewis, E. Hinton, and P. Bettess

Optimum Structural Design
Edited by E. Atrek, R. H. Gallagher, K. M. Ragsdell, and O. C. Zienkiewicz

Finite Elements in Fluids
Vol. 5
Edited by R. H. Gallagher, J. T. Oden, O. C. Zienkiewicz, T. Kawai and M. Kawahara

Mechanics of Engineering Materials
Edited by C. S. Desai and R. H. Gallagher

Numerical Analysis of Forming processes
Edited by J. M. Alexander, J. F. T. Pittman, R. D. Wood, and O. C. Zienkiewicz

Mechanics of Geomaterials: Rocks, Concretes, Soils
Edited by Z. P. Bažant

Mechanics of Geomaterials
Rocks, Concretes, Soils

Edited by
Zdeněk P. Bažant
Center for Concrete & Geomaterials,
Northwestern University, Evanston,
Illinois, USA

in collaboration with
The International Union of Theoretical and
Applied Mechanics

A Wiley–Interscience Publication

JOHN WILEY AND SONS
Chichester · New York · Brisbane · Toronto · Singapore

Copyright © 1985 by John Wiley & Sons Ltd.

All rights reserved.

No part of this book may be reproduced by any means, nor transmitted, nor translated into a machine language without the written permission of the publisher.

Library of Congress Cataloging in Publication Data:

IUTAM William Prager Symposium (1983: Northwestern University)
  Mechanics of geomaterials.
  (Wiley series in numerical methods in engineering)
  'A Wiley–Interscience publication.'
  Includes index.
  1. Rock mechanics—Congresses. 2. Concrete—Congresses. 3. Soil mechanics—Congresses. 4. Prager, William, 1903–1980—Congresses. I. Bazant, Z. P. II. Prager, William, 1903–1980. III. International Union of Theoretical and Applied Mechanics. IV. Title. V. Series.
  TA706.I94   1983   624.1′513   84-10448
  ISBN 0 471 90541 0

British Library Cataloguing in Publication Data:

IUTAM William Prager Symposium (*1983: Evanston*)
  Mechanics of geomaterials.—(Wiley series in numerical methods in engineering)
  1. Materials 2. Mechanics, Applied
  I. Title II. Bažant, Z.P.
  620.1′12   TA404.8

ISBN 0 471 90541 0

Photoset at Thomson Press (India) Limited, New Delhi
Printed in Great Britain by Page Bros. (Norwich) Ltd

Dedicated to

William Prager

1908–1980

Contributing Authors

C. A. ANDERSON — Los Alamos National Laboratory,
MS-J576, Los Alamos,
New Mexico 87545, USA.

M. BARON — Weidlinger Associates,
333 Seventh Avenue,
New York, NY 10001, USA.

Z. P. BAŽANT — Center for Concrete & Geomaterials,
The Technological Institute,
Northwestern University,
Evanston, Illinois 60201, USA.

P. G. BERGAN — Division of Structural Mechanics,
The Norwegion Institute of Technology,
N-7034 Trondheim-NTH, Norway.

L. CEDOLIN — Dipartimento di Ingegneria Strutturale,
Politecnico di Milano,
Piazza Leonardo da Vinci 32,
20133 Milano, Italy.

W. F. CHEN — School of Civil Engineering,
Purdue University,
West Lafayette, Indiana 47907, USA.

M. P. CLEARY — Massachusetts Institute of Technology,
77 Massachusetts Avenue,
Cambridge, Massachusetts 02139, USA.

S. C. COWIN — Department of Biomedical Engineering,
Tulane University,
New Orleans, Louisiana 70118, USA.

Y. F. DAFALIAS — Department of Civil Engineering,
University of California at Davis,
Davis, California 95616, USA.

Contributing Authors

J. W. DOUGILL	Department of Civil Engineering, Imperial College of Science and Technology, London SW7 2BU, UK.
D. C. DRUCKER	Graduate Research Professor of Engineering Sciences 231 Aerospace Engineering Building University of Florida Gainesville, Florida 32611
A. EHRLACHER	Laboratoire de Mécanique des Solides Ecole Polytéchnique, 91128, Palaiseau Cedex, France.
B. EVANS	Department of Earth, Atmospheric and Planetary Science 54–718 Massachusetts of Technology Cambridge, Massachusetts 02139
D. GRADY	Sandia National Laboratories Division 1534, PO Box 5800, Albuquerque, New Mexico 87185, USA.
G. GUDEHUS	Institut für Bodenmechanik und Felsmechanik, D7500 Karlsruhe, 1 Richard–Willstatter–Allee, Postfach NR 6380, West Germany.
P. G. HODGE	Department of Mechanics, 107 Akermon University of Minnesota, Minneapolis, Minnesota 55406, USA.
D. J. HOLCOMB	Sandia National Laboratories, Geomechanics Division 5532 Albuquerque, New Mexico 87185, USA.
G. HORRIGMOE	Sivilingenior Ravlo Multiconsult A. S., Droningens GT51, PO Box 381, 8501 Narvik, Norway.
A. R. INGRAFFEA	School of Civil Engineering, Hollister Hall, Cornell University, Ithaca, New York 14853, USA.
Z. MRÓZ	Institute of Fundamental Technological Research, Polish Academy of Sciences, Swietokrzyska 21, 00–049 Warsaw, Poland.

Contributing Authors

V. N. NIKOLAEVSKY	Institute of Physics of the Earth U.S.S.R. Academy of Sciences 117526 Moscow 3–526 Prospekt Vernadokovo 101 USSR.
F. OUCHTERLONY	Swedish Detonic Research Foundation, PO Box 32058, S-12611 Stockholm, Sweden.
K. S. PISTER	College of Engineering, University of California, Berkeley, California 94720, USA.
J. H. PREVOST	Department of Civil Engineering, Princeton University, Princeton, New Jersey 08544, USA.
J. R. RICE	Division of Applied Sciences, Pierce Hall 224, Harvard University, Cambridge, Massachusetts 02138, USA.
J. W. RUDNICKI	Department of Civil Engineering, Northwestern University, Evanston, Illinois 60201, USA.
A. L. RUINA	Department of Theoretical and Applied Mechanics, Cornell University, Ithaca, New York 14853, USA.
I. S. SANDLER	Weidlinger Associates, 333 Seventh Avenue, New York, NY 10001, USA.
S. STURE	Department of Civil Engineering, University of Colorado, Campus Box 428, Boulder, Colorado 80302, USA.
L. W. TEUFEL	Sandia National Laboratories, Geomechanics Research 5532, PO Box 5800, Albuquerque, New Mexico 87185, USA.
F. H. WITTMANN	Ecole Polytéchnique Féderale de Lausanne, Département des Matériaux, 32 ch. de Bellerive, CH-1107 Lausanne, Switzerland.

T.-F. WONG   Department of Earth and Space Sciences,
                  State University of New York at Stony Brook,
                  Long Island, New York 11794, USA.

Yu. V. ZAITSEV   All-Union Polytechnical Institute (VZPI),
                  Pavla Korčagina 22
                  129805 Moscow, USSR.

O. C. ZIENKIEWICZ Department of Civil Engineering,
                  University College of Swansea,
                  Singleton Park, Swansea SA2 8PP, West Glamorgan, UK.

Preface

Recent years have witnessed an intensified interest in the mechanics of soils and rocks, as well as concrete. Applications in geotechnical and structural engineering, mining and petroleum engineering, nuclear power plant safety, underground excavation, earthquake predictions, structures in the ocean, and the mechanics of celestial bodies abound with mechanics problems which call for more realistic and more accurate solutions.

In response to these needs, the International Union of Theoretical and Applied Mechanics (IUTAM) decided at its meeting in Toronto in August 1980 to hold at Northwestern University a symposium dealing with the mechanics of geomaterials, and to dedicate this symposium to the memory of William Prager, a man whose major contributions to continuum mechanics and inelastic material behaviour have had a profound impact on our present understanding of geomaterials. Work in this field has proceeded to a large extent separately within various professional groups, and communication among civil engineers, geophysicists, mining engineers, and theoretical mechanicists have been quite limited, even though their problems are similar from the viewpoint of mechanics. Therefore, the primary objective of the William Prager Symposium was to bring together specialists from these various professional groups and disciplines, and thus promote interdisciplinary cooperation.

The William Prager Symposium on "Mechanics of Geomaterials: Rocks, Concretes, Soils", organized under the auspices of The International Union of Theoretical and Applied Mechanics, was held at Northwestern University, Evanston, Illinois, U.S.A., during September 11–15, 1983. The present volume constitutes the Proceedings of this Symposium.

The Symposium featured a series of principal lectures by eminent specialists who were invited to report on certain preassigned themes carefully selected with the help of the International Scientific Committee. These lectures were enhanced by invited principal discussers who prepared their reports after receiving complete or condensed preliminary texts of the principal lectures. Having benefited from the extensive and stimulating discussions at the Symposium, the principal lecturers and the invited discussers were given the opportunity to update, revise, and expand their reports after the Symposium. The aggregate of all the principal lectures and discussers' reports, collected in the present volume, provides an up-to-date exposition of the mechanics of rocks, concretes, and soils,

Preface

unique for its comprehensive coverage of this broad field and its timely analysis of important issues. Furthermore, reputed experts in various topics were invited to summarize the general discussions which took place at the Symposium. Their summaries, providing valuable critical insights, are included in the Appendix of the present volume.

Grateful acknowledgment is due to the US National Science Foundation for financially sponsoring the Symposium under Grant No. CEE-8208208, which was provided jointly from the Geotechnical Engineering Program and the Structural Mechanics Program directed by C. A. Babendreier and M. P. Gaus, respectively. Without this generous support the Symposium could not have taken place. Additional financial support for the travel of foreign participants was received with appreciation from IUTAM.

I would like to thank the members of the International Scientific Committee for all the advice they gave me; they included:

J. R. Rice (Co-Chairman), Harvard University, USA;
M. P. Cleary, Massachuseets Institute of Technology, USA;
A. Drescher, Institute of Fundamental Technological Research, Poland;
P. Habib, Ecole Polytéchnique, France;
H. Lippmann, Technical University of Munich, Germany;
G. Maier, Politenico di Milano, Italy;
V. N. Nikolaevsky, Institute of Physics of the Earth, USSR;
H. Okamura, University of Tokyo, Japan.

I am especially grateful to Jim Rice for his extensive help in choosing the principal themes and the names of the invited speakers and participants. The local organization of the Symposium was principally an endeavour of the Department of Civil Engineering and the Center for Concrete and Geomaterials in The Technological Institute of Northwestern University. Many people contributed to this endeavour. I would like to thank all the members of the Organizing Committee who assisted me in organizing the Symposium; they included James R. Rice (Associate Chairman of the Symposium), Mary Hill (Symposium Secretary), Ta-Peng Chang (Program Manager), J. D. Achenbach, T. B. Belytschko, R. J. Krizek, J. Rudnicki, S. P. Shah, and J. Weertman. Especially, Mary Hill's expertise and her relentless, dedicated effort was most instrumental in the success of the Symposium. I wish to thank also Bruno A. Boley, Dean of the Technological Institute, and Raymond J. Krizek, Chairman of the Department of Civil Engineering, for their useful suggestions and persistent supporting spirit.

Seeing this book become a reality, I feel gratified by my co-authors' success in producing results worthy of the illustrious name of the man to whom the Symposium was dedicated.

December 27, 1983 *Zdeněk P. Bažant*
Evanston, Illinois, USA *Symposium Chairman*

Contents

Part I. WILLIAM PRAGER AND HIS LEGACY

1. The Impact of William Prager on the Evolution of Inelastic Continuum Mechanics   3

   D. C. Drucker

2. William Prager: Personal Recollections   13

   P. G. Hodge, Jr.

Part II. CONSTITUTIVE MODELLING OF NONLINEAR TRIAXIAL BEHAVIOUR

3. Constitutive Relations for Concrete and Rock: Applications and Extensions of Elasticity and Plasticity Theory . . .   21

   J. W. Dougill

4. Requirements for Constitutive Relations for Soils   47

   G. Gudehus

5. Constitutive Relations for Concrete, Rock and Soils: Discusser's Report.   65

   W. F. Chen

Part III. BEHAVIOUR OF SOLIDS WITH A SYSTEM OF CRACKS

6. Inelastic Properties of Solids with Random Cracks   89

   Yu. V. Zaitsev

7. The Mechanics of Fracture under High-Rate Stress Loading   129
   D. Grady

8. Behaviour of Solids with a System of Cracks: Discusser's Report.   157
   A. Ehrlacher

Part IV. SHEAR LOCALIZATION, FAULTING, AND FRICTIONAL SLIP

9. Constitutive Relations for Frictional Slip   169
   A. L. Ruina

10. Shear Localization in Rocks Induced by Tectonic Deformation   189
    B. Evans and T. F. Wong

11. Shear Localization, Faulting, and Frictional Slip: Discusser's Report.   211
    J. R. Rice

Part V. FRACTURE PROPAGATION AND FRACTURE ENERGY

12. Fracture Propagation in Rock.   219
    A. R. Ingraffea

13. Fracture in Concrete and Reinforced Concrete   259
    Z. P. Bažant

14. Fracture Propagation and Fracture Energy: Discusser's Report.   305
    M. P. Cleary

Part VI. FLUID INFILTRATED GEOMATERIALS

15. Effect of Pore Fluid Diffusion on Deformation and Failure of Rock   315
    J. W. Rudnicki

Contents

16. **Effect of Pore Water and Its Diffusion in Concrete** . . . 349
    G. Horrigmoe

17. **Mechanics of Fluid-Saturated Geomaterials: Discusser's Report.** 379
    V. N. Nikolaevsky

Part VII. CREEP, SHRINKAGE, AND AGEING

18. **Creep and Thermal Effects in Ageing Solids** 405
    C. A. Anderson

19. **Deformation of Concrete at Variable Moisture Content** . . 425
    F. H. Wittmann

20. **Some Remarks on Constitutive Equations for Concrete and Geomaterials: Discusser's Report.** 461
    K. S. Pister

Part VIII. NUMERICAL MODELLING

21. **Numerical Modelling and Geomechanics (Soil–Rock–Concrete)** 471
    O. C. Zienkiewicz

22. **Numerical Models for Dynamic Loading** 501
    I. Sandler and M. Baron

23. **Numerical Aspects in Modelling of Inelastic Materials: Discusser's Report.** 527
    P. G. Bergan

Part IX. MODERN TRENDS AND NEW DIRECTIONS

24. **Current Problems and New Directions in Mechanics of Geomaterials** 539
    Z. Mróz

APPENDIX. SUMMARIES OF DISCUSSIONS FROM THE SYMPOSIUM

A. Y. F. Dafalias—Part II	569
B. D. J. Holcomb—Part III	571
C. L. W. Teufel—Part IV	575
D. F. Ouchterlony—Part V	577
E. S. C. Cowin—Part VI	581
F. L. Cedolin—Part VII	583
G. J. H. Prevost—Part VIII	587
H. S. Sture—Part IX	591

List of Symposium Participants   597

List of Contributed Papers   603

Index   605

William Prager Symposium on Mechanics of Geomaterials: Rocks, Concretes, Soils
under the auspices of International Union of Theoretical and Applied Mechanics, sponsored by U.S. National Science Foundation
Department of Civil Engineering and Center for Concrete and Geomaterials, The Technological Institute, Northwestern University, September 11–15, 1983

UPPER ROW: 1. van Mier, 2. Stout, 3. Ottosen, 4. Marchertas, 5. Margolin, 6. Kusters, 7. Cowin, 8. Bergan, 9. Cedolin, 10. Horrigmoe, 11. Galloway, 12. Papadopoulos, 13. Jonasson, 14. Marti, 15. Hillerborg, 16. Reinhardt, 17. Blaauwendraad, 18. Ehrlacher, 19. Gambarova, 20. Dafalias, 21. Jeter, 22. Wittmann, 23. Wu, 24. Powell, 25. Chen, 26. Buyukozturk, 27. Krajcinovic, 28. Dundurs, 29. Saada, 30. Malvern, 31. Lazic, 32. Okamura, 33. Crawford, 34. Maekawa, 35. Darwin, 36. Wecharatana, 37. Levine, 38. Sandler, 39. Holcomb, 40. Lippmann, 41. Schofield, 42. Teufel. LOWER ROW: 1. Desai, 2. Nilsson, 3. Shah, 4. Willam, 5. Pister, 6. Sture, 7. Carroll, 8. Murray, 9. Ballesteros, 10. Badillo, 11. Braestrup, 12. Ruina, 13. Madsen, 14. Cervenka, 15. Cuellar, 16. Hill (Symposium Secretary), 17. Bažant (Symposium Chairman), 18. Ziegler, 19. Drucker (IUTAM President), 20. Dougill, 21. Astill (NSF), 22. Higgs, 23. Mróz, 24. Jenike, 25. Darve, 26. Ingraffea, 27. Rudnicki, 28. Li, 29. Drescher, 30. Babendreier (NSF), 31. Gerstle, 32. Vardoulakis, 33. Katsube, 34. Warner, 35. Cleary, 36. Goldberg (NSF), 37. Ghaboussi, 38. Mura, 39. Atkinson, 40. Grady, 41. Jaeger, 42. Ouchterlony, 43. Uittenbogaard, 44. Rice (Associate Chairman), 45. Zienkiewicz, 46. Gudehus, 47. Wong, 48. Wawersik, 49. Prasannan, 50. John.
Photo: V. S. Gopalaratnam and T. P. Chang (Programme Manager)

PART I

WILLIAM PRAGER AND HIS LEGACY

Mechanics of Geomaterials
Edited by Z. Bažant
© 1985 John Wiley & Sons Ltd

Chapter 1

The Impact of William Prager on the Evolution of Inelastic Continuum Mechanics†

Daniel C. Drucker

It is almost impossible to encompass the scope of the great influence Professor William Prager exerted in so many ways on the development of theoretical and applied mechanics throughout the world. Restriction to the USA alone helps only a little, as does consideration of solid mechanics alone or just inelastic behaviour. As a teacher, as a researcher, and as a prolific writer and speaker of unsurpassed clarity in English, French, German, and Turkish, he became and remains the personification of modern mechanics everywhere in the world. Yet he played a very special role in the USA beyond all of that. Directly and through his students and all of the many who were privileged to come under the influence of his thinking, Professor Prager created a revolution in the education of civil and mechanical engineers. When he first came to the USA in 1941, engineering educators respected mathematics and required their students to take formal courses in the calculus and simple ordinary differential equations. Yet educators shared the feeling of practitioners that rarely if ever would this cultural requirement be a useful tool after graduation. With a few exceptions here and there engineering research was separate from engineering instruction. Only a few engineering educators engaged in fundamental research. Those who did were viewed as indulging in a somewhat disreputable hobby, not really acceptable even if it did not interfere with the normal teaching load of four courses each semester. That remained the prevailing view in most schools of engineering through the 1940s. Professors of civil and mechanical engineering with doctorates were few and far between until the mid-1950s. However, by 1950 it had become acceptable to do research and by 1955 research was a real plus in the consideration of salary, tenure, and promotion. Electrical Engineering was driven in this direction by the needs and lessons of World War II. Chemical Engineering had been given an earlier impulse by World War I as well. Civil

†Presented as the Opening Lecture at the Symposium.

Engineering and Mechanical Engineering moved almost as rapidly following World War II, pulled by the example established so visibly at Brown University by Professor Prager. With ONR support and diverse Army support prior to the establishment of NSF, the remarkably productive and fully collaborative research mode of Prager's group already had a profound psychological effect on civil and mechanical engineering in the late 1940s that continued to grow rapidly with time through the 1950s and 1960s. Fundamental changes were made in codes and specifications to take inelastic behaviour into account as more and more engineers became convinced of the validity and the practical value of the research. Engineering educators in these fields were intrigued by the power of general approaches and theorems, by the advantages of thinking in multidimensional spaces and in function spaces. Research integrated into teaching and teaching integrated into research with graduate students as colleagues became the established way of life to the great benefit of students at all levels and resulted in continual great advances in civil and mechanical engineering design.

Stimulated by Bill Prager's lectures and writings, almost all engineering educators took the first giant step from a rather primitive, primarily one-dimensional, strength-of-materials mode of thinking to a genuine three-dimensional geometry and the concepts of stress and strain as tensors. Many also took the more difficult conceptual step of going full circle and thinking effectively of the multidimensional quantities of displacement and strain and the corresponding quantities of force and stress in analogy to one-dimensional strength-of-materials. The clear separation of field equations and constitutive relations was adopted for the elementary strength-of-materials approach. Linear elasticity then took its proper place as one important possibility among many. Students were provided with the proper basis for advanced work in mechanics of solids.

The world-wide research community active in mechanics of solids and the user community competent to apply the results of the research expanded enormously. The very small group of experimentalists and applied mathematicians along with an occasional engineer grew spectacularly with the addition of a very large group of civil and mechanical engineers engaged in education and practice. Thousands of papers a year now appear in the technical literature when before there were tens. Understanding developed of the need for constitutive relations to solve problems of soil and rock mechanics as well as metal deformation processing and the usual elasto–plastic problems of interests to structural and mechanical engineers. The inelastic behaviour of metals, polymeric materials, soils, rock, and concrete now is contained in computer codes employed widely in engineering practice.

Perhaps you will permit me, before I go on to mention more of the specific contributions of Bill Prager to tell you of his influence on me. In the summer of 1947, I jumped at the opportunity offered by Bill Prager to immerse myself in the Brown atmosphere, to become a part-time applied mathematician while building experimental facilities needed for both the then *Graduate* Division of Applied

Mathematics and the Division of Engineering. The intellectual jump that occurred almost immediately under Bill's tutelage and the friendly spirit of cooperation among so many students and staff opened a whole new world for me as it had done for the staff and students already there and would do for those who came afterwards. We spent much of our time helping each other as best we could. We read each others' papers in draft form overnight, did our best to provide written and oral constructive criticism the next day. In return we expected no joint authorship, and most often no acknowledgement unless we made a major contribution, just the same treatment from everyone else. That was Bill Prager's approach, set by example, not by words, in so many ways. Bill served as editor of a variety of important journals, was founding editor of the *Quarterly of Applied Mathematics*, and did much much more for the mechanics community as a matter of course.

Always very pleased to give full credit to those who originated or developed ideas and methods, he searched out and brought important results on inelastic behaviour obtained elsewhere in the world to the attention of US researchers and vice versa. Visitors from many countries were brought to Brown University for short and for long periods. Seminars elsewhere were arranged. People at Brown were encouraged to go abroad. Plasticity symposia were organized without publication to present current thinking and to debate and clarify the important interesting issues of the day. Correspondence flowed across international boundaries. The intellectual ferment within this extended community of researchers centred around Bill Prager led to remarkable progress that accelerated throughout the decade of the 1950s and has grown steadily since.

Just a representative few of the students, staff, visitors from abroad, and visitees in the late 1940s and early 1950s gives a picture of the extent of that informal network in plasticity alone of those early years: J. F. Baker, L. Finzi, H. Ford, J. Heyman, H. G. Hopkins, W. Koiter, C. Massonnet, J. L. M. Morrison, F. K. G. Odqvist, W. Olszak, Y. N. Rabotnov, V. V. Sokolovsky, H. Ziegler, B. Budiansky, G. F. Carrier, H. J. Greenberg, G. H. Handelman, R. M. Haythornthwaite, P. G. Hodge, Jr., J. Kestens, E. H. Lee, C. C. Lin, L. E. Malvern, F. Niordson, E. T. Onat, J. L. Sanders, F. S. Shaw, R. T. Shield, P. S. Symonds, B. Thürlimann, A. J. (Ren) Wang, H. J. Weiss, A. Winzer. In addition, just about everyone elsewhere in the USA interested in plasticity and wishing to exchange ideas was sought out and welcomed into this extended family.

Bill's range of research topics always was broad and continued to include the mechanics of fluids and other branches of mechanics of solids even during the time of his greatest concentration on plasticity. His interest in scheduling, transportation, and management problems began to emerge during this period.

His role as inspirational intellectual leader on the international scene and great expositor would alone have been enough to warrant his extensive recognition. He held membership in both the National Academy of Engineering and the National Academy of Sciences, and in the Polish and the French academies, was awarded honorary degrees from universities in Europe and the USA, and was showered with prizes and medals and honorary memberships from societies and organizations around the world.

However, had he been a stay-at-home working away quietly with students and colleagues, his contributions to the literature alone would have more than warranted all of his many honours. At age 30 he was already a well-established professor in Karlsruhe with one book and over thirty papers to his credit on a variety of topics including reinforced concrete, torsion in the elastic and plastic range, fluid flow, dynamics, and vibrations of plane and space structures. His book with Hohenemser *Dynamik der Stabwerke*, published by Springer in 1933, and related papers set a very high standard in generality of approach, understanding achieved, and specific usefulness to the engineer.

Rodney Hill in his treatise *The Mathematical Theory of Plasticity* published by the Clarendon Press, Oxford, in 1950 refers frequently to Bill Prager's original work in plasticity during the 1930s and 1940s. Included are: his concepts of consistency and a neutral change in stress (one traversing the yield surface in stress space), and of a means of facilitating computation by a gradual transition from elastic to plastic response (or as we would say now, a zero diameter yield surface with or without a limit surface); his extremum and variational principles; warping and stress functions for plastic torsion; geometric properties of slipline fields for plane strain; and the first systematic study of stress discontinuities.

His analysis of experimental results on the plastic deformation of steel included the beginning of many geometric or kinematic analogues and descriptions for plasticity that intrigued and stimulated researchers everywhere. Use of the hodograph plane for plasticity problems of plane strain and the picture of kinematic hardening were two of his subsequent proposals that swept around the world, spawning books and papers everywhere. His easily understood geometric constructions are so logical and appealing that they take on a physical reality for many people far beyond the useful simplifications they were so clearly stated to portray. Ideal locking materials with a limit surface in strain space, the counterpart to ideal or perfect plasticity with its fixed yield or limit surface in stress space, is another one of Bill Prager's fertile and valuable inventions. He elucidated the need for constitutive relations in soil mechanics as in all branches of mechanics of solids and discussed suitable idealizations of the behaviour of soils, composites, and living tissues. Variational and extremum principles for elastic and for plastic response appear in profusion from his work with Synge that introduced the solid mechanics community in the US to the advantages of working in function space. Elastic stability, plastic limit analysis and design, shakedown, incremental principles, design for minimum weight, optimization of many kinds in many fields all show this common thread. Time and temperature effects under quasi-static and dynamic loading were given attention over the years.

When presenting his ideas to the research community, Bill's objective was always to be simple, clear, and fully understandable to those unfamiliar with the new concepts advanced. As a matter of policy, whenever it was possible, the first publication would describe the solution to a practical problem of one-dimensional or equivalently simple form. The second, sometime later to allow time for the new thoughts to be absorbed, would be in two dimensions or the next level up in complexity. Only later still would the general approach be advanced encompassing the special cases as obvious, almost trivial examples to an audience well-prepared to accept and use the new concept or approach. His books also show this bottom-up technique. The *Theory of Perfectly Plastic Solids* co-authored with Phil Hodge, published by Wiley in 1951, is a good example, as is his *Probleme der Plastizitätstheorie* published by Birkhäuser in 1955. Often the level reached stops short of the most general approaches that he had developed and the subtleties of the most difficult problems he had solved, to avoid confusing and discouraging a student or other newcomer to the field. His success in bringing the research community and the user community along with him was spectacular.

With his strong feeling for geometry, for analysis, and for relating theory to experiment, Bill Prager enthusiastically embraced the computer in education and research. As he stated in the closing paragraph of his 'Introductory Remarks' to the Symposium on the Future of Applied Mathematics published in the April 1972 issue of the *Quarterly of Applied Mathematics* 'With your indulgence, I should like to conclude on a personal note. In my student days, descriptive geometry, graphical statics, and kinematic geometry were still mandatory courses in applied mathematics, and differential geometry and projective geometry were not regarded as esoteric electives. I have regretted the gradual disappearance of these subjects from the curriculum and I am delighted at the prospect that, with the enormously powerful tool of computer graphics at our disposal, the geometric approach will become fashionable again, for I believe with Descartes that "nothing enters the mind as readily as geometric figures."' (Reproduced by permission of Brown University.) He brought the computer age to Brown University and served as the first Director of the Computer Laboratory there in 1960. His increasing attention to the use of computers in his later years was another manifestation of his flexibility and his clear vision of the future throughout his entire life.

A truly remarkable teacher, researcher, administrator, editor, and all-round human being of enormous direct and indirect influence on so many engineers and applied mathematicians, it is most fitting that we pay our respect to him on this occasion.

APPENDIX

Professor William Prager's impact on inelastic continuum mechanics appears clearly in the titles of the following representative but very small sample, up to 1969, of the many papers he published from 1926 to his death in 1980.

'Beitrag zur Mechanik des bildsamen Verhaltens von Flusstahl', with K. Hohenemser, *Zeit. angew. Math. Mech.*, **12** (1932) pp. 1–14.

'Mechanik isotroper Koerper im plastichen Zustand', with H. Geiringer, *Ergebnisse der exakten Naturwissenschaften*, **13** (1934) pp. 310–363.

'Der Einfluss der Verformung auf die Fliessbedingung zaehplastischer Koerper', *Zeit. angew. Math. Mech.*, **15** (1935) pp. 76–80.

'On isotropic materials with continuous transition from elastic to plastic state', *Proc. 5th Int'l Congress Appl. Mech.*, Cambridge (1938) pp. 234–237.

'A new mathematical theory of plasticity', *Rev. Fac. Sci. Istanbul, Ser. A*, **5** (1941) pp. 215–226; alos *J. of Appl. Math. and Mech. Acad. Sci., USSR*, **5** (1941) pp. 419–430.

'Strain hardening under combined stresses', *J. Appl. Phys.*, **16** (1945) pp. 837–840.

'Approximations in elasticity based on the concept of function space', with J. Synge, *Quart. Appl. Math.*, **5** (1947) pp. 241–269.

'On the mechanical behaviour of metals in the strain-hardening range', with G. H. Handelman and C. C. Lin, *Quart. Appl. Math.*, **4** (1947) pp. 397–407.

'A variational principle for plastic materials with strain-hardening', with P. G. Hodge, *J. Math. and Phys.*, 27 (1948) pp. 1–10.

'The stress–strain laws of the mathematical theory of plasticity—a survey of recent progress', *J. Appl. Mech.*, **15** (1948) pp. 226–233.

'Discontinuous solutions in the theory of plasticity', *Courant Anniversary Volume* (1948) pp. 289–300.

'On the interpretation of combined torsion and tension tests of thin-wall tubes', NACA Tech. Note 1501 (1948) 11 pp.

'Elastic–plastic analysis of structures subjected to loads varying arbitrarily between prescribed limits', with P. S. Symonds, *J. Appl. Mech.*, **17** (1950) pp. 315–323.

'Limit design of beams and frames', with H. J. Greenberg, *Proc. ASCE*, **77**, Sep. No. 59 (1951); *Trans. ASCE*, **117** (1952) pp. 447–484 (with discussion).

'The safety factor of an elastic–plastic body in plane strain', with D. C. Drucker and H. J. Greenberg, *J. Appl. Mech.*, **18** (1951) pp. 371–378.

'Recent contributions to the theory of plasticity', *Appl. Mech. Reviews*, **4** (1951) pp. 585–588.

'Extended limit design theorems for continuous media', with D. C. Drucker and H. J. Greenberg, *Quart. Appl. Math.*, **9** (1952) pp. 381–389.

'Limit design of full reinforcement for a circular cutout in a uniform slab', with H. J. Weiss and P. G. Hodge, Jr., *J. Appl. Mech.*, **19** (1952) pp. 397–401.

'Soil mechanics and plastic analysis of limit design', with D. C. Drucker, *Quart. Appl. Math.*, **10** (1952) pp. 157–165.

'Limit design of plates', with W. H. Pell, *Proc. 1st Nat'l Congress Appl. Mech.*, Chicago, Ill. (1951) pp. 547–550.

'Limit analysis of arches', with E. T. Onat, *J. Mech. Phys. Solids*, **1** (1953) pp. 77–89.

'On the use of singular yield conditions and associated flow rules', *J. Appl. Mech.*, **20** (1953) pp. 317–320.

'A geometrical discussion of the slip line field in plane plastic flow', *Trans. Royal Inst. Technology, Stockholm*, No. 65 (1953); also *Acta Polytechnica*, **123** (1953).

'The influence of axial forces on the collapse load of frames', with E. T. Onat, *Proc. 1st Midwest. Conf. Mech. Solids*, Urbana, Ill. (1953) pp. 40–42.

'On the dynamics of plastic circular plates', with H. G. Hopkins, *Zeitschrift f. angew. Mathematik und physik*, **5** (1954) pp. 317–330.

'On the theory of the bulge test', with E. W. Ross, Jr., *Quart. Appl. Math.*, **12** (1954) pp. 86–91.

'The bursting speed of a rotating plastic disc', with H. J. Weiss, *J. Aero. Sci.*, **21** (1954) pp. 196–200.

'The necking of a tension specimen in plane plastic flow', with E. T. Onat, *J. Appl. Phys.*, **25** (1954) pp. 491–493.

'On the kinematics of solids', *Colloque Junius Massau 1952 Mémoires des Sciences, Ac. Roy. Belgique*, **28** (1954) fasc. 6.

'Thermal and creep effects in work-hardening elastic–plastic solids', with A. J. Wang, *J. Aeronautical Sci.*, **21** (1954) pp. 343–344.

'Discontinuous fields of plastic stress and flow', General Lecture, *Proc., 2nd U.S. 'l Congress of Appl. Mech.*, Ann Arbor (1954) pp. 21–32.

'Limit analysis of shells of revolution', with E. T. Onat, *Proc. Koninkl. Nederl. Akad. Wetens. (B)*, **57** (1954) pp. 534–548.

'The theory of plasticity: a survey of recent achievements', James Clayton Lecture to the Institution of Mechanical Engineers, London, *Proc. Inst. Mech. Engrs.*, **169** (1955) pp. 41–57.

'The sign of plastic power in the graphical treatment of problems of plane plastic flow', *Quart. Appl. Math.*, **13** (1955) pp. 333–335.

'Plastic twisting of thick-walled circular ring sectors', with W. Freiberger, *J. Appl. Mech.*, **23** (1956) pp. 461–463.

'Minimum-weight design of a portal frame', *J. Engg. Mech. Div., Proc. Amer. Soc. Civil Engrs.*, **82** (1956) 1073, 1–10; also *Trans. ASCE*, **123** (1958) pp. 66–76 (with discussion).

'Shakedown in elastic, plastic media subjected to cycles of load and temperature', *Symposium sulla Plasticita nella Scienza delle Costruzioni*, N. Zanichelli, Bologna (1957) pp. 239–244.

'Total creep under varying loads', *J. Aeronautical Sci.*, **24** (1957) pp. 153–155.

'On ideal locking materials', *Trans. Soc. Rheology*, **1** (1957) pp. 169–175.

'Non-isothermal plastic deformation', *Proc. Koninkl. Nederl. Akad. Wetens. (b)*, **61** (1958) pp. 176–182.

'Automatic minimum weight design of steel frames', with J. Heyman, *J. Franklin Inst.*, **266** (1958) pp. 339–364.

'Minimum weight design of circular plates under arbitrary loading', with R. T. Shield, *Zeitschrift angew. Math. & Physik*, **10** (1959) pp. 421–426.

'On the plastic analysis of sandwich structures', contribution to *Problems of Continuum Mechanics*, published by the Soc. of Indus. and Appl. Math., Philadelphia (1961) pp. 342–349.

'Linearization in visco-plasticity', *Oesterreichisches Ingenieur-Archiv*, **15** (1961) pp. 152–157.

'Minimum weight design of beams subjected to fixed and moving loads', with M. Save, *J. Mech. Phys. Solids*, **20** (1963) pp. 255–267.

'A method of optimal plastic design', with P. V. Marcal, *Journal de Mécanique*, **3** (1964) pp. 509–530.

'A general theory of optimal plastic design', with R. T. Shield, *J. of Appl. Mech.*, **34** (1967) 184–186.

'Models of plastic behavior', *Proceedings, 5th U.S. Nat'l Congress of Appl. Mech.*, Minneapolis, Minn. (1966), New York (1966) pp. 435–450.

'On the formulation of constitutive equations for soft living tissues', *Quart. Appl. Math.*, **27** (1969) 128–132.

'Plastic failure of fiber-reinforced materials', *J. Appl. Mech.*, **36** (1969) 542–544.

Mechanics of Geomaterials
Edited by Z. Bažant
© 1985 John Wiley & Sons Ltd

Chapter 2

William Prager: Personal Recollections†

Philip G. Hodge, Jr

Professor Bažant, Professor Rice, Professor Drucker, distinguished colleagues, welcome guests, or—to put it concisely—friends.

Shortly after I accepted the kind invitation from Professor Bažant to deliver a 'Banquet Speech related to Prager' I obtained a copy of Jay Barry's (1982) book *Gentlemen under the Elms*, with essays on eleven of Brown University's famous professors, including, of course, William Prager. 'Aha', I thought, 'this talk will be easy to prepare'.

I already had quite a collection of articles on Prager. These included two ASME documents: an account of his acceptance speech for the Timoshenko Medal from ASME in 1966 (Paul and Hodge, 1963) and a copy of his nomination form for Honorary Membership in 1970. I had a 1968 article from the *Brown Alumni Monthly* (Anon., 1968) about his triumphant return to Brown after five years of 'exile' at IBM in Zurich and the University of California, San Diego. I had my own introduction of him for the Timoshenko medal, and H. G. (Geoffrey) Hopkins had sent me a copy of his urbane presentation when Prager was awarded an *honoris causa* by the University of Manchester. I could refer to two 'birthday cards': one written by Dan Drucker (Drucker, 1973) for his 70th birthday, and Jim Rice's beautiful 'Laudatio' for his 75th (Rice, 1979; Prager, 1979). And I had two bulging folders of personal correspondence extending over thirty-some years. All I had to do was to put all this material together in a talk.

I began my preparation by carefully reading Barry's essay. And I found that everything I wanted to say is already there in well-written detail. How Prager worked with Prandtl in Goettingen, and how he accepted a professorship in Karlsruhe in 1933 only to be fired by Hitler. How he went to Istanbul, and what he accomplished in eight years there. How Henry Wriston brought him to Brown in 1941, and what Prager did there during and after World War II. The statistics are there, too: twenty-some books, over 200 papers, lists of awards, proficiency in four languages, the journal editorships. And finally, the essay is replete with

†Text of a speech presented at symposium banquet.

personal glimpses and anecdotes contributed by such people as Dan Drucker, Harry Kolsky, Paul Maeder, Jim Rice, Paul Symonds, and his secretaries Eleanor Addison and Frances Gadjowski.

It was soon obvious that the only way to literally adhere to the publisher's original title of 'William Prager—the scientist and the man', would be to spend the next 75 minutes reading Chapter 5 of *Gentlemen under the Elms* to you, a task which would tax your patience and my vocal chords—and probably provoke a suit for copyright infringement.

So, let me refer you to these other sources for most of the public record which constitutes William Prager's illustrious career. I will make passing reference to a few well-know facts, but for the most part I want to share with you some of the personal glimpses I had of this exceptional man who was my thesis adviser, mentor, and friend.

A prime requisite for achieving success in life is to be in the right place at the right time. For me the right place was the Graduate Division of Applied Mathematics (GDAM) at Brown University at the end of World War II, and the right time was seven o'clock in the morning—any morning.

I was a conscientious graduate student, and fully expected to work sixty hours or more a week at my studies and research. However, unlike normal people, I preferred to start work before sunrise and to eschew burning the midnight oil.

According to Barry (1982, p. 92), Prager's rigidly regular schedule called for 10.30 bedtime, up at 4.30, and not later than 6.30 to work. Thus, for an hour or so in the early morning we were the only two people in GDAM. Usually we each worked in our separate offices. Occasionally I would have a question related to my work, and he would always interrupt his pursuits to discuss it with me. And on rare mornings when he didn't feel like working, he would come in my office, perch on a desk corner, and reminisce.

Thus, I knew some of his history in the late 1940s—well before the laudatory articles and speeches spread the knowledge. And there were little differences.

For example, Barry (1982, p.79) says that Hitler had him dismissed from Karlsruhe 'because of Prager's strong anti-Nazi views'. As I heard the story from him early one morning, Professor Prager was singled out because he was the youngest full professor in Germany and the object of some jealousy. Therefore, the universities didn't protest as strenuously as they might have over a more established professor, and Hitler had been able to set a precedent.

The harrowing details of how Bill and Ann Prager and their 12-year-old son, Stephen, came from Turkey to the United States in 1941 by way of Baghdad, Karachi, and Capetown may be found in many places. But to me, the tension they were under was brought home most sharply when he talked about how for over two months they had lived fully packed and ready to leave on 24-hour's notice.

Sometimes he talked about Prandtl. His favourite story, of course, was about the student who plotted some numerically obtained data and suggested that the points lay on a semi-ellipse. Prandtl took one look and said, 'No. An ellipse is symmetric; this problem is not. Try a cycloid'. Looking for the cycloid, the student returned to his equations and was able to prove that it was indeed the correct solution.

A balance between intuition and rigour, and a gift for, in Einstein's words, being 'as simple as possible—but no simpler' were Prager's trademarks. For my first assignment as a research assistant Professor Prager asked me to write up, as a paper, a theorem he had presented in our plasticity class. It took me weeks to flesh out the proof that he had intuitively sketched in a few minutes, but eventually I did it for the general three-dimensional case. He suggested I apply the method to a simple beam problem. I did, and in the process discovered a mathematical error in my proof which was rather subtle in general 3D tensor notation, but glaringly obvious in a single scalar dimension. It was a lesson in the value of simplicity which I never forgot.

One August morning when I was working in shorts and T-shirt (which I still do), Professor Prager, who could look formal in slacks and a short-sleeved, open-collar shirt, commented on the changes he had witnessed. Prandtl had once complained about a graduate student who had come to his office on a hot day without putting on his waistcoat. Stiff-collared shirt with tie and jacket were considered insultingly informal! There was no censure in Prager's telling of the tale. He had his own standards of dress and conduct, but he had no desire to force them on anyone else. He was just amused at the contrast.

A lot of nonsense has been written about Prager's formality. It was present on the surface, but his humanity easily shone through it. One afternoon a burly delivery-man stamped into the GDAM with a package and bellowed 'Is there a guy named Prager here?' (He pronounced it 'pray-ger'.) We, the students and secretaries, were horrified at the sacrilege, but Prager wasn't fazed. He came out of his office, signed as requested, and thanked the man.

Speaking of formality, formal calls had not gone completely out of style in 1948. One spring Sunday my wife Thea, year-old daughter Sue, and I all dressed in appropriate finery and set out for the Prager's house. In our anxiety not to be late we appeared at the door about 15 minutes before the agreed-upon hour of 2 p.m. Professor Prager was ready for us, but Mrs. Prager was doing some last-minute things in the kitchen. (This was a lesson to us not to be early for social calls!).

Now, many people have mentioned that William Prager was not noted for 'small-talk' but preferred to sit in comfortable silence for any length of time. However, after a minute or so he sensed that Thea's efforts to start a conversation were becoming desperate. So he picked up Sue and told us what a beautiful baby she was. I think we were all relieved when Ann Prager came in, and quickly put us at ease.

Apparently the event also made an impression on Bill. Twenty-some years later he was giving a talk at Washington University in St. Louis where Sue was a graduate student. When she introduced herself after the talk, he seemed very pleased as he reminded her of the day he'd held her in his lap.

When I left Brown in August 1949 we began a lengthy correspondence over the book we were writing. His first letter began 'Dear Hodge' and was signed 'W. Prager'. I responded with 'Dear Professor Prager' and signed 'Phil'. He immediately switched to 'Dear Phil' and in letters of September 21 and October 3, signed himself 'Bill'. When it happened twice I assumed he meant it, and from then on we were 'Phil' and 'Bill' to each other.

Let me tell you about our book (Prager and Hodge, 1951). To the mechanics community it says a lot about plasticity; to me it says a lot about William Prager.

In 1948 the Office of Naval Research asked Prager to prepare a survey report on the status of plasticity, and he assigned the job to me. It was a typical professor–student relationship: he'd say generally what should be in a chapter, I'd write a draft, he'd red-pencil it all over, I'd write another draft, and eventually we'd converge.

Chapter 9 started out with a typical history. I submitted a first draft. Prager suggested numerous changes, corrections, and additions, and he also completely rewrote the introductory paragraph. By the time I responded to his ideas the manuscript was a mess so I recopied it all, in my own handwriting—including the opening paragraph exactly as he had written it. Draft 2 was returned with several further changes, but on the first page Prager had written only 'An excellent first paragraph'. I reread that paragraph the other day; it still sounds good!

After many months, eleven chapters containing 396 pages were issued as GDAM Report A11-S2: 'An Introduction to the Mathematical Theory of Perfectly Plastic Solids', by P. G. Hodge, Jr. Period. Prager insisted that only my name belonged on the cover and that there should not even by an acknowledgement page.

As we neared the final chapters, Prager, ONR, and John Wiley decided that the material should have the wider distribution of a published book. Prager *completely* rewrote what I had done, and showed it to me for suggestions. Here was I, with the ink still wet on my Ph.D. diploma, telling William Prager how to improve his writing! I may have suggested a couple of things to be included, possibly clarified a mathematical point or two, and spotted various typographical errors. I did prepare the problems and the index, but it would be exaggerating to say that ten per cent of the final book was mine. But Prager insisted that we share equally in the infinitesimal royalties. In fact, he sent the manuscript in as 'Hodge and Prager', and only agreed to have his name first when Wiley pointed out it might sell better that way!

Even though I left Brown to go to UCLA in 1949, my relation with Bill Prager continued to grow. He was always willing to make suggestions about my work, and he also helped my in other ways. I don't think the words had been invented then, but William Prager was both my 'mentor' and my 'role model'.

California had a state law that required all state-employed professors to swear that they didn't belong to any subversive organization. Now, although I had once voted for Norman Thomas, I had never joined anything more subversive than the Boy Scouts, so I could sign the oath without fear of perjury. Still, I wasn't happy with the implications of the requirement. I visited Brown while I was still mulling over what, if anything I should do. Not only did Prager talk to me at length about the subject, but he arranged an appointment for me with Brown's president, Henry Wriston, to hear his views.

William Prager is, of course, as famous in Europe as in the United States. One of the few real regrets in my life is that I did not go to the 1952 International Congress of Theoretical and Applied Mechanics in Istanbul. Bill had offered to show me around the city—and who was better qualified? —but with a salary of $4,600, two children and a third on the way, and no contract support, I didn't go. I should have.

However, I did go and present a paper at the 1956 Congress in Brussels. I sat down next to Bill after my talk and a Frenchman got up. Judging from his gestures and rapid French, he was irate about something, but darned if I knew what. Prager leaned over and explained that the Frenchman wasn't criticizing my paper but complaining that it had gotten a better billing than his own had. When Bill, who'd been a member of the papers committee, asked me if I'd like him to respond, I murmured a heartfelt 'Thank you'.

A week earlier, also in Brussels, Prager and I had both attended an AGARD meeting, chaired by von Karman. This meeting was held in English or French with simulcast translation available for the other language. Prager, of course, spoke four languages including French and English extremely well. Von Karman, on the other hand, spoke many languages, including French and English in his own inimitable fashion. Von Karman's introductory remarks were equally divided between the two languages, and Prager kept his earphones *on* the whole time, whispering to me, 'I'm letting the translator do the work'.

I'm not quite sure how it happened, but once long ago in Chicago I took Bill to Minsky's Burlesque—for a matinee, yet! As we left, he said, 'Sometime I must take you to the Folies Bergère'. And in 1960 he did. I had to admit that in Paris the visible flesh was much more artistically presented.

But my chief memory of that evening was near the end of the show when the attractive mistress of ceremonies called on a gentleman from the audience to join her on the stage. Most of the audience was highly amused as she asked the man personal questions, kissed him, and made suggestive remarks and gestures, but Prager turned to me and said in a voice which conveyed absolute horror and utter disbelief, 'The young man seems to be enjoying it!'

When he was 60, Prager and several other people left Brown, and two years later he accepted a position at the University of California, San Diego. He went with promises of great support, many of which did not materialize. I visited him there in 1966, and he confessed he was somewhat frustrated. He told me, 'Older people are supposed to move more slowly and be more patient. I find I am less patient because I realize I have less time to get things done.'

William Prager visited Minneapolis on October 26, 1979, less than six months before he died. His colloquim lecture that afternoon was typical Prager: a profound idea in minimum-weight design, clearly presented without obfuscating mathematical detail, and applied to a trivial example to bring out its essential quality. That evening a group of six of us went to dinner—Bill, his son Stephen, Thea and I, and a young visiting couple from our department. Bill seemed to really enjoy the evening with family, old friends, and new friends. Somehow it conveyed a perfect balance and a sense of continuity.

I didn't know, of course, that it would be the last time I'd see him, but I couldn't ask for a better last memory of a great man and a good friend.

REFERENCES

Anon. (1968) 'Dr. Willi Prager: Applied math's clean desk man', *Brown Alumni Bulletin*, December.

Barry, J. (1982) 'Gentlemen under the elms', *Brown Alumni Monthly*, Brown University, Providence.

Drucker, D. C. (1973) 'Preface' (to special issue, edited by A. C. Pipkin), *SIAM Journal of Applied Math.*, **25**, 3.

Paul, B., and Hodge, P. G. Jr (1963) 'Presentation of Timoshenko Medal to Professor William Prager', *Applied Mechanics Division News*, ASME, New York.

Prager, W. (1979) 'Reply', FENOMECH '78 Special Banquet, North-Holland, Amsterdam.

Prager, W. and Hodge, P. G. Jr (1951) *Theory of Perfectly Plastic Solids*, John Wiley, New York.

Rice, James R. (1979) 'Laudatio', FENOMECH '78 Special Banquet, North-Holland, Amsterdam.

PART II

CONSTITUTIVE MODELLING OF NONLINEAR TRIAXIAL BEHAVIOUR

Mechanics of Geomaterials
Edited by Z. Bažant
© 1985 John Wiley & Sons Ltd

Chapter 3

Constitutive Relations for Concrete and Rock: Applications and Extensions of Elasticity and Plasticity Theory

J. W. Dougill

3.1 INTRODUCTION

A variety of approaches have been used in describing the behaviour of materials such as rock and concrete. At one extreme, attempts are made to generate rules to reproduce the results of experiments but without dependence on any general principles of mechanics. The resulting equations can be exceedingly useful. However, there can be no guarantee of general utility outside the range of behaviour covered by the data on which the rules are based. At the other extreme, attention is focused on a class of ideal materials defined by elementary postulates that are sufficient to provide a general theory of behaviour. Of course, this generality is concerned with the ideal behaviour so that the question remains as to how closely this can be made to correspond to that of any particular physical material. The two approaches are complementary. Both have attractions in particular circumstances. The experimentalist's view can provide precision over a narrow range. The mechanician's, broader brush, treatment may be less responsive to the fine detail of behaviour of a given material, but has the potential for greater generality in applications.

In practice, neither extreme is followed to the exclusion of the other. In designing and interpreting experiments, the range of variables may be constrained using mechanics arguments that reflect the consequences of assessing isotropy, linearity, elasticity, etc. Similarly, the choice of initial postulates, central to the development of a more general theory, is conditioned by knowledge of the physical phenomena obtained from experiments on real materials.

In what follows, we review briefly the phenomena to be described by constitutive relations for rock and concrete. Following this, various approaches to materials categorization are discussed with particular reference to time independent behaviour. Some theories developed particularly for rock and concrete are then discussed from the view of path dependence and coupling of degradation and plastic deformation. There is no attempt to be comprehensive in this review. The books by Chen and Saleeb (1982) and Chen (1982) provide a detailed account of particular theories which need not be repeated.

3.2 STRUCTURAL FEATURES

3.2.1 Physical behaviour

Concrete is a composite material. Particles of aggregate are contained in a continuous matrix of mortar which itself comprises a mixture of cement paste and smaller aggregate particles. The manner in which the properties of cement paste evolve with time and the structure of the composite material itself largely determine its physical behaviour.

The processes of hydration and curing of the cement paste produce incompatible strains in the matrix causing minor breakdown in the form of microcracking even in the absence of load (Hsu et al., 1963). Under applied loading, the material as a whole deforms, but again, there are significant incompatibilities between the aggregate and the matrix which promote further breakdown. At the macroscopic level, breakdown is accompanied by both loss in stiffness and the accumulation of irrecoverable deformation. At the structural level, breakdown appears as micro-cracking and possibly slip at the aggregate–cement paste interfaces, although this too could be described as a breakage of bonds at a yet smaller dimensional level.

The dimensions of the larger aggregate particles are significant and may be comparable to the size of the region of interest in structural problems of a local character. For instance; with removal of material by spalling, the depth of the spalled layer is of the same order as the maximum aggregate size. Similarly, the diameter of bar reinforcement is typically not too different from that of the aggregate. In both these situations, local behaviour is inevitably conditioned by the detailed structure of the material. More generally, it is still an open question to what extent the structure of the material or the grain size may need to be considered in developing theories for material behaviour. Recent work links aggregate size with the width of crack band to be used in analysis when failure becomes localized (Hillerborg et al., 1976; Bažant, 1982). On the other hand, theories for concrete behaviour developed from a structural viewpoint do not include any influence of grain size or structure (Brandtzaeg, 1927; Ortiz and Popov, 1982). This is an area where further work is required.

Almost invariably, however, the structure of the material is ignored and rules of behaviour are developed for an amorphous continuum. A similar situation applies with rock although here attention is paid to directional properties whereas, with concrete, it is customary to assume the material is initially isotropic.

The occurrence of micro-cracking and slip leads to nonlinearity and softening in the stress–strain response and a marked dependence on the mean normal stress

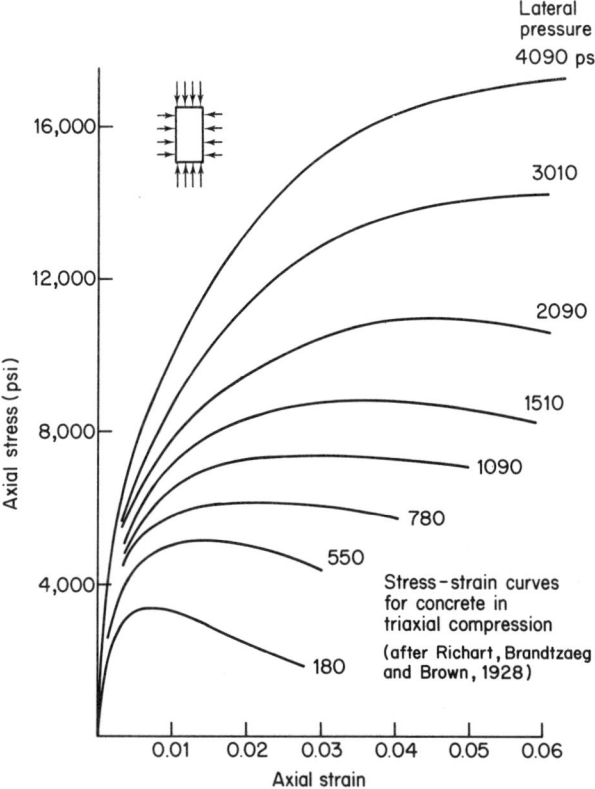

Figure 3.1 Behaviour of concrete in triaxial compression after Richart *et al.* (1928)

(Figure 3.1, after Richart *et al.*, 1928). Both concrete and rock are dilatant materials. In uniaxial compression, the test specimen decreases in volume at first but the tendency is reversed with increasing strain. Most early work to compare strength under different combinations of stress used proportional loading and took the maximum or peak values of stress to be the strength of the material. Typical strength envelopes obtained in this way for biaxial states of stress are shown in Figure 3.2. At low strains, behaviour is very nearly linear elastic but the peak stresses are achieved only after a significant amount of degradation and damage. Because of this, much work has been devoted to determining when damage becomes significant. Different techniques have been used to monitor the incidence of cracking each giving different results. Currently, most attention is given to the point of departure from linearity in the volumetric strain/axial strain plot and also to the point of minimum volumetric strain. These give envelopes for the first occurrence of significant damage (referred to by Kotsovos and Newman (1977), as the onset of slow crack propagation) and for an intermediate state at

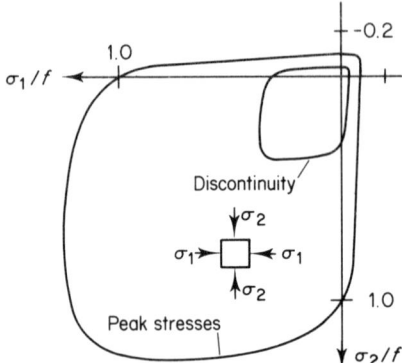

Figure 3.2 Strength envelopes from tests on concrete under proportional loading for biaxial states of stress after Kupfer et al. (1969)

which the material is just about to dilate. The similarity between these various envelopes has led a number of workers to use them as loading surfaces in theories of behaviour based on hardening plasticity (e.g. Reid, 1970; Murray et al., 1979). This general approach would be helped if it were found possible to associate some quantitative measure of damage with the various envelopes and so define the state of the material more precisely.

So far, there have been few attempts to define a quantitative measure of damage in concrete or rock in terms that would be useful in linking experiment with theory. A limited view was adopted by Spooner and Dougill (1975) in studying the behaviour of concrete in uniaxial compression. In their tests, concrete samples were subjected to a series of loading and unloading cycles in which the load was always reduced to zero but the strain range increased in each cycle as shown in

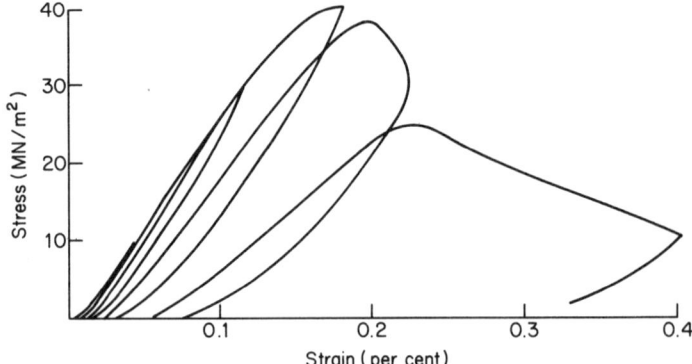

Figure 3.3 Stress-strain curves obtained in tests reported by Spooner and Dougill (1975)

Figure 3.3. Interestingly, significant acoustic emission was recorded from the samples only during first loading and not during subsequent unloading or reloading over any given strain range. This suggested that there were at least two distinct mechanisms operating to dissipate energy in the sample. One of these is confined to first loading, and is presumably responsible for micro-cracking and loss of stiffness, whilst the other applies at all times and leads to hysteresis and damping. This view of the energy dissipation processes allowed the work done in loading to a given strain to be separated into components: W_γ, the energy dissipated in damage by the first mechanism; W_δ, the dissipation under repeated loading after the first cycle and W_ε, the energy that can be recovered on unloading.

Typical results for the three energy components are given in Figures 3.4 and 3.5. These do depend on the particular view of the energy dissipation process adopted by Spooner and Dougill (1975) and it should be recognized that different criteria could be devised to determine the individual energy components. Nevertheless, the form of the results fits the observed pattern of behaviour for concrete and rock rather well. The W_γ component mirrors the change in stiffness of the material as measured by the reloading modulus (Figure 3.6) whilst the W_δ component can be closely related to damping at very low frequencies and large amplitude.

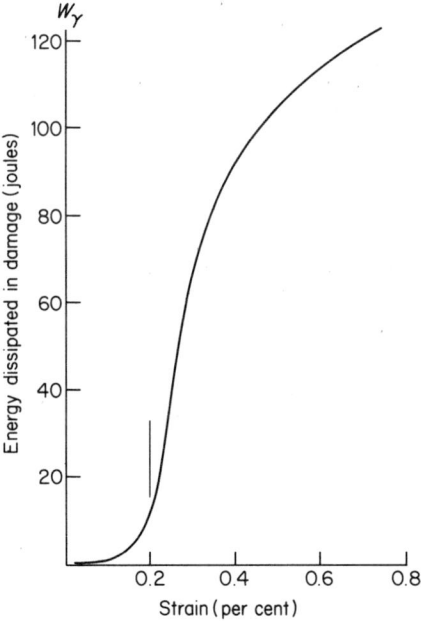

Figure 3.4 Energy dissipated in damage in uniaxial compression W_γ from Spooner and Dougill (1975)

26 Mechanics of Geomaterials

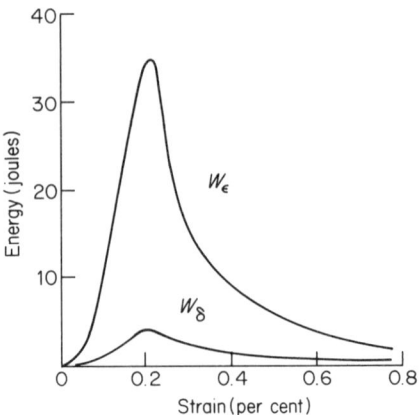

Figure 3.5 Energy dissipated in damping W_δ from Spooner and Dougill (1975)

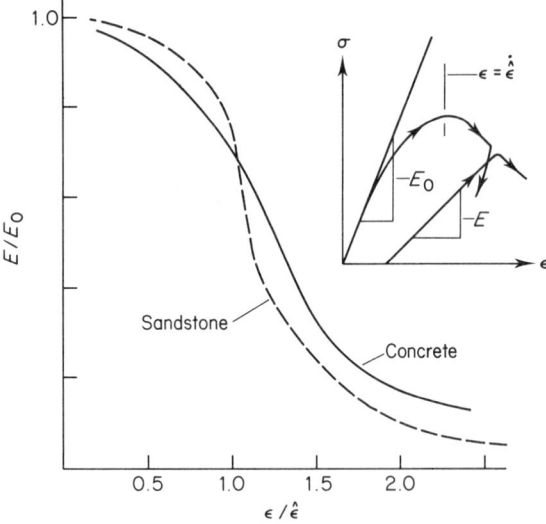

Figure 3.6 Variation of reloading modulus for concrete and rock in uniaxial compression data from Spooner and Dougill (1975) and Bieniawski (1971)

The principal conclusion from the results shown in Figures 3.4 and 3.6 is that damage occurs continuously during loading and is initiated at very low strains. This last point is not surprising when it is remembered that concrete may contain micro-cracks before any load is applied. However, the result is in conflict with the concept of discontinuity and the idea of a state surface to separate initially elastic behaviour from nonlinear behaviour controlled by stable crack propagation. In practice, this difference is not important. Damage at low strains is slight and the elastic approximation is useful for a significant range of strain.

As yet, the quantitative analysis of energy dissipation has not been extended to other stress states. It would be informative to do so and in particular to observe how the quantities W_γ and W_δ are influenced by the mean normal stress. It might be supposed that, at a given strain, W_γ would be decreased and W_δ increased by an increase in the volumetric component of applied loading. The reduction in the rate of softening with σ_{kk} is clearly apparent from Figure 3.1 and an increase in W_δ could be expected if friction played some part in the relevant mechanism. Information of this sort would be directly useful in constructing constitutive relations based on the physical processes causing nonlinearity in concrete and rock. Current information is limited, however, and so provides only a guide to the kinds of theory that may be most appropriate. Before discussing these, it is useful to look at elastic behaviour and the extensions of elasticity that have been used to derive total stress, rather than incremental, theories for material behaviour.

3.3 ELASTICITY

With time-independent materials, a distinction is made between those for which the order in which loads are applied does not affect the deformations in the final loaded state and those for which the sequence of loading does influence the final deformations. Materials in the first category are termed elastic. This class of materials thus comprises those for which the work done on loading depends only on the initial and final states. This leads to a definition of elastic materials based on the existence of a Potential Energy Function $W(\varepsilon_{ij})$ and the Complementary Potential Energy Function $W^*(\sigma_{ij})$ having the properties

$$\sigma_{ij} = \frac{\partial W}{\partial \varepsilon_{ij}} \quad \text{and} \quad \varepsilon_{ij} = \frac{\partial W^*}{\partial \sigma_{ij}} \tag{3.1}$$

It will be useful to examine the implication of these relations. In this, we will deal only with the Potential Energy Function $W(\varepsilon_{ij})$ and note that results derived from W^* can be obtained by an exchange of static and kinematic variables.

Using the chain rule for differentiation, Equation (3.1) gives relations between increments of stress and strain, i.e.

$$\delta\sigma_{ij} = \frac{\partial^2 W}{\partial \varepsilon_{ij} \partial \varepsilon_{km}} \delta\varepsilon_{km} \tag{3.2}$$

Here, it is convenient to introduce an arbitrary but monotonically increasing 'time' scale t; then to divide both sides of Equation (3.2) by δt and take the limit as δt tends to zero. This gives relations between the stress and strain 'rates' in the form

$$\dot{\sigma}_{ij} = \frac{\partial^2 W}{\partial \varepsilon_{ij} \partial \varepsilon_{km}} \dot{\varepsilon}_{km} \tag{3.3}$$

where the superior dot indicates differentiation with respect to t. It will be noted that the equation is homogeneous in time and so reflects truly time independent behaviour despite the introduction of 'rates' of stress and strain.

The quantity $\partial^2 W/\partial\varepsilon_{ij}\partial\varepsilon_{km}$ is the tensor of tangent moduli. It is clearly symmetric for all kinds of elastic materials.

Equation (3.1) can also be written in the form

$$\sigma_{ij} = \frac{\dfrac{\partial W}{\partial \varepsilon_{ij}} \dfrac{\partial W}{\partial \varepsilon_{km}}}{\dfrac{\partial W}{\partial \varepsilon_{pq}} \varepsilon_{pq}} \varepsilon_{km} \quad (3.4)$$

showing that a symmetric form can also be found for the tensor of secant moduli for any elastic material. If the material is isotropic, the potential energy can be taken to be a function of the invariants of strain. In these circumstances

$$W = W(I_1, I_2, I_3) \quad (3.5)$$

where

$$I_1 = \varepsilon_{pp},\ I_2 = \tfrac{1}{2}\varepsilon_{pq}\varepsilon_{pq},\ \text{and}\ I_3 = \tfrac{1}{3}\varepsilon_{pq}\varepsilon_{qr}\varepsilon_{rp} \quad (3.6)$$

so that

$$\sigma_{ij} = \left[\delta_{ij}\delta_{km}\frac{1}{I_1}\frac{\partial W}{\partial I_1} + \delta_{ik}\delta_{jm}\frac{\partial W}{\partial I_2} + \frac{1}{2}(\varepsilon_{ik}\delta_{jm} + \varepsilon_{jm}\delta_{ik})\frac{\partial W}{\partial I_3}\right]\varepsilon_{km} \quad (3.7)$$

In this equation, (3.7), the only connection between the shear stresses and direct strains occurs from the term involving $\partial W/\partial I_3$. If the cubic invariant has no influence ($\partial W/\partial I_3 = 0$) Equation (3.7) has the same form as Hooke's Law for isotropic linear elasticity but with the two moduli depending on stress and strain, i.e.

$$\sigma_{ij} = [\tfrac{1}{3}\delta_{ij}\delta_{km}(3K - 2G) + 2\delta_{ik}\delta_{jm}G]\varepsilon_{km} \quad (3.8)$$

in which the shear modulus

$$G = \frac{1}{2}\frac{\partial W}{\partial I_2} \quad (3.9a)$$

and the bulk modulus

$$K = \frac{1}{I_1}\frac{\partial W}{\partial I_1} + \frac{1}{3}\frac{\partial W}{\partial I_2} \quad (3.9b)$$

With linear elasticity, G and K are independent material constants, linked loosely through the requirements for material stability. However, with nonlinear materials, the shear and bulk moduli are clearly related by

$$\frac{\partial}{\partial I_2}\left[\frac{3K - 2G}{3}\right] = \frac{2}{I_1}\frac{\partial G}{\partial I_1} \quad (3.10)$$

a result following formally from Equation (3.9) and providing the condition for the tensor of tangent moduli to be symmetric as required by (3.3). The condition (3.10) implies that there are difficulties in matching behaviour which is nonlinear, but not truly elastic, with an extension of elasticity based on two strain or stress dependent moduli.

Concrete is nonlinear and clearly not elastic. In spite of this, a number of workers have devised total stress—total strain relations for concrete using stress dependent shear and bulk moduli (Kupfer and Gerstle, 1973; Cedolin et al., 1977). In general, it might be supposed that approaches of this sort have the disadvantage that the material description is no longer path independent so that total stress–strain relationships are not really appropriate. Nevertheless, considerable success has been achieved in this way in reproducing results from experiments using mainly proportional loading in triaxial and biaxial compression. In doing this, the principal deficiency is the absence of dilation in the materials description. This was dealt with by Kotsovos and Newman (1978) by introducing an additional term σ_o for the reduction in volumetric stress occurring at a given strain due to internal cracking and dilation. In this way,

$$\sigma_{ij} = 2G\varepsilon_{ij} + \frac{3K-2G}{3}\delta_{ij}\varepsilon_{kk} - \delta_{ij}\sigma_o \qquad (3.11)$$

where σ_o is a function of the octahedral shear stress which is found from experiment. Very good agreement with experimental results has been achieved using this equation, both for biaxial and triaxial compression and a variety of stress paths. This degree of path independence for compressive stress regimes is possibly surprising but would be essential for successful use of the total stress theories in analysis.

Rather than depend on the path independence property of elasticity, some workers (Coon and Evans, 1972, and see Bažant, 1979) have preferred to develop incremental theories of the form:

$$\dot{\sigma}_{ij} = T_{ijkm}\dot{\varepsilon}_{km} \qquad (3.12)$$

where the tensor of tangent moduli T_{ijkm} is a function of the current stress and strain. The conditions for the material to be isotropic initially and for the T_{ijkm} to be tensorially invariant (Bažant, 1979) impose restrictions on the form of the T_{ijkm} but still leave a large number of coefficients to be found by curve fitting experimental results. A quite common simplification has been to adopt an orthotropic form for T_{ijkm}. The three functions then required can be found from experimental data without too much difficulty. With this approach, experimental results for common situations, in which the directions of principal stress do not change during loading, can be reproduced satisfactorily. This restriction must be noted as the orthotropic form for T_{ijkm} is not tensorially invariant and so leads to ambiguities in analysis of more general load histories involving rotation of principal directions (Bažant, 1979).

Despite the success, for simple load paths, of stress–strain relations based on elasticity and hypoelasticity, all these approaches fail to distinguish between loading and unloading. From the earlier discussions of physical behaviour it is clear that quite different processes are involved during loading and unloading for rocks and concrete. If these are to be separated for the purpose of analysis, additional criteria need to be introduced so that the stress or strain paths associated with loading or unloading can be identified. The existence of a loading function of this sort is the principal feature of the most common forms of plasticity theory. These then provide a basis for a more general and physically acceptable description of behaviour.

3.4 HARDENING PLASTICITY WITH A REGULAR LOADING FUNCTION

In considering plasticity, it is appropriate first to focus attention on the form of theory developed by Prager and his colleagues working mainly at Brown University. This involves relations between increments of stress and strain; the so-called incremental theory existing in parallel with the deformation and flow theories which link total stress and strain.

The theory involves the formal division of the strain rate $\dot{\varepsilon}_{ij}$ caused by changes in stress $\dot{\sigma}_{ij}$ into elastic ($\dot{\varepsilon}'_{ij}$) and plastic ($\dot{\varepsilon}''_{ij}$) components so that

$$\dot{\varepsilon}_{ij} = \dot{\varepsilon}'_{ij} + \dot{\varepsilon}''_{ij} \tag{3.13}$$

Here the elastic component is recovered on unloading whilst the plastic component remains as shown in Figure 3.7. In order to distinguish between loading involving plastic deformation and elastic unloading, a loading function

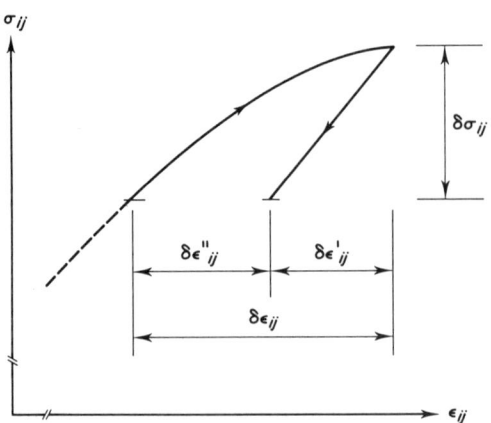

Figure 3.7 Elastic and plastic components of the strain increment as used in plasticity theory

$F(\sigma_{ij}, H_k)$ is introduced in which the H_k are measures of plastic deformation and so change only when plastic deformation occurs. The loading function provides the criteria for yield in the form

$$F(\sigma_{ij}, H_k) = 0 \tag{3.14}$$

which can be visualized as the equation of a surface—the yield surface—in stress space (σ_{ij}). The sign of F is chosen so that interior points ($F < 0$) represent stress states for which the next increment of stress will cause purely elastic behaviour. When the stresses are such that $F(\sigma_{ij}, H_k) = 0$, a subsequent increment of stress may cause either purely elastic behaviour or lead to plastic deformation depending on the direction of the increment. The choices available are shown in Figure 3.8. From this, it is evident that elastic loading or unloading occurs when

$$F < 0 \quad \text{or} \quad F = 0 \quad \text{and} \quad \frac{\partial F}{\partial \sigma_{ij}} \dot{\sigma}_{ij} < 0 \tag{3.15}$$

Alternatively, behaviour is accompanied by plastic deformation when

$$F = 0 \quad \text{and} \quad \frac{\partial F}{\partial \sigma_{ij}} \dot{\sigma}_{ij} > 0 \tag{3.16}$$

The possibility remains that loading may be along the yield surface. Here, Prager (1949) noted that this behaviour must be capable of description in terms of elasticity so that no plastic deformation can occur. On the basis of this continuity condition, the criteria for purely elastic behaviour can be revised and become

$$F < 0 \quad \text{or} \quad F = 0 \quad \text{and} \quad \frac{\partial F}{\partial \sigma_{ij}} \dot{\sigma}_{ij} \leqslant 0 \tag{3.17}$$

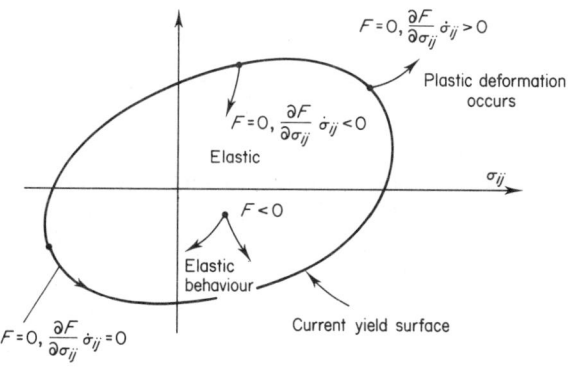

Figure 3.8 Loading and unloading in plasticity

A major consequence of the continuity condition is that the flow rule for the plastic strain rate can be written in the form

$$\dot{\varepsilon}_{ij}'' = \alpha_{ij} \frac{\partial F}{\partial \sigma_{km}} \dot{\sigma}_{km} \qquad (3.18)$$

The theory then proceeds by noting that the yield condition (3.14) must be satisfied at all times when plastic deformation is in progress. This leads to Prager's condition for consistency, i.e. during plastic deformation $\dot{F} = 0$ and

$$\frac{\partial F}{\partial \sigma_{km}} \dot{\sigma}_{km} + \frac{\partial F}{\partial H_k} \dot{H}_k = 0 \qquad (3.19)$$

At this stage, we note that the rate work is done in unit volume of material is

$$\dot{W} = \sigma_{ij}\dot{\varepsilon}_{ij} = \sigma_{ij}\dot{\varepsilon}_{ij}' + \sigma_{ij}\dot{\varepsilon}_{ij}'' \qquad (3.20)$$

In this, the last term will be recognized as being the rate that energy is dissipated in plastic deformation so,

$$\dot{D}^p = \sigma_{ij}\dot{\varepsilon}_{ij}'' \qquad (3.21)$$

On using this result with the flow rule (3.18) and the consistency condition (3.19), we find

$$\alpha_{ij}\sigma_{ij} = -\frac{1}{\dfrac{\partial F}{\partial H_k}\dfrac{dH_k}{dD^p}} \qquad (3.22)$$

Further definition of the flow rule requires additional information concerning material behaviour. For metals, Drucker's (1950) definition of hardening is appropriate and leads to the condition that

$$\dot{\sigma}_{ij}\dot{\varepsilon}_{ij}'' \geq 0 \qquad (3.23)$$

To meet this conditon, the multipliers, α_{ij}, in the flow rule must be such that

$$\alpha_{ij}\dot{\sigma}_{ij} \geq 0 \qquad (3.24)$$

An interesting result follows when the α_{ij} are taken to be independent of the stress rates so that the material is linear in the small. In such a case, the condition (3.23) can be satisfied for arbitrary $\dot{\sigma}_{ij}$ only if the α_{ij} have the same form as a vector in the direction of the normal to the yield surface at the current stress point. In this way,

$$\alpha_{ij} = K \frac{\partial F}{\partial \sigma_{ij}} \qquad (3.25)$$

where the scalar K can be found directly from the flow rule (3.18) and the consistency condition (3.19). Alternatively, using Equation (3.22)

$$K = - \frac{1}{\dfrac{\partial F}{\partial H_k} \dfrac{dH_k}{dD^p} \dfrac{\partial F}{\partial \sigma_{pq}} \sigma_{pq}} \tag{3.26}$$

so completing the flow rule.

This completes the theory for a model of hardening plasticity based on Prager's continuity condition and Drucker's definition of hardening. Because of this last definition and the assumption of linearity in the small, the flow rule is associated with the loading function through the normality principle. In applications, most attention has been given to isotropic hardening in which a single measure is taken to describe the history of plastic deformation and this determines the scale of the yield surface in stress space. In this way, the loading function

$$F(\sigma_{ij}, H_k) \to F(\sigma_{ij}) - H \tag{3.27}$$

where the hardening parameter H can be taken to be a function of the energy dissipation in plastic deformation or, more usually, a scalar measure of the plastic strains themselves.

Theories of this sort tend to take $F(\sigma_{ij})$ to correspond with the shape of the discontinuity surface or the surface for peak stresses obtained in tests under proportional loading (Chen and Chen, 1975; Murray et al., 1979). However, there are no convincing experimental data to show that normality obtains or that the material is linear in the small. Experiments of this sort would be extra-ordinarily difficult both to devise and interpret, especially for materials like rock and concrete where more than one mechanism operates to induce inelastic behaviour.

3.5 DUAL RELATIONSHIPS FOR HARDENING PLASTICITY

There is one feature of hardening plasticity that is not often exploited. At any stage, the yield surface in stress space can be mapped, using Hooke's Law, to provide a corresponding yield surface in deformation space, ε_{ij}. This suggests separating the stress rate into elastic and inelastic components

$$\dot{\sigma}_{ij} = \dot{\sigma}'_{ij} + \dot{\sigma}''_{ij} \tag{3.28}$$

where the elastic component is derived from the total strain increment $\dot{\varepsilon}_{ij}$ by the current tensor of moduli S_{ijkm}, i.e.

$$\dot{\sigma}'_{ij} = S_{ijkm} \dot{\varepsilon}_{km} \tag{3.29}$$

A flow rule for the inelastic component $\dot{\sigma}''_{ij}$ of the stress rate can now be derived by analogy with the established theory (Equations 3.14–3.26) and an exchange of kinematic and static variables. In this, Il'iushin's (1961) postulate of plasticity replaces Drucker's (1950) definition of work hardening as the principal determinant of material behaviour. The analogy is shown in Table 3.1: it again leads

Table 3.1 Static kinematic relations for alternative formulations of hardening plasticity theories

Stress space	Formulation	Strain space
$\dot{\varepsilon}'_{ij}$ $\dot{\varepsilon}''_{ij}$	**Working Variables** elastic components inelastic components	$\dot{\sigma}'_{ij}$ $-\dot{\sigma}''_{ij}$
$\dot{D} = \sigma_{ij}\dot{\varepsilon}''_{ij}$	**Dissipation measure** for inelastic behaviour	$\dot{D} = -\dot{\varepsilon}''_{ij}\dot{\sigma}''_{ij}$
$F = F(\sigma_{ij}, H_k)$ H_k is a measure of ε''_{ij} history	**Loading Function**	$F = F(\varepsilon_{ij}, H_k)$ H_k is a measure of σ''_{ij} history
from Drucker $\dot{\sigma}_{ij}\dot{\varepsilon}''_{ij} \geqslant 0$	**Material Classification**	from Il'iushin $\dot{\varepsilon}_{ij}\dot{\sigma}''_{ij} \leqslant 0$
$\dot{\varepsilon}''_{ij} = K \dfrac{\partial F}{\partial \sigma_{ij}} \dfrac{\partial F}{\partial \sigma_{km}} \dot{\sigma}_{km}$	**Flow Rule**	$\dot{\sigma}''_{ij} = -K \dfrac{\partial F}{\partial \varepsilon_{ij}} \dfrac{\partial F}{\partial \varepsilon_{km}} \dot{\varepsilon}_{km}$

to an associated flow rule and normality of the inelastic stress rate, $\dot{\sigma}''_{ij}$, again follows the assumption of linearity in the small.

3.6 A COMMENT ON PERFECT PLASTICITY

Although not directly relevant to descriptions of rock and concrete, perfect plasticity has some points of interest in the present discussion. Here, the loading function of perfect plasticity $F(\sigma_{ij}, H_k)$ is replaced by a yield function providing a yield surface fixed in stress space,

$$F(\sigma_{ij}) = 0 \tag{3.30}$$

As it is not possible to achieve stress states outside the yield surface and the yield criterion $F = 0$ must be satisfied during yield, the necessary conditions for plastic deformation to occur are

$$F = 0 \quad \text{and} \quad \frac{\partial F}{\partial \sigma_{ij}}\dot{\sigma}_{ij} = 0 \tag{3.31}$$

Clearly, Prager's continuity condition does not apply and the theory is of a different sort to that conventionally adopted to describe hardening. Moreover, as plastic strain is induced by stress increments that lie on the yield surface it is not possible to map the yield surface into deformation space as can be done with hardening plasticity. This does not mean that dual relationships do not exist. For hardening materials, relationships derived by the exchange of kinematic and static variables can be used to describe the same material and phenomenon. On the other hand, with perfect plasticity, the exchange leads to the description of a completely different class of material as recognized by Prager (1949) in developing his theory for perfectly locking materials.

3.7 HARDENING PLASTICITY: EXTENSIONS AND ALTERNATIVES

The theory of hardening plasticity is attractively compact. Also, there is no restriction to any particular physical mechanism for the occurrence of the nonlinear strain components. Because of this, the effects of both slip and fracture are equally amenable to description by the theory. The restriction to hardening may seem to be an embarrassment but this is eliminated if the theory is developed using Il'iushin's postulate of plasticity rather than Drucker's definition of work hardening. Alternatively, the view could be taken that the material is path independent in the small in the sense explored by Drucker (1968) and Dougill (1975). Here it is supposed that the value of the second order work term

$$W_D = \int_t^{t+\delta t} \delta\sigma_{ij} \dot{\varepsilon}''_{ij} dt \qquad (3.32)$$

is independent of the path in stress space used to effect the increment $\delta\sigma_{ij}$. With elasticity, path independence arguments lead to symmetry of the tensor of moduli as indicated in Equations (3.3) and (3.4). A similar implication follows from this form of path independence leading to symmetry of the flow rule (3.18) and normality. It is of interest that the assumption of linearity in the small is not required in this argument. Less positively, although it may be attractive to adopt path independence in the small as a means of classifying ideal materials in mechanics terms, the concept also introduces the difficulty of recognizing the corresponding behaviour in real materials.

It is useful to note that the theory of hardening plasticity is usually considered in the context of the behaviour of metals for which it can be assumed that plastic deformation occurs without any accompanying change in the moduli that determine the elastic response. This is not a necessary restriction, however: the form of the flow rule is unaffected by degradation. In these terms, the nonlinearities observed in concrete and rock should be amenable to description by the theory. Additionally, though, there is the possibility that a theory could be developed to provide information on the changes of stiffness that occur during loading. This possibility leads to theories which concentrate on degradation to the exclusion of any other process that might lead to nonlinearity (Dougill, 1976; Dougill and Rida, 1980).

3.8 DEGRADATION AND THE PROGRESSIVELY FRACTURING SOLID

We consider a material in which stress is related to strain at all times by Hooke's Law, i.e.

$$\sigma_{ij} = S_{ijkm} \varepsilon_{km} \qquad (3.33)$$

However, the tensor of moduli S_{ijkm} changes due to progressive fracture during loading when

$$\dot{\sigma}_{ij} = S_{ijkm}\dot{\varepsilon}_{km} + \dot{S}_{ijkm}\varepsilon_{km}$$

This suggests a division of the stress rate tensor into an elastic component

$$\dot{\sigma}'_{ij} = S_{ijkm}\dot{\varepsilon}_{km} \qquad (3.34)$$

and a component due to degradation of stiffness

$$\dot{\sigma}''_{ij} = \dot{S}_{ijkm}\varepsilon_{km} \qquad (3.35)$$

Use of these components allows a theory to be developed by analogy with the theory of hardening plasticity using the relationships given in Table 3.1. To do this, we introduce a loading function $F(\varepsilon_{ij}, H_k)$ to define a fracture (or yield) surface, $F = 0$, in deformation space in order to distinguish between deformation paths associated with elastic behaviour and those for which degradation occurs. Then, on the basis of Il'iushin's postulate, the assumption of linearity in the small and Prager's continuity condition, the flow rule is obtained i.e.

$$\dot{\sigma}''_{ij} = K\frac{\partial F}{\partial \varepsilon_{ij}}\frac{\partial F}{\partial \varepsilon_{km}}\dot{\varepsilon}_{km} \qquad (3.36)$$

Alternative formulations are possible involving incrementally nonlinear behaviour and departures from normality (Dougill, 1983). However, for the purpose of illustration, we continue with the associated flow rule (3.36) and consider, particularly, loading functions giving behaviour analogous to isotropic hardening (cf. Equation (3.27)). Accordingly, we put

$$F(\varepsilon_{ij}, H_k) = F(\varepsilon_{ij}) - H(D) \qquad (3.37)$$

and take the energy dissipated per unit volume of material to be the single measure of damage D. It follows that

$$\dot{D} = -\tfrac{1}{2}\varepsilon_{ij}\dot{\sigma}''_{ij}, \qquad (3.38)$$

so that use of Prager's consistency condition with the 'yield' criteria $F = 0$, gives,

$$\dot{\sigma}''_{ij} = -\frac{1}{M}\frac{\partial F}{\partial \varepsilon_{ij}}\dot{G} \qquad (3.39)$$

where

$$M = \frac{\partial F}{\partial \varepsilon_{pq}}\varepsilon_{pq} \qquad (3.40)$$

and

$$\dot{G} = 2\frac{\mathrm{d}D}{\mathrm{d}H}\frac{\partial F}{\partial \varepsilon_{km}}\dot{\varepsilon}_{km} \qquad (3.41)$$

The flow rule gives no information on the change of stiffness. For this, additional assumptions concerning material behaviour are required. One approach is to assume that the change in stiffness $\dot{S}_{ijkm}\delta t$ caused by an increment of deformation $\dot{\varepsilon}_{ij}\delta t$ is independent of the deformation path. This is a different form of path independence in the small than discussed earlier. In applying this condition, the two alternative deformation paths shown in Figure 3.9 are considered. Attention is therefore focused on the difference in the change of stiffness that would occur by loading to a common point from two closely neighbouring points ε_{ij} and $\varepsilon_{ij} + \delta\varepsilon_{ij}$ situated on the fracture surface. The conditions for path independence are thus

for
$$\frac{\partial \dot{S}_{ijkm}}{\partial \varepsilon_{rs}}\delta\varepsilon_{rs} = 0 \tag{3.42}$$

$$\frac{\partial F}{\partial \varepsilon_{rs}}\delta\varepsilon_{rs} = 0 \tag{3.43}$$

and $\dot{G}$ constant for each path.

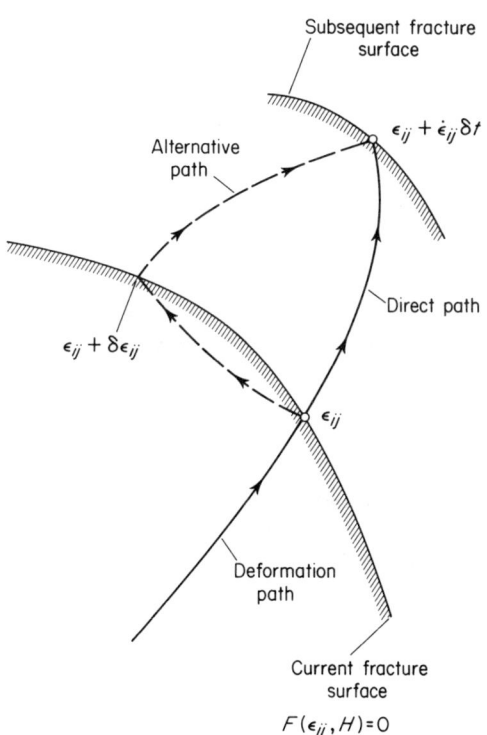

Figure 3.9 Alternative deformation paths producing the same change in stiffness

The conditions (3.42) and (3.43) can be combined in the form

$$\frac{\partial \dot{S}_{ijkm}}{\partial \varepsilon_{rs}}\varepsilon_{km} + \lambda_{ij}\dot{G}\frac{\partial F}{\partial \varepsilon_{rs}} = 0 \tag{3.44}$$

where the unknown multipliers remain to be determined. This is done using Equations (3.35) and (3.39) which, after some manipulation, leads to the result,

$$\begin{aligned}\dot{S}_{ijkm} = \frac{\dot{G}}{M}&\left[\frac{\partial^2 F}{\partial \varepsilon_{ij}\partial \varepsilon_{km}} + \frac{1}{M}\frac{\partial F}{\partial \varepsilon_{ij}}\frac{\partial F}{\partial \varepsilon_{km}}\right.\\ &- \frac{1}{M}\left[\frac{\partial F}{\partial \varepsilon_{ij}}\frac{\partial^2 F}{\partial \varepsilon_{km}\partial \varepsilon_{pq}}\varepsilon_{pq} + \frac{\partial F}{\partial \varepsilon_{km}}\frac{\partial^2 F}{\partial \varepsilon_{ij}\partial \varepsilon_{pq}}\varepsilon_{pq}\right]\\ &+ \left.\frac{1}{M^2}\frac{\partial F}{\partial \varepsilon_{ij}}\frac{\partial F}{\partial \varepsilon_{km}}\frac{\partial^2 F}{\partial \varepsilon_{pq}\partial \varepsilon_{rs}}\varepsilon_{pq}\varepsilon_{rs}\right] \end{aligned} \tag{3.45}$$

Equations (3.39) and (3.45) are sufficient to describe the process of degradation. It is of interest that a useful simplification is obtained when $F(\varepsilon_{ij})$ is a homogeneous function of the strains of degree n. In these circumstances

$$\dot{S}_{ijkm} = -\frac{2}{nH}\frac{dD}{dH}\dot{H}\left[\frac{\partial^2 F}{\partial \varepsilon_{ij}\partial \varepsilon_{km}} + \frac{2-n}{nF}\frac{\partial F}{\partial \varepsilon_{ij}}\frac{\partial F}{\partial \varepsilon_{km}}\right] \tag{3.46}$$

$$\dot{\sigma}''_{ij} = -\frac{2}{nH}\frac{dD}{dH}\frac{\partial F}{\partial \varepsilon_{ij}}\dot{H} \tag{3.47}$$

One further comment is necessary. It was convenient to present the approach to degradation as an extension to a theory for softening based on Il'iushin's postulate or a direct assumption of normality. On more detailed examination, (Dougill, 1983) it turns out that this is not necessary. The assumption of path independence for the change in stiffness is sufficient on its own to establish the change in stiffness and the flow rule for this class of material.

3.9 EXAMPLE—A 'NO TENSION' MATERIAL

The results (3.46) and (3.47) allow a simple theory to be developed providing behaviour approximating to a 'no-tension' material. To do this, consider an isotropic softening model described by the loading function (3.37) with

$$F(\varepsilon_{ij}) = -I_2/I_1 \tag{3.48}$$

and where the invariants I_1 and I_2 are as described in Equations (3.6). During progressive fracture

$$\varepsilon_{ij}(\varepsilon_{ij} + 2\delta_{ij}H) = 0 \tag{3.49}$$

so that the fracture surface is a hypersphere passing through the points $\varepsilon_{ij} = 0$ and

Constitutive Relations for Concrete and Rock

$\varepsilon_{ij} = -2\delta_{ij}H$. The fracture surface thus remains captured to the origin whilst it grows in size as H increases due to the fracture process. The change in stiffness follows from Equation (3.46) with $n = 1$, i.e.

$$\dot{S}_{ijkm} = -2\dot{D}\left[\frac{\delta_{ik}\delta_{jm}}{I_2} + \frac{3\delta_{ij}\delta_{km}}{I_1^2} - \frac{2(\varepsilon_{ij}\delta_{km} + \varepsilon_{km}\delta_{ij})}{I_1 I_2} + \frac{\varepsilon_{ij}\varepsilon_{km}}{I_2^2}\right] \quad (3.50)$$

the result being illustrated in Figures 3.10–3.13.

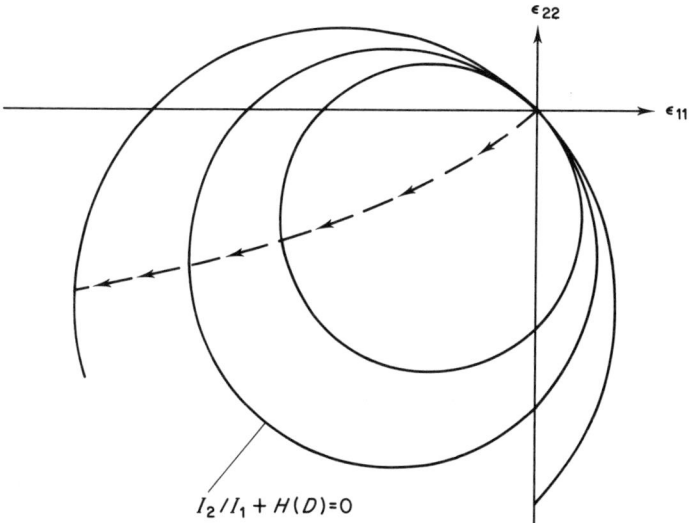

Figure 3.10 No-tension material: fracture surface and isotropic softening

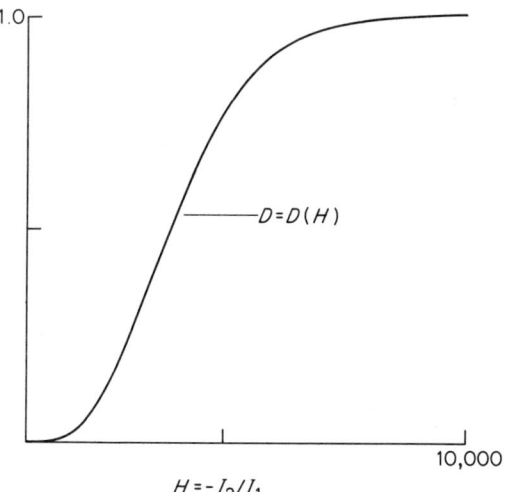

Figure 3.11 Energy dissipation function $H(D)$

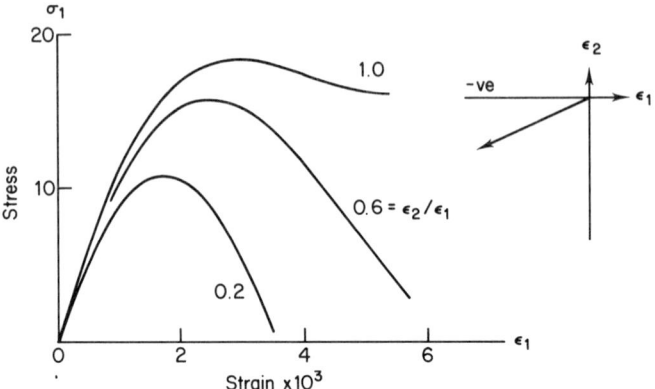

Figure 3.12　No-tension material: stress–strain curves for proportional straining

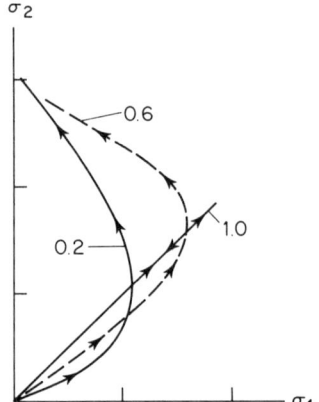

Figure 3.13　No-tension material: stress paths corresponding to proportional straining

3.10 MULTIPLE LOADING SURFACES

So far, theories have been derived on the assumption that the loading function used for either plasticity or progressive degradation leads to a single smooth surface in stress or strain space. This may be too much of a simplification. Certainly, theories constructed from a micro-structural background lead to yield surfaces that have corners or vertices. At such points, the direction of the outward normal changes abruptly and the conditions for loading (3.17) no longer apply.

Koiter (1953) suggested one way of dealing with this problem in formulating a theory using more than one loading function to define the conditions for yield.

Each loading function has its corresponding yield surface. The individual yield surfaces may intersect so that the innermost boundary, which defines the region of elastic behaviour, is a composite surface with segments from several loading functions. The conditions for yield (3.17) are then taken to apply separately to each loading function so that, at any instant, some are active and involved in plastic deformation whilst others are not.

A theory of plasticity of this sort was developed by Sanders (1954) using independent linear loading functions. A similar result was obtained by Batdorf and Budiansky (1949), on physical grounds, by considering the possibility of slip on a number of distinct and differently orientated slip planes each with its own loading function. This approach is the formal plasticity based equivalent of Brandtzaeg's (1927) classical treatment of nonlinearity in concrete based on the consequences of slip. The analogue to Batdorf and Budiansky's model was used by Dougill (1976) to illustrate the behaviour of the fracturing solid with linear loading functions being used to describe the behaviour of individual oriented fibre bundles comprising an ideal fibrous composite. Some results from a model of this type for a sheet-like fibrous material are shown in Figure 3.14. A similar approach is implicit in Gambarova's (1983) use of a lattice element to represent local behaviour of concrete subject to shear.

This approach has great flexibility for describing materials behaviour and it is comparatively simple to develop models for combined fracturing and plastic behaviour using a series of appropriate loading functions in deformation space. If this is done, the results are identical to those obtained using the 'layering' technique employed by Zienkiewicz et al. (1972) for direct modelling of materials

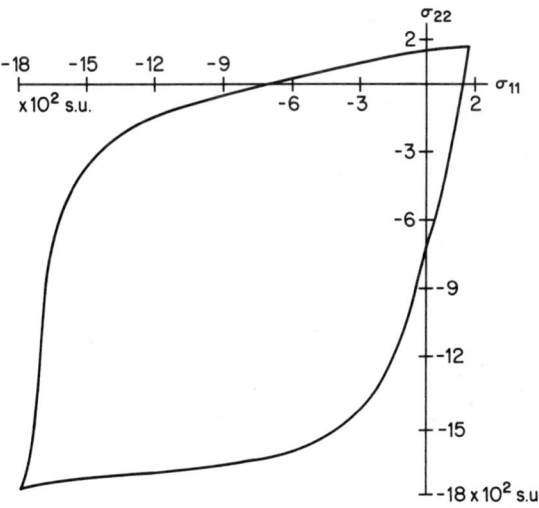

Figure 3.14 Results for progressive fracture using linear loading function from Dougill and Rida (1980)

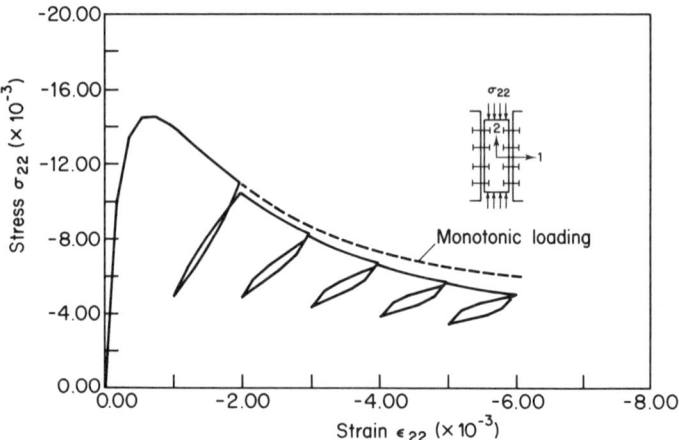

Figure 3.15   Combined plasticity and fracture (Rida, 1981)

in finite element theory. An example of material behaviour modelled by a number of loading functions for fracture and a Von Mises' type kinematic hardening model for plasticity is given in Figure 3.15. The result shows the loss of stiffness and hysteretic behaviour typical of rocks, concrete and similar real materials.

The use of a range of loading functions adds to the computation and storage requirements and is only attractive for structural calculations when the number of functions can be severely limited. This is a feature of the so-called 'cap' models now used extensively in soil and rock mechanics. These all emanate from an idea from Drucker, Gibson and Henkel (1957). The conditions at yield for a rock or soil are conveniently given by the Mohr–Coulomb or Drucker–Prager (1952) yield criterion. In these terms, though, adoption of normality and an associated flow rule can give anomalous results for the plastic component of the volumetric strain. To meet this difficulty, Drucker *et al.* (1957) introduced an isotropic hardening cap to close the region enclosed by the Coulomb–Mohr yield surface. This dual criterion model involving a combination of perfect plasticity and isotropic hardening has subsequently been extensively developed (Resende and Martin, 1982). In general, though, the computational difficulties accompanying the use of multiple loading functions have been such that alternative approaches have been preferred.

3.11   VERTEX HARDENING

Ideally, it might be supposed that the effects caused by the presence of vertices in the yield surface might be represented by a suitable choice of smooth loading function and hardening rule. However, it is soon evident that the simplest idealization—isotropic hardening—offers little prospect of success. As can be

Constitutive Relations for Concrete and Rock

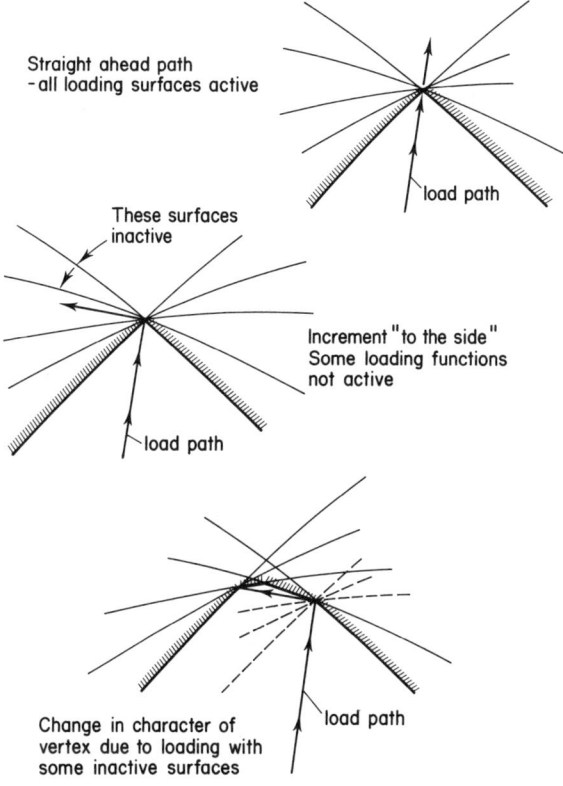

Figure 3.16 Vertex hardening

seen from Figure 3.16, the range of stress increments causing plastic deformation is limited when using a continuous yield surface to those increments outside the tangential plane at the point corresponding to the current stress point. With a sharp vertex at the same point, the range of stress increments causing plastic deformation is much greater so that inelastic strains can occur for increments that would be regarded as unloading with the continuous surface. All this promotes the idea of using a special form of flow rule based on behaviour at a vertex.

We imagine the vertex to be formed by a number of yield surfaces drawn together to the current stress point by the preceding stress path. It is evident that subsequent loading 'straight ahead' is likely to involve all the loading functions and so conserve the form of the vertex. Alternatively, if the next stress increment causes a sharp change in the direction of the stress path, a number of the loading functions are inactive and not involved in plastic deformation. Because of this, any flow rule describing the full range of behaviour associated with a vertex is likely to be nonlinear in the small. This was recognized by Rudnicki and Rice (1975) when formulating their model for vertex hardening incorporating the effects of dilatancy and, in particular, in proposing a linear form for stress paths that are not too far removed from straight ahead loading. In the more general case, the tangential moduli for the inelastic components are not symmetric (Bažant, 1978) and the material described is accordingly path-dependent in the small. This may well be a feature of the behaviour of some real materials but it must introduce difficulties into numerical computations as the precise path used to effect an increment $\delta\sigma_{ij}$ or $\delta\varepsilon_{ij}$ becomes influential.

3.12 CONCLUDING REMARKS

A number of features of theories used to describe time-independent behaviour in rocks and concretes have been reviewed. This discussion has been concentrated on path dependence and theories based on elasticity, plasticity, and their various extensions. The variety of approaches possible means that certain ranges of observed physical behaviour can be well represented by more than one model. This means that there is a choice to be made when considering applications. As yet, there is little objective guidance on the precision required in materials models when these are used in the solution of engineering problems of practical interest. It may be that a lack of detailed correspondence can be accepted provided that the theory does not obscure recognition of the physical phenomena involved. Some research on these matters is clearly needed.

In applications, the quality of the model depends on the quality of the experimental data used to evaluate the materials constants or functions used in the theory. In most instances, results are available only for the total stress and total strain for particular stress paths, it being very difficult to isolate the nonlinear components. In spite of this, attempts are needed to investigate experimentally questions of path dependence and the influence of loading paths that show sharp changes in direction. An extension of the method already used by Spooner and Dougill (1975) to quantify damage in concrete in uniaxial compression may be useful in this.

REFERENCES

Batdorf, S. B., and Budiansky, B. (1949) 'A mathematical theory of plasticity based on the concept of slip', NACA Technical Note, TN 1871.

Bažant, Z. P. (1978) 'Endochronic inelasticity and incremental plasticity', *Int. Journal of Solids and Structures*, **14**, 691–714.

Bažant, Z. P. (1979) 'Critique of orthotropic models and triaxial testing of concrete and soils', Structural Engineering Report No. 79-10/640C, Department of Civil Engineering, Northwestern University, Evanston, Illinois.

Bažant, Z. P. (1980) 'Work inequalities for plastic fracturing materials', *Int. Journal of Solids and Structures*, **16**.

Bažant, Z. P. (1982) 'Crack band model for fracture of geomaterials', *Proceedings of 4th International Conference on Numerical Methods in Geomechanics, Edmonton, Alberta, Canada*.

Bieniawski, Z. T. (1971) 'Deformational behaviour of fractured rock under triaxial compression, Paper 50, Vol. 1, *Structure, Solid Mechanics and Engineering Design* (Ed. M. Teeni), Wiley, pp. 589–598.
Brandtzaeg, A. (1927) 'Failure of a material composed of non-isotropic elements, *Nor. Vidensk. Selsk. Skr*, No. 2.
Cedolin, L., Crutzen, Y. R. J., and Dei Poli, S. (1977) 'Triaxial stress–strain relationship for concrete, *Journal Eng. Mech. Divn. ASCE*, **103**, EM3, 423–439.
Chen, A. C. T., and Chen, W. F. (1975) 'Constitutive relations for concrete', *Journal Eng. Mech. Divn. ASCE*, **101**, EM4, 465–481.
Chen, W. F. (1982) *'Plasticity in Reinforced Concrete*, McGraw-Hill, pp. 474.
Chen, W. F. and Saleeb, A. F. (1982) *Constitutive Equations for Engineering Materials*, Vol. 1, *Elasticity and Modelling*, Wiley-Interscience, pp. 579.
Coon M. D., and Evans, R. J. (1972) 'Incremental constitutive laws and their associated failure criteria with application to plain concrete, *Int. Journal of Solids and Structures*, **8**, 1169–1180.
Dougill, J. W. (1975) 'Some remarks on path-independence in the small in plasticity, *Quarterly of Applied Mathematics*, **33**, 3, 233–243.
Dougill, J. W. (1976) 'On stable progressively fracturing solids, *Zeitschrift fur angewandte Mathematik und Physik (ZAMP)*, **27**, 4, 423–437.
Dougill, J. W. (1983) 'Path dependence and a general theory for the progressively fracturing solid'. *Proc. Royal. Soc. Lond.* A 390, 341–351.
Dougill, J. W., and Rida, M. A. M. (1980) 'Further consideration of progressively fracturing solids', *Journal of the Engineering Mechanics Division, ASCE* **106**, EM5, 1021–1038.
Drucker, D. C. (1950) 'Some implications of work-hardening and ideal plasticity, *Quart. App. Maths.*, **7**, 411–418.
Drucker, D. C. (1968) 'On the continuum as an assemblage of homogeneous elements or states', *Proc. IUTAM Symposium on Irreversible effects in elasticity, plasticity and fluid dynamics* (Eds. ABIR, D., and Reiner, M.), Pergamon Press, Oxford, pp. 331–350.
Drucker, D. C., Gibson, R. E., and Henkel, D. J. (1957) 'Soils mechanics and work hardening theories of plasticity', *Trans. Am. Soc. of Civil Engineers*, **122**, 338–346.
Drucker, D. C., and Prager, W. (1952) 'Soil mechanics and plastic analysis or limit design', *Quart. of Applied Maths.*, **10**, 157–175.
Gambarova, P. G. (1983) 'Analytical models for cracking of concrete subject to shear', contribution to IUTAM Symposium on Mechanics of Geomaterials, rocks, concretes and soils, Northwestern University, USA, 1983.
Hillerborg, A., Modeer, M., and Petersson, P. E. (1976) 'Analysis of crack formation and crack growth in concrete by means of fracture mechanics and finite elements,' *Cement and Concrete Research*, **6**, 773–782.
Hsu, T. C., Slate, F. O., Sturman, G. M., and Winter, G. (1963) 'Microcracking of plain concrete and the shape of the stress–strain curve', *Proc. American Concrete Institute*, **60**, 2, 209–224.
Il'iushin, A. A. (1961) 'On the postulate of plasticity', *Applied Maths. and Mech.*, **25**, 746–752, translation of *Prikladnaya Matematika i Mekhanika*, **25**, 503–507.
Koiter, W. T. (1953) 'Stress strain relations, uniqueness and veriational theorems for elastic–plastic materials with a singular yield surface', *Quart. Appl Maths.*, **II**, 350–354.
Kotsovos, M. D., and Newman, J. B. (1977) 'Behaviour of concrete until multiaxial stress', *Journal of the American Concrete Institute, Proc.*, **74**, 9, 443–446.
Kotsovos, M. D., and Newman, J. B. (1978) 'Generalised stress–strain relationships for concrete', *Journal Eng. Mech. Divn. ASCE*, **104**, EM4, 845–856.
Kupfer, H. B., and Gerstle, K. H. (1973) 'Behaviour of concrete under biaxial stresses', *Journal Eng. Mech. Divn. ASCE*, **99**, EM4, 852–866.

Kupfer, H., Hilsdorf, H. K., and Rusch, H. (1969) 'Behaviour of concrete under biaxial stresses', *Journal A. C. I. Proc.*, **66**, 8, 656–666.

Murray, D. W. Chitnuyanondh, L., Rijub-Agha, K. Y., and Wong, C. (1979) 'Concrete plasticity theory for biaxial stress analysis', *Journal Eng. Mech. Divn. ASCE*, **105**, EM6 989–1006.

Ortiz, M., and Popov, E. P. (1982) 'Plain concrete as a composite material', *Mechanics of Materials*, **1**, 2, 139–150.

Prager, W. (1949) 'Recent developments in the mathematical theory of plasticity', *J. Applied Physics*, **20**, 235–241.

Prager, W. (1966) 'On elastic, perfectly locking materials', *Proc. 11th International Conference on Applied Mechanics, Munchen, 1964*, Springer, pp. 538–544.

Reid, S. R. (1970) 'On the construction of mathematical model for the inelastic deformation of concrete', Central Electricity Research Laboratories, Lab. Note RD/L/N 106/70, Leatherhead, UK.

Resende, L., and Martin, J. B. (1982) 'A consistent Drucker–Prager cap model for geotechnical materials', Non-linear Structural Mechanics Research Unit Technical Report, No. 15, University of Cape Town, S.A.

Richart, F. E., Brandtzaeg, A., and Brown, R. L. (1928) 'A study of the failure of concrete under combined stress', Univ. of Illinois Expt. Station Bulletin No. 185, 102 pp.

Rida, M. A. M. (1981) 'The use of linear loading functions in progressive fracture, plasticity and combined behaviour', Ph.D. Thesis, University of London, 277 pp.

Rudnicki, J. W., and Rice, J. R. (1975) 'Conditions for the localization of deformation in pressure-sensitive dilatant materials', *J. Mech. and Physics of Solids*, **23**, 371–394.

Sanders, J. L. Jnr. (1954) 'Plastic stress strain relations based on linear loading functions', *Proc. 2nd U.S. Nat. Congress on Applied Mechanics*, pp. 455–460.

Spooner, D. C. and Dougill, J. W. (1975) 'A quantitative assessment of damage sustained in concrete during compressive loading', *Magazine of Concrete Research*, **27**, 92, 151–160.

Zienkiewicz, O. C., Nayak, G. C., and Owen, D. R. J. (1972) 'Composite and 'overlay' models in numerical analysis of elasto–plastic continuua', *Proc. Int. Symposium on Foundations of Plasticity, Warsaw*, (Ed. A. Sawczuk), Noordhoff, pp. 107–123.

Mechanics of Geomaterials
Edited by Z. Bažant
© 1985 John Wiley & Sons Ltd

Chapter 4

Requirements for Constitutive Relations for Soils

G. Gudehus

4.1 INTRODUCTION

Constitutive relations are needed in almost every method of soil mechanics: planning and evaluation of laboratory and field tests, analytical and numerical prediction or back analysis of forces and displacements within soil bodies. Until about 20 years ago, one was rather satisfied with linear elastic and ideally plastic formulations. The need of more sophisticated approaches arose from

- large deviations among thus predicted and observed soil behaviour,
- an enormous increase of computer capacity allowing voluminous numerical approaches.

Considerable efforts have been made by building and using new experimental set-ups, notably truly triaxial apparatus, by mathematically formulating various constitutive assumptions, and by adapting them to finite difference and finite element calculations. The result is, in the author's opinion, not very satisfying as yet. Promising constitutive relations of increasing complexity, frequently with experimental and/or numerical applications, have been presented in numerous papers. A close inspection by trying to repeat experiments and calculations frequently leads to a disappointment.

This state of affairs has initiated an international workshop on constitutive relations which was held in Grenoble in September 1982 (it was not the only workshop of this kind). I had hoped that it would enable me to write an overview paper for the William Prager Symposium in Evanston. The editorial work (Gudehus *et al.*, 1984) has convinced me that the time is not ready for such an attempt, and for the following reasons:

- many of the available papers are not fully tractable or almost unreadable,
- various of the proposed equations turn out to be incomplete or inconsistent,
- the necessary procedures for determining material constants are not outlined in many papers, or not technically feasible,

48   *Mechanics of Geomaterials*

- the authors are rather far from an agreement on material properties which should be covered by any constitutive relation,
- many proposals can only serve as a fit to triaxial test results and do not suit for typical boundary value problems of foundation engineering.

It is stressed that these flaws are not characteristic of all publications, and do not prove incompetence of the authors. There are fully tractable, mathematically sound papers with constitutive assumptions, however, which are rather too restrictive as against observed behaviour. Approaches in other papers are so sophisticated (in order to describe some details more precisely) that the consequences can hardly be overlooked. The soil seems to defeat its investigators again and again. It is understandable then that users revert to very simple formulations. More sophisticated theories are not accepted if they are only recommended as black boxes. The requirements outlined in the sequel are intended to better avoid the above shortcomings. They are thought as a kind of check-list for judging constitutive relations. Negative examples from my own experience are not given as it would be unfair to blame a random group of authors. We should keep in mind that science can only proceed by ideas and subsequent selection. Selection procedures should not discourage those who try to reveal the intricate mechanical behaviour of soils.

The requirements are grouped from an engineering point of view. If constitutive relations are fit for solving practical boundary value problems of soil mechanics they should cover frequently occurring processes (section 4.2), reflect repeatedly observed properties (section 4.3), and be manageable (section 4.4). The well-known physical and mathematical requirements are implied. The representation is mainly graphical in order to avoid symbols that have not generally been stipulated. Some notes have been added after the Prager Symposium (section 4.5).

4.2 PROCESSES TO BE COVERED

Constitutive relations of soils should mainly serve to achieve safe and economic designs in foundation engineering. Thus they should cover, with an adequate accuracy, stress and strain histories typically occurring in the ground, in field, model and element tests. This may be illustrated by some examples.

Uniaxial compression and extension occurs in uniformly layered, eroded and loaded grounds (Figure 4.1). Consider a soil element (a) with principal effective stress σ_1 and $\sigma_2 = \sigma_3$, and strains ε_1 and $\varepsilon_2 = \varepsilon_3 = 0$ (pressure and compression positive). Strain and stress paths are numbered to indicate subsequent times; time magnitudes could be represented by associating them with these numbers (not done in the figure). The soil may be first compressed by a superimposed layer, then partly extended by partial erosion, and recompressed (b). The stress path is first proportional, then curved with σ_2/σ_1 increasing towards its possible

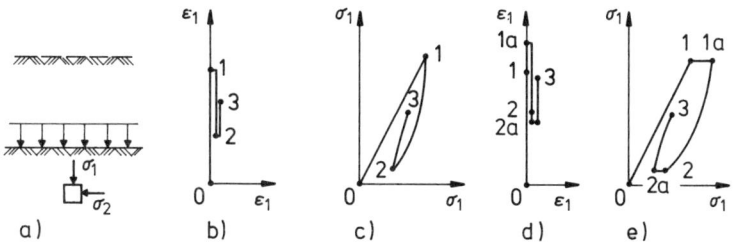

Figure 4.1 Strain (b,c) and stress paths (d,e) under vertical compression and extension (a)

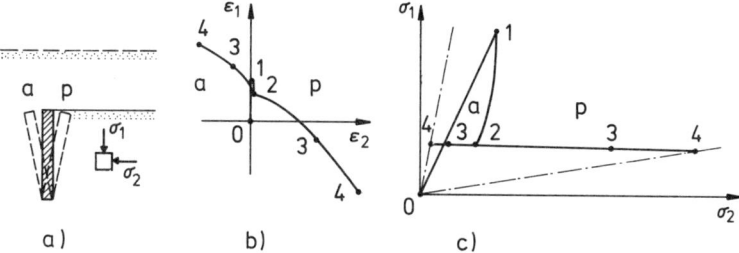

Figure 4.2 Strain (b) and stress paths (c) in case of active or passive earth pressure in dry sand (a)

maximum, and finally tending towards the first one without necessarily reaching it (c). In order to represent the influence of viscosity let us assume the same amounts of σ_1 at the points 1 and 2 of reversal, but delays at these stages (d and e). There is an additional increase of ε_1 and σ_2 during the time 1–1a. With the same thickness of eroded layer a reduction of σ_2 and ε_1 occurs approximately as before (1a–2). During the second delay (2–2a) ε_1 and σ_2 decrease further. The subsequent recompression can be stronger or weaker than before; this is the decisive information for a settlement calculation.

A plane strain earth pressure problem with dry granular soil is depicted in Figure 4.2. The wall (a), placed after sedimentation and partial erosion of the soil, is assumed, for simplicity's sake, to rotate around its foot up to the active (a) or passive (p) state. Vertical strains increase and horizontal strains decrease in the active case; the opposite process occurs in the passive case (b) (state 2 serves as reference state). The volume decreases first, and later increases. During the wall rotation σ_1 remains fairly constant for reasons of equilibrium (c). σ_2 decreases (increases in the passive case) and remains constant for a certain amount of further wall rotation. The limiting states (4) are accompanied by narrow inclined shear zones called slip surfaces. Delays are of minor importance in case of sand if cementation is excluded.

Figure 4.3 shows the corresponding earth pressure problem (a) with saturated clay, which is more difficult. The strain path (b) is essentially the same as in

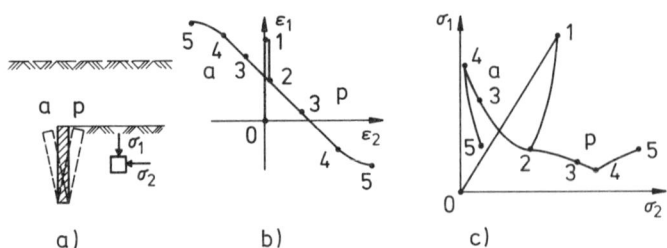

Figure 4.3 Strain (b) and stress paths (c) in case of active and passive earth pressure in saturated clay (a)

Figure 4.1(b) for stages 0–1–2, practically volume-conserving during a rapid wall rotation 2–3–4, and accompanied by volume changes during a subsequent fully drained slow rotation. The stress path (c) deviates, with the onset of wall rotation, from $\sigma_1 =$ const. until the excess pore pressure is fully dissipated. In the active case, the pore pressure decreases so that σ_1 increases; the opposite happens in the passive case. A volume increase (4–5) is accompanied by an increase of σ_2 in the active case, and a volume decrease with σ_2-increase in the passive case. Close to the limiting states (4) narrow shear zones arise. Delays reveal a marked influence of pore-water flow and viscosity of the grain skeleton.

A bearing capacity problem with dry sand is represented by Figure 4.4. It is helpful to consider at least one soil element under the centre of a strip foundation (A) and one under the neighbouring free surface (B). Sedimentation and

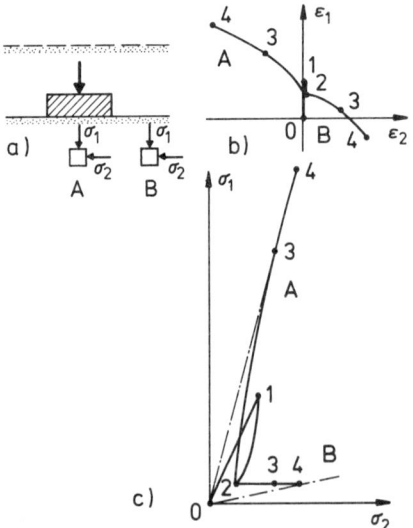

Figure 4.4 Strain (b) and stress paths (c) beneath a strip foundation upon dry sand (a)

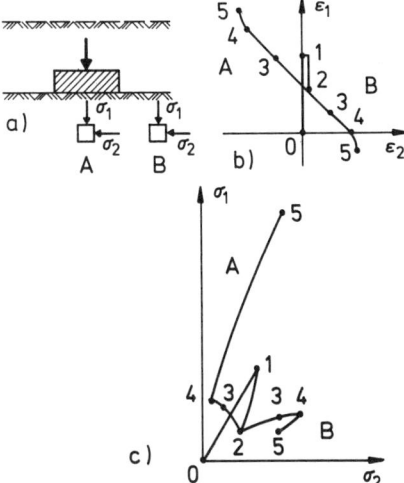

Figure 4.5 Strain (b) and stress paths (c) beneath a strip foundation upon saturated clay (a)

unloading by erosion are the same for A and B, the paths corresponding to Figure 4.1. The strains (b) due to strip loading are first contractant, later dilatant. The stress paths (c) tend towards limiting states earlier in A than in B. For equilibrium reasons σ_1 is nearly constant in B and σ_2 is roughly equal in A and B. In strain space A corresponds to the active and B to the passive case of Figure 4.2. Narrow shear zones occur in the vicinity of limit states (4).

Figure 4.5 shows the corresponding problem with saturated clay. After uniform precompression the strains (b) are nearly volume-conserving during rapid load application, and afterwards, with sufficient drainage time, contractant in A and dilatant in B. The effective stress paths (c) are governed by pore pressure increase in A and decrease in B which is dissipated slowly so that the paths tend to the same statical conditions as in Figure 4.4. Slip surfaces can occur.

Figure 4.1 to 4.4 may suffice to represent typical boundary value problems of foundation engineering. Other and more complicated problems can only be mentioned here:

- dams, slopes, cuts and cavities imply other shapes of boundaries;
- walls and foundations of other shapes and other foundation bodies imply other boundaries and kinematical conditions;
- layers, faults and tectonic stresses produce other initial states than the ones of Figure 4.1 to 4.5;
- loads and displacements imposed to the earth body can be very rapid, transient, repeated, or cyclic;
- hydraulic, chemical, thermal or electrical processes can play an important role.

We leave these aspects aside not as they are unimportant but as the existing constitutive relations do not even cover yet all the cases described by Figures 4.1 to 4.5 in a satisfactory manner. Another class of boundary value problems has to be dealt with here, however. It arises when soil bodies are solicited in order to detect their material properties. Numerous field and laboratory tests can serve, without being element tests (i.e. implying uniform stress and strain in the soil body), to check constitutive assumptions and to determine material constants defined by one or the other constitutive relation. Preferably such boundary value problems should yield closed-form solutions.

The expansion of a vertically clindrical cavity by a pressuremeter is represented by Figure 4.6. The strain path (b) of a surrounding soil element is practically volume-conserving with saturated clay (c), and first contractant than dilatant with dry sand (s). The effective stress path (c) tends to a limit state with decreasing σ_2 and σ_1 with soft clay, whereas σ_1 increases with sand (and hard clay). Unfortunately, the initial state (1 in Figure 4.6) is not well-defined and at best produced by a process as in Figure 4.1 which implies a plane of strains orthogonal to the one of Figure 4.6. Thus even the best pressuremeter and associated theory cannot yield more than crude estimate of material parameters.

Figure 4.7 shows a sample in a triaxial apparatus. Even with a perfect end lubrication and non-rotating end plates bulging (a) can occur. This bifurcation conceals the true stress–strain curve (dashed) from a certain point (B) onwards. It can be prevented by use of flat samples, whereas necking under triaxial extension

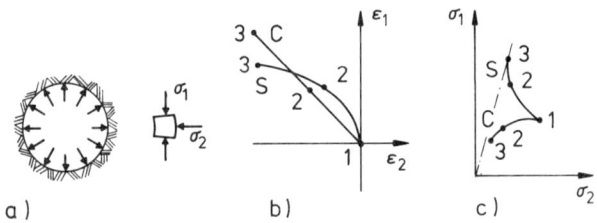

Figure 4.6 Stress (b) and strain paths (c) around an expanding cavity (a)

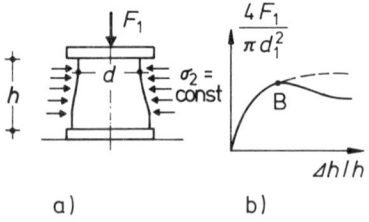

Figure 4.7 Load–displacement curve (b) of a cylindrical sample (a)

is always occurring if the confining stress σ_2 is produced by fluid pressure. Localized modes of bifurcation (narrow shear bands and cracks) cannot even be avoided with purely strain-controlled tests. Evaluation of test results as if diffuse or localized modes of bifurcation had not occurred leads to wrong conclusions with respect to constitutive relations and parameters.

Another important group of boundary value problems is associated with so-called undisturbed samples. Extracting a soil element from the ground and inserting it into a testing apparatus adds a complicated and rarely well-defined history to the one of Figure 4.1. If a constitutive relation does not reflect this procedure its practical application can be rather unrealistic.

Physical model tests, possibly with artificially increased gravity, can reproduce such processes as depicted in Figures 4.1 to 4.6. Constitutive relations are the clue to model laws which are needed for the transfer of results to prototypes. Thus constitutive relations should represent similarity properties of the material. (Another use of model tests—and large scale tests—will be discussed in section 4.4.)

4.3 PROPERTIES TO BE REPRESENTED

Strictly speaking, material properties are defined by constitutive relations (irrespective of their analytical or graphical representation). We here denote by properties relations for certain very restricted processes of soil elements represented by certain types of graphs or functions as far as they have been corroborated by element tests. Such relations do not suffice to solve boundary value problems but serve as a kind of frame for constitutive relations.

The stress and strain paths represented in Figures 4.1 to 4.6 can be decomposed into monotonic sections separated by sharp bends or even reversals. A monotonic section can be straight in stress or strain space, but not generally both. We consider paths with 0, 1 or 2 sharp bends. The influence of viscosity is suppressed by assuming constant magnitudes of strain rates first; this simplification will be dropped at the end of this section.

Monotonic paths *without sharp bends* occur during sedimentation and, more generally, in the laboratory. They have been studied in strain-controlled cuboidal tests with sand (Goldscheider, 1975) and biaxial tests with clay (Kuntsche, 1982). When starting with zero stress, proportional strain and stress paths are associated (Figure 4.8). We consider three distinguished strain path directions (a): isotropic (A), uniaxial (B), and so that the soil element remains practically unstressed (C). The latter cannot precisely be determined and is therefore replaced, in Figure 4.8, by a closely neighbouring path. It is contractant for clay, nearly isochoric for loose sand, and dilatant for dense sand. The physically possible sector of strain paths is deliminated by type C directions.

The associated stress paths (b) are also hydrostatic in case A, and have an inclination $\sigma_2/\sigma_1 = K_0$ in case B; K_0 is the so-called earth pressure coefficient at rest. The nearly stress-free path is close to the inclination $\sigma_2/\sigma_1 = K_a$, K_a being

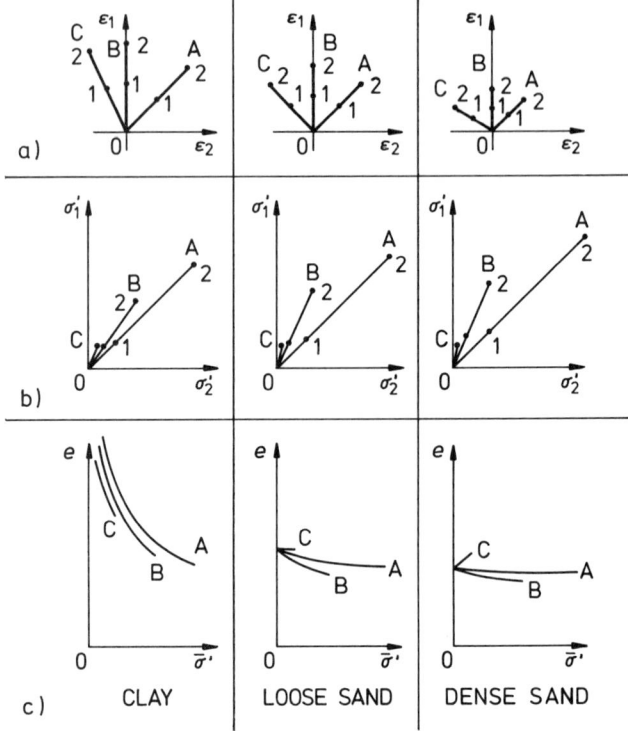

Figure 4.8 Stress paths (b) and changes of void ratio (c) under proportional stress paths (a)

the coefficient of active earth pressure. The relationship among stress and strain path directions is nearly independent of the material (not depicted). K_0 and K_a in case of clay are independent of density, but increase with initial density in case of sand.

The representation is completed by a plot of void ratio e versus mean effective stress $\bar{\sigma}$ (c). The curves for clay do not strongly depend on the path direction. They can scarcely be determined for $\bar{\sigma} \to 0$ and evidently show $e \to 0$ for $\bar{\sigma} \to \infty$. The void ratio of sand at $\bar{\sigma} = 0$ can be produced by placement of the grains. With increasing $\bar{\sigma}$, the change of e is markedly different from the one of clay. With type A and B paths, the compressibility is drastically reduced by an increase of grain hardness and density (although the asymptote $e \to 0$ for $\bar{\sigma} \to \infty$ remains valid). Type C paths are abnormal by leaving e constant with loose sand and implying an increase of e with dense sand. These relations are partly veiled by the scatter of results: it is not possible to exactly reproduce a certain initial void ratio.

We now turn to paths *with one sharp bend*. The first sections may be as in Figure 4.8, the second ones straight either in strain or stress space. Owing to their

Requirements for Constitutive Relations for Soils 55

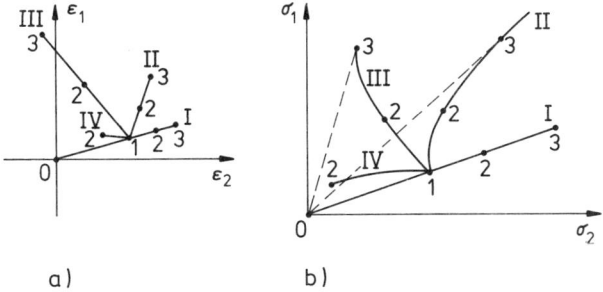

Figure 4.9 Stress paths (b) due to bilinear strain paths (a)

wide variety only narrow subgroups of such paths have been studied experimentally. The material properties outlined in the sequel are subject to some doubt as they have not generally been checked. Consider four bilinear strain paths (Figure 4.9(a)). The associated stress paths (b) of the second sections are curved. They tend towards those stress paths which had been produced by the strain path direction of the second section from the very beginning. This property of both sand (Gudehus et al., 1977) and clay (Kuntsche, 1982) holds for strain paths within the possible sector of Figure 4.8(a); path II of Figure 4.9 is one example. States during proportional stress and strain paths are called swept-out-memory- or SOM-states as they do not bear the memory of more complicated histories. It appears that SOM-states are produced by any monotonic strain path of sufficient amount. If the strain path direction is outside the physically possible sector (cf. Figure 4.8(a)) the stress path comes back to the origin (e.g. IV), i.e. the soil element tends to decay. Certain path directions (e.g. III) lead to the SOM-states associated with the sector limits of Figure 4.8 and leave the stress state unchanged then. These are limit states both in the statical and the kinematical sense (Goldscheider, 1975).

Figure 4.10 illustrates that identical strain paths are associated with geometrically similar stress paths, if the latter start close to zero stress. This kind of self-similarity is characteristic of sand and clay. It holds also for unstressing paths, as

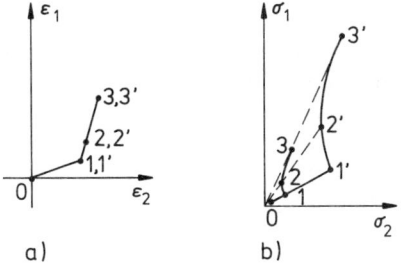

Figure 4.10 Strain paths (a) associated with similar stress paths (b)

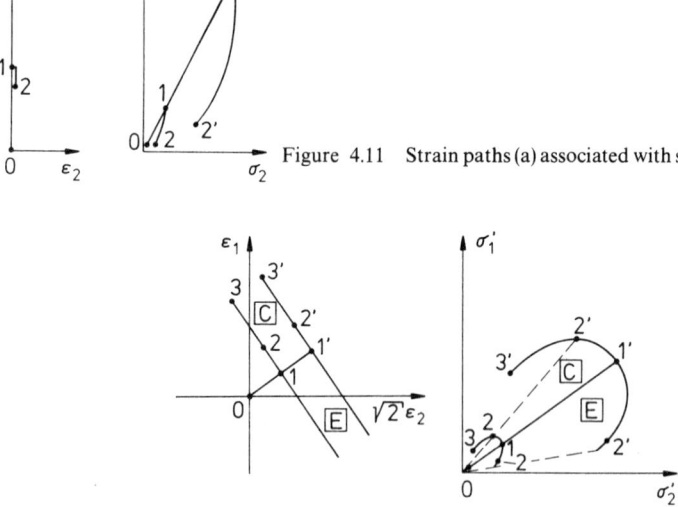

Figure 4.11 Strain paths (a) associated with similar stress paths (b)

Figure 4.12 Strain paths (a) associated with similar stress paths (b)

shown in Figure 4.11 with an initial uniaxial compression (cf. Figure 4.1). (Zero stress has to be excluded in order to avoid contradictions.)

Figure 4.12 shows a special case of the properties outlined by means of Figure 4.9 and 4.10: constant-volume cylindrical compression (C) and extension (E) of clay after isotropic compression; such processes occur in undrained triaxial tests. The stress paths (b) imply a reduction of mean stress up to the limit states (in this case called critical states). Limit states are not clearly reached under extension with confining fluid pressure due to necking of the sample. Note that such stress paths are very special as compared with the ones in the field (cf. Figures 4.3 and 4.5). It appears that only the SOM and similarity rules as outlined above remain valid.

Similar properties are revealed with bilinear stress paths. We will refer here to results of triaxial compression tests only as with other triaxial tests of this kind uniform sample deformations have rarely been secured. Typical sand results by Hettler and Vardoulakis (1984) are plotted in Figure 4.13. Two geometrically similar stress paths (b) produce identical strain paths (a). After an isotropic compression of negligible amount there is first uniaxial compression ($\dot{\varepsilon}_2 = 0$). The relative dilation $\dot{\varepsilon}_2/\dot{\varepsilon}_1$ at the limit state is the same as the one which produces a limit state in the sense of Figure 4.9 if the density is equal.

Figure 4.14 shows corresponding results for clay (drained triaxial compression). The stress paths (b) differ from the ones for sand only by the stress ratio

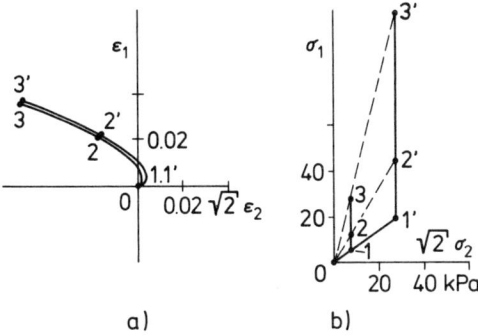

Figure 4.13 Strain paths (a) associated with similar stress paths (b)

σ_1/σ_2 reached at limit states. The strain paths (a) are nearly vertical ($\hat{\varepsilon}_2 \simeq 0$) first; the relative volume change $\dot{\varepsilon}_2/\dot{\varepsilon}_1$ comes close to $-1/2$, i.e. constant volume, when reaching the limit states.

A generalization of these properties to other bilinear stress paths requires some caution. The observation of $\dot{\varepsilon}_2 = 0$ at the onset of cylindrical compression (not made with conventional tests) is certainly restricted to cylindrical symmetry. A change from contractancy to dilatancy is typical of the strain paths. More generally, it has been found that immediately after any sharp bend of a bilinear stress path there is a volume decrease (Goldscheider, 1975; Kuntsche, 1982).

The similarity property has to be relaxed for large amounts of pressure, presumably due to deformation and fracture of grains. Strain paths with higher initial stress reach limit states with lower dilatancy and stress ratios and higher amounts of strain. Conversely, stress paths with higher pressure level produce higher strain and compression amounts. However, this effect is practically negligible if the stress components do not exceed ca. 1 MN/m².

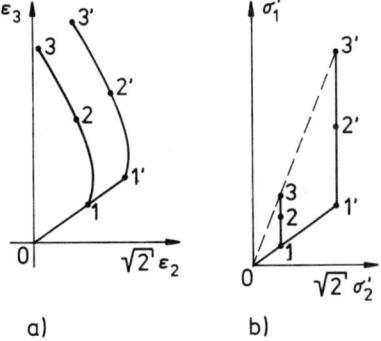

Figure 4.14 Strain paths (a) associated with similar stress paths (b)

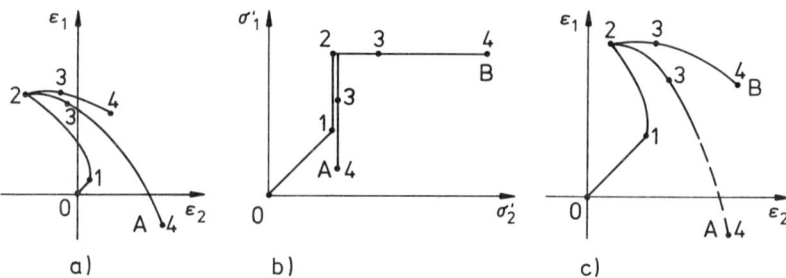

Figure 4.15 Trilinear stress paths (b) and associated typical strain paths of sand (a) and clay (c)

Rather little is known about paths with *two sharp bends* because of their wide variety. Some features of sand behaviour under plane strain are illustrated by Figure 4.15(a) and (b). After the second sharp bend (point 2) a straight stress path section may turn back (A) or be orthogonal to the previous section (B, sometimes called loading to the side). With any density the strain path is contractant immediately after the bend (Goldscheider, 1975). From a certain stress point (3) onwards the subsequent strains develop as if the previous stress path had been proportional (i.e. 0–3 straight). The then required path length (2–3) increases with the deviation of the path directions before and after the bend. Observations indicate that the similarity rule as outlined above holds again.

It appears that sufficiently long monotonic paths after the second bend lead to SOM-states again. There are states in between (2–3), however, which cannot only be characterized by stress and density. (In a rather vague sense, not yet defined for arbitrary paths with two sharp bends, they are frequently called overconsolidated or preloaded states.)

Similarity and memory properties can also explain the repeatability of test results with not more than one scheduled sharp bend: remnants of more path bends at low stress level inevitably encountered at the beginning of the test are swept out by subsequent monotonic paths.

Stress paths of the same kind imposed upon clay produce strain paths as shown in Figure 4.15(c). The volume changes differ from the ones of sand both in quality and quantity. Volume decrease is more prenounced, and increase is almost inevitably localized in shear bands (dashed path section). It appears that larger monotonic strains than with sands are needed to reach SOM-states. The similarity rules seem to hold again.

Finally, some notes on *soil viscosity* are suited. Its influence can be represented by associated plots of strain and stress paths with absolute time labels. The experimental results are restricted to very few types of paths. The importance of viscous effects increases with the relative amount of water in the diffuse double layers around the soil particles. The following restriction to clay should not veil

Requirements for Constitutive Relations for Soils

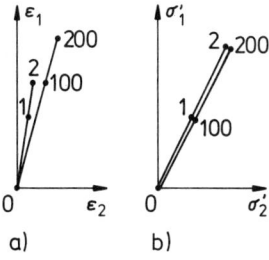

Figure 4.16 Proportional strain (a) and stress paths (b) with time labels

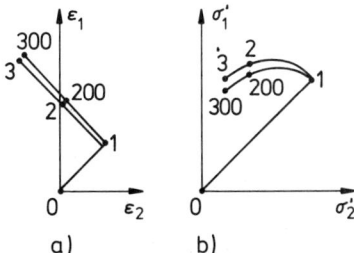

Figure 4.17 Bilinear strain (a) and associated stress paths (b) with time labels

the fact that qualitatively the same viscous properties are observed with sand and silt.

Proportional stress paths with different rates are associated with different proportional strain paths (Figure 4.16). A reduction of stress rates by orders of magnitude causes only slight changes of strain ratios and amounts. Isochoric strain paths with different rates are associated with different effective stress paths of the same type (Figure 4.17, cf. Figure 4.12). The change is only noticeable if the rates vary by orders of magnitude.

It appears that the similarity and memory properties outlined above are not affected if the rates of stress or strain remain within one order of magnitude. Stronger changes of rates affect both directional and intensity (stiffness) properties, however.

4.4 MANAGEABILITY

A constitutive relation should be conceived and represented in such a manner that one can reasonably work with it. As this seemingly trivial requirement is by no means always satisfied it will be specified in the sequel. To sum up, the representation should be tractable, the determination of the required material constants should be technically and economically feasible, and the equations should be analytically and/or numerically well-posed and economically feasible. All these requirements are only necessary, not sufficient.

Tractability of a paper can effectively be tested by having it worked through by a reader who is willing and able to understand but not familiar with details of the concept. If he obtains the same analytical relations from the given set of assumptions, and the same figures from given sets of data, the paper is tractable. As such tests, made by the author and some of his co-workers, produced negative results with more than three-quarters of the papers tested, it is of reason to give some hints here. (Their necessity reflects how complicated the field of constitutive relations has become.)

The set of assumptions used should be completely outlined. Logical completeness, as with a set of axioms, is only desirable. More important is a hierarchical order of presentation so that the reader can more easily look for possible inconsistencies. Following K. Popper, one should strictly distinguish contradictions of the assumptions among each other and versus observations. Even oversimplified assumptions may be justified for engineering approaches, if they are clearly stated as such. It is helpful, though not indispensable in the sense of tractability, to justify the assumptions by observations (this is not yet a strict validation).

All the quantities occurring in the proposed relations should be completely defined. Of course, stress and strain definitions can be taken from the literature on continuum mechanics. Notations and sign conventions are not yet standardized, unfortunately, but this is only a minor source of confusion. New quantities should be introduced by showing how to work with them analytically and numerically, i.e. by working definitions. Functionals, functions, variables, parameters and constants should be clearly discerned. The reader can learn their meaning with a well-organized typology and some significant examples. (It is sometimes required that all the quantities should have 'physical meaning'; this will be achieved if the reader can get familiar with the proposed relations by working with them.)

A complete procedure for determining the *material constants* should be given. This implies a description of

- required sampling and testing equipment and procedure,
- evaluation of test results and typical range of constants.

The tractable description should enable the reader to repeat the procedure and to judge the range of applicability of the proposed relations. Thus it will turn out whether the proposal is technically and economically feasible.

Some sophisticated constitutive relations imply material constants which can only be determined by element tests with rarely available equipment (such as a truly triaxial apparatus) or not-yet-existing machines (such as for general shear deformation). This has to be clearly stated together with recommended values of such constants. This shortcoming cannot be avoided: simpler constitutive relations tacitly imply fixed values of such constants (e.g. by neglecting the effects of third and mixed invariants). If the mathematical formulation is well-posed, such assumptions can be checked by comparing solutions of boundary value problems with model test results.

The *economical aspect* cannot easily be judged as simplicity is not a measurable notion. The number of constants is certainly not a relevant measure, except for the required human and artificial memory. A rough economical measure is the required sum of expenses for sampling, testing, and evaluation. If constitutive relations are thought to be for practical application, required working duration and qualification of equipment and staff should be given (note that with a non-tractable description the duration would be infinite). As with any data acquisition in the light of safety and economy, such information cannot be sufficient, however; engineering judgement is also needed.

Constitutive relations are analytically and numerically *well-posed* if

- they lead to solutions of boundary value problems which can be physically realized (existence);
- they lead to one solution if there is only one in reality (uniqueness);
- the solutions with small deviations of material constants show small deviations only if this is also the case in reality (stability).

As relevant constitutive relations of soils are represented by nonlinear or sector-wise linear sets of differential equations these requirements cannot generally be checked by the currently available methods of functional analysis. They can as yet only be tested by numerical experiments which will be indicated in the sequel. Note that the above requirements are only necessary and imply some rather general hypotheses of continuum mechanics (local action, simple material).

The numerical tests should begin with soil elements under uniform stress and strain, i.e. with simulated element tests. One has a set of ordinary (not partial) differential equations then. Within the intended range of application, characterized by path directions and strain rates, histories should be systematically scanned by means of numerical integration. Graphical representations (as by response envelopes, e.g., Gudehus, 1979) are recommended for rapid discrimination. The reader can easily follow these tests if flow charts or even programs are given.

This procedure should precede any publication. It will help the respective author to eliminate improper formulations and to deliminate the range of applicability. The class of histories must not be restricted to element tests available to the author (such as triaxial compression only, say). The examples of Figures 4.1 to 4.7 show that the complete range has to be tested. Numerical stability should be checked first. If this has been achieved (and tractably demonstrated), the range of uniqueness should be ascertained. Note that non-uniqueness of the response of soil elements has not been observed experimentally (which is not to be confused with bifurcation of samples). Non-existence should only occur with certain continuations from limit states.

If the proposed constitutive relation has passed this numerical element test it should be further tested with boundary value problems implying only one or two spatial derivatives. Examples are

- expansion and compression of a cylindrical or spherical cavity;
- expansion and compression of a cylinder or cuboid with deviations from uniform strain;
- rotation of a wall with uniform backfill around its feet;
- shear of a thin layer between two rigid plates.

Some reference experiments are available for such cases. A few numerical and physical aspects will be outlined now.

The moment of truth during such tests arises with sudden changes of path directions. They occur in reality if imposed boundary stresses or displacements are reversed, and in the case of bifurcations. A sensitive change of intensity response to change of path direction, especially in the vicinity of limit states, is characteristic of soils and has thus to be modelled by relevant constitutive relations. If the equations are sectorwise incrementally linear (as elasto–plastic laws, for example), the switch conditions needed for changes of sectors can destroy numerical stability. Completely nonlinear foundations (as some rate-type laws) have to be linearized by a kind of Newton–Raphson method, the convergence of which is by no means guaranteed. Many of the published constitutive relations cannot pass such tests even with only one spatial derivative (cavity expansion and compression, for example).

Economy of the proposed equations should be tested as outlined above for material constants. Although the sum of expenses for hardware and staff is not a sufficient measure estimated figures for it are helpful; they should rely upon the solution of real boundary value problems.

It is noted once more that the criteria of manageability are necessary and not sufficient altogether. As with any theory, comparison of predictions with acceptable experimental results is an indispensable part of validation.

4.5 NOTES ADDED AFTER THE SYMPOSIUM

The discussion paper by Professor Chen can be used to illustrate the requirements given here by means of the well-established 'cap-laws'. There is no doubt that my Figures 4.2 and 4.3 are covered by this kind of elasto–plastic relations; the same could be shown for Figures 4.4, 4.5, and 4.6. These figures are semi-quantitative and intentionally compatible with some earlier elasto–plastic theories. Problems arise if *all* these cases are to be covered quantitatively with the *same* constitutive relation. This is unfortunately not achieved with isotropic hardening; for details see Gudehus *et al.* (1984).

The simple 'cap-models' also meet most of the requirements of manageability, as outlined by Professor Chen. Thus it is justified to use them numerically for well-deliminated classes of problems. One must not overlook then that the constitutive parameters are at best valid only for *one* class, e.g. active earth pressure, passive earth pressure, vertically loaded shallow foundation, triaxial test. The transfer of the same parameters from one class to another one can lead to gross errors. This lack of consistency may be acceptable from an engineering point of view but cries out for improved constitutive relations.

The response envelopes associated with elasto–plastic relations are continuous, but not smooth (Gudehus, 1979). Chambon (1981) has pointed out that smoothness can be necessary to achieve numerical stability. He has proposed tensor-valued smooth interpolation functions by means of which stability is obtained. This procedure is advisable for all sectorwise incrementally linear relations such as elasto–plastic ones.

Finally, it may be worth mentioning that our requirements of manageability assure objectivity in a rather wide sense. If any constitutive relation is user-invariant, i.e. if different users obtain the same solutions with the same data sets, it must be

- logically consistent and language-independent;
- independent of the choice of reference systems, coordinates, and units;
- insensitive to minor numerical deviations.

REFERENCES

Chambon, R. (1981) 'Contribution à la modelisation numérique non-linéaire des sols', Thèse Docteur d'Etat, Univ. Grenoble.
Goldscheider, M. (1975) 'Dilatanzverhalten von Sand bei geknickten Verformungswegen', *Mech. Res. Comm.*, **2**, 143–148.
Gudehus, G., Goldscheider, M., and Winter, H. (1977) 'Mechanical properties of sand and clay and numerical integration methods: some sources of errors and bounds of accuracy', in *Finite Elements in Geomechanics* (Ed. G. Gudehus), Wiley, Chichester, pp. 121–150.
Gudehus, G. (1979) 'A comparison of some constitutive laws for soils under radially symmetric loading and unloading', *Proc. 3rd Int. Conf. Num. Meth. Geomech.* (Ed. W. Wittke), Balkema, Aachen, pp. 1309–1323.
Gudehus, G., Darve, F., and Vardoulakis, I. (1984) *Constitutive Relations of Soils—Results of a Workshop*, Balkema, Rotterdam.
Hettler, A. and Vardoulakis, I. (1984) 'Behaviour of dry sand tested in a large triaxial apparatus', *Géotechnique* **34**, No. 2, pp. 183–198.
Kuntsche, K. (1982) 'Materialverhalten von wassergesättigtem Ton bei ebenen und zylindrischen Verformungen', Veröffentl. Inst. F. Bodenmech. u. Felsmech. Karlsruhe, Heft 91.
Vardoulakis, I. (1979) 'Bifurcation analysis of the triaxial test on sand samples', *Acta Mechanica*, **32**, 35–54.

Mechanics of Geomaterials
Edited by Z. Bažant
© 1985 John Wiley & Sons Ltd

Chapter 5

Constitutive Relations for Concrete, Rock and Soils: Discusser's Report

W. F. Chen

Section I Constitutive Relations for Concrete and Rock

5.1 INTRODUCTION

Although the applications of mechanics to reinforced concrete structures and to rock engineering are old and in many respects well established, the nonlinear deformation and ultimate load analysis of triaxially loaded concrete structures and mining problems related to rocks by means of finite element techniques is relatively recent (Chen, 1983). Different aspects of these advances were reported in several recent books, conference proceedings, and state-of-the-art reports. This includes the books by Chen (1982), Chen and Saleeb (1982, 1985), the Conference Proceedings by IABSE (1979, 1981), by US Defence Nuclear Agency (Schreyer *et al.*, 1981), and the state-of-the-art reports by ASCE (1981) and by Politecnico di Milano (1978), among others.

Most of the finite element studies consider concrete and rock to behave as an elastic or elastic–plastic solid in compression and as a brittle material in tension. Simple constitutive models have been proposed and used extensively in the early studies of concrete and rock mechanics by modelling cracking in the form of tension cut-off criteria via Rankine theory of maximum tensile stresses for tension concrete and by combining this with either a nonlinear elasticity theory or with a plasticity theory via von Mises or Coulomb yield criterion and its associated flow rule.

With the present rapid developments in computational techniques and computing capability, more general theory of continuum mechanics like hyper- or hypoelasticity, classical and endochronic plasticity, and viscoelasticity and plasticity must be developed to describe the very complex behaviour of concrete and rock materials involving phenomena like inelasticity, cracking, time dependency, and discontinuity. In Professor Dougill's paper, he has confined his presentation to the most fundamental aspects of concrete and rock mechanics, that is, the general technique used in the discussion of stress–strain laws based on the theories of elasticity and plasticity leading from modelling concrete and rock materials in the pre-failure regime to modelling the progressively fracturing solids in the post-failure range. In Professor Dougill's presentation, he has placed his emphasis on the critical subject of path dependence requirements and their implications for plasticity and fracturing and also the dual formulation in stress and strain space. In the following, I will concentrate my discussion on a number of topics that have not been covered in Professor Dougill's lecture (preceding chapter). This includes a number of constitutive modelling techniques that were introduced in recent years to describe the nonlinear deformation as well as failure behaviour of concrete or rock under triaxial conditions. The scope is restricted to time-independent material behaviour in the pre- and post-failure range which can be described on the continuum level.

5.2 CONSTITUTIVE MODELLING

Within the framework of continuum mechanics, the behaviour of real materials is generally idealized as time-independent or time-dependent. In the time-independent idealization, such as elastic and elastic–plastic models described in Professor Dougill's lecture, time effects are neglected. Time does not appear explicitly as a variable in the constitutive relations; phenomena like rate sensitivity, ageing effects, and creep are not included in these modelling techniques. Further, for an ideal elastic material, the behaviour is reversible and independent of the loading path, while it is irreversible and load path dependent in a plasticity-based model. On the other hand, in the time-dependent material idealization such as the viscoelastic and viscoplastic models, time effects are considered, and, therefore, they are generally capable of describing rate- and history-dependent behaviour.

In concrete and rock mechanics, constitutive modelling of these materials under triaxial stress conditions is of central importance to the analyses and engineering design of triaxially loaded concrete structures and mining systems. Elastic modelling has been used most widely and is well understood, but irreversible deformation is not. This is an area of great importance for concrete and rock materials. The principles of continuum mechanics provide the needed general guidelines for characterization of these materials, a field of increasing importance for sophisticated analyses.

It must be emphasized here that the previous idealizations and subsequent classification of the constitutive modelling techniques are only for mathematical convenience in describing the actual complex behaviour of real materials. Nothing can compel the material to behave according to any of these idealized models. Indeed, for concrete and rock materials, the actual material response will exhibit the behaviour characteristics of most of these models under certain conditions of stresses, temperatures, vibration, and strain rates. Therefore, in any practical problem, it is essential that we determine the limits and conditions under which the material can sensibly be assumed to exhibit the dominant

Constitutive Relations for Concrete, Rock and Soils

characteristics of a particular type of the idealized models. Furthermore, since any idealized model has its own shortcomings, all the results obtained must be interpreted carefully in terms of these shortcomings.

5.3 FAILURE CRITERIA

The failure of concrete or rock in three-dimensional state of stress is extremely complicated. Numerous criteria have been devised to explain the conditions for failure of a material under such a loading state. These models can be classified as one-parameter models including the Rankine or Griffith criterion of maximum tensile stress failure and the Tresca criterion of failure at maximum shear stress, two-parameter models including the Mohr–Coulomb criterion of shear failure, three-parameter models including the well-known Mohr–Coulomb criterion

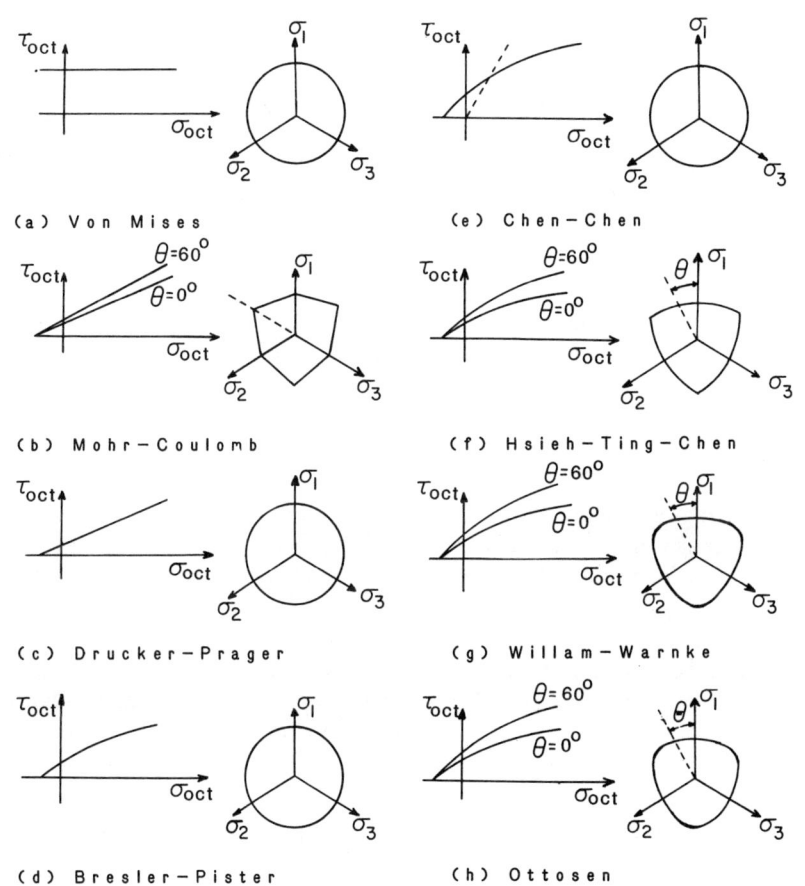

Figure 5.1 Failure models

with a small tension cut-off, and four- and five-parameter models with a nonlinear relation between octahedral normal and shear stresses ($\sigma_{oct} - \tau_{oct}$) as well as noncircular cross-sections on the deviatoric plane (Figure 5.1).

In all these strength models, two basic postulates are adopted: isotropy, and convexity in the principal stress space. The first assumption is mainly introduced because of the inherent simplification of the failure model. It is certainly true some concrete especially rocks exhibit significant anisotropy with respect to their strength which requires the formulation of the failure surface in the six-dimensional stress space instead of the three-dimensional space of principal stresses. For concrete and many rocks, however, the assumption of isotropy is reasonable. On the other hand, convexity is an assumption which is supported by global stability arguments in plasticity (see for example, Chen and Saleeb, 1982). Clearly, there are some questions on the validity of this postulate, and in fact, there is strong indication that the failure envelope for rocks, for example, over a wide range of hydrostatic (confining) pressures may be non-convex with respect to the hydrostatic axis.

Most of these three-, four-, and five-parameter models give a close estimate of the relevant experimental data, contain all the three stress invariants for an isotropic material, reflect all the required characteristics concerning smoothness, convexity, symmetry, curved meridians, etc. as shown in Figure 5.1. But Willam-Warnke's five-parameter model includes most of the earlier one-, two-, and three-parameter models as special cases and therefore becomes increasingly popular in recent years (Chen and Han, 1983a). The five-parameter failure model of Willam-Warnke may be adopted as the basic surface for further development of elastic-plastic-fracture model for concrete (Chen and Han, 1983b).

5.4 PRE-FAILURE RANGE

Linear elasticity for isotropic and transversely isotropic materials constitutes the oldest and simplest approach to modelling the stress–strain behaviour of concrete and rocks under low deforming loads. However, for higher loads or for rocks with large pore space such as the weaker sedimentary rocks, the stress–strain curve is generally nonlinear, and any analysis based on linear elasticity would not be realistic. Such nonlinear behaviour may be characterized by variable stress–strain moduli. The simplest approach to formulate such nonlinear models is to simply replace the elastic constants in the linear stress–strain relations with secant moduli dependent on the stress or strain invariants.

These models are mathematically and conceptually very simple. The models account for two of the main characteristics of concrete and rock behaviour: nonlinearity and the dependence on the hydrostatic stress. The main disadvantage of the models is that they describe a path-independent behaviour. Therefore, their application is primarily directed towards monotonic or proportional loading regimes.

A more rigorous approach in formulating secant stress–strain models for concrete and rocks can be developed on the basis of hyperelasticity theory. This type of formulation can be quite accurate for concrete and rock in proportional loading. They satisfy the rigorous theoretical requirements of continuity, stability, uniqueness, and energy consideration of continuum mechanics. However, here, as noted previously, this type of model fails to identify the inelastic character of concrete and rock deformations, a shortcoming that becomes apparent when the material experiences unloading. The main objection to the hyperelastic formulations is that it often contains too many material parameters. For instance, a third-order isotropic model requires nine constants; while 14 constants are needed for a fifth-order isotropic hyperelastic model. A large number of tests is generally required to determine these constants, which limit the practical usefulness of the models.

The path-independent behaviour implied in the previous secant type of stress–strain formulation can be improved by the hypoelastic formulation in which the incremental stress and strain tensors are linearly related through variable tangent material response moduli that are functions of the stress or strain state. In the simplest case of hypoelastic models, the incremental stress–strain relations are formulated directly as a simple extensions of the isotropic linear elastic model with the elastic constants replaced by variable tangential moduli which are taken to be functions of the stress and/or strain invariants. Models of this type are attractive from both computational and practical viewpoints. They are well suited from both computational and practical viewpoints. They are well suited for implementation of finite element computer codes. The material parameters involved in the models can be easily determined from standard laboratory tests using well defined procedures; and many of these parameters have broad data base. However, the application of this type of hypoelastic models should be confined to loading situations which do not basically differ from the experimental tests from which the material constants were determined or curve-fitted. Thus, the isotropic models should not be used in cases such as non-proportional loading paths or cyclic loadings.

A general description of hyper- and hypoelastic formulations has been given in Professor Dougill's paper. Here, I would like to add the following two comments on problems associated with the hypoelastic modelling. The first problem is that, in the nonlinear range, the hypoelastic models exhibit stress induced anisotropy. This anisotropy implies that the principal axes of stress and strain are different, introducing coupling effect between normal stresses and shear strains. As a result, a total of 21 material moduli for general triaxial conditions have to be defined for every point of the material loading path. This is a difficult task for practical application. The second problem is that under the uniaxial stress condition, the definition of loading and unloading is clear. However, under multiaxial stress conditions, the hypoelastic formulation provides no clear criterion for loading or unloading. Thus, a loading in shear may be accompanied by an unloading in some of the normal stress components. Therefore, assumptions are needed for defining loading–unloading criterion. Furthermore, the material tangent stiffness matrix for a hypoelastic model is generally unsymmetric which results in a considerable increase in both storage and computational time. As a result of this, uniqueness of the solution of boundary value problems cannot generally be assured.

The mechanics of concrete and rocks in the elastic range is well developed. Predictions of acceptable accuracy based on some of these models mentioned previously have been made to many practical problems. However, elastic analyses do not account for the important phenomenon of irreversible deformations of materials. In general, it is this inelastic, irreversible range of deformation that is of great concern in a nonlinear deformation and ultimate load analysis of concrete structures and mining problems. Plasticity theory is well established with a long history of successful application to metals. For concrete or rock materials, the internal micro-cracks developed and grown during loading can be considered as a irreversible behaviour. Figure 5.2 shows a typical uniaxial stress–strain diagram for plain concrete in the compression range. The material exhibits an almost linear behaviour up to the proportional limit at point A, from which the material constitution is progressively weakened due to internal microcracking up to failure at point D. The nonlinear deformations are essentially inelastic, since upon unloading only the portion of ε^e can be recovered from the total deformations $\varepsilon = \varepsilon^e + \varepsilon^p$. It is clear that this phenomenon corresponds exactly to the behaviour of a work-hardening elastic–plastic solid.

The plasticity-based models can account for in principle, path-dependent irreversible inelastic deformations. Using a properly choosing pressure-dependent yield function, a hardening rule, and a flow rule, the models developed can fit rather well to the triaxial stress–strain test data of concrete. The material parameters involved can also be determined relatively easily.

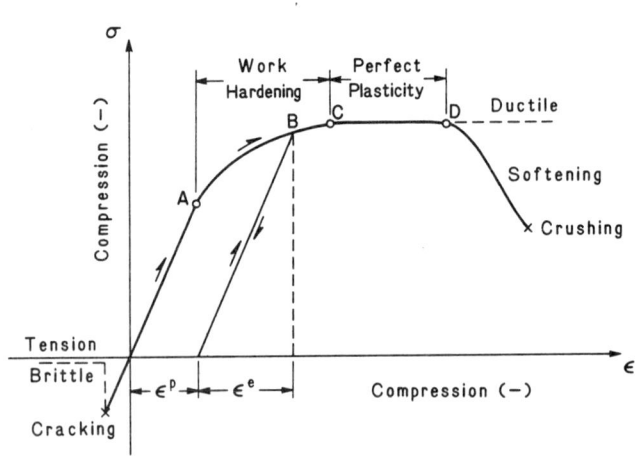

Figure 5.2 Schematic uniaxial stress–strain curve for plain concrete

The theory of plasticity enables one to go beyond the elastic range in a time-independent but theoretically consistent way, because the theory satisfies the conditions of uniqueness, continuity, stability, and thermodynamic laws. Consistency is admirable and the practicability of the theoretical situation is something that cannot be avoided. Thus, where it is possible, efforts should be made to introduce theoretical concepts and mathematics to concrete and rocks. If this is not possible, an empirical and engineering treatment may be introduced. However, experiments for determining the stress–strain diagrams beyond the elastic range for general two- and three-dimensional states of stress are difficult to perform, especially for concrete and rock type of materials. Little has been done in the area of cyclic loading, rate-dependency, non-proportional loading, and post-failure response. The question of modelling strain softening in the post-failure range using plasticity theory is highly controversial. This can, to some extent, be rectified by introducing the concept of progressively fracturing solids, a subject very much emphasized in Professor Dougill's lecture. The formulation of a purely fracturing material is completely analogous to that of flow plasticity.

As apparent from the preceding discussion, it is necessary to include many complex effects in the material model in order to describe the behaviour of concrete and rock properly. An important step in the direction of developing a more unified and comprehensive material model for concrete or rock is the generalization of the theory of viscoplasticity by introducing the measure of intrinsic time (Valanis, 1971). This is known as the endochronic theory of viscoplasticity. This theory offers a possibility to accurately model concrete or rock behaviour in a wide range of loading conditions. The theory has been adopted with good success to concrete materials (Bažant, 1976), but there seems to be no application to rock yet either in static or dynamic loading conditions.

The concept of intrinsic time provides a good measure of the irreversible damage caused by the internal micro-cracks that contribute directly to the inelastic strains. In this theory the degradation of material stiffness is obtained directly from the evolution of the damage function without recourse to the loading–unloading conditions. This type of model can represent many features of concrete behaviour. However, the models generally involved many functions and parameters which are obtained by a rather complicated fitting procedure. Also, some of these models are incrementally nonlinear, and thus they may require excessive programming and computer effort.

5.5 POST-FAILURE RANGE

Concrete fails or fractures in extremely complex modes. It is known that concrete or rock material can be roughly as brittle at one extreme and ductile at another extreme, depending on the confining pressure or hydrostatic pressure. Thus, strictly speaking, we should speak of the 'brittle state' or 'ductile state' of materials, rather than of 'brittle materials' or 'ductile materials'. Nevertheless, in the following discussion, we shall consider concrete or rock as a brittle material under ordinary load condition and it approaches a ductile material under extreme high hydrostatic pressure condition.

To this end, we classify the mode of failure into three types, namely, the cracking, crushing and a mixture of cracking and crushing. Documented test results for tension–tension or tension–compression biaxial conditions show the cause of fracture is primarily a brittle splitting in the plane normal to the maximum tensile strain direction. This phenomenon should be studied within the framework of fracture mechanics correlating the propagation of discrete cracks to the stress concentration at the crack tip. Here, in the concrete community, a contrary continuum point of view is generally adopted, in which distinct cracks are smeared or distributed evenly within a finite element. It is then assumed that the cracked concrete element becomes an orthotropic (or more accurately, transversely isotropic) elastic material, with one of the material axes being oriented along the direction of cracking, such formulations easily allow for gradual build-down of strength in the direction of tension (tension softening or stiffening). Also shear-strength reserves due to aggregate interlocking and dowel action of reinforcement can be accounted for by retaining a positive shear modulus. The continuous model with a tension cut-off criterion (maximum tensile stress or strain) for cracking has been used in most of the computational models for analysis of reinforced concrete structures.

It is known (Bažant, 1976) that the strength criterion for crack formation over a finite element is unobjective and leads to incorrect results when the stress concentrations near the front of the fracture are calculated with more and more refined finite element meshes. It has been found that the overall stiffness of a structure and the predicted load-carrying capacity reduced with the decrease of the size of the element meshes. If the characteristic length of the problem is large (say, 0.34 m, as suggested by Nilsson, 1982), the brittle fracture model based on a strength criterion can be used. For smaller size structures or for small element meshes, the softening cracking model based on an energy criterion should be used.

In the mixed failure range, the material fails in the form of cracking and sliding due to excessive tension and shear. Since micro-cracks are not oriented regularly, but rather randomly in this region, softening of materials would occur in all stress components. Such a multiaxial softening behaviour may be treated by the fracturing theory (Dougill, 1976). A further improvement may be made by combining the plasticity and fracturing theories to model the softening behaviour of fractured concrete due to cracking and sliding in the mixed failure zone (Bažant and Kim, 1979).

In summary, there are no brittle or ductile materials, the medium is only subjected to loading conditions which cause brittle or ductile response in the post-failure range. For concrete or rock type of materials, hydrostatic confinement plays the key role in distinction between different failure modes by the state of stress. Other factors such as the rate of loading, the composition of concrete, the amount of reinforcements have also important effects on the post-failure response of this material. Therefore, the perfectly brittle and the perfectly ductile models for the two extreme hydrostatic pressure cases combined with the plastic-fracturing model for the transition range form a good modelling basis for the actual post-failure behaviour of fractured concrete or rock material. The difficulty of this approach is that good softening data for concrete or rock are lacking. So, the shape and the expansion of the fracturing surface required in the progressively fracturing theory can not be easily defined. Further investigations are therefore urgently needed on the following topics: post-failure behaviour, unconventional loading paths including cyclic response phenomena, rate effects, composite bahaviour and numerical algorithms.

REFERENCES TO SECTION I

ASCE Committee on Concrete and Masonry Structures (1981) *A State-of-the-Art Report on Finite Element Analysis of Reinforced Concrete*, Task Committee on Finite Element Analysis of Reinforced Concrete Structures, ASCE, Spec. Publication, New York.

Bažant, Z. P. (1976) 'Instability, ductility, and size-effect in strain-softening concrete', *J. Engrg. Mech. Div., ASCE*, **102** (EM2), April, 331–344.

Bažant, Z. P. (1978) 'Inelasticity and failure of concrete: a survey of recent progress', *Proc. Spec. Sem. Anal. Reinforced Concr. Struct. Means Finite Element Method*, Milan, pp. 5–59.

Bažant, Z. P., and Kim, S. S. (1979) 'Plastic-fracturing theory of concrete, *J. Engrg. Mech. Div., ASCE*, **105** (EM3), June, 407–428.

Chen, W. F. (1982) *Plasticity in Reinforced Concrete*, McGraw-Hill, New York.

Chen, W. F. (1983) 'The continuum theory of rock mechanics', in *Mechanics of Oil Shale*, (K. P. Chong and J. W. Smith, Eds.), Applied Science, Amsterdam.

Chen, W. F., and Han, D. J. (1983a) 'Failure criteria for concrete materials', *Proc. of Colloq. Failure Criteria of Structured Media*, Grenoble, France.

Chen, W. F., and Han, D. J. (1983b) 'A five-parameter mixed-hardening model for concrete materials', *Proc. Int. Sym. Plasticity Today*, Udine, Italy.

Chen, W. F., and Saleeb, A. F. (1982, 1985) *Constitutive Equations for Engineering Materials*, Vol. 1—*Elasticity and Modeling*, Vol. 2—*Plasticity and Modeling*, Wiley, New York.

Dougill, J. W. (1976) 'On stable progressively fracturing solids', *Zeitschrift Angewandte Mathematik und Physik*, **27**, 4, 423–437.

IABSE (1979) *Proc. Int. Assoc. Bridge Structr. Engrg. Colloq. Plasticity Reinforced Concrete*, Vol. 28, Introductory Report and Final Report, Lyngby, Copenhagen, Zurich.

IABSE (1981) *Proc. Int. Assoc. Bridge Structr. Engrg. Colloq. Adv. Mech. of Reinforced Concr.*, Vol. 34, Introductory Report and Final Report, Delft, Zurich.

Nilsson, L. (1982) 'Nonlinear wave propagation in plastic fracturing materials—a constitutive modelling and finite element analysis, *Proc. IUTAM Symposium Nonlinear Deformation Wave*, Tallin, August.

Politecnico di Milano (1978) *Proc. Spec. Sem. Anal. Reinforced Concr. Struct. Means Finite Element Method*, Milan.

Schreyer, H. L., and Jeter, Jr., J. W. (Eds.) (1981) *Proc. of Workshop on Constitutive Relations for Concrete*, US Defense Nuclear Agency, Albuquerque.

Valanis, K. C. (1971) 'A theory of viscoplasticity without a yield surface', *Archiwum Mechaniki Stossowanej (Archives of Mechanics)*, Warszaw, **23**, 517–551.

Section II Constitutive Relations for Soils

5.6 INTRODUCTION

Before I go into the discussion of Professor Gudehus's paper on constitutive relations for soils, I think it is appropriate to say a few words about the historical developments of constitutive equations in soil mechanics. This discussion has to start with the historical development of analytical methods for solving problems in soil mechanics in the past. The analysis of problems in soil mechanics is generally divided into two distinct groups—the stability problems and the elasticity problems. They are then treated in two separate and unrelated ways. The stability problems deal with the condition of ultimate failure of a mass of soil: problems of earth pressure, bearing capacity, and stability of slopes most often are considered in this group. The most important feature of such problems is the determination of the loads which will cause failure of the soil mass. Solutions to these problems can often be obtained by simple statics by assuming failure surface of various simple shapes—plane, circular, or logspiral—and by using Coulomb failure criterion. This is known as the limit equilibrium method in soil mechanics.

The earliest contribution to this method was made in 1773 by Coulomb who proposed the Coulomb criterion for soils and also established the important concept of limiting equilibrium to a continuum and applied it to determine the pressure of a fill on a retaining wall. Later, in 1857, Rankine investigated the limiting equilibrium of an infinite body and developed the theory of earth pressure in soil mechanics. In this historical development, the introduction of stress–strain relations or constitutive relations of soils was obviated by the restriction to the consideration of limiting equilibrium and the appeal to the extremum principle. Subsequent developments by Fellenius (1926) and Terzaghi (1943), among many others, have made the limit equilibrium method a working tool with which many engineers develop their own practical solutions. Perhaps the most striking feature of this approach is that no matter how complex the geometry of a problem or loading condition, it is always possible to obtain some approximate but realistic solution.

The elasticity problems on the other hand deal with stress and deformation of the soil at working load level when no failure of the soil is involved. Stresses at points in a soil mass under a footing, or behind a retaining wall, deformations around tunnels or excavations, and all settlement problems belong in this group. Solutions to these problems are often obtained by using the theory of linear elasticity. This approach is rational for problems at short-term working load level, but limited by the assumed elasticity of the soils whose properties approach most nearly those of a time-independent elastic material. While time-dependent effects are significantly large; introducing long-term working stresses over a given period, it is obviously wrong to design a structure on the basis of this time-independent Hooke's law for soils. In this case the design must consider the influence of time on the deformations. This is known as creep. Such a behaviour may be modelled as viscoelastic and the theory of viscoelasticity may be applied to obtain solutions.

Intermediate between the elasticity problems and the stability problems mentioned above are the problems known as progressive failure. Progressive failure problems deal with the elastic–plastic transition from the initial linear elastic state to the ultimate failure state of the soil by plastic flow. The essential constituent in obtaining the solution of a progressive failure problem is the explicit introduction of stress–strain or constitutive relations of soils which must be considered in any solution of a solid mechanics problem.

As mentioned previously, for a long time, solutions in soil mechanics have been based upon Hooke's law of linear elasticity for describing soil behaviour under working loading condition and Coulomb's law of perfect plasticity for describing soil behaviour under collapse state because of simplicity in their respective applications. It is well known that soils are not linearly elastic and perfectly plastic for the entire range of loading of practical interest. In fact, actual behaviour of soils is known to be very complicated and it shows a great variety of behaviour when subjected to different conditions. Drastic idealizations are therefore essential in order to develop simple mathematical constitutive models for practical applications. For example, time-independent idealization is necessary in order to apply the theories of elasticity and plasticity to problems in soil mechanics.

It must be emphasized here that no one mathematical model can completely describe the complex behaviour of real soils under all conditions. Each soil model is aimed at a certain class of phenomena, captures their essential features, and disregards what is considered to be of minor importance in that class of applications. Thus, a constitutive model meets its limits of applicability where a disregarded influence becomes important. This is why Hooke's law has been used so successfully in soil mechanics to describe the general behaviour of soil media under short-term working load conditions, while the Coulomb's law of perfect plasticity providing good predictions of soil behaviour near ultimate strength conditions, because plastic flow at this ultimate load level attains a dominating influence, whereas elastic behaviour becomes of relatively minor importance.

As we can see from this historical sketch, constitutive modelling of soils has come a very long way, not only through the hundreds of years of historical developments in the older theory of earth pressure by Coulomb's and by Rankine's work, but also the establishment of the classical theory of soil mechanics by Terzaghi. During the last 15 years, the theory of soil plasticity has been intensively developed. The modern development of soil plasticity has been strongly influenced by the modern development of somewhat older theory of metal plasticity. It is therefore appropriate to mention here the important works of Roscoe and his students (1958–63) on work-hardening theory of soil plasticity (Palmer, 1973, Parry, 1972) and also the subsequent developments and applications that mark the beginning of the modern development of a consistent theory of soil plasticity (Chen, 1975).

As a result of these and allied developments coupled with the rapid development of finite element computer programs, there are exaggerated hopes in the soil mechanics field that soil problems could soon be solved on a sound theoretical basis similar to that existing with regard to problems relating to steel structures. This limitless faith, developed in recent years, especially by the younger generation of soil mechanicians, does give a strong indication of recent progress in soil mechanics today. The high hope on the application of mechanics to soil leads to the conception by some young soil mechanicians, who were trained by their curriculum in modern continuum mechanics in general and soil mechanics in particular, to insist upon the prediction of computer calculation of the settlement of a foundation with the field measurement to within an exactness of say 1 mm. Otherwise, it is the constitutive model of soil that is to be blamed for the wrong prediction.

On the other hand, the soil engineer, usually an elder member of the profession with his analysis and design based primarily on experience and case history, knows well the fact that no matter how thorough a site investigation is attempted, there is always incomplete exploration and knowledge of the original conditions. It is not possible to explore, detect, and measure every sand seam, loose spot, or weak plane. Thus, no actual site could be completely characterized by any mathematical models, simple or complex. Even a most sophisticated theory cannot describe fully the details of an actual geotechnical problem—an account of the imperfectness of the laboratory investigations and the incompleteness of site investigations.

As demonstrated convincingly by Professor Leonards in his 1980 Terzaghi lecture (1982) on several failure case studies, the actual failures of many geotechnical problems were often the result of the existence of some weak seams or loose spots that could not be detected in the exploration or that were not considered in the theories of soil mechanics, even though the defects have been detected. These painful experiences revealed clearly that they were due neither to the deficiencies of the theories, nor to the imperfectness of the laboratory investigations, but to the fact that the lack of adequate knowledge and understanding of some physical phenomena at the site that requires the establishment of a proper concept, that guides the exploration before construction, that helps to decide an appropriate rational approach and mathematical theory during the analysis and design stage, that gives a deeper insight into the working conditions of the problem, and that provides, at the end, the engineer with a clear physical picture of his problem and the range of validity of his theory and design.

As mentioned previously, the early studies in soil mechanics were based on the mathematical theory of linear elasticity and limit equilibrium of perfect plasticity.

This theoretical mechanics approach has led to the rational establishment of a number of good design rules for practical applications. With the present developments in computational techniques like the finite element method, more general theory of continuum mechanics like hyper- or hypo-elasticity, classical and endochronic plasticity, and visco-elasticity and -plasticity (Chen and Saleeb, 1982) have been developed to describe the very complex behaviour of soils involving phenomena like inelasticity, soil–water interaction, time dependency, and dynamic and cyclic loading conditions (Chen, 1984b, 1984c). This advanced development may produce the danger of a possible separation between the practical soil engineer and the theoretical academic engineer. This danger has been brought out to the open by Professor Gudehus's paper. In his overview on constitutive relations for soils, he pointed out clearly the shortcomings of the existing theories and the inadequacies of the existing methods. As a result, on the one hand, the academic soil engineers or mechanicians tend to generate more expert opinions that will drift farther and farther away from the reality of construction and their propositions and refined approaches may not be put into practice. On the other, discouraged partly by the negative contributions of the recent achievements of modern soil mechanics and partly by the complexity of modern mathematics, continuum mechanics, and computational methods, the practising soil engineer may be forced and driven away from this fundamental theoretical approach and try to rely more and more upon his experiences and case histories.

In the preceding discussion, I may exaggerate the trends and factors as perceived by Professor Gudehus's viewpoint on the disordered state of the present state of soil mechanics that may be detrimental for the further development of soil mechanics. However, it can be stated here that the present state of constitutive modelling in the field of soil mechanics is a fairly satisfactory one and the efforts made by the soil engineers and soil mechanicians in recent years have led to a more fundamental understanding of soil behaviour under different conditions. As a result, some of the constitutive models and methods that have been developed and refined in recent years to such a completeness that many practical applications have been made and good predictions with field measurements have been observed. In these later cases, it can be appreciated that these concepts, theories, and methods have become working tools with which every modern engineer should be conversant. Therefore, in the following sections, some of these positive developments will be briefly summarized. Here, my emphasis is placed on a special class of simple soil plasticity models and the way in which they affect the practical solutions.

5.7 CRITERIA OF MODEL EVALUATION

There exist a large variety of models which have been proposed in recent years to characterize the stress–strain and failure behaviour of soils. All these models have certain inherent advantages and limitations which depend to a large degree on their particular application. Professor Gudehus has proposed some basic requirements for evaluating these models. They include tractability, material constants, economy, and numerical considerations. Alternatively, we may consider the following three basic criteria for model evaluation:

(1) *Theoretical evaluation* of the models with respect to the basic principles of continuum mechanics to ascertain their consistency with the theoretical requirements of continuity, stability, and uniqueness.
(2) *Experimental evaluation* of the models with respect to their suitability to fit experimental data from a variety of available tests, and the ease of the determination of the material parameters from standard test data.
(3) *Numerical and computational evaluation* of the models with respect to the facility with which they can be implemented in computer calculations.

In general, the criterion for model evaluation should always consider the balance between the requirements of rigour from the continuum mechanics viewpoint, the requirements of realistic representation of soil behaviour from the experimental-testing viewpoint, as well as the requirements for simplicity in application from the computation viewpoint.

For the most part, the concept of perfect plasticity has been used extensively in conventional soil mechanics in assessing the collapse load in stability problems. The standard and widely known technique used in conventional soil mechanics is the limit equilibrium method. However, it neglects altogether the important fact that the stress–strain relations constitute an essential part in a complete theory of continuum mechanics of deformable solids. Modern limit analysis method, however, takes into consideration, in an idealized manner, the stress–strain relations of soils. This idealization, termed normality or flow rule, establishes the limit theorems on which limit analysis is based. Within the framework of perfect plasticity and the associated flow rule assumption, the approach is rigorous and the techniques are competitive with those of limit equilibrium approach. In several instances, especially in slope stability analysis and bearing capacity calculations, such a level and completeness has been achieved and firmly established in recent years that the limit analysis method can be used as a working tool for design engineers to solve everyday problems (Chen, 1975).

Most of the early applications of limit analysis of perfect plasticity to geotechnical problems have been limited to soil statics. Recent works attempt to extend this method to soil dynamics, in particular to earthquake-induced stability problems. Recent results show convincingly that the upper bound analysis method can be applied to soils for obtaining reasonably accurate solutions of slope failures and lateral earth pressures subjected to earthquake forces (Chen, 1980, 1984a; Chang and Chen, 1982, Chen and Chang, 1981).

As a further example, the Drucker–Prager type of elastic–perfectly plastic models were discussed and evaluated in the book by Chen (1975), among others. These models are computationally simple. With the proper selection of the material constants, the Drucker–Prager model can be matched with Coulomb condition. This simple model reflects some of the important characteristics of soil behaviour such as: elastic response at lower loads, small material stiffness near failure, failure condition, and elastic unloading after yielding. A simple model of this type can be considered a fair first approximation in the progressive failure analysis of soil media and soil–structure interaction problems.

5.8 ROLE OF STRAIN-HARDENING PLASTICITY IN SOIL MECHANICS

There are several important features mentioned in the first part of Professor Gudehus's paper (Leonards, 1983)

(1) There is a clear review of the stress and strain paths associated with typical boundary value problems in soil mechanics. These paths are generally non-proportional, irreversible and nonlinear.
(2) It is shown that element (uniform strain distribution) tests are:
    (a) rare
    (b) often do not simulate those stress paths for common boundary value problems
(3) The difficulties of interpreting non-element tests close to, or beyond, the initiation of failure are emphasized.

From an academic point of view, strain-hardening models are the most attractive to model the soil, because they are inherently capable of treating conditions of unloading, stress path dependency, and dilatancy as required in item (1). Furthermore, use of these models usually satisfy the rigorous theoretical requirements of continuity, uniqueness, and stability. On the other hand, it is rather difficult to correlate these plasticity models with data from conventional tests as described in items (2) and (3). Thus, the difficulty is extended to the proper determination of specific values for the parameters involved.

Because real materials tested in various manners at different laboratories often do not appear to behave in any consistent and unique way, no practical model can be expected to represent such materials in full details. The Drucker–Prager perfectly plastic model may represent a first but crude attempt to use a plasticity theory. The 'cap model' represents a more refined attempt which can fit most available experimental data reasonable well (Chen, 1984c). Herein, the cap model will be used to handle and predict the complicated situations such as nonlinear, non-proportional loading and unloading, stress path dependency, and dilatancy as reviewed by Professor Gudehus on some typical boundary value problems in soil mechanics. From this numerical demonstration, it is shown clearly that the 'cap' type of plasticity model not only reflects some important features of soil behaviour under laboratory condition, but also provides a fairly accurate prediction of some detailed soil stress–strain histories typically occurring in the ground and in field as sketched in Figures 4.2, 4.3 and 4.9 of Professor Gudehus's paper.

5.9 DEMONSTRATION OF A SIMPLE CAP MODEL

The schematic shape of this simple model is drawn in Figure 5.3. The model is composed of a linearly elastic region bounded by a perfectly plastic failure surface of Drucker–Prager type taking the simple form, in the usual notation

$$\sqrt{J_2} + \alpha I_1 = k \tag{5.1}$$

and a strain-hardening cap taking the form of a quarter of an ellipse

$$(I_1 - L)^2 + R^2 J_2 - (X - L)^2 = 0 \tag{5.2}$$

where

α, k = material constants related to c, ϕ, of Coulomb criterion,
L = value of I_1 at centre of elliptic cap,
R = ratio of major to minor axis of elliptic cap, taking a constant aspect ratio,
X = hardening function that effectively controls material compaction and/or dilatancy, taking the simple form

$$\varepsilon_{kk}^p = W(e^{DX} - 1) \tag{5.3}$$

The following values of the material constants were used in the present calculation made by McCarron (1983) since they were already available from

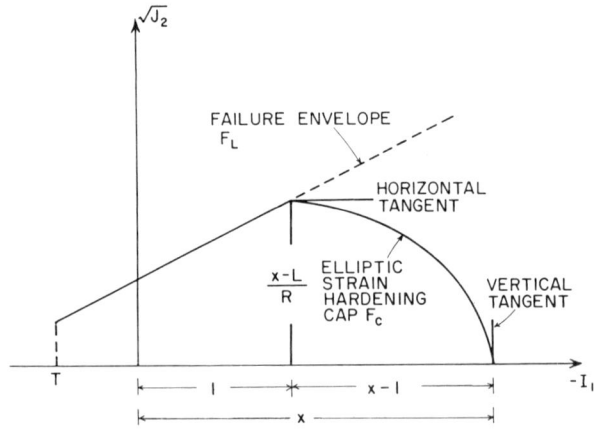

Figure 5.3 Elliptic cap model in $I_1 - \sqrt{J_2}$ space

Constitutive Relations for Concrete, Rock and Soils

previous work for a specific material (Baladi and Rohani, 1979).

$\phi = 49.1°$ $R = 4.33$
$c = 0$ $W = 0.0075$
$v = 0.2736$ $D = 6.78 \times 10^{-5} \text{ ft}^2/\text{lb}$
$E = 841400 \text{ lb/ft}^2$

For the plane strain condition, we have:

$$\alpha = 0.2309$$
$$k = 0$$

The initial cap position is assumed at the origin of the I_1, J_2 space (Figure 5.3). The numerical results are summarized below.

The plane strain earth pressure problem with granular soil as depicted in Figure 4.2 of Professor Gudehus's paper was selected and solved in the following manner and the numerical results are shown in Figure 5.4.

(1) The material was loaded in uniaxial strain until a vertical stress of 30 psi was achieved (path 0–1, Figure 5.4). During this interval the stress state is on the cap for the path.
(2) Unloading (uniaxial strain) then occurred until the vertical stress was reduced to a value of 8.5 psi (path 1–2, Figure 5.4). The stress–strain relation was linearly elastic along this path.
(3) For the passive case a iterative procedure was used to follow the stress path (constant σ_1). The material first is linearly elastic and then elastic–plastic when loading begins on the cap (path 2–3–4, Figure 5.4).
(4) For the active case the stress path was also followed. The material behaved elastically until the failure surface was reached (path 2–3–4, Figure 5.4). At this point the model was unable to follow the stress path.

The corresponding earth pressure problem of Figure 4.2 with saturated clay as shown in Figure 4.3 of Professor Gudehus's paper was also selected here and solved, and the results are summarized in the following (Figure 5.5).

(1) The initial loading and unloading paths are identical to those in Figure 4.2 (path 0–1–2, Figure 5.5). To simulate the saturated condition, strain paths which were volume-constant were prescribed ($\varepsilon_1 = -\varepsilon_2$, $\varepsilon_3 = 0$).
(2) For the passive case the response is initially elastic until the failure surface is reached (path 2–3–4, Figure 5.5). If loading continues in a constant-volume manner the state of stress eventually becomes constant (unchanged) (path 4–5, Figure 5.5). For the present case loading continues along the failure surface until the state of stress coincided with the intersection of the cap and failure surface (corner loading). At this point the direction of the strain path was altered so that the volume decreased.

82   *Mechanics of Geomaterials*

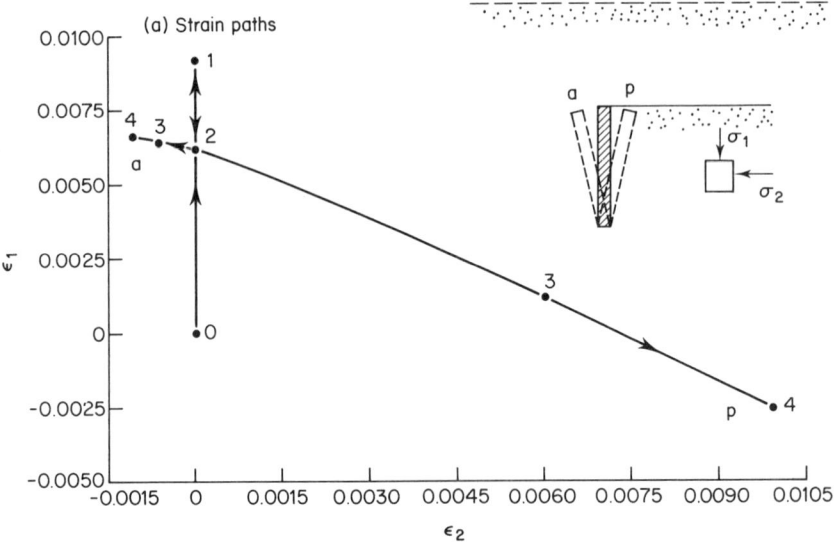

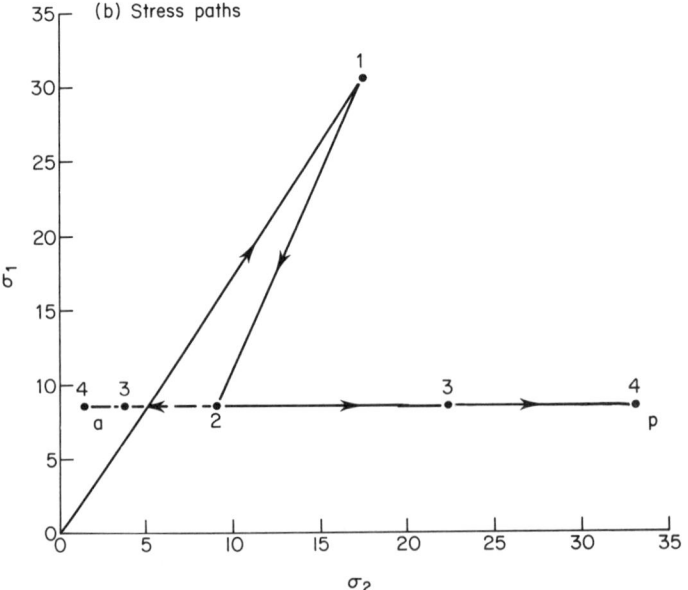

Figure 5.4 Earth pressure problem with dry granular soil corresponding to (a) Gudehus's Figure 4.2(b) and (b) Gudehus's Figure 4.2(c)

Constitutive Relations for Concrete, Rock and Soils 83

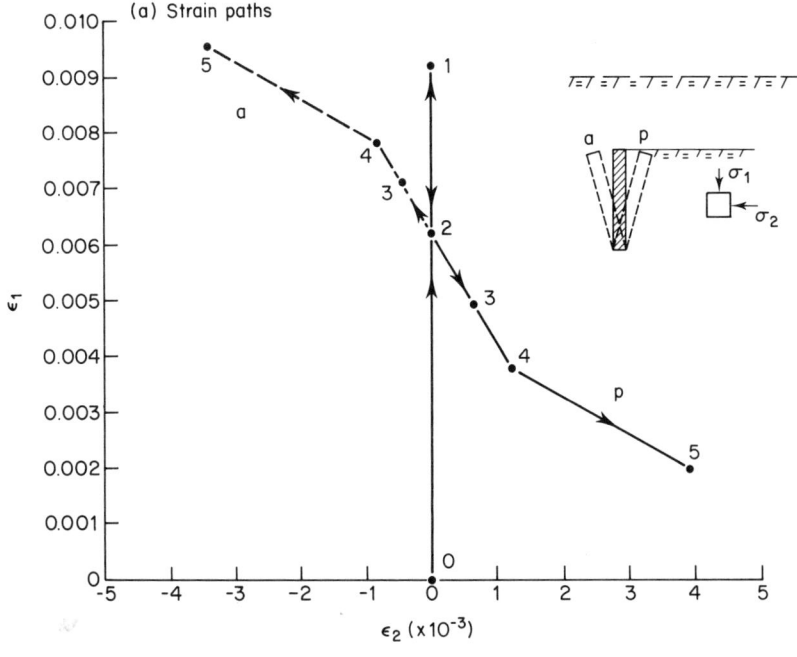

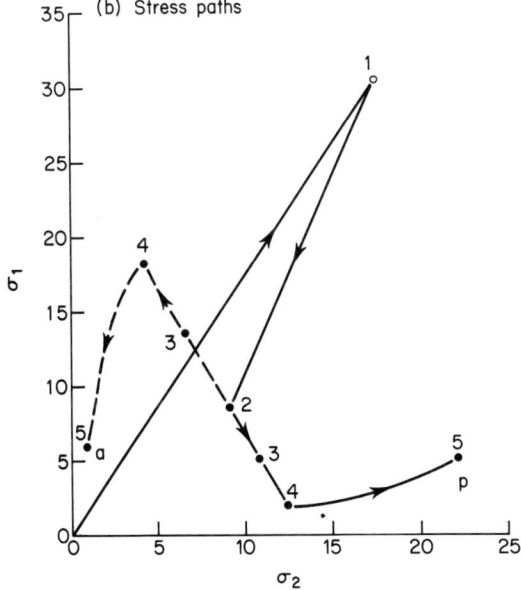

Figure 5.5 Earth pressure problem with saturated clay corresponding to (a) Gudehus's Figure 4.3(b) and (b) Gudehus's Figure 4.3(c)

84   Mechanics of Geomaterials

(3) For the active case the same procedure was followed except the volume-conserving strain path was terminated before reaching the failure surface (path 2–3–4, Figure 5.5). A volume-increasing path was then followed (path 4–5, Figure 5.5).

If the original volume-constant path has been followed along the failure surface, the stress path would change directions so that both σ_1 and σ_2 would increase and stress path would be opposite to the curve 4–5 shown in Figure 5.5(b).

As a last example, the material response given by the cap model for the four bilinear strain paths described in Figure 4.2(a) of professor Gudehus's paper is shown in Figure 5.6. It exhibits the type of behaviour that Professor Gudehus discussed, namely, that sufficiently large monotonic strain path may remove the memory of a previous stress/strain state.

The following four strain paths were used for the calculation of the corresponding stress paths shown in Figure 5.6

For Path i: The strain ratio $(\Delta\varepsilon_1 : \Delta\varepsilon_2 : \Delta\varepsilon_3)$.
For Path ii: The strain ratio was $(1:3:0)$.
For Path iii: The strain ratio was $(3: -2:0)$
For Path iv: The strain ratio was $(1: -4:0)$.

Paths i, ii, and iii have continuous loading on the cap.

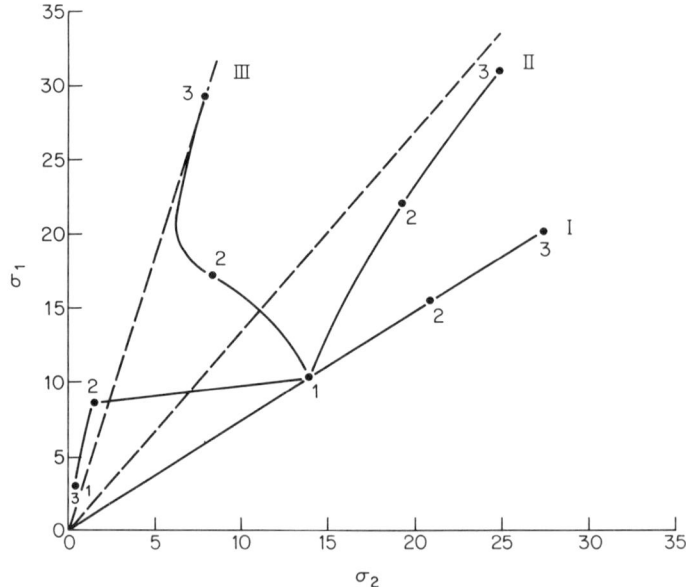

Figure 5.6   Stress paths corresponding to the four bilinear strain paths of Gudehus's Figure 4.9(a)

Path iv loads first on the cap when the strain path is parallel to Path i. When the incremental strain direction changes, the path is first elastic, then on the failure surface, then at the intersection of the failure surface and cap (or corner loading).

From these example calculations, it can be concluded that the cap model is capable of reproducing the stress–strain behaviour discussed by Professor Gudehus. However, as Professor Gudehus mentioned, several practical and numerical problems may cause difficulties when solving boundary value problems. Professor Gudehus seemed particularly concerned with the following three problems:

(1) Viscosity (time-effects).
(2) Saturated soils (water).
(3) Correct representation of the soil load history.

The first item presents considerable difficulties. Considerable effort should be expected in developing the necessary constitutive relationships, computer code, and determination of the required material parameters. Perhaps a more basic problem is the requirement of a deterministic load history.

The analysis of saturated soils presents an interesting problem. The normal procedure appears to be to consider the total medium as incompressible ($d\varepsilon_{kk} = 0$). For an elastic material this may be approximated by a Poisson's ratio of ≈ 0.5 or an extremely large bulk modulus. If, however, the material is in an elastic–plastic state, it appears that the appropriate treatment to enforce incompressibility is to have $d\varepsilon_{kk}^e = - d\varepsilon_{kk}^p$. At the present time it seems unclear as to the best method to satisfy this relation.

Finally, the concern with correctly representing the load history of a soil history is addressed by Professor Gudehus. This aspect presents considerable practical problems in determining the past geological history of a site as well as the influence that construction activity may have in disturbing the soil mass. Professor Gudehus's concern seem to focus on the validity of solutions obtained by say the finite element methods if the 'history' of a soil is not well represented. However, the theoretical solution does provide a rational and useful understanding of the behaviour of geotechnical problems, just one of the steps that is needed with any practical applications.

It is therefore very important to conclude here that soil mechanicians and soil engineers must have both adequate understanding and appreciation in the field of the other party. This will certainly help to contribute to the sound development of constitutive relations in soil mechanics and, as a consequence, a certain overlap in each other's development, rather than a certain separation to occur in the future.

REFERENCES TO SECTION II

Baladi, G. Y., and Rohani, B. (1979) 'An elastic–plastic constitutive model for saturated sands subjected to monotonic and/or cyclic loadings', *3rd Int. Conf. Numer. Methods Geomech. Aachen*, 1979, pp. 389–404.

Chang, M. F., and Chen, W. F. (1982) 'Lateral earth pressures on rigid retaining walls subjected to earthquake forces', *Solid Mechanics Archives*, **7**, 315–362.
Chen, W. F. (1975) *Limit Analysis and Soil Plasticity*, Elsevier, Amsterdam.
Chen, W. F. (1980) 'Plasticity in soil mechanics and landslides', *J. of Engineering Mechanics Division, ASCE*, **106**(3), 443–464.
Chen, W. F. (1984a) 'Soil mechanics, plasticity and landslides', in *Mechanics of Inelastic Materials* 31–58 (Eds. G. J. Dvorak and R. T. Shield), Elsevier Science Publishers, Amsterdam.
Chen, W. F. (1984b) 'The continuum theory of rock mechanics', in *Mechanics of Oil Shale* (Eds. K. P. Chong and J. W. Smith), Applied Science Publishers, London.
Chen, W. F. (1984c) 'Constitutive modeling in soil mechanics', in *Mechanics of Engineering Materials* 91–120 (Eds. C. S. Desai and R. H. Gallagher), Wiley, Chichester.
Chen, W. F., and Chang, M. F. (1981) 'Limit analysis in soil mechanics and its applications to lateral earth pressure problems', *Solid Mechanics Archives*, **6**(3), 331–399.
Chen, W. F., and Saleeb, A. F. (1982) *Constitutive Equations for Engineering Materials*, Vol. 1—*Elasticity and Modeling*, Wiley, New York.
Fellenius, W. O. (1926) *Mechanics of Soils*, Statika Gruntov, Gosstrollzdat, 1933.
Leonards, G. A. (1982) 'Investigation of failures', *J. Geotechnical Engineering Division, ASCE*, **108**(2), 187–246.
Leonards, G. A. (1983) Private communication.
McCarron, W. (1983) Private communication.
Palmer, A. C. (Ed.) (1973) *Proceedings of the Symposium on the Role of Plasticity in Soil Mechanics*, Cambridge University Press, London.
Parry, R. H. G. (Ed.) (1972) *Roscoe Memorial Symposium: Stress-Strain Behaviour of Soils*, Henley-on-Thames, Cambridge University Press, London.
Terzaghi, K. (1943) *Theoretical Soil Mechanics*, Wiley, New York.

PART III

BEHAVIOUR OF SOLIDS WITH A SYSTEM OF CRACKS

Chapter 6
Inelastic Properties of Solids with Random Cracks*

Yu. V. Zaitsev

6.1 INTRODUCTION

Geological materials often have microstructures composed of coherent grains (monocrystals), sometimes with only weak bonding forces between them, or of granules in an unbonded aggregate. When nominally coherent, they typically have pores or cracks (or both) which are randomly distributed through the volume of the material (Nikolaevskii and Rice, 1979).

Concrete is one of the most complex construction materials, consisting of randomly distributed inclusions embedded in a continuous, relatively soft porous matrix (hardened cement paste), the properties of which depend on time. The effect of this heterogeneity is to cause non-uniform internal strains, and the processes of both cracking and failure in such a material become complex and discontinuous (Swamy and Sriravindrarajah, 1982). The aggregate–matrix interface in such a system becomes critical and represents the weakest link; pre-existing micro-cracks in the interface create nuclei of potential crack propagation, inelastic behaviour, and failure of concrete.

There are many models which make it possible to describe the behaviour of random cracks in materials such as rocks and concrete under various load conditions. Mathematical treatment of the problem of random cracks in a three-dimensional body is very complicated. Therefore, in most theories this problem is reduced to the problem of random cracks in a two-dimensional body.

Let us at first analyse models in which a material with random cracks is supposed to be quasi-homogeneous. These models are mainly used to describe the behaviour of rocks and hardened cement paste (hcp). After that, the models which take the significantly inhomogeneous structure of the material (such as concrete) into consideration will be analysed.

---

*On the Symposium programme but not presented at the Symposium because of travel difficulties beyond the author's control.

6.2 MODELS OF QUASI-HOMOGENEOUS SOLIDS (ROCKS, HCP) WITH RANDOM CRACKS

6.2.1 Uniaxial tension, biaxial loading (tension + tension, tension + compression)

For such loading conditions, opening mode cracks (Mode I) are typical. Carpinteri, Di Tomasso, and Viola (1978) have analysed a structural part represented by a thin plate with random cracks. The plate is subjected to biaxial tension, or to compression with tension; σ_1 and σ_3. The length of cracks is assumed to be small as compared to the plate dimensions. The interaction among individual cracks is neglected. It is assumed that propagation of the most dangerous crack leads to failure of the solid. To estimate the load value when the crack starts to propagate, formulae of fracture mechanics are used. For various values of σ_3/σ_1 different values of crack inclination β are found to be decisive for strength of the structure (see Figure 6.1). A similar model was analysed by Panasjuk *et al.* (1976).

It has been shown that the interaction among n existing random cracks can in

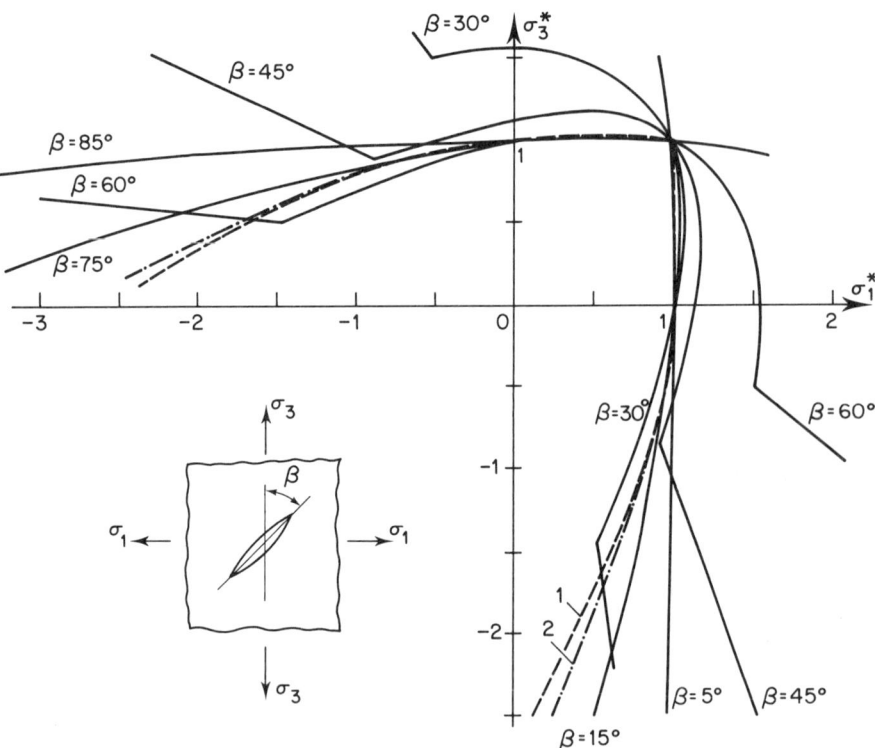

Figure 6.1 Failure criteria in biaxial tension for a solid with random cracks

principle be taken into consideration with the help of a system of integral equations. However, difficulties of computation prevent the estimation of the interaction effect for more than two arbitrarily located cracks. Results for two such cracks according to Panasjuk et al. (1976) are given in Figures 6.2–6.5. Here the dashed lines correspond to the left tips of the cracks, and the solid lines to the right tips. Curves 1 and 2 correspond to cracks No. 1 and No. 2. From Figure 6.2 one can see that for $0 < \beta < 7\pi/30$ fracture begins from the internal tips, and for $\beta > 7\pi/30$ from the external tips. The critical load, p_*, reaches a minimal for $\beta \cong 5\pi/36$. For $\beta = \pi/3$, the critical load is the same as for an isolated crack $p_* = p_0$), oriented perpendicularly to the applied stress.

From Figure 6.2 one can see that the angles θ_* of the initial direction of crack propagation are negative for $0 < \beta < \pi/6$ and $5\pi/12 < \beta \leqslant \pi/2$. This means that fracture propagates not in the direction towards the nearby crack but in the direction away from it.

In Figure 6.3, the minimum critical load corresponds again to $\beta \cong 5\pi/36$. Fracture begins from the left (internal) tip of the crack No. 2. With an increase of the β-value, the initiation of fracture goes over to the right tip (and then to the left one) of crack No. 1. Figure 6.3 also shows the corresponding angle of the initial crack propagation direction.

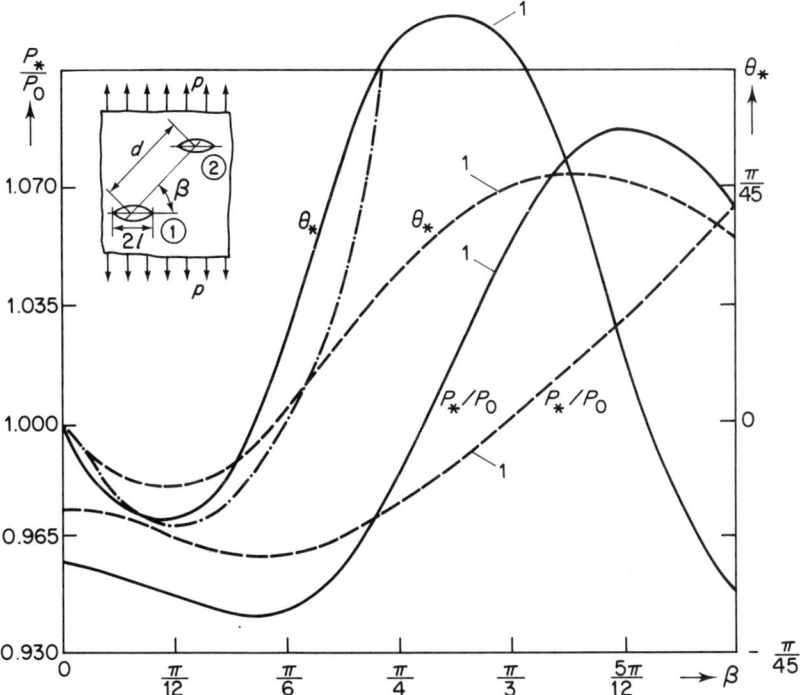

Figure 6.2 Interaction of two random cracks

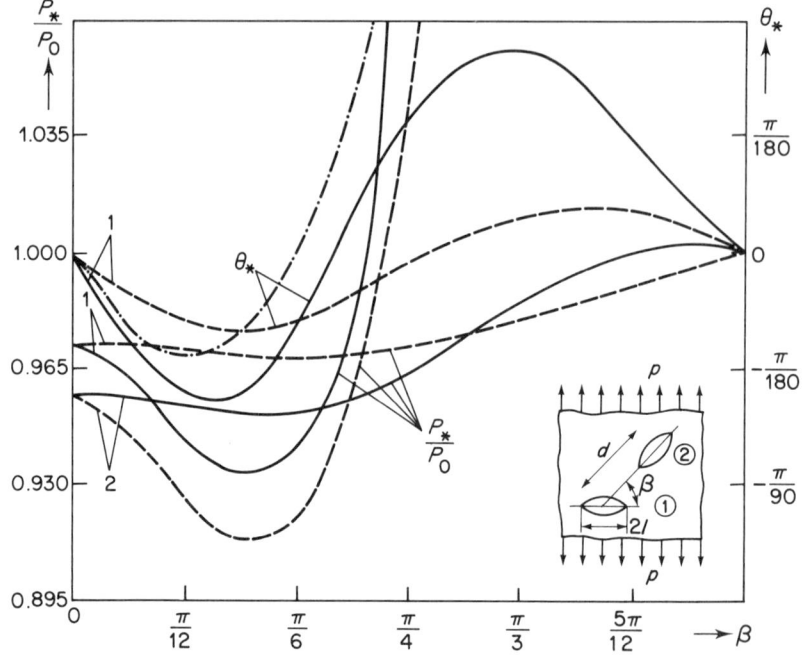

Figure 6.3 Interaction of two random cracks

From Figure 6.4 one can see that for $\alpha < 7\pi/36$ fracture initiates from the left (internal) tip of the second crack, and for $\alpha > 7\pi/36$ from the right (internal) tip of the first crack.

Figure 6.4 shows also the effect of orientation of the second crack on the direction of initial propagation of the first crack. If the second crack ceases to be co-planar, the fracture from the right (internal) tip of the first crack will propagate in the direction to the more distant tip of the second crack.

From Figure 6.5 one can see that for a crack orientation near that of the second crack, the critical load is higher than for an isolated crack $(p_*/p_0 > 1)$ that is oriented perpendicularly to the applied stress. It is remarkable that the curves for p_*/p_0 in Figures 6.2–6.5 differ qualitatively only little from the curves for an isolated crack of arbitrary orientation (dotted lines).

Failure criteria for a solid with random cracks subjected to biaxial tension, or to compression with tension, have been also investigated by Panasjuk et al. (1976). It was assumed that all random cracks interact but their interaction takes place only in pairs. It means that in a solid there are some pairs of interacting cracks; moreover, distance between the pairs is so large that the pairs may be considered as isolated pairs. For each pair, the crack length $2l$ and the distance d between centres of the cracks are given. For such a model, Figure 6.1 shows the failure criterion for a plane state of stress with $\lambda = 2l/d = 0.5$ (curve 1). Curve 2 shows the

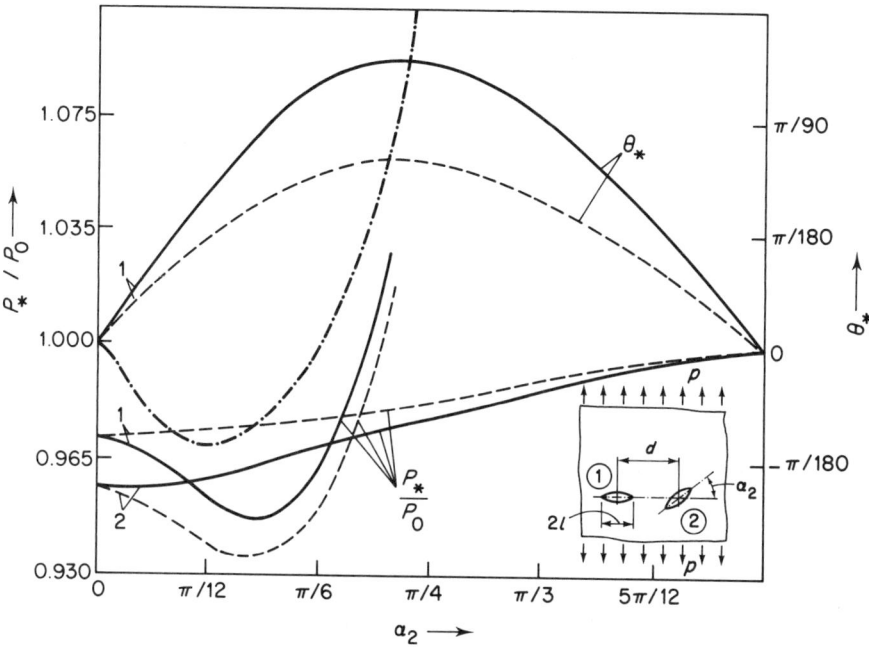

Figure 6.4 Interaction of two random cracks

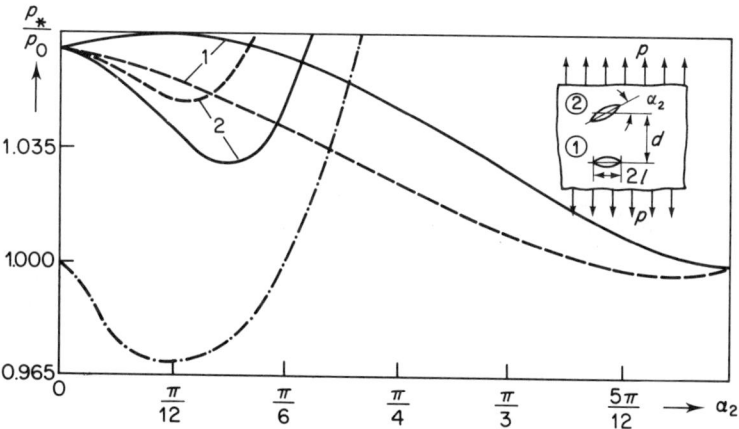

Figure 6.5 Interaction of two random cracks

same failure criterion for a body with only one isolated crack of a certain related length; this length is chosen from the condition that the uniaxial tensile strengths of the body with one isolated crack and another body with pairs of interacting cracks are the same. One can see from Figure 6.1 that the interaction of cracks affect qualitatively only slightly the failure criteria for a plane state of stress.

There exist also some theories that pay attention not so much to the geometrical arrangement of random cracks as to probabilistic concepts of crack formation and crack propagation. Mihashi (1983) has assumed that the failure processes of brittle materials may be regarded as a kind of Markov process because the probability of cracking is influenced only by the state at the last previous moment but is independent of the history preceding this moment, t. The Markov process is characterized by a transition probability $\bar{p}_{01}(t)$ representing the probability that the state of the element changes from state 0 (no fracture) to state 1 (fractured) (Figure 6.6(a)). Since fracture of brittle materials is caused by cracking, the transition probability might be equal to the rate of crack initiation, $p(t)$. When an element consists of n units, the rate of crack initiation which leads to the fracture of the element is given by the following equation:

$$\bar{p}(t) = n\bar{\mu}(t) \tag{6.1}$$

where $\bar{\mu}(t)$ is the mean value of $\mu(t)$ for a large number n, and $\mu(t)$ is the probability of fracture initiation at the moment t.

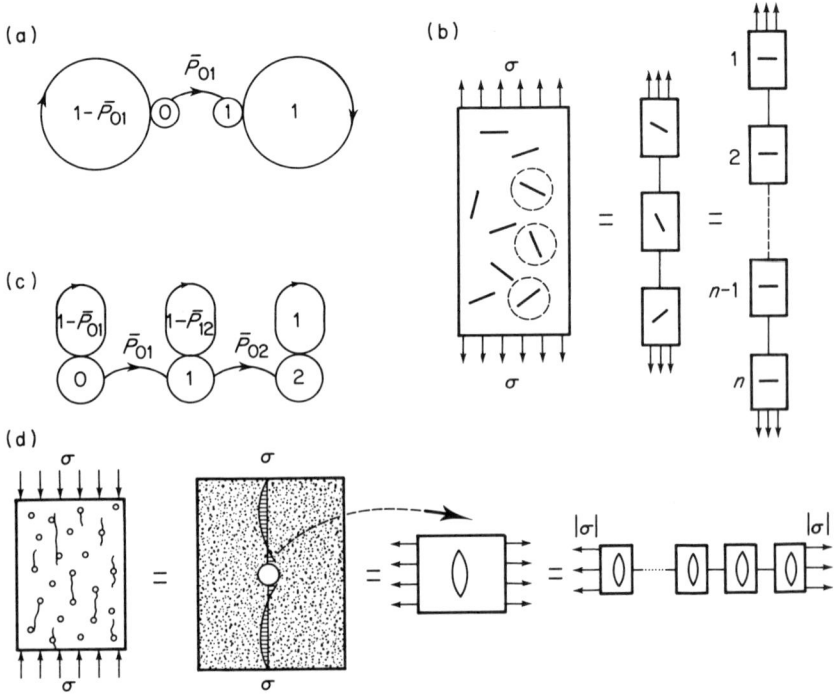

Figure 6.6 Stochastic model for hardened cement paste; (a) transition line graph of a stochastic process for tension; (b) scheme of fracture in tension; (c) transition line graph for compression; (d) scheme of fracture in compression

According to the theory of stochastic processes,

$$\frac{dp_0(t)}{dt} = -\bar{p}_{01}(t)P_0(t) \tag{6.2}$$

where $P_0(t)$ means the probability that no fracture occurs before time t(given in any units). The variable t may have other meanings than time—for example, stress under a monotonically increasing load, or the number of cycles under a repeated load.

The failure process of a solid consists of a series of local brittle fractures (which are assumed to represent cracking). The number of units, n, might be connected with the number of relatively large micro-cracks of some type (Figure 6.6(b)). A large number of 'n' may describe a very porous solid. The theory of Mihashi provides a basis for mathematical formulation of the influence that some factors have on strength; the rate of loading, environmental temperature, size of the specimen, age of the specimen, and the upper limit stress for repeated load.

However, for this model, as well as for all the models described above, the following question still remains open: how to calculate the tensile macrostrain of a solid on the basis of crack formation and crack propagation, i.e. how to estimate the real inelastic behaviour of a solid with random cracks.

There have been some attempts to determine the deformation of a body from its defects (pores and cracks) using an effective modulus E of elasticity and an effective Poisson's ratio v. Their values may be expressed as follows (Salganik, 1973):

$$E = E_0(1 - k\pi\Omega/4); \quad v = v_0(1 - k\pi\Omega/4) \tag{6.3}$$

where Ω is a small parameter, $\Omega = Nl^2$, N = number of defects in a unit volume, l = defect size, E_0 and v_0 = values of E and v in a body without defects; $k = 3$ for pores and $k = 1$ for cracks. It is assumed in this approach that the defects are isotropically distributed throughout the body and that defects are small as compared to the dimension of the body. It must be noted that such assumptions in fact present using this model for concrete and rocks, because propagating cracks in these materials have a certain distinct orientation with respect to the applied load being parallel to load for compression, or perpendicular to it for tension, and also because the crack dimensions near the ultimate load are comparable to the dimension of the specimen. Therefore, it seems to be appropriate for the calculation of inelastic deformation to take into account cracks of real length and of real orientation.

Budiansky and O'Connel (1976) have proposed an analytical method for estimating effective elastic moduli of a solid containing random cracks of different shapes. Calculations are made for the case of elliptical cracks.

The effect that (orientation and) the number of cracks have on strength and other properties of a solid was analysed by Ikeda, Kabayashi, and Sakurai (1973). A model of a microstructure with cracks which uses a statistical approach and

96 Mechanics of Geomaterials

includes a generalization of Walsh's model for multiaxial states of stress, has been developed by Brady (1969, 1970, 1973).

A more precise estimation of inelastic behaviour of a solid with random cracks can be made with the help of the model shown in Figure 6.7(a); it was assumed that the centres of cracks are statistically uniformly distributed over the area in Figure 6.7(a) which represents the specimen size (Zaitsev, 1980, 1982; Zaitsev and Wittmann, 1981). The lengths of the cracks are uniformly distributed over the interval $[a, b]$. The angle of crack orientation α with respect to the external load is also uniformly distributed, over the interval $[0, 2\pi]$. For every crack, the interaction with only one nearby crack was taken into account; this was made according to the results given in Figures 6.2–6.5. The random structure has been created by a computer program. The possibility of crack arrest due to crack crossing was taken into consideration. Taking the opening of cracks into account, the contribution of cracks to the longitudinal strain ε can be found if one assumes

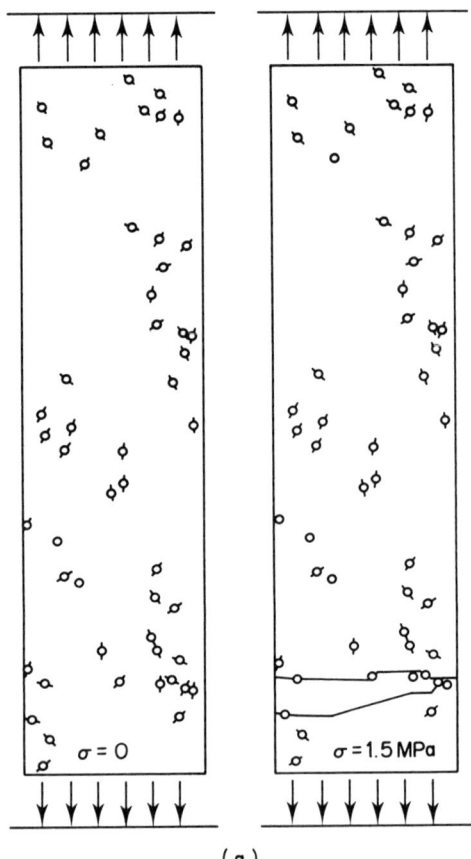

(a)

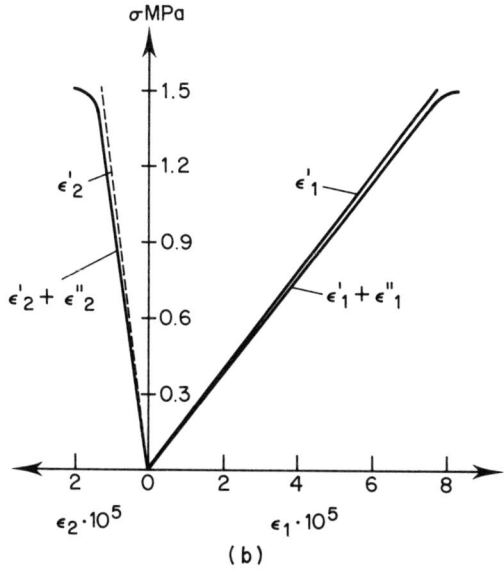

Figure 6.7 Consequent stages of crack propagation in a quasi-homogeneous porous solid under tension and stress–strain curves according to results of crack propagation

linear superposition of two strain components, i.e.

$$\varepsilon_1 = \varepsilon_1' + \varepsilon_1'' \qquad (6.4)$$

where ε_1' is the contribution of the elastic unfractured material, ε_1'' is the inelastic component caused by the displacement of crack edges. According to Panasjuk, Savruk and Dazyshin (1976), the half width v of a crack of length $2l$ in a material with modulus of elasticity E subjected to tensile stress p that is perpendicular to the cracks ($\alpha = \pi/2$), can be found as follows:

$$v(x, 0) = \frac{2p}{E}\sqrt{l^2 - x^2} \qquad (6.5)$$

whereby the maximum half-width is equal to

$$v_{\max} = 2pl/E \qquad (6.6)$$

If the angle between the crack direction and the load direction $\alpha \neq \pi/2$, i.e., if the crack is inclined with respect to the load direction, then these expressions may be rewritten as follows:

$$v(x, 0) = 2p\sin^2\alpha\sqrt{l^2 - x^2}/E \qquad (6.7)$$

$$v_{\max} = 2pl\sin^2\alpha/E \qquad (6.8)$$

The effect of the shear stress $\tau = p\sin\alpha\cos\alpha$ on the crack opening is neglected.

98   *Mechanics of Geomaterials*

The component ε_1' is equal to σ/E, while the component ε_1'' can be found if one assumes that the elongation caused by the displacements of the crack edges is 'smeared' over the volume of the sample. Thus, for one crack we have

$$\varepsilon_1'' = 2 \int_0^{2l} v(\xi)\mathrm{d}\xi / bh \qquad (6.9)$$

and for all the n cracks

$$\varepsilon_1'' = 2 \sum_{i=1}^{n} \int_0^{2l_i} v(\xi)\mathrm{d}\xi = \frac{4p}{E} \sum_{i=1}^{n} \sin^2 \alpha_i \int_0^{2l_i} \sqrt{l_i^2 - \xi^2}\, \mathrm{d}\xi \qquad (6.10)$$

Figure 6.7(b) shows the stress–strain curves in uniaxial tension according to the results of simulation of crack propagation in the specimen (see Figure 6.7(a)).

Damage approach for concrete in uniaxial tension has been developed by Løland (1980), and Lorrain and Løland (1983). Damage, ω, denotes the relative portion of the nominal fracture area (A in Figure 6.8(a)) which is not mechanically intact. Damage includes pores (1) and cracks of all types (2). It is suggested that the initiation of cracking depends more on strains than on stresses.

As a specimen is being strained, damage will occur within the total length of the strained body when the strain (ε) is less than the strain capacity ($\varepsilon_{\mathrm{cap}}$). Above this strain level, damage will develop only in the fracture zone, as shown in Figure 6.8(b). Figure 6.8(c) shows a summary of the damage approach for concrete in uniaxial tension.

6.2.2 Uni- and multiaxial compression

The models with opening-mode cracks (Mode I) can describe the behaviour of solids subjected mainly to uniaxial tension, and sometimes also to biaxial tension or tension with compression. For uniaxial compression (typical for concrete), and especially for multiaxial compression (which is typical for rocks), propagation of shear cracks (Mode II) must be also taken into account.

Chen, Yao and Xie (1979) have studied the behaviour of gabbro under triaxial compression. They observed the development of microcracks under the microscope, and they found that, under low stresses, the micro-cracks occur within the crystal grains of the rock minerals and their orientation is determined principally by the configuration of the crystals. When the applied stress reaches about 70–80 per cent of the ultimate strength, the cracks become more concentrated in the central part near one diagonal of the sample. At this time there appear a few micro-faults which pass through several crystal grains. The orientation of these micro-faults depends obviously on the direction of the applied stress, which makes an angle less than 40° with the direction of the maximum stress. The volume changes in the rock specimens were also measured and compared with the observations of the micro-cracks. It was found that the development of the

Inelastic Properties of Solids with Random Cracks 99

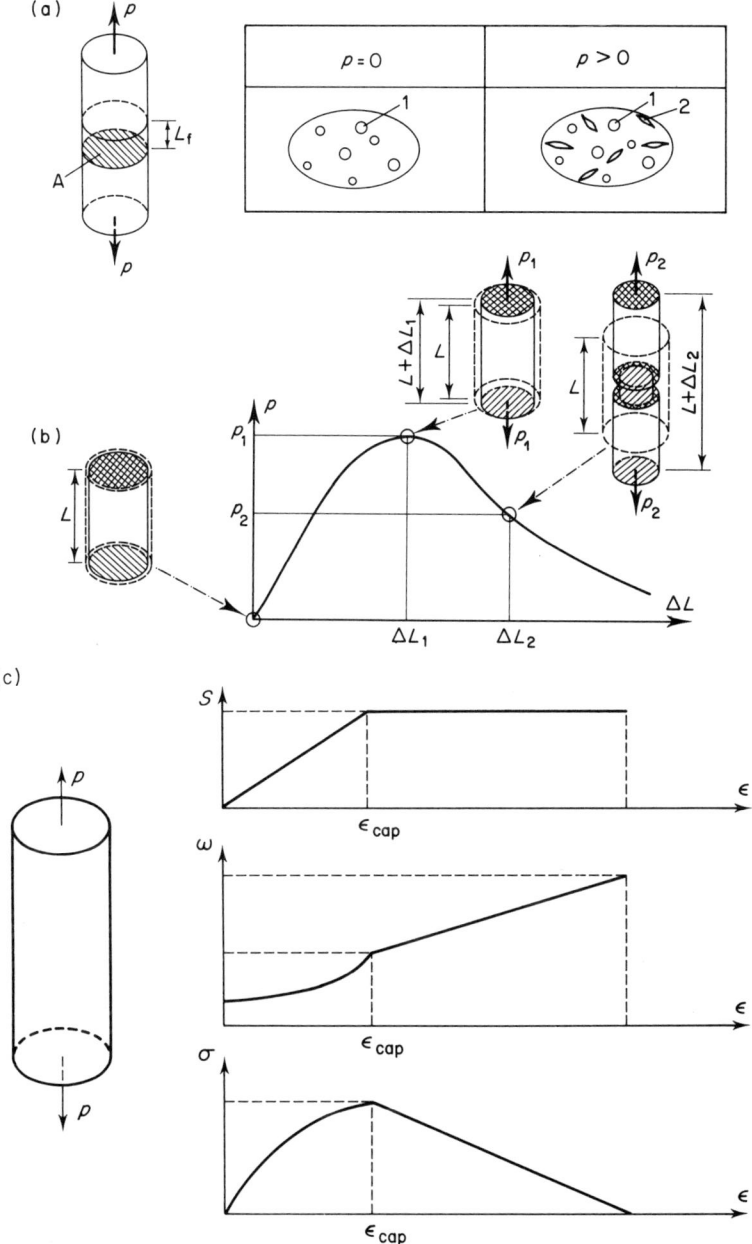

Figure 6.8 Damage approach for hardened cement paste and concrete in tension; (a) fracture zone with pores (1) and cracks (2); (b) net contours (net area A_n) of a specimen with reference to the load-elongation curve; (c) net stress S, damage ω and nominal stress σ, related to strain ε

micro-cracks, as well as their influence on other physical properties of the rock, depends on the dimensions of the crystal grains.

Simmons and Richter (1976) present results of petrographical studies of microcracks in rock. The length of micro-cracks is usually of 10^2 mm range but may reach even a few metres. Data on the structure and propagation of pores and micro-cracks are given also by Brace et al. (1972), Gash (1971) and other investigators.

Micro-crack closure in rocks under increasing stress was observed directly by Batzle et al. (1980) with a scanning electron microscope. Uniaxial stresses up to 300 bars were applied to specimens of granite, both unheated and previously heat-cycled up to 500°C, and diabaze, heat cycled up to 700 °C. It was assumed that the behaviour of cracks under stress is influenced by numerous factors including orientation, shape, source, and the proximity of other fractures. In general, fractures oriented perpendicularly to the maximum stress direction tend to close, and those oriented in parallel tend to open. These trends, however, are often changed by the other factors. Natural cracks have walls that are irregular, etched and pitted, and poorly matched. Closure is incomplete, with many portions of the crack remaining open and interconnected. A single fracture may partially close, forming numerous smaller cracks.

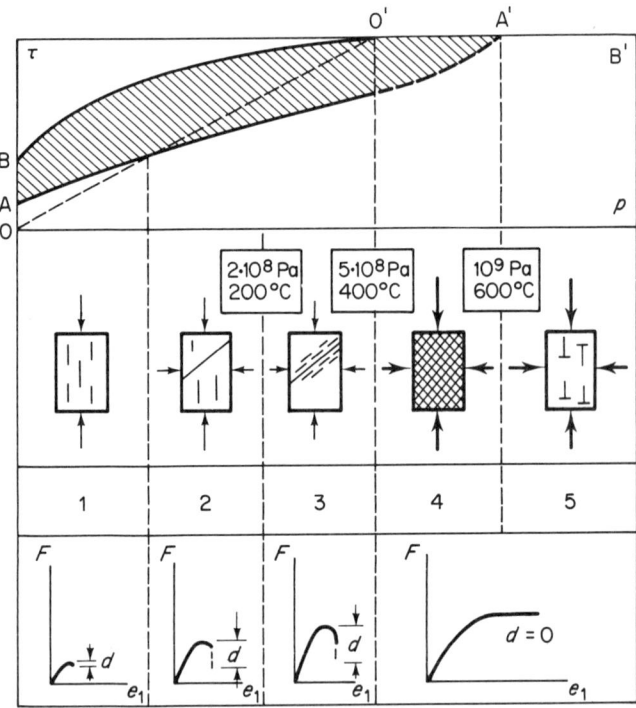

Figure 6.9 Classification of fracture for rocks

Experimental results have been summarized by Nikolaevskii (1982) in the diagram shown in Figure 6.9, giving a classification for fracture of rocks. One can see in Figure 6.9, fracture surface BB', surface of the onset of dilatancy (inelastic volume change due to micro-cracking) AA', zone of dilatancy (shaded area), and line OO' representing the line of the Coulomb condition $\tau = \mu\sigma$, with friction coefficient μ.

The numbers in Figure 6.9 correspond to granite. To the left from the point O' (0.5 GPa, 400°C) rocks fail in a brittle manner, and to the right in a plastic manner. At low confining pressure (I), cracks propagate in parallel to the axial (vertical) compression as the opening mode cracks. At increasing confining pressure, crack edges close (2), cracks now propagate as inclined cracks, and the angle of inclination with respect to the axial load depends on the principal stress ratio and some other factors.

In the interval from the point of 0.2 GPa and 200°C to the point O', shear bands (3) are formed due to the localization of dilatancy cracks. Then one can see pseudoplastic (cataclastic) fracture (4) (due to formation of a large number of small cracks), and plastic (dislocation) fracture (5). To the right of the point 1 GPa (or 1.5 GPa, 500°C; or 0.7 GPa, 800°C), there are no cracks, and the rock becomes impervious.

An idealized model for rocks (see Figure 6.10(a)) consists, after Nikolaevskii

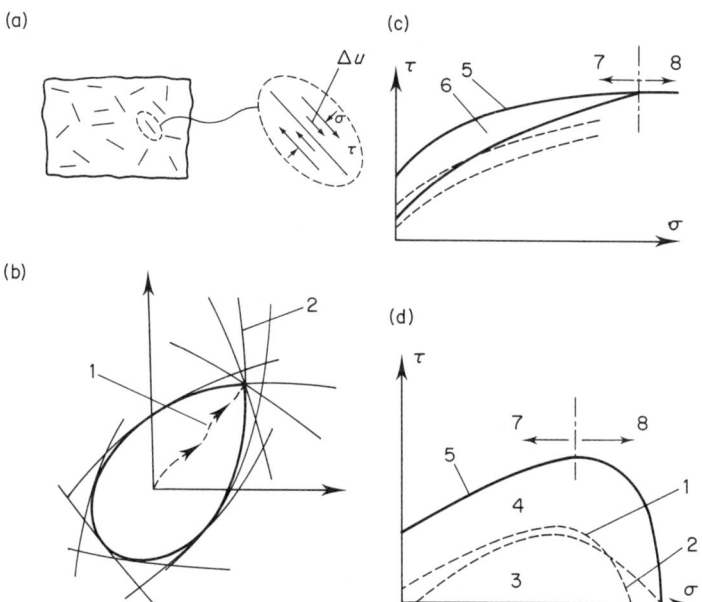

Figure 6.10 Model of rocks with random cracks. 1—stress path; 2—yield surface for individual fissure; 3—elastic zone; 4—post-loading limit surfaces; 5—failure; 6—dilatancy zone; 7—brittle, 8—ductile

and Rice (1979), of a solid with a random array of flat Griffith cracks that are closed under compressive principal stresses and can begin to slip when the Coulomb condition $\tau = \mu\sigma$ is met based on locally resolved shear (τ) and normal (σ) stresses. If the crack surfaces are flat so that slippage is not accompanied by dilatant uplift at asperities, and if no local tensile fissures are opened at the crack tips, there is no plastic dilatancy.

The idealized model of Figure 6.10(a) suffices also for an examination of subsequent yield surfaces. In particular, as illustrated in Figure 6.10(b), the macroscopic yield surface can be considered as the inner envelope of an essentially infinite family of yield surfaces for individual fissures of all possible orientations (Nikolaevskii and Rice, 1979). Initially, these individual yield surfaces have the form $\tau = \mu\sigma$ where the orientation of an individual crack enters the calculation of τ and σ from the macroscopic stress state. Later, as slips Δu develop, a kind of 'hardening' occurs since not all of the nominally applied shear stress τ is actually transmitted across the crack surface.

In Figure 6.10(c), limit surfaces representative of coherent rock, e.g. granite, are given after Nikolaevskii and Rice (1979). Here the zone of dilatancy is located between the line of the onset of dilatancy and the limit failure curve. Inside, the response can be modelled by the rule of non-associative plasticity, although on the micro-level the cracking is brittle.

In Figure 6.10(d), the case of porous limestone is illustrated. Under high σ, the response is similar to that of porous metals, due to the effects of plastic flow around the pores. This gives the capped form of the overall yield surface of the material, where the region of lower σ corresponds to inter-pore shearing. At limit conditions, these shears combine into a macroscopic surface of localization.

The localization of previously homogeneous deformation into a narrow shear band is a common feature of geological materials that are loaded to failure under multiaxial compression. Rudnicki and Rice (1975) explained the onset of localization as a bifurcation into a localized mode that is predictable in terms of the non-elastic constitutive relations prevailing up to the moment of localization.

The onset of localization may be important for the inception of earth faulting and also for the concentration of deformation into narrow shear zones in a variety of geotechnical problems, e.g. landslides of slopes. What is missing in the theoretical description developed thus far is an explanation of what happens as deformations proceed within the localized zone.

Using Geniev, Kissjuk, and Tjupin's method (1974), Kozachevskij (1981), and Kozachevskij and Zjazin (1982), develop a dilatancy model for the theory of plasticity. This model is based on the nonlinear behaviour exhibited by concrete (not only at shear deformation, but also at volume deformation) and also takes into account the cracking connected with dilatancy.

Dilatancy that results from cracking can be explained with the help of the model by Stavrogin (1969), Stavrogin and Protasenja (1979). This model involves statistical ideas explained in Figure 6.11 (a, b, c, d) which shows four different

Inelastic Properties of Solids with Random Cracks 103

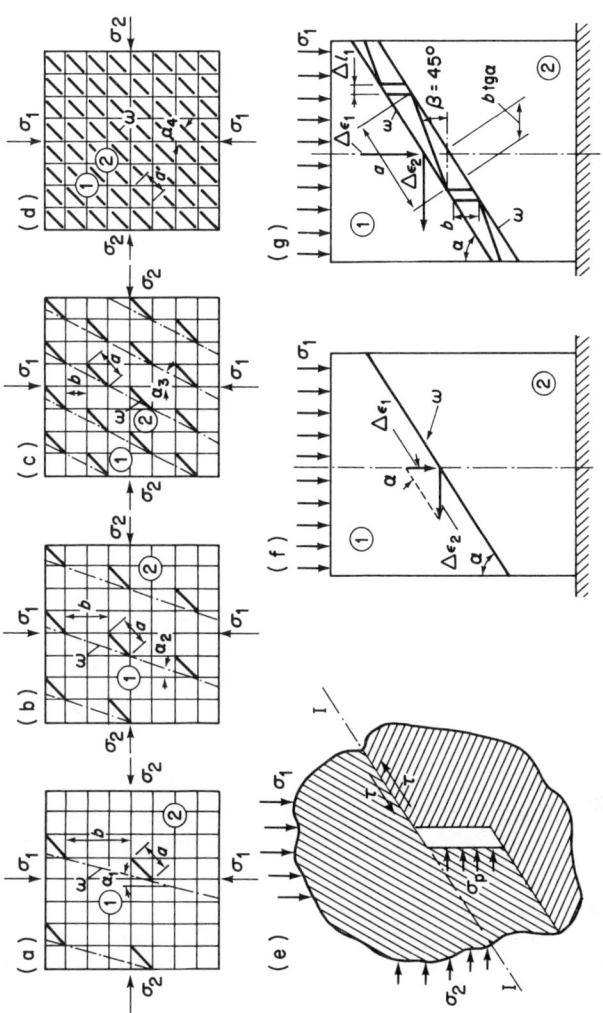

Figure 6.11 Model of rocks with a geometrical arrangement of cracks

cases of loading with stresses σ_1, σ_2, whereby the value of $C = \sigma_2/\sigma_1$ increases at the transition from 'a' to 'd'. Statistically distributed defects produced by loading of the body cause shear on the micro-areas which coincide with defects. These defects are distributed uniformly, with mean distance 'b'. Density of defects N increases as C increases. If the shear stress on these micro-areas has reached a critical value, shear deformation occurs, which leads to fracture in vertical micro-area (see Figure 6.11(e)). Macroscopically, a rough shear surface ω is formed, and part 1 of the body moves with respect to part 2. Fracture in the vertical micro-areas causes an increase of the volume of the body.

It can be calculated from this model (see Figure 6.11(f), (g)) that, with respect to crack formation, $v = \varepsilon_1/\varepsilon_2 = (1/2)\tan\alpha$.

The influence of cracks on the effective elastic moduli in compression has been studied by many investigators. In the papers by Walsh (1965a, 1965b, 1980) and Walsh and Brace (1968, 1972), rock is considered to be elastic and isotropic, containing pores or elliptical cracks of random orientation. Nonlinear stress–strain behaviour and hysteresis due to cracks are analysed. Cracks close under increasing compressive stress, causing an increase in Young's modulus. This modulus need not be equal to that of a solid material since frictional sliding at crack faces can occur. If fluid is not permitted to enter the pores due to a plastic barrier, the effective compressibility must also include a term to account for the resulting decrease in porosity. Expressions for the effective compressibility of rocks having various shapes of pores, including spherical, penny-shaped, and elliptical shapes, are given. The pressure required to close a crack is of the order of $E\alpha$ where E is Young's modulus and α is the aspect ratio. Cracks having an aspect ratio of 1/1000 can be closed by moderate pressure.

Results of an investigation of all elastic moduli have been published by Warren and Nashner (1976). Mechanical properties of rocks and rock masses have been considered by Bernaix (1974) to depend on the presence and the behaviour of cracks (fissures). Fissure reaction mechanisms are investigated using models which range from plane fissures to fissures with random asperities. Rock behaviour is analysed with reference to these basic mechanisms and it is found that, under certain confining stresses, fissures develop gradually under load from stable to unstable.

The models mentioned above make it possible to describe inelastic deformation of a quasi-homogeneous material under multiaxial compression, which is typical for rocks. The second quasi-homogeneous material considered in this chapter is the hardened cement paste, which in most cases is subjected to the state of stress close to uniaxial compression. For such conditions, the propagation of longitudinal opening cracks (splitting cracks) is typical—see Figure 6.9, region 1.

Mihashi has modified his model mentioned above (see Figure 6.6) for this case. The case of compressive failure (Figure 6.6(d)) may be described by two different types of models. If the failure process is modelled by two states (i.e., a non-fractured state and a fractured state), Type A model is applicable even to compressive fracture. (see Figure 6.6(a)). When the first step of the failure process does not involve fracture of the specimen, Type B model should be used. (Figure 6.6(c)). It is also possible to construct more complex models just by increasing the number of states. However, the equation becomes very complex and it is not easy to determine the coefficients rationally.

It must be noted that the theory of Mihashi does not allow estimating the macrostrain of a solid due to crack formation and crack propagation.

Bažant (1980) has suggested that there are basically two types of inelastic strains—plastic and fracturing. The plastic strains, de_{ij}^{pl}, are due to plastic slip which takes place at constant stress and does not affect the elastic stiffness, and the fracturing strains, de_{ij}^{fr}, are due to microfracturing or micro-cracking which is accompanied by a stress drop and results in a degradation of material stiffness; accordingly

$$de_{ij} = \frac{ds_{ij}}{2G} + de_{ij}''; \qquad de_{ij}'' = de_{ij}^{pl} + de_{ij}^{fr} \qquad (6.11)$$

$$de_{ij}^{fr} = \frac{\partial \Phi}{\partial e_{ij}} dk; \qquad de_{ij}^{pl} = \frac{1}{2G} \frac{\partial \Psi}{\partial e_{ij}} d\zeta \qquad (6.12)$$

in which Ψ and Φ are the plastic and fracturing loading functions; G = elastic shear modulus, introduced for dimensional convenience; $s_{ij} = \sigma_{ij} - \delta_{ij}\sigma$ = stress deviator; $\sigma = \sigma_{kk}/3$ = volumetric stress; $e_{ij} = \varepsilon_{ij} - \delta_{ij}\varepsilon$ = strain deviator; $\varepsilon = \varepsilon_{kk}/3$ = volumetric strain, δ_{ij} = Kronecker delta; de_{ij}^{pl} = plastic strain increments, de_{ij}^{fr} = fracturing strain increments.

For data fitting, a computer program has been developed. Small loading steps are used to integrate the constitutive relation numerically for specified forms of materials functions and given material parameters.

Problems of application of fracture mechanics to concrete and hcp have been studied by many authors, for instance, Bažant (1979), Carpinteri, Di Tomasso, and Viola (1978), Desay (1977), Dias and Hilsdorf (1973), Hillerborg (1983), Mindess (1983), Moavenzadeh and Kuguel (1969), Saouma, Ingraffea and Catalano (1980), Slate, Jaquot, Lierse, Ringkamp, Rastogi, and Terrien (1983), Shah (1979), Shah and McGarry (1971), Wittmann (1983), Ziegeldorf (1983) and others.

Nonlinear behaviour of concrete has been experimentally investigated by many authors, for instance, Bascoul and Maso (1982), Desay (1977), Dias and Hilsdorf (1973), Evans and Marathe (1968), Gerstle, Aschl, Belotti, Bertacchi, Kotsovos, Ko, Linse, Newman, Rossi, Schickert, Taylor, Traina, Winkler, and Zimmerman (1980), Hsu, Slate, Sturman, and Winter (1963), Johnston (1970), Lott and Kesler (1966), Stroeven (1973), Ziegeldorf (1983) and others.

In a model of Zaitsev (1969 to 1982), Zaitsev and Wittmann (1971 to 1981), Wittmann and Zaitsev (1972 to 1981) it was assumed that hardened cement paste has pre-existing defects in the form of pores and cracks. First of all, the interaction of two pores with cracks (Figure 6.12) has been studied with the help of fracture mechanics methods. The main results are shown in Figure 6.13. The first stages of crack propagation are stable. The presence of neighbouring pores causes an unstable manner of crack propagation (at $\Psi = \Psi_2$ and $q_*^0 = q_*^{max}$, see curves 2 in Figure 6.13); curves 2a correspond to irreversible cracks, and curves 2b to reversible cracks. The calculations for different c/r-values have shown that the relative crack length Ψ_2 (corresponding to the end of stable crack propagation) is nearly constant ($\Psi_2 = 0.69$ to 0.73 for internal cracks, and $\Psi_2 = 0.51$ to 0.55 for external cracks). The relative crack length Ψ_1, which corresponds to q_*^{max} on the line 1a (one pore with two coplanar cracks), can also be regarded as a constant ($\Psi_1 = 0.46$ to 0.47).

The process of crack propagation discussed above will lead to joining of the two internal cracks and formation of a new crack, crossing both pores (see the dotted lines on Figure 6.12). The results, which were obtained in a similar way, are given in Figure 6.13 (line 4). A comparison of curves 2 and 4 shows that if the load has reached the value q_*^{max} (i.e., the cracks have entered the zone of unstable propagation), both interacting cracks join at once and the λ-value 'jumps' from curve 2 to curve 4 (wavy line).

When three, four or more coplanar cracks interact, the subsequent stage of crack propagation, is a process having a statistical nature, because the interaction depends significantly on the distribution of the distances between the pores. This process can be simulated using Monte Carlo methods. The results of such a simulation, described by Zaitsev (1971, 1974), and Zaitsev and Wittmann (1971, 1974), show a satisfactory agreement with the existing experimental results.

In the model described above it is assumed that all the cracks are coplanar. It is, however, well known that the real hardened cement paste has cracks oriented at random. Therefore, a more complicated model has been analysed. It was assumed that all pores are statistically uniformly distributed over the area in Figure 6.14(a), which represents the specimen size. Each pore has two pre-existing cracks, the length of which is uniformly distributed within the range $0 < l/r < 2$. The angle of crack orientation α_i with respect to the applied load is also uniformly distributed, within the limits of 0 and 2π. The interaction of cracks was taken into account by assuming that two cracks ($i = 1, 2$) will interact and coalesce if α_1 and α_2 are both below $\pi/6$ (or above $5\pi/6$). The random structure has been created by a computer program.

Simulation of crack propagation as described above makes it possible to estimate the effect of the pore size distribution on the fracture mechanism, mechanical strength, and strain behaviour of hardened cement paste. In particular, it was found (Zaitsev, 1980) that by increasing the mean size of pores while keeping their number constant, the ultimate load will decrease. Increasing the maximum size of pores but keeping a constant mean size and a constant number of pores will decrease the ultimate load too.

Figure 6.14(b) shows the stress–strain curves for uniaxial compression of a

Inelastic Properties of Solids with Random Cracks 107

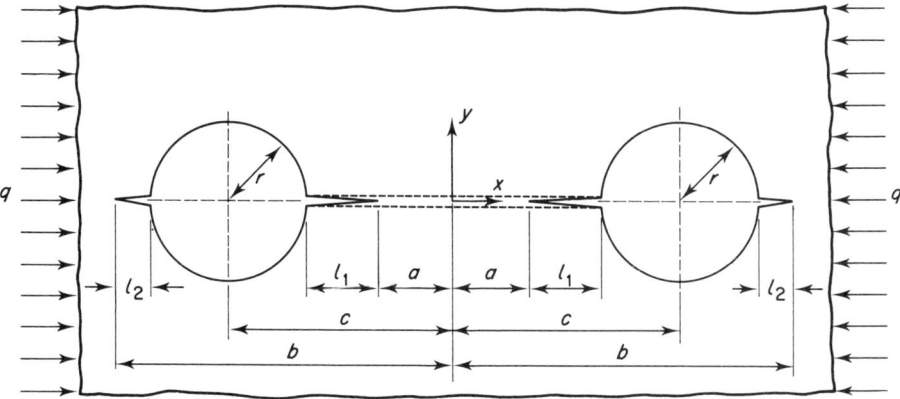

Figure 6.12 Schematic representation of two pores and coplanar cracks in hardened cement paste

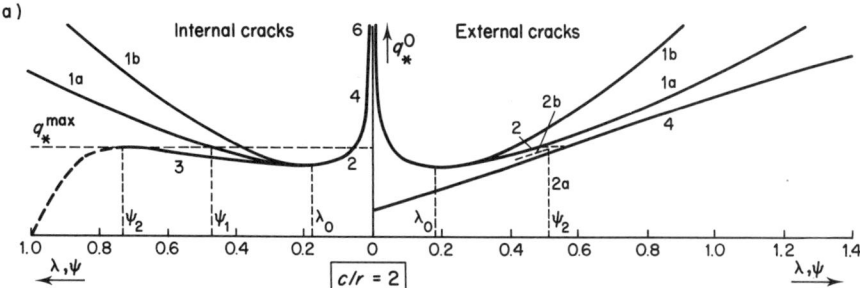

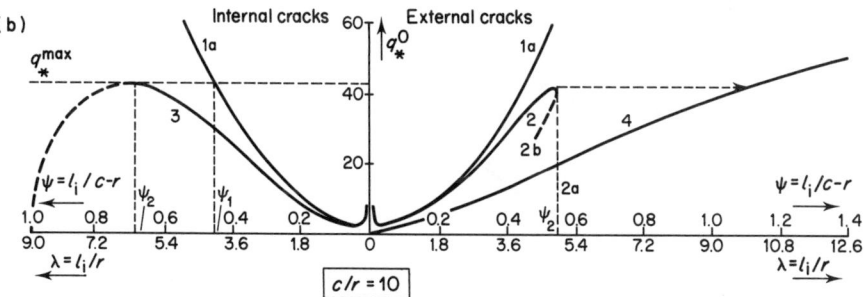

Figure 6.13 Relationship between related external load and related crack length near two pores:
(a) $c/r = 2$, (b) $c/r = 10$.
1—relationship for two cracks (1a) and one (1b) isolated crack (without interaction); 2—relationship for interacting external cracks; 3—relationship for interacting internal cracks

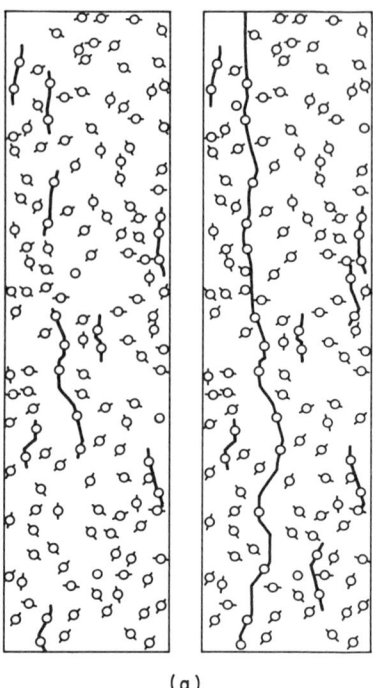

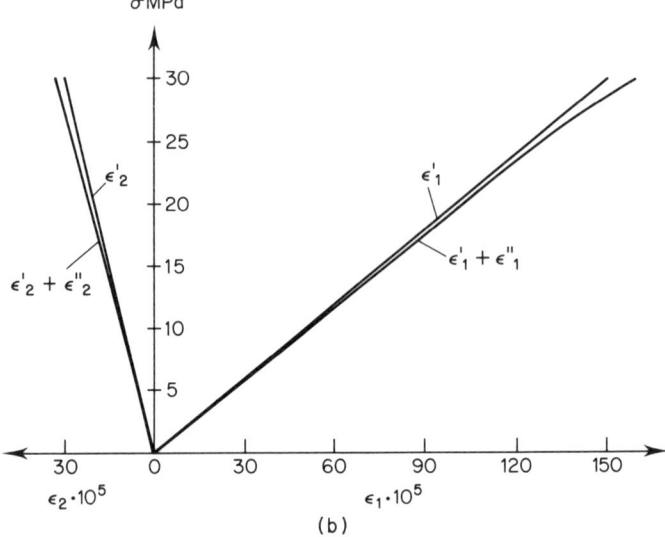

Figure 6.14 Consequent stages of crack propagation in a quasi-homogeneous porous solid under compression and stress–strain curves according to results of simulation of crack propagation

specimen (see Figure 6.14(a)), plotted according to the results of simulation of crack propagation for ε_1'' based on the results of section 6.2.1.

6.3 MODELS OF SIGNIFICANTLY INHOMOGENEOUS SOLIDS WITH RANDOM CRACKS (CONCRETES)

6.3.1 Uniaxial tension

The phenomenological aspects of the inelastic properties of concrete in uniaxial tension may be summarized as follows. The σ–ε diagram is linear up to stresses of about 80 per cent of the ultimate stress. The deviation of the σ–ε diagram from the straight line is associated with enlargement of pre-existing bond cracks. At higher loads, continuous cracks (i.e. cracks comprising some bond cracks and mortar cracks) are formed. Some propagating cracks can be arrested by aggregate particles. Different views exist about the origin of pre-existing bond cracks. Usually such crack formation is explained by shrinkage of the hydrating cement paste. However, according to Ziegeldorf (1983), shrinkage cracks must emanate radially from aggregates. The second point of view is related to the observation that water lenses develop under coarse aggregate pieces during setting of the fresh concrete (bleeding), and the crack density is therefore the greatest in the horizontal direction at all stress levels. It seems that both effects (shrinkage and bleeding) are responsible for pre-existing cracks, and the crack pattern according to Figure 6.15(a) must be expected in concrete (Ziegeldorf, 1983).

Mihashi's model described above can also be applied to the case of concrete in tension. Application of the methods of fracture mechanics may be very useful for such problems. However, Bažant (1979) has shown that methods of linear

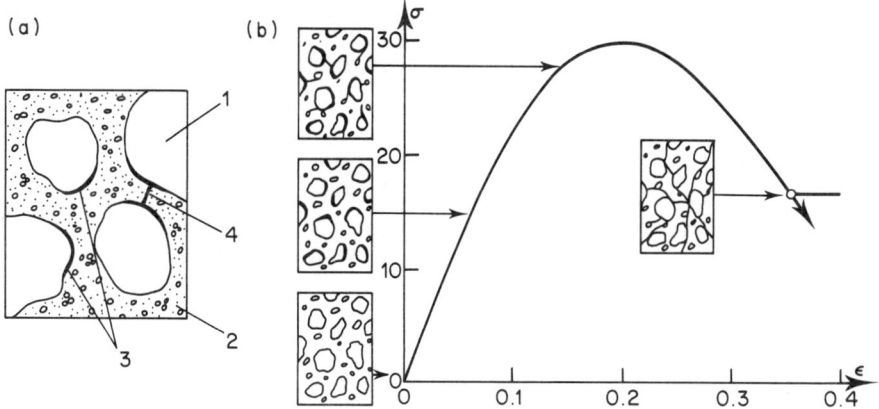

Figure 6.15 Cracks in concrete. (a) cracks in unloaded concrete; (b) consequent stages of crack propagation and stress–strain relationship; in compression. 1—aggregate; 2—mortar; 3—interface cracks; 4—cracks between aggregates

fracture mechanics can be applied to concrete (regarded as a quasi-homogeneous solid) only for crack dimensions whose order of magnitude is several metres, which is significantly more than the dimensions of real cracks in concrete structures. Therefore, investigation of crack propagation in concrete with help of methods of fracture mechanics necessitates analysing concrete as a multiphase system consisting of a quasi-homogeneous matrix (hardened cement paste) and inclusions (aggregate particles). Such an approach has also an important advantage: it gives the possibility of evaluating the effect of the quantity and shape of the particles, and of K_{IC}- and E-values of aggregate, on inelastic behaviour of concrete.

Let us now analyse crack propagation in concrete subjected to uniaxial tension, according to the model of Wittmann and Zaitsev. We shall investigate the problem of crack propagation in an elastic plate of relative thickness 1 containing circular inclusions (coarse aggregate particles). Each inclusion has one pre-existing bond crack; the dimension of the most dangerous crack in the matrix (hardened cement paste), defining its strength, is small as compared to the dimension of the inclusions. We shall assume also that the value of K_{IC} for the inclusions is greater than the value of K_{IC} for the matrix, and the value of K_{IC} for the matrix is greater than the value of K_{IC} for the interface, i.e.

$$K_{IC}^{INCL} > K_{IC}^{M} > K_{IC}^{IF} \tag{6.13}$$

Under these assumptions, the problem has been solved by Cherepanov. He has found the following relationship between the applied tensile load q_* and the angle, θ, which defines the dimension of the crack:

$$q_* = K_{IC}^{IF} F(\theta) \sqrt{R} \tag{6.14}$$

$$F(\theta) = \frac{4(3 - \cos\theta)/\sqrt{\pi}}{\sqrt{\sin\theta(44 + 12\cos\theta + 12\cos^2\theta - 4\cos 4\theta + \sin^4\theta)}} \tag{6.15}$$

Line 1 in Figure 6.16 shows this relationship; the y-axis gives values of the relative load $q_{IF}^{0} = q_* \sqrt{2R/\pi}/K_{IC}^{IF}$.

Where $\theta < \theta_0 (\theta_0 = \pi/4)$, a crack propagates in an unstable manner (descending part of the curve), but when $\theta > \theta_0$ the crack becomes stable (ascending part of the curve). The crack can propagate in a stable manner only until the applied load reaches a certain critical value, corresponding to the appearance of crack branches within the matrix, illustrated on top of Figure 6.16, which will again lead (for $\theta > \theta_2$) to an unstable manner of crack propagation. The precise value of θ_2 depends on the value of K_{IC}^{IF}/K_{IC}^{M}; the greater this value, the less the angle θ_2 for $K_{IC}^{IF}/K_{IC}^{M} = 0.6$ (which is close to the experimental results of Alexander, Hillermaier and Hilsdorf, Ziegeldorf and Hilsdorf, etc.), $\theta_2 = \pi/2$. This situation is shown in Figure 6.16, where line 3 gives the values of the relative load for the interfacial part of the crack $q_{M}^{0} = q_* \sqrt{2R/\pi}/K_{IC}^{M}$ (i.e. $q_{M}^{0} = 0.6 q_{IF}^{0}$), and line 2

Inelastic Properties of Solids with Random Cracks

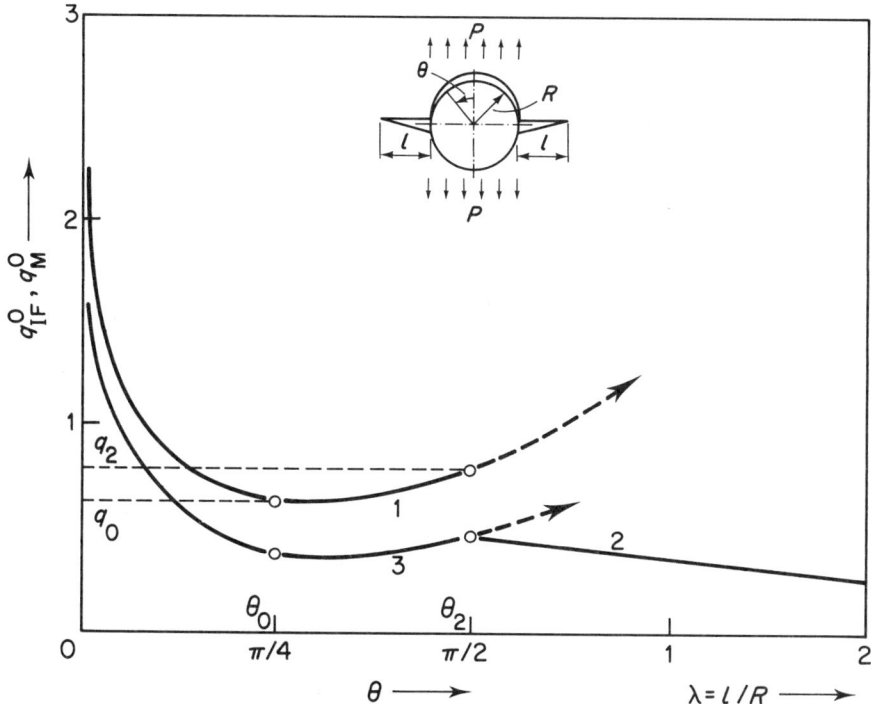

Figure 6.16 Relationship between related external tensile load and related length of an interfacial crack; 1—interfacial crack (load is related to K_{IC}^{IF}); 2—interfacial crack (load is related to K_{IC}^{M}); 3—interfacial crack and additional matrix (load is related to K_{IC}^{M})

gives the values of the relative load for the crack in the matrix. As an approximation, the latter values have been determined as the values for a Griffith's crack of length $2(l + R)$. The load value q_0 at the beginning of the stable stage of crack propagation along the interface is equal to approximately 80 per cent of the value of q_2 at the beginning of the unstable stage of crack propagation in the matrix. The subsequent stages of crack propagation in concrete can be analysed in an analogous way.

Now we can simulate the structure of concrete and the crack propagation using Monte Carlo methods. Typical examples of one of the computer realizations are shown in Figure 6.17. For the simulation, 50 circular inclusions have been produced, and each of them is assumed to have one pre-existing interfacial crack. It is also supposed that there are pre-existing cracks in the matrix. It is assumed that the centres of cracks in the matrix and the centres of aggregate particles are uniformly distributed over the area of the sample. The lengths of pre-existing matrix and interfacial cracks as well as the diameters of aggregate particles are assumed to have a Gaussian distribution.

Taking the opening widths of cracks into consideration, the contribution of

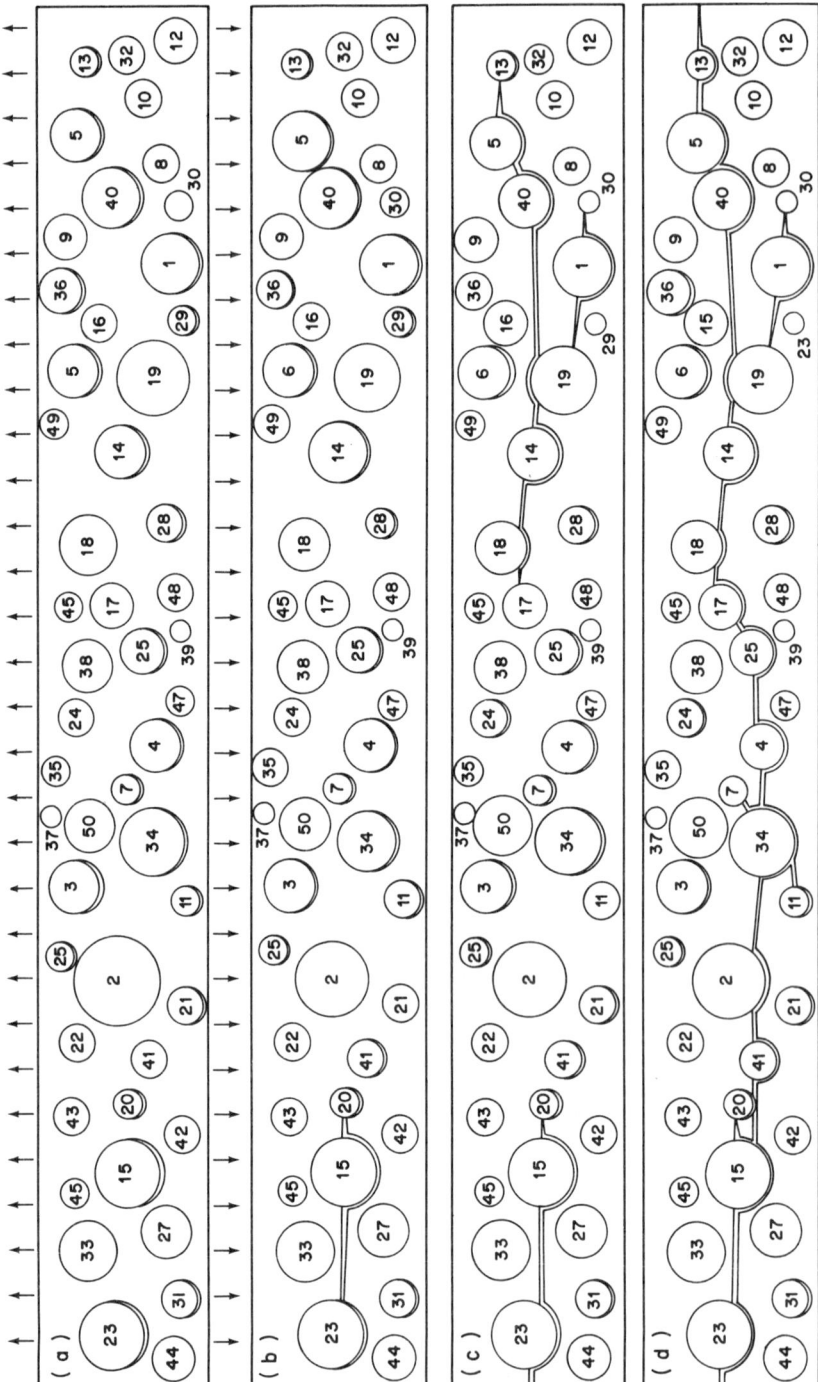

Figure 6.17 Crack pattern for different load level in concrete. Cracks originating from interfaces run around aggregates

cracks to the longitudinal deformation ε_1 can be found if we assume linear superposition of the two components which appear in Equation (6.4).

Figure 6.18 shows the stress–strain curves for concrete according to the results of simulation of crack propagation. The values of the relative strain $\varepsilon/\varepsilon_{el}^*$ are plotted on the x = axis ($\varepsilon_{el}^* = \sigma_{ult}/E$, σ_{ult} is the fracture stress), and the values of the relative load $q_*^0 = q_* \sqrt{2\bar{R}/\pi}/K_{IC}^M$ are plotted on the y-axis; $\bar{R}$ is mean value of the diameter of aggregate particles.

The simulated σ–ε curves for concrete in tension may be divided into the three

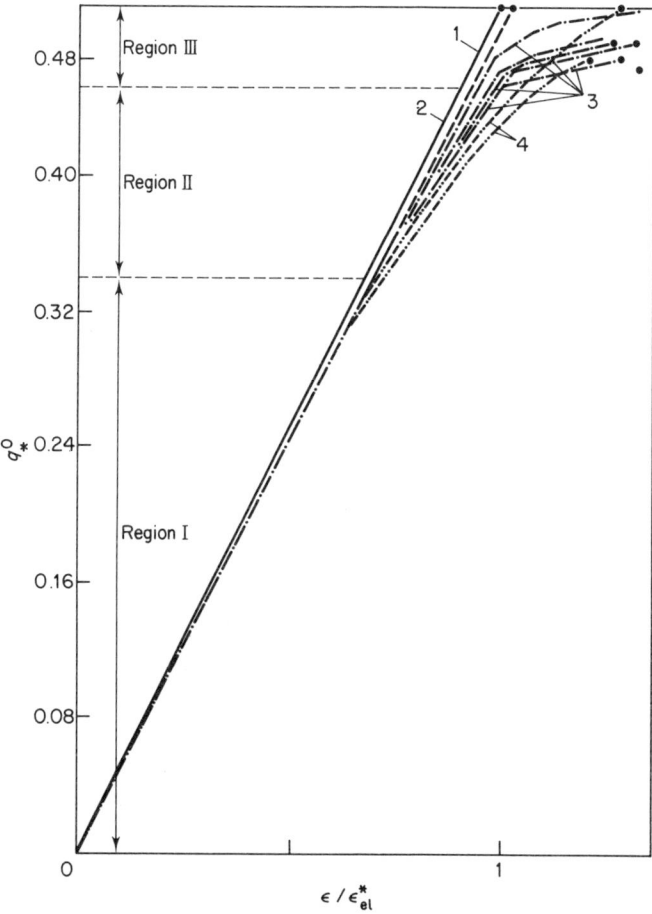

Figure 6.18 Stress–strain curves for concrete in tension according to results of simulation of crack propagation; 1—linear for a material without cracks; 2—linear for a material with pre-existing cracks which do not propagate; 3—nonlinear taking into account crack propagation; 4—experimental curves for concrete in tension

regions shown in Figure 6.18. In region 1, which corresponds to the crack pattern shown in Figure 6.17(a), the applied load is less than the q_0-value (see Figure 6.16) for each of the pre-existing cracks. This means that the cracks do not propagate and there is only very little extension (widening) of these cracks. The σ–ε diagram (line 2 in Figure 6.18) is linear, but it has a slightly steeper declining slope than line 1, which corresponds to a material without cracks. This difference is due to the crack extension (widening) mentioned above.

In region 2, cracks according to Figure 6.17(a) begin to propagate in a stable manner which corresponds to applied load values greater than the q_0-value (see Figure 6.16) for the most dangerous cracks. The corresponding σ–ε curves become slightly nonlinear (see lines 3 in Figure 6.18, where each line corresponds to one realization by the Monte Carlo method.)

In region 3 (see Figure 6.17(b,c)), cracks begin to propagate through the matrix between the aggregate particles. The σ–ε curves become significantly nonlinear (see line 3 in the upper part of Figure 6.18). Finally, failure of the sample occurs (see Figure 6.17(d)). It was found that region 1 corresponds to $\sigma < 0.65\sigma_{ult}$; region 2 to $0.65\sigma_{ult} < \sigma < 0.9\sigma_{ult}$, and region 3 to $\sigma > 0.9\sigma_{ult}$. It was also found that, for assumed parameters of crack redistribution, the critical load for pre-existing matrix cracks (according to paragraph 6.2.1) was higher than σ_{ult} for concrete. Thus, the failure of concrete depends in this case only on the propagation of pre-existing interfacial cracks.

Figure 6.19 shows the mean value (from 20 realizations) of the sum of the lengths of cracks (relative to the area of the specimen) as a function of the relative strain $\varepsilon/\varepsilon_*$. Lines 1 and 2 correspond to interfacial and matrix cracks. Lines 3 and 4 represent the experimental results. All the results of simulation of crack

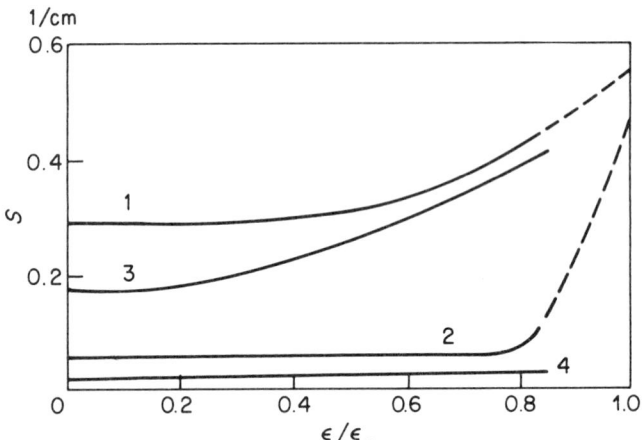

Figure 6.19 Summarized length of cracks as a function of related strain $\varepsilon/\varepsilon_*$; 1—simulated (interface cracks); 2—simulated (matrix cracks); 3—experimental (interface cracks); 4—experimental (matrix cracks)

propagation in concrete as described above are in reasonable agreement with existing experimental results.

6.3.2 Uniaxial compression

The phenomenological aspects of inelastic properties of concrete subjected to uniaxial compression may be summarized as follows. The pre-existing bond cracks initially do not propagate. At applied stress of 30–40 per cent of the ultimate stress, the bond cracks begin to propagate in a stable manner along the interface, and mortar cracks then appear. At 70–80 per cent of the ultimate stress, there is a significant increase in the number of mortar cracks, and due to their joining with the nearby bond cracks continuous cracks are formed. Their orientations are mainly parallel to the direction of the applied load. At all stress levels, bond cracks predominate, and so failure of the contact zone of sand particles rarely occurs. For higher strength concretes, the stress–strain curve is linear up to a higher stress–strength ratio than it is for normal concretes because of a decrease in the amount and extent of bond cracking (Ziegeldorf, 1983).

The phenomenological description of the fracture process of concrete has furnished valuable information. But many fundamental problems are still unsolved. In particular, the experimental results concerning the effect of aggregate concentration on concrete strength are rather contradictory, and the effect of the aggregate–paste bond strength on the concrete strength is not well understood. What is generally missing is a theoretical basis capable of explaining such effects.

Mihashi's model, and Zaitsev and Wittmann's model described above, can be applied to the case of concrete in compression as well. In the model of Zaitsev and Wittmann for the case of concrete in compression, the problem of crack propagation in an elastic plate of thickness 1, containing inclusions (coarse aggregate particles), is analysed. Inclusions having in this case random polygonal shapes are randomly distributed through out the matrix. The size and shape distribution and the volume content of the inclusions can be varied in order to simulate different concrete mix proportions. Each inclusion has one pre-existing bond crack. The dimension of the most dangerous crack in the matrix (hardened cement paste), governing its strength, is small as compared to the dimension of the inclusions.

We begin with the simplest case of a randomly inclined crack in an elastic plate loaded at infinity. Figure 6.20 shows an initial crack having length $2l$ and inclination α with respect to the direction of applied compressive load $q(q<0)$. This crack might propagate along the same inclined line as a shear crack (Mode II). It can be shown, however, that in our example two branching cracks of Mode I (splitting, or opening cracks) are created at the tips of the initial crack (Dias and Hilsdorf, 1973; Desay, 1977). By introducing certain simplifying assumptions, crack length l_2 according to Zaitsev (1977), and Zaitsev and Wittmann (1977), can

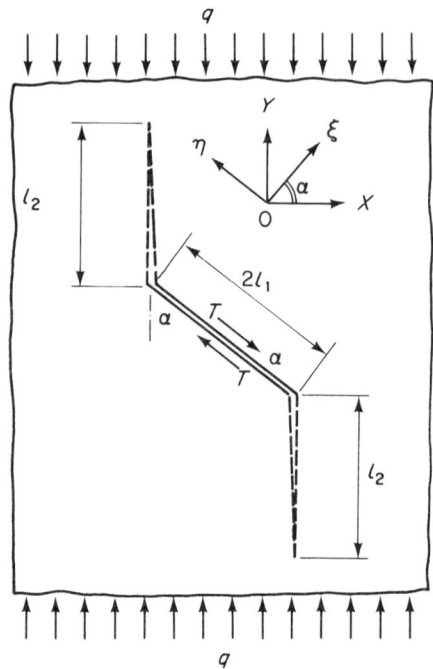

Figure 6.20 Schematic representation of the development of branching cracks and definition of symbols used in corresponding equations

be expressed as follows:

$$P/\sqrt{\pi l^2} = K_{IC} \quad (6.16)$$

where $P = T \sin \alpha$ and T is the resulting force of shear stress $\tau_{\xi\eta}$ which causes sliding of two opposite sides of the inclined crack. Taking the coefficient of friction ρ into account, T can be expressed as follows:

$$T = 2l_1 \tau_{\xi\eta} = 2l_1 q(\sin \alpha \cos \alpha - \rho \sin^2 \alpha) \quad (6.17)$$

From these equations, we can obtain (crack of Mode I in the matrix):

$$q_I^M = -\frac{\sqrt{\pi l_2}}{2l_1} \frac{K_{IC}^M}{A(\alpha, \rho)} \quad (6.18)$$

where $A(\alpha, \rho) = \sin^2\alpha \cos\alpha - \rho \sin^3\alpha$. From the last equation one can see that the propagation of such a crack is stable, i.e. l_2 will steadily increase as $|q|$ increases. A similar stable propagation of such cracks has been observed experimentally.

After analysing this simplest case, we shall now consider a homogeneous matrix with one polygonal inclusion, representing an aggregate particle in an infinite matrix. An initial interfacial crack with length $2l_1$ is assumed to be located along one side AB (see Figure 6.21(a)). This problem can be treated similarly to the one with an inclined crack in a homogeneous matrix, taking into consideration, however, the concentration of shear and normal stress in the interface. This can be done by introducing coefficients of stress concentration k_τ, k_σ. It can be shown that the initial shear crack extends (Mode II) in an unstable manner as soon as the critical load q_{II}^{IF} (Mode II, interface) is reached:

$$q_{II}^{IF} = -\frac{K_{IIC}^{IF}}{\sqrt{\pi l_1}\, D_{IF}(\alpha,\rho)} \qquad (6.19)$$

where $D = k_\tau \sin\alpha\cos\alpha - k_\sigma \rho \sin^2\alpha$. The shear crack reaches length $2L_1$ (see Figure 6.21(b)) and stops because further crack propagation in the same inclined direction would take place through the matrix, provided that $K_{IIC}^M \gg K_{IIC}^{IF}$. If, however, the external load is increased to a higher critical value q_I^M:

$$q_I^M = -\frac{\sqrt{3/4}\, K_{IC}^M}{\sqrt{\pi L_1}\, D_{IF}(\alpha,\rho)} \qquad (6.20)$$

then branching cracks will develop in the matrix (see Figure 6.21(c)). The actual crack length in the matrix can be expressed as a function of the load in a manner that is analogous to the case of a homogeneous material treated alone:

$$q = -\frac{\sqrt{\pi l_2}}{2L_1}\frac{K_{IC}^M}{A_{IF}(\alpha,\rho)} \qquad (6.21)$$

where $A_{IF}(\alpha,\rho) = D_{IF}(\alpha,\rho)\sin\alpha$, and l_2 corresponds to the distance AA′ as shown in Figure 6.21(c).

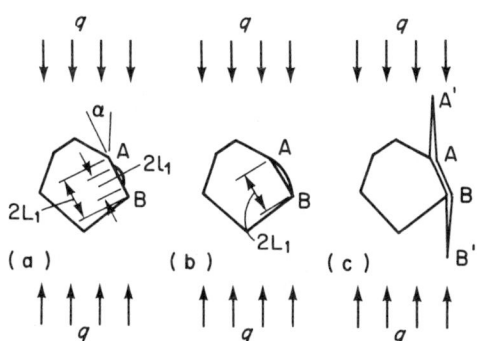

Figure 6.21 An initial crack with length $2l_1$ (a) grows in an unstable fashion along an interface AB (b) and finally stable branching cracks AA′ and BB′ are created as the load is increased

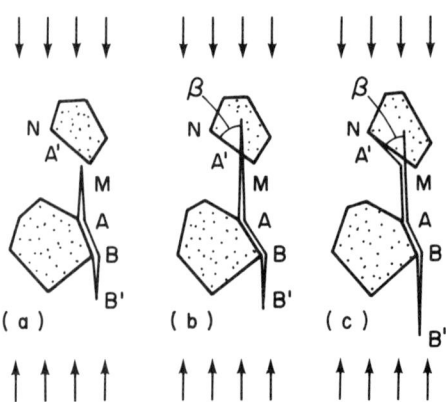

Figure 6.22 A crack path as shown in Figure 6.21, in the vicinity of the second inclusion (a); the crack meets the second inclusion (b); finally the crack will propagate along the interface MN (c)

In Figure 6.22, the situation shown in Figure 6.21(c) is repeated, but now it is assumed that the branching crack AA' meets a second inclusion as it propagates. Further crack growth will either take place through the inclusion maintaining the same direction, or the crack must follow the same interface MN, but as a crack of Mode II (see Figure 6.22(c)). To decide which path will be followed, three critical load values must be compared:

$$q_I^{INCL} = -\frac{\sqrt{\pi l_2}}{2L_1} \frac{K_{IC}^{INCL}}{A_{IF}(\alpha,\rho)} \qquad (6.22)$$

$$q_I^{INT} = -\frac{2K_{IC}^{IF}\sqrt{\pi l_2/L_1}}{A_{IF}(\alpha,\rho)\left\{3\cos\frac{\beta}{2}+\cos\frac{3\beta}{2}\right\}-3B_{IF}(\alpha,\rho)\left\{\sin\frac{\beta}{2}+\sin\frac{3\beta}{2}\right\}} \qquad (6.23)$$

$$q_{II}^{INT} = -\frac{2K_{IC}^{IF}\sqrt{\pi l_2/L_1}}{A_{IF}(\alpha,\rho)\left\{\sin\frac{\beta}{2}+\sin\frac{3\beta}{2}\right\}+B_{IF}(\alpha,\rho)\left\{\cos\frac{\beta}{2}+\cos\frac{3\beta}{2}\right\}} \qquad (6.24)$$

The index INCL denotes that values are valid for the crack propagation through the second inclusion; l_2 corresponds to the distance AA' as shown in Figure 6.22(b) and $B_{IF}(\alpha,\rho)$ is given by

$$B_{IF}(\alpha,\rho) = D_{IF}(\alpha,\rho)\cos\alpha \qquad (6.25)$$

From the expressions for q_I^{INCL}, q_I^{INT}, and q_{II}^{INT} it is evident that the further crack path depends on the relationship between the interface and the inclusion characteristics K_{IC}^{INCL}, K_{IC}^{IF}, and K_{IIC}^{IF}, as well as on the geometry of the crack path

A' ABB' (see expressions for $A_{IF}(\alpha, \rho)$, $B_{IF}(\alpha, \rho)$) and on the inclination of the interface MN.

For further discussion of these theoretical predictions, we shall look at three different cases:

—normal concrete (where $K_{IC}^{IF} \ll K_{IC}^{INCL}$)
—high strength concrete (where $K_{IC}^{IF} \approx K_{IC}^{INCL}$)
—lightweight concrete (where $K_{IC}^{IF} \gg K_{IC}^{INCL}$)

These three cases with be dealt with separately.

In the case of normal concrete, a crack will propagate mainly along the interface MN, because the critical q_I^{INCL}-value will be too high as a result of high K_{IC}^{INCL}-values. Whether crack propagation will take place according to Mode I or Mode II, depends significantly on the sign of β (see Figure 6.22). If $\beta < 0$, a crack of Mode I is to be expected but if $\beta > 0$ (as in Figure 6.22), a crack of Mode II is more probable.

It should be noted that shear cracks (Mode II) are facilitated by the shear components of applied compression, whereas the presence of normal confining components of applied compression makes the formation of opening cracks in the interface (Mode I for $\beta < 0$) less likely as compared to the formation of shear cracks (Mode II for $\beta > 0$). This theoretical prediction has been verified experimentally (Zaitsev and Wittmann, 1981). As a consequence, new interfacial cracks in a material with randomly distributed inclusions (for which the probabilities of occurrence of positive and negative values of β are equal) will propagate mainly according to Mode II if $\beta > 0$. This means that a final crack running through the whole specimen will contain some interfacial parts, which deviate predominantly in the same direction ($\beta > 0$!) from the applied load direction (see also Figure 6.22). Thus, the final crack will be slightly inclined and not exactly parallel to the external load direction.

In the case of high-strength concrete, a crack as shown in Figure 6.22 will either propagate according to the mechanism described above (i.e., along the interface), or it will penetrate into the inclusion. With increasing values of β, crack propagation along an interface occurs at higher and higher loads.

Above a critical value β_* of the angle of inclination of the interface side β, cracks prefer to pass through the inclusions. The critical value, β_*, is essentially independent of α or ρ or both, and depends only on the ratio χ_1 of K_{IC}^{INCL} and K_{IC}^{IF}. In particular, for $\chi_1 = K_{IC}^{INCL}/K_{IC}^{IF} = 1$ we obtain $\beta_* = \pi/3$. Crack propagation through the inclusions also depends on the ratio χ_1. If $\chi_1 < 1$, cracks grow faster, and if $\chi_1 > 1$, they grow slower, as compared to the homogeneous matrix. Thus, for high-strength concrete the probability that the crack would deviate from the direction of the external load is much less as compared to the case of normal strength concrete.

In the case of lightweight concrete, for which the interface strength is much higher than the matrix strength, fracture surfaces run across the matrix and aggregate particles. For this case, the model of crack initiation at pores according to section 6.3.1 can be used. It has been shown that such cracks propagate in lightweight concrete in a stable manner analogous to the case of hardened cement paste discussed above. After a crack has reached an inclusion, it propagates into the inclusion. This propagation becomes unstable because the K_{IC}^{INCL}-values are small compared to the corresponding K_{IC}^{M}-values. When the crack, after having passed through the inclusion, reaches the matrix again, it stops. After further increase of the applied load (depending on the $\chi_1 = K_{IC}^{INCL}/K_{IC}^{M}$-value), the crack will propagate through the matrix again as if it were a homogeneous material.

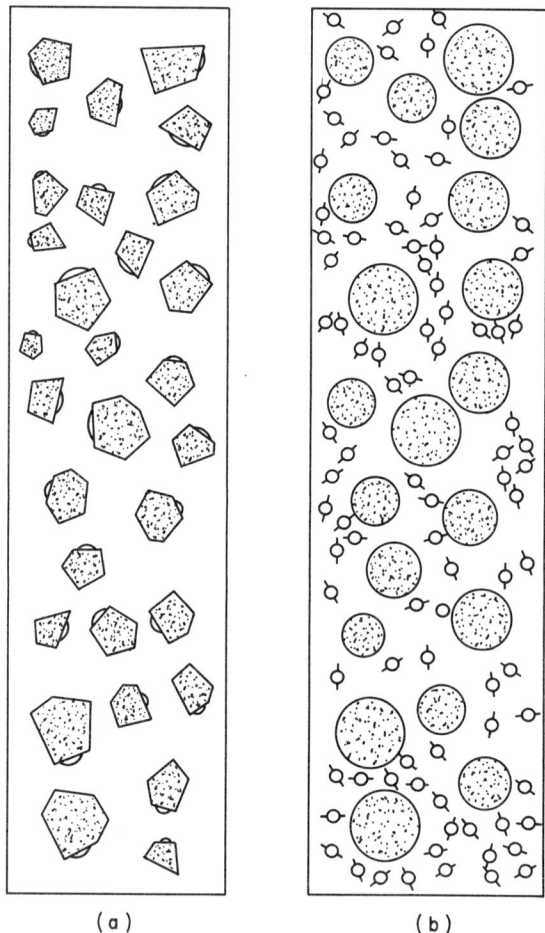

Figure 6.23 Typical computer realization of random structure; (a) normal concrete; (b) lightweight concrete

After describing all essential elements of crack propagation in a two-phase material, we can simulate the structure of concrete and the crack propagation using Monte Carlo methods. Typical examples of computer realizations of the structure of normal and lightweight concrete are shown in Figure 6.23. In the case of normal concrete, 30 polygonal inclusions have been produced. Each particle is supposed to have one interfacial crack. The structure can also be used to study crack propagation in high strength concrete. In this case, the geometrical arrangement is maintained, but the fracture mechanics parameters are modified. The random structure of lightweight concrete is simulated by 20 round inclusions and small pores spread over the matrix.

As the load in the computer experiment is increased, the most critical cracks will propagate first. Further increase of the load produces a characteristic crack pattern and finally one crack will run through the whole specimen. This is defined to be the failure of the material.

Figure 6.24 Crack pattern for two different load levels in normal concrete. Cracks are running around inclusions

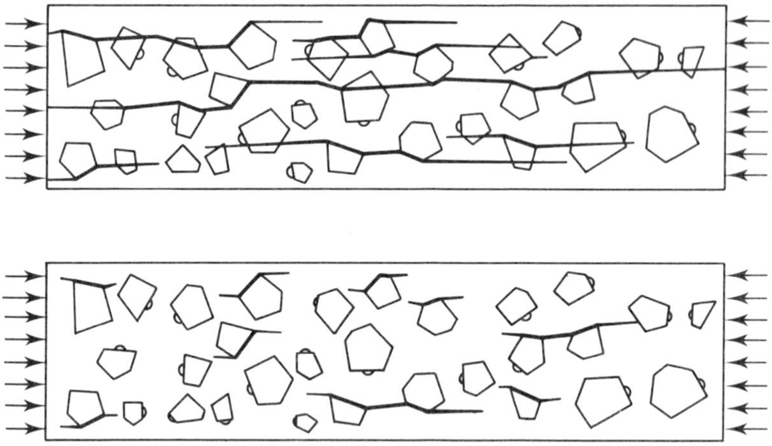

122 *Mechanics of Geomaterials*

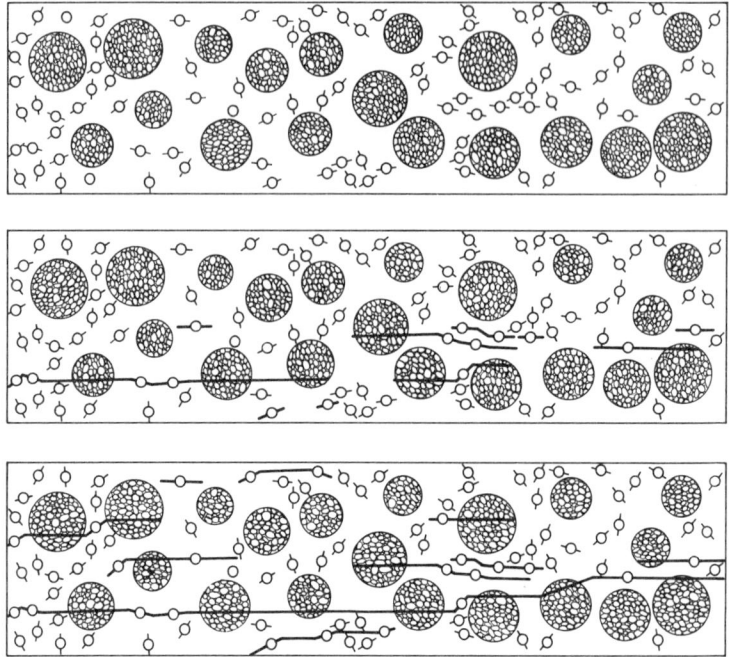

Figure 6.26 Crack pattern for two different load levels in lightweight concrete. Cracks originating from pores run through inclusions

Figure 6.25 Crack pattern for two different load levels in high strength concrete. Some cracks penetrate through inclusions

Different load levels and corresponding crack patterns are found by means of a computer experiment, as shown in Figures 6.24–6.26. A discussion of this result is published by Zaitsev and Wittmann (1981), and Wittmann and Zaitsev (1981). Cracks in normal concrete do not penetrate the aggregate pieces; thus, they contain some interfacial parts which mostly have an angle of inclination of the same sign, and so the resulting overall crack is slightly inclined. To the author's knowledge, this often observed behaviour is now theoretically explained for the first time.

Figure 6.27 shows the stress–strain curves for uniaxial compression based on Zaitsev and Kazatskij's (1982) results of simulation of crack propagation for normal concrete. The effects of K_{IC}- and E-values of the aggregate on the shape of the σ–ε curves of concrete have been studied. The K_{IC}- and E-values for the matrix were equal to 0.3 MPa m$^{1/2}$ and 27 GPa, respectively. The K_{IC}- and E-values for the aggregate were equal to 0.33 MPa m$^{1/2}$ and 29.7 GPa (curve 17), 0.45 MPa m$^{1/2}$ and 40.5 GPa (curve 2), 0.57 MPa m$^{1/2}$ and 51.3 GPa (curve 3), 0.81 MPa m$^{1/2}$ and 72.9 GPa (curve 4). The geometrical arrangement of the simulated structure of concrete, including the pre-existing cracks, was the same for all realizations. As we can see from Figure 6.27, computer experiments give a possibility to evaluate the effect of aggregate properties on the inelastic behaviour and strength of concrete and provide a solid basis for further systematic investigations. Various material structures as well as failure under a multiaxial

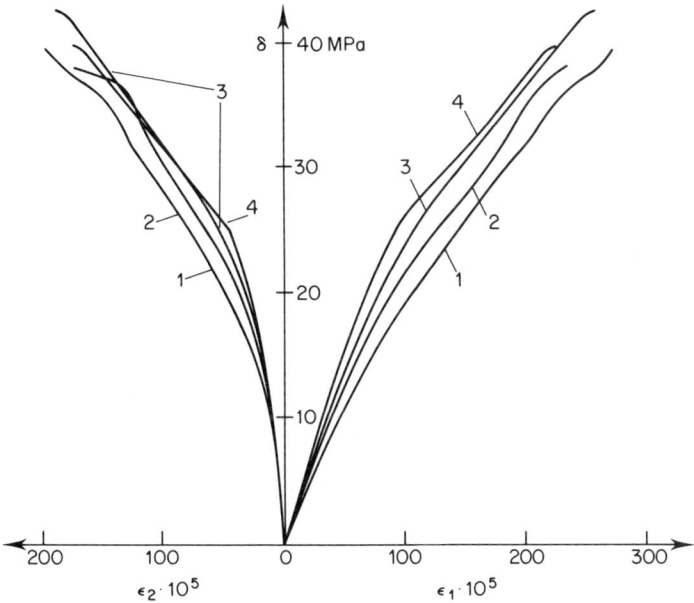

Figure 6.27 Stress–strain curves for concrete according to results of simulation of crack propagation

state of stress can be studied in a similar way. Time-dependent processes, such as failure under a high sustained load, can be included in this type of investigation, too.

REFERENCES

Ashrabov, A. A., and Zaitsev, Y. V. (1981) 'Elements of fracture mechanics of concrete (in Russian), Ukituvchi, Tashkent.

Bascoul, A., and Maso, J. C. (1982) 'Influence of the spatial strain gradient on the cracking limit state of concrete', in *Proc. Int. Symp. Bonds between Cement Paste and Other Materials*, INSA, Toulouse, 17–19 Nov. 1982, pp. C.47–C.45.

Batzle, M. L., Simmons, G., and Seigfried, R. W. (1980) 'Microcrack closure in rocks under stress: direct observation', *J. of Geophysical Research*, **85**, 7072–7090.

Bažant, Z. P. (1974) (A new approach to inelasticity and failure of concrete, sand and rock: endochronic theory', *Proc. 11th Ann. Meet. Soc. of Engng. Sci.*, Duke University, Durham, N. C., Nov. 1974 (Ed. G. J. Dvorak), pp. 158–159.

Bažant, Z. P. (1979) 'Material behaviour under various types of loading', in *High Strength Concrete* (Ed. S. P. Shah), Proc. of a Workshop held at the University of Illinois at Chicago Circle, Chicago, 1979, pp. 79–92.

Bažant, Z. P. (1980) 'Hysteretic fracturing endochronic theory for concrete', *Journal of the Engineering Mechanics Division, Proceedings of the ASCE*, **106**, EM5.

Bažant, Z. P. (1982) 'Mathematical models for creep and shrinkage of concrete', in *Creep and Shrinkage in Concrete Structures* (Eds. Z. P. Bažant and F. H. Wittmann), Wiley, pp. 163–256.

Bernaix, J. (1974) 'Properties of rock and rock masses', *Advances in rock mechanics*, Vol. 1, Part A, Proceedings of the Third Congress of the International Society for Rock Mechanics. Washington, D. C., 1974, pp. 9–38.

Brace, W. F., Silver, E., Hadley, K., and Goetze, C. (1972) 'Cracks and pores: a closer look', *Science*, **178**, 162–164.

Brady, B. T. (1969) 'The non-linear mechanical behaviour of brittle rock', *Int. J. Rock Mech. and Mining Sci.*, **6**, 211–225 and 301–310.

Brady, B. T. (1970) 'A mechanical equation of state for brittle rock', Part I, *Int. J. Rock Mech. and Mining Sci.*, **7**, 385–421.

Brady, B. T. (1973). 'A mechanical equation of state for brittle rock', Part II, *Int. J. Rock Mech. and Mining Sci.*, **10**, 291–309.

Budiansky, B., and O'Connel, J. (1976) 'Elastic moduli of a cracked solid', *Int. J. Solids and Struct.*, **12**, 2, 81–97.

Carpinter, A., Di Tomasso, A., and Viola, E. (1978) 'Stato limite di frattura nei materiali fragili. Modelli meccanici teorici per l'interpretazione della fratura nei calcestruzzi e nelle rocce', Publicazione n. 38, Università di Bologna, Istituto di scienza delle costruzioni, Bologna, pp. 1–24.

Desay, P. (1977) 'Fracture of concrete in compression', *Matériaux et constructions*, **10**, 139–143.

Dias, S., and Hilsdorf, H. K. (1973) 'Fracture mechanisms of concrete under compressive loads', *Cem. Concr. Res.*, **3**, 4.

Dougill, J. W. (1982) 'Mechanics of concrete systems: current approaches to assessing material behaviour and some possible extensions', in *Creep and Shrinkage in Concrete Structures* (Eds. Z. P. Bažant and F. H. Wittmann), Wiley, pp. 23–49.

Dougill, J. W., and Rida, M. A. (1980) 'Further consideration of progressively fracturing solids', *J. Engng. Mech. Div. ASCE*, 1021–1038.

Evans, R. H., and Marathe, M. S. (1968) 'Microcracking and stress–strain curves for concrete in tension', *Materiaux et Constructions*, **1**, 1, 61–64.

Gash, P. J. (1971) 'A study of surface features related to brittle and semibrittle fracture', *Tectonophysics*, **12**, 349–391.

Geniev, G. A., Kissjuk, V. N., and Tjupin, G. A. (1974) 'Theory of plasticity of concrete and reinforced concrete' (in Russian), Strojizdat, Moscow.

Gerstle, K. H., Aschl, H., Belotti, R., Bertacchi, P., Kotsovos, M. D., Hon-Jim, Ko, Linse, D., Newman, J. B., Rossi, P., Schickert, G., Taylor, M. A., Traina, L. A., Winkler, H., and Zimmerman R. M. (1980) 'Behaviour of concrete under multiaxial stresses', *Journal of the Engineering Mechanics Division, ASCE* **106**, EM6, 1383–1403.

Hillerborg, A. (1983) 'Analysis of a single crack', in *Fracture Mechanics of Concrete* (Ed. F. H. Wittmann), Elsevier, Amsterdam.

Hsu, T. T. C., Slate, F. O., Sturman, G. M., and Winter G. (1963) 'Microcracking of plain concrete and the shape of the stress–strain curve', *J. Amer. Concr. Inst.*, **60**, 2, 209–224.

Ikeda, K., Kabayashi, Y., and Sakurai, T. (1973) 'Effect of flows on the strength of rocks', *Quart. Repts. Railway Techn. Res. Inst.*, **14**, 2, 61–64.

Johnston, C. D. (1970) 'Deformation of concrete and its constituent materials in uniaxial tension', *Highway Res. Rec.*, 324.

Kozachevskij, A. I. (1981) 'About relations of deformation theory of plasticity for concrete with respect on structural anisotrophy', in *Resistance of Materials and Theory of Structures* (in Russian), **39**, Budivelnik, Kiev, pp. 84–88.

Kozachevskij, A. I., and Zjazin, A. M. (1982) 'Study of non-linearity of dilatancy model of deformation theory of plasticity for concrete', in *Static and Dynamic of Sophisticated Building Structures* (in Russian), LISI, Leningrad, pp. 38–45.

Løland, K. E. (1980) 'Continuous damage model for load-response estimation of concrete', *Cem. Concr. Res.*, **10**, 395–402.

Lorrain, M., and Løland, K. E. (1983) 'Damage theory applied to concrete', in *Fracture Mechanics of Concrete* (Ed. F. H. Wittmann), Elsevier, Amsterdam.

Lott, J., and Kesler, C. E. (1966) 'Crack propagation in plain concrete', in *Symp. on Structure of Portland Cement Paste and Concrete*, Spec. Rep. 90, Highw. Res. Board, Washington, D. C., pp. 204–218.

Mihashi, H. (1983) 'A stochastic theory for fracture of concrete', in *Fracture Mechanics of Concrete* (Ed. F. H. Wittmann), Elsevier, Amsterdam.

Mindess, S. (1983) 'Application of fracture mechanics to hardened cement paste and concrete. A historical review', in *Fracture Mechanics of Concrete* (Ed. F. H. Wittmann), Elsevier, Amsterdam.

Mindess, S., and Diamond, S. (1980) 'A preliminary SEM study of crack propagation in mortar', *Cem. Concr. Res.* **10**, 4, 509–519.

Moavenzadeh, F., and Kuguel, R. (1969) 'Fracture of concrete', *J. of Materials*, **4**, 3, 497–519.

Nikolaevskii, V. N. (1982) 'Deformation of geomaterials and porous media', *Proc. USSR Acad. Sci., Mechanics of Solid*, 2, 96–109 (in Russian).

Nikolaevskii, V. N., and Rice, J. R. (1979) 'Current topics in the nonelastic deformation of geological materials', in *High pressure Science and Technology*, vol. **2**, Plenum, New York, pp. 455–464.

Panasjuk, V. V. (1968) 'Limit equilibrium of brittle bodies with cracks', Naukova Dumka, Kiev (in Russian).

Panasjuk, V. V., Savruk, M. P., and Dazyshin, A. P. (1976) 'Stress distribution around cracks in plates and shells' Naukova Dumka, Kiev, (in Russian).

Rice, J. R., and Rudnicki, J. W. (1980) 'A note on some features of the theory of localization of deformation', *Internat. J. Solids Struct.*, **16**, 597–605.

Chen, R., Yao, X. X. and Xie, H. S. (1979) 'Studies of the fracture of gabbro', *International Journal of Rock Mechanics and Mining Sciences and Geomechanics Abstracts*, **16**, 187–193.

Rudnicki, J. W., and Rice, J. R. (1975) 'Conditions for the localization of deformation in pressure-sensitive dilatant materials', *J. Mech. Phys. Solids*, **23**, 6, 371–394.

Ruetz, W. (1966) 'Das Kriechen des Zemensteins im Beton uns seine Beeinflussung durch gleichzeitiges Schwinden', *Dtsch. Ausschuss Stahlbeton, Schriftenr*, Heft 183, Wilhelm Ernst & Sohn, W. Berlin.

Salganik, R. L. (1973) 'Mechanics of bodies with great number of cracks', *Proc. USSR Acad. Sci., Mech. of Solid*, 4, 149–158 (in Russian).

Saouma, V. E., Ingraffea, A. R., and Catalano, D. M. (1980) 'Fracture toughness of concrete—K_{IC} revisited', *Rep.* 80-9, Department of Structural Engineering, Cornell University, Ithaca, New York.

Sellevold E. J. (1976) 'Low frequency internal friction and short-time creep of hardened cement paste: an experimental correlation', *Proc. Conf. on Hydraulic Cement Pastes: Their Structure and Properties*, Sheffield, pp. 330–334.

Shah, S. P. (1979) 'Whither fracture mechanics in concrete design? *Proc. Engng. Foundation Conf. on Cement Production and Use*, Rindge, June 25–29, 1979, pp. 187–199.

Shah, S. P., and McGarry, F. J. (1971) 'Griffith fracture criterion and concrete', *J. Engng. Mech. Div., ASCE*, **97**, EM6, 1663–1676.

Simmons, G., and Richter, D. (1976) 'Microcracks in rocks', *Physics and Chemistry of Minerals and Rocks*, London, pp. 105–137.

Slate, F. O., Jaquot, P., Lierse, J., Ringkamp, M., Rastogi, P. K., and Terrien, M. (1983) 'Experimental methods for detection and analysis of cracks in concrete', in *Fracture Mechanics of Concrete* (Ed. F. H. Wittmann), Elsevier, Amsterdam.

Stavrogin, A. N. (1969) 'Research on limit state and deformation of rocks', *Proc. USSR Acad. Sci., Physics of the Earth*, 12, 3–17.

Stavrogin, A. N., and Protasenja, A. G. (1979) *Plasticity of rocks*, Nedra, Moscow (in Russian).

Stroeven, P. (1973) 'Some aspects of the micromechanics of concrete', Ph.D. thesis, Delft.

Swamy, K. N. (1983) 'Experimental methods to determine fracture mechanics parameters—mortar and concrete', in *Fracture Mechanics of Concrete* (Ed. F. H. Wittmann), Elsevier, Amsterdam.

Swamy, R. N., and Sriravindrarajah, R. (1982) 'Influence of time on the aggregate-matrix bond under sustained load', *Proc. Int. Symp. Bonds between Cement Paste and Other Materials*, INSA, Toulouse, 17–19 Nov. 1982, pp. C.66–C.78.

Walsh, J. B. (1965a) 'The effect of cracks on the compressibility of rock' *J. Geophys. Res.*, **70**, 2, 381–389.

Walsh, J. B. (1965b) 'The effect of cracks on the uniaxial elastic compression of rocks', *J. Geophys. Res.*, **70**, 2, 399–411.

Walsh, J. B. (1965c) 'The effect of cracks on Poisson's ratio', *J. Geophys. Res.*, **70**, 20, 5249–5257.

Walsh, J. B. (1980) 'Static deformation of rock', *J. Engng. Mech. Div. ASCE*, 1005–1019.

Walsh, J. B., and Brace, W. F. (1968) 'Elasticity of rock: a review of some recent theoretical studies', *Felsmech. und Ingenieur-geol.* **44**, 4, 283–297.

Walsh, J. B., and Brace, W. F. (1972) 'Elasticity of rock in uniaxial strain', *Int. J. Rock. Mech. and Mining Sci.*, **9**, I, 7–15.

Warren, N., and Nashner R. (1976) 'Theoretical calculation of compliences of a porous medium', in *Physics and Chemistry of Minerals and Rocks*, London.

Wang, P. T., Shah, S. P., and Naaman, A. E. (1978) 'Stress–strain curves of normal and lightweight concrete in compression', *J. of Amer. Concr. Inst.*, **75**, 11, 603–611.

Wittmann, F. H. (1982) 'Creep and shrinkage mechanisms', in *Creep and Shrinkage in Concrete Structures* (Eds. Z. P. Bažant and F. H. Wittmann), Wiley, pp. 129–161.

Wittmann, F. H. (1983) 'Structure of concrete with respect to crack formation', in *Fracture Mechanics of Concrete* (Ed. F. H. Wittmann), Elsevier, Amsterdam.

Wittmann, F. H., and Zaitsev, Y. V. (1972) 'Behaviour of hardened cement paste and concrete under high sustained load', *Mechanical Behaviour of Materials, Proc. of the 1971 Int. Conf. on Mech. Behaviour of Materials*, Kyoto, Japan, pp. 84–95.

Wittmann, F. H., and Zaitsev, Y. V. (1976) 'Concrete—a viscoelastic porous body', *Proc. 2nd Int. Conf. on Mech. Behaviour of Materials*, Boston, USA, August 16–20, 1976, pp. 163–167.

Wittmann, F. H., and Zaitsev, Y. V. (1981) 'Crack propagation and fracture of composite materials such as concrete', *Proc. 5th int. Conf. on Fracture*, Cannes, March 29–April 3, 1981.

Zaitsev, Y. V. (1969) 'Limit equilibrium of a compressed plate containing a circular hole and cracks' (in Russian), *USSR Acad. Sci., Journ. Appl. Mech. and Techn. Phys.*, **5**, 100–101.

Zaitsev, Y. V. (1971a) 'Deformation and failure of hardened cement paste and concrete subjected to short-term load', *Cem. Concr. Res.*, **1**, 123–137.

Zaitsev, Y. V. (1971b) 'Deformation and failure of hardened cement paste and concrete under sustained load', *Cem. Concr. Res.*, **1**, 329–344.

Zaitsev, Y. V. (1971c) 'Experimental investigation to determine the behaviour of hardened cement paste and plaster of Paris under high load', *Cem. Concr. Res.*, **1**, 437–447.

Zaitsev, Y. V. (1974) 'Propagation of opening mode cracks in brittle materials subjected to compression' (in Russian), *Proc. USSR Acad. Sci., Mech. of Solid*, 4, 118–125.

Zaitsev, Y. V. (1975) 'Consideration of micro- and macro-structure of a material and its physical non-linearity in problems of crack propagation in concrete' (in Russian), *Izvestija Vuzov, Stroitelstvo i Architectura*, 11, 15–20.

Zaitsev, Y. V. (1977) 'Fracture mechanism of concrete under compression' (in Russian), *Beton i Zhelezobeton*, 7, 35–37.

Zaitsev, Y. V. (1980) 'Influence of structure on fracture mechanism of hardened cement paste', *Proc. 7th Int. Congr. on Chemistry of Cement*, Paris, June 30–July 4, 1980, pp. VI-176–VI-180.

Zaitsev, Y. V. (1981) 'Fracture mechanism and strength of concrete under triaxial compression', *Proc. 5th Int. Conf. on Fracture*, Cannes, March 29–April 3, 1981.

Zaitsev, Y. V. (1982) Simulation of deformation and strength of concrete with help of methods of fracture mechanics' (in Russian), Strojizdat, Moscow.

Zaitsev, Y. V., and Wittmann, F. H. (1971) 'Zur Dauerfestigkeit des Betons unter konstanter Belastung', *Der Bauingenieur*, 46 Jahrg., 84–90.

Zaitsev, Y. V. and Wittmann, F. H. (1973) 'Fracture of porous viscoelastic materials, *Proc. 3rd Int. Conf. on Fracture*, Munich, FRG, IX-323.

Zaitsev, Y. V., and Wittmann, F. H. (1974) 'Verformung und Bruchvorgang poröser Baustoffe bei kurzzeitiger Belastung und Dauerlast', *Deutcher Ausschuss für Stahlbeton*, Schriftenr., Heft 232, Wilhelm Ernst und Sohn, W. Berlin, pp. 65–145.

Zaitsev, Y. V., and Wittmann, F. H. (1977) 'Crack propagation in a two-phase material such as concrete', *Proc. 4th Int. Conf. on Fracture*, Waterloo, Canada, 19–24 June 1977, pp. 1197–1203.

Zaitsev, Y. V., and Wittmann, F. H. (1981) 'Simulation of crack propagation and failure of concrete', *Materiaux et Constructions* **14**, 357–365.

Zaitsev, Y. V., and Kazatskij, M. B. (1982) 'Evaluation of crack resistance of concrete under uniaxial compression using methods of fracture mechanics', in *Evaluation and Security of Safety of Hydrotechnical Structures*, Proc. of Conferences and Meetings on hydraulic engineering, Energija, Leningrad.

Ziegeldorf, S. (1983) 'phenomenological aspects of the fracture of concrete', in *Fracture Mechanics of Concrete* (Ed. F. H. Wittmann), Elsevier, Amsterdam.

Mechanics of Geomaterials
Edited by Z. Bažant
© 1985 John Wiley & Sons Ltd

Chapter 7

The Mechanics of Fracture Under High-Rate Stress Loading*

D. Grady

7.1 INTRODUCTION

The response of a single crack to both static and impulsive loading has received considerable attention over the past several decades and is reasonably well understood. The mechanics of a system of such cracks under impulsive or stress-wave loading and how this cooperative response relates to the transient strength and ultimate failure of a solid body is less well understood, and has been under study in several laboratories over the previous several years. Experimental studies of fracture under high rate loading have revealed unusual features associated with the phenomena such as greatly enhanced material strength and fracture stress dependence on loading conditions. Although such observations have led to the postulation of rate dependent material properties, most of these features can be understood in terms of fundamental fracture concepts when considered in terms of a system of interacting cracks.

In applications of the concepts of dynamic fracture, or the theories which purport to describe such fracture, there are a number of features worthy of prediction. Perhaps the first, and most fundamental, is the transient strength, or ability to support an impulsive load, either without sustaining fracture damage, or else sustaining fracture damage within some tolerable level without permitting total failure. In partially fractured bodies the spacing or fabric of the cracking may be of importance along with the void volume and extent of intersection, which will relate to the permeability of the crack system. In completely failed bodies the degree of fragmentation is of interest in many applications. The size and velocity of ejected fragments is of concern, and the distribution in fragment sizes and how this relates to the conditions of loading is also of importance.

The present chapter reviews several of the features of dynamic fracture and fragmentation which have come to light over the past several years. In the second section, the concept of dynamic fracture strength is treated. Experimental studies

*This work was performed at Sandia National Laboratories supported by the US Department of Energy under contact number DE-AC04-76DP00789.

have shown that solids subjected to high rate or impulsive loading exhibit dramatically enhanced material strength. A number of criteria have been proposed to account for this effect over the years. Recently, the strength properties of a single crack subjected to stress-wave loading have been explored and found to relate well with the behaviour of a system of cracks within a body as a whole. Earlier criteria appear to be similar statements of the same behaviour.

In the third section the properties of the material and the conditions of loading which lead to the number of fractures participating in the fracture process and the number and size of fragments resulting in the failed body are considered. Two interrelated concepts are important here. First is the inherent distribution of flaws or sites of weakness in the body which constitute the points of fracture activation. The second relates to the energy or the rate of energy application required to sustain the system of growing cracks. Although the inherent flaw distribution, a material property, has been used to determine fracture number and fragment size in most previous work, it has been found that in numerous applications a kinematic energy condition appears to govern the fracture fabric.

The following section is concerned with the mechanical and statistical conditions which determine the distributions in fragment size resulting from catastrophic fracture events. This is an interesting, diverse, and extremely complex topic about which very little is currently known. Two statistical concepts which have been proposed in the earlier literature for determining fragment size distributions are considered. The first is based on Poisson statistics while the second is a unique application of Boltzmann energy statistics. They lead to vastly different distributions and each appears to have application in certain circumstances. There is evidence that both the mode and the multiplicity of the dynamic fracture event is important in determining the shape of the distribution.

In the final topical section a brief review of the various approaches to continuum modelling of dynamic fracture currently under consideration is presented. Such modelling represents a necessary final goal in that the complexities of stress loading, geometry, and the interaction of stress and relief waves necessitate the use of wave-propagation codes to address realistic problems.

7.2 DYNAMIC FRACTURE STRENGTH

The dependence of the dynamic fracture strength of rock on the rate of loading can be studied through the response of an isolated crack under the action of constant tensile strain-rate loading. Illuminating studies on the strength of a body due to stress concentrating effects of cracks under static loading have emerged from the concepts initiated by Griffith (1920). The impact response of an elastic solid containing a crack and subjected to an abrupt tensile loading normal to the crack surface has also been well characterized and is extensively discussed by Chen and Sih (1977), while response of a solid to more general loading functions applied to the crack has been considered by Freund 1973. Application of these methods to constant tensile strain-rate loading have been explored in detail by Kipp et al. (1980) and provide an understanding of many of the features observed in the dynamic fracture of rock.

7.2.1 The dynamic stress intensity factor

The theory of linear elastic fracture mechanics applied to the dynamic loading of an isolated crack has provided a clear understanding of the response of cracks to transient tensile loads. The response to various types of Heaviside loading has been outlined by Chen and Sih (1977) and the Heaviside response function may be employed as a Green's function for other dynamic pulse shapes (Freund, 1973).

In particular, if a Heaviside tensile stress of magnitude σ_0 is applied to a crack with a characteristic dimension a, then the functional form of the stress intensity factor, K_1 at the crack tip, is

$$K_1(a,t) = \sigma_0 \sqrt{\pi a} f(c_s t/a) \tag{7.1}$$

where c_s is the shear wave velocity. The response to an arbitrary stress loading function, $\sigma(t)$, may then be expressed as (Freund, 1973),

$$K_1(a,t) = \sqrt{\pi a} \int_0^t \sigma'(s) f(c_s(t-s)/a) ds \tag{7.2}$$

A convenient loading function for comparison with experimental data on the dynamic fracture strength of rock is that of constant strain rate loading (Kipp et al., 1980). For linear elastic response the constant strain rate, $\dot{\varepsilon}_0$, and stress rate, $\dot{\sigma}_0$, are related through the elastic modulus of the material. Under this special loading condition Equation (7.2) becomes,

$$K_1(a,t) = \dot{\sigma}_0 \sqrt{\pi a} \int_0^t f(c_s/a) ds \tag{7.3}$$

Fracture is expected to initiate at a time, t_c, when the stress intensity factor in Equation (7.3) achieves the critical stress intensity factor. At this time the applied stress will have achieved the fracture stress level, σ_c, related to t_c through $\sigma_c = \dot{\sigma}_0 t_c$. Thus,

$$K_{1c} = \dot{\sigma}_0 \sqrt{\pi a} \int_0^{\sigma_0/\dot{\sigma}_0} f(c_s s/a) ds \tag{7.4}$$

provides an implicit relation for the fracture initiation stress and the dependence on both the crack size and the critical stress intensity factor.

7.2.2 A strain rate dependent fracture criterion

Before discussing the dynamic fracture process it is important to recall the form of the function, $f(c_s s/a)$. An elastic analysis of the Heaviside loading has been considered by Chen and Sih (1977). Solutions for both the penny-shaped crack and the plane crack exhibit the familiar square root singularity in time at the crack tip for the stress intensity factor, followed by overshoot and oscillating convergence to the static value.

To achieve an explicit relation for the strain rate dependence of the fracture stress, a solution expected to be valid at high loading rates has been used (Grady and Kipp, 1979a; Kipp et al., 1980). Assuming the solution for penny-shaped crack response to Heaviside loading valid for small normalized time,

$$K_1(t) = N \frac{2}{\sqrt{\pi}} \sigma_0 \sqrt{a} \sqrt{c_s t/a} \tag{7.5}$$

and, as before, using Equation (7.5) as a Green's function for constant strain-rate loading of a crack results in the crack-size independent relation for the stress intensity factor,

$$K_1(t) = N \frac{2}{\sqrt{\pi}} \sqrt{c_s} \dot{\sigma}_0 \tfrac{2}{3} t^{3/2} \tag{7.6}$$

Here N is a geometric coefficient equal to 1.12 for the penny-shaped crack. As in Equation (7.4), a relation among the critical stress intensity factor, the fracture stress, and the fracture time can be established and results in the strain-rate dependent fracture stress,

$$\sigma_c = \left(\frac{9\pi E K_{Ic}^2}{16 N^2 c_s} \right)^{1/3} \dot{\varepsilon}_0^{1/3} \tag{7.7}$$

Equation (7.7), of course, corresponds to the stress level required to initiate fracture on an isolated, normally-oriented circular crack which is sufficiently large relative to the loading conditions specified by the strain rate, $\dot{\varepsilon}$. In a body with a distribution of flaws under similar loading conditions, fracture initiation within some population of this distribution would be expected, providing flaws spanning a characteristic size of the order $a \sim (c_s K_{Ic}/E\dot{\varepsilon}_0)^{2/3}$, were contained within the distribution. Assuming that the material is sufficiently flawed, Equation (7.7) might be expected to provide a reasonable measure of the dynamic strength of the body.

7.2.3 Experimental measurements of the fracture strength of rock

Explosive or percussive rock breakage involves dynamic fracture in the sense that transient waves or the interaction of transient waves with local free surfaces carry regions of rock into tension, initiating fracture and fragmentation. It has been known since the early studies of Rinehart (1965) that the dynamic fracture strength of rock can exceed the static strength by as much as one order of magnitude. This observation has been substantiated by later studies and has attained fairly wide acceptance. In rock blasting calculations, the known static fracture strength is often arbitrarily increased by a factor of seven or eight to account for dynamic conditions.

A fairly substantial body of experimental data exists in which loading rates comparable to impulsive rock breakage applications have been simulated. These data tend to support the discrepancy between static and dynamic strength. Dynamic laboratory measurements using compressive (Kumar, 1968; Green and Perkins, 1968; Lindholm et al., 1974; Lundberg, 1976; Lankford, 1976), torsional (Lipkin et al., 1977), and tension (Birkimer, 1971) Hopkinson bar techniques have identified rate sensitive rock fracture. Plate impact induced spall in rock has also been studied (Shockey et al., 1974; Grady and Hollenbach, 1979; Grady and Kipp, 1984; Cohn and Ahrens, 1981). Rock fracture through magnetic stress loading methods has also been pursued (Forrestal et al., 1978).

Fracture toughness measurements (Oucherlony, 1980) have been obtained on some of the rocks for which dynamic strength as measured by plate impact spall experiments is also available. In Figure 7.1, comparison of dynamic spall data with the dynamic fracture relation in Equation (7.7) illustrates a reasonable correlation, and the practical applicability of this relation in providing estimates of dynamic rock strength. In the spall experiments a nominal strain rate of $\dot{\varepsilon}_0 \simeq 10^4/s$ was achieved and used in the calculation. Figure (7.2) illustrates a similar comparison of the strain rate dependence of Equation (7.7) with data obtained with both Hopkinson bar and plate impact experiments (Grady and Lipkin, 1980). Although in reasonable agreement in both rate dependence and magnitude

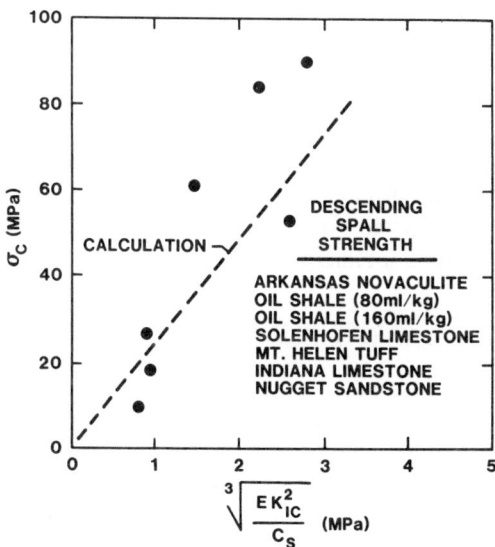

Figure 7.1 Spall fracture strength of selected rocks

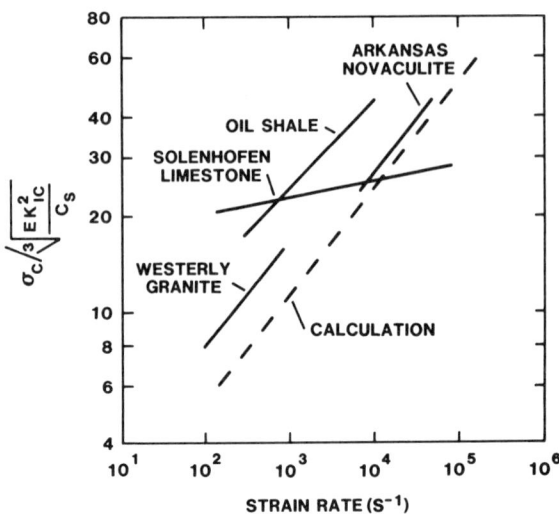

Figure 7.2 Strain-rate dependent fracture strength of several rocks

for most rocks, the relatively rate independent response of Solenhofen limestone, which has also been observed by others (Green and Perkins, 1968), is noted. It is believed that the relatively small flaw size and homogeneity of this rock leads to the observed rate independence over the $10^2/s$ to $10^4/s$ strain rate regime. The behaviour of Solenhofen limestone emphasizes that caution should be exercised in generally applying Equation (7.7) to dynamic rock strength predictions.

7.2.4 Comparison with other dynamic fracture criteria

In past years other dynamic fracture criteria have been established to describe observed dynamic strength behaviour. In general, these studies have lead to criteria similar in form to the expression in Equation (7.7). Tuler and Butcher (1968) proposed a criteria for incipient dynamic fracture damage of the form

$$\int_0^\tau (\sigma - \sigma_t)^\lambda dt = c \qquad (7.8)$$

where σ and τ are the tensile pulse amplitude and duration, respectively, and σ_t is the stress threshold for damage growth. They found that $\lambda = 2$ provided the best agreement with spall data on aluminium. Assuming that σ is significantly larger than σ_t and considering constant strain rate loading, Equation (7.8) leads to a cube root dependence on strain rate similar to Equation (7.7).

Birkimer (1971), in an attempt to explain Hopkinson bar fracture data on quartzite and concrete proposed that fracture occurred when the work on the fracture plane,

$$w = A \int \sigma \, dx, \tag{7.9}$$

achieved a critical value. σ is the stress and A is the area of the bar. Using the relation $dx = (c\sigma/E)dt$ (c is the wave velocity), and assuming constant strain rate loading results in the relation.

$$\sigma_c = \left(\frac{3E^2 W}{AC}\right)^{1/3} \dot{\varepsilon}_0^{1/3} \tag{7.10}$$

Steverding and Lehnigk (1976) developed a relation described as a least action law for fracture initiation,

$$\int \sigma^2 dt = \frac{\pi \gamma E}{c} \tag{7.11}$$

Here, γ is the specific surface energy for crack growth. It is apparent that their proposed fracture criterion will also lead to a cube root dependence on strain rate. A similar spall fracture criterion has been discussed by Ivanov and Mineev (1980).

Because of the importance of the state of fracture damage created in rock breakage, it is useful to consider a seemingly different criteria related to rock fracture due to Von Rittinger (1867). This relation states that the energy required to fragment a body is inversely proportional to the size modulus, x, of the resulting fragment distribution, namely;

$$U = \frac{b}{x} \tag{7.12}$$

where b is a constant. As far as can be determined, no physical basis has been put forth for Rittinger's law. Application of this relation has been actively debated in the field of grinding and crushing (e.g. Charles, 1957; Faddeenkov, 1975) and has given rise to alternative relations such as those of Kick (1885) and Bond (1952). Conditions of multiple particle breakage during crushing and grinding are complex. Bergstron et al., (1961), however, has shown that in single breakage Equation (7.12) is obeyed for numerous brittle materials.

In single breakage, the energy to fracture in the material is initially stored as elastic energy and if the specimen supports a stress σ at fracture then the energy in Rittinger's relation (Equation (7.12)) is of the order $\sigma^2/2E$. In brittle catastrophic fracture, crack propagation will rapidly approach a terminal growth velocity, c_g. If cracks coalesce to form fragments at a time, t_c, then the nominal fragment size will be on the order of $x \sim c_g t_c$. Applying these relations to Equation (7.12) and assuming constant strain rate loading results in

$$\sigma_c = \left(\frac{2bE^2}{c_g}\right)^{1/3} \dot{\varepsilon}_0^{1/3} \tag{7.13}$$

136 *Mechanics of Geomaterials*

suggesting that the fragmentation behaviour noted by Rittinger may have a similar physical basis.

7.3 ENERGY CONCEPTS IN CATASTROPHIC FRAGMENTATION

The damage created in severe impulsive tensile fracture events in rock or rock like material is a complexity of intersecting cracks which may reflect both the intensity and the orientation of the applied tensile stress field. As an example, the fracture damage of an explosion placed in a deep borehole consists of severe comminution to small particle sizes in the rock immediately adjacent to the explosion and grading to larger fragments with increasing distance. Further from the charge the fracture damage reduces to a small number of radial cracks which may propagate a considerable distance before arresting. In many applications of explosive or percussive rock breakage the production of optimum and uniform fragment sizes is either the objective or is important to the efficiency of the process. Consequently an understanding of the fracture effects governing the intensity of cracking and fragment size is needed.

7.3.1 Inherent flaw effects in the dynamic fracture process

As pointed out in the previous section, by adjusting the rate of loading, different levels of tensile stress can be achieved in the material before catastrophic failure occurs. This behaviour is understood by hypothesizing that the virgin material contains a distribution of flaws with various sizes and orientations (Grady and Kipp, 1980). The flaws become active and grow when subjected to tensile loads of various magnitudes. The active flaws grow during the period of load application and eventually coalesce, causing material failure. When a load is applied slowly, only those flaws which become active at low stress levels actually contribute to the fragmentation process because these flaws grow and coalesce, and failure occurs, before the applied load reaches a level of stress high enough to activate other flaws. This results in a low apparent threshold for material failure and comparatively large fragments because the number of contributing flaws is small. When the load is applied more rapidly, a higher level of stress is achieved before flaw coalescence occurs; thus, a greater number of flaws participate, causing the fragment dimensions to be smaller and the apparent threshold for material failure to be higher.

It is also recognized that a specific amount of energy is required to create a new fracture surface during the fracture event, and this energy must come from the loading stress wave. This surface energy is of the order $K_{Ic}^2/2E$, where K_{Ic} is a measure of the fracture toughness and E is the elastic modulus. If the material is loaded at a rate such that a stress σ_c is achieved before failure occurs then an elastic energy of order $\sigma_c^2/2E$ is stored in the body and available for fragmentation. A consideration of energy balance suggests that a fragment size of approximately $d \sim (K_{\mathrm{Ic}}/\sigma_{\mathrm{c}})^2$ might be achieved in the dynamic fragmentation event with a corresponding fragment surface area per unit volume of $A \sim 6/d$. Direct observation has shown that this energy balance is never achieved, and measured fragment surfaces are typically one or more orders of magnitude below this energy limit.

Consequently, based on this observation, previous studies have assumed that fragment sizes achieved in dynamic fracture events are flaw rather than energy governed and considerable effort has been focused on characterizing the flaw distribution in rock material, through direct observation (Shockey et al., 1974), or other indirect methods (Grady and Kipp, 1980; Margolin, 1983). Usually the flaw distribution has been related to the stress level of flaw activation with $N(\sigma)$ providing the number of flaws per unit volume which activate at or below a stress, σ. Thus, if a threshold stress σ_{c} is achieved before failure, the nominal fragment size predicted is $d \sim N(\sigma_{\mathrm{c}})^{-1/3}$.

7.3.2 Kinetic energy considerations

Although accurate characterization of the flaw structure appears to be important, there is evidence that such characterization alone is not sufficient to explain all of the observed effects in dynamic fracture and fragmentation. Energy balance principles still appear to play a significant role. As described earlier, the dynamic fracture stress appears to be governed largely by fracture energy considerations. Also, the fragmentation studies of Von Rittinger (1867), Charles (1957), and Bergstrom et al. (1961) reflect, at least qualitatively, energy aspects in the breakage process.

There is another way to model the dynamic fracture process which leads to an energy balance governing the average fragment size (Grady, 1982a). In this approach kinetic energy rather than elastic strain energy is considered to be the important energy fueling the fracture process. The model is thought to be most reliable in extremely catastrophic fragmentation events, however, in application it has been found quite useful over a fairly broad range of fracture rates.

Consider a body which has previously been compressed by some means and is currently in a state of rapid expansion. The instantaneous kinematic state will be determined by the density, ρ, and the density rate, $\dot{\rho}$, which will be assumed uniform over a sufficiently large region encompassing the point of interest. Conceptually, one might consider an elastic sphere which has been compressed uniformly and suddenly released and is currently in a state of rapid outward expansion. The kinetic energy associated with the outward motion is responsible for the fracturing forces and surface tension associated with newly created fractures resists the fracture process. It is equally intuitive that, after fragmentation, particles will continue to fly apart at high velocities. Thus, the production of fragment surface area cannot simply be governed by a balance of kinetic energy and surface energy, since a large portion of the kinetic energy remains after fragmentation.

An expression for the kinetic energy available for fragmentation can be determined by considering an element of the expanding body with a volume of the order of the fragment size expected from the event. With reference to a specific coordinate system, the kinetic energy of this element can be decomposed into a centre-of-mass kinetic energy, T_{cm}, and the kinetic energy relative to a coordinate system referenced to the centre of mass, T'. Assuming average response, forces acting on the element during fragmentation should, due to symmetry, exert no net impulse and, consequently, both the centre-of-mass velocity and kinetic energy of the particle should remain invariant during fragmentation. Thus in a decomposition of the kinetic energy, the centre-of-mass kinetic energy, T_{cm}, of the fragments must be conserved during fragmentation and only the kinetic energy relative to the centre of mass, T', is available to fuel the breakage process. This latter energy may be regarded as a local kinetic energy which is available for fragmentation without violating local momentum conservation.

An explicit expression for the local kinetic energy can be obtained by considering a spherical mass element of radius, a, expanding uniformly at a density rate, $\dot{\rho}$. The kinetic energy about the centre of mass for this single sphere is,

$$T' = \frac{2\pi}{45} \frac{\dot{\rho}^2}{\rho} a^5 \tag{7.14}$$

Dividing by the volume of the spherical element and expressing in terms of the fragment surface area to volume ratio, $A = 3/a$, a measure of the local kinetic energy density, in terms of the surface area created by fragmentation is obtained

$$T = \frac{3}{10} \frac{\dot{\rho}^2}{\rho A^2} \tag{7.15}$$

The new fragment surface energy density is simply,

$$\Gamma = \gamma A \tag{7.16}$$

and the total energy is given by,

$$U = \frac{3}{10} \frac{\dot{\rho}^2}{\rho A^2} + \gamma A \tag{7.17}$$

We will assume that during the catastrophic fragmentation process, forces brought about will seek to minimize the energy in Equation (7.17) with respect to the fracture surface area. This approach assumes that the coordinate, A, is available or free to vary during the fracture process and, consequently, requires a sufficient supply of flaws or sites of fracture initiation during the process, although precise requirements are not yet clearly understood.

The Mechanics of Fracture Under High-Rate Stress Loading

At equilibrium, $dU/dA = 0$, and the equilibrium fracture surface area is,

$$A = \left(\frac{3\dot{\rho}^2}{5\rho\gamma}\right)^{1/3} \tag{7.18}$$

Equation (7.18) provides a quantitative measure of the fragment surface area created in the fracture process in terms of fundamental thermodynamic and kinematic properties. Typically a nominal fragment size, rather than surface area, is of interest. Assuming spherical fragments of equal size, the fragment diameter is related to the surface area through, $d = 6/A$.

If Equation (7.18) is used along with, $\dot{\varepsilon} = \dot{\rho}/3\rho$, and $\gamma = K_{Ic}^2/2\rho c^2$, an expression for the nominal fragment diameter for dynamic fragmentation in a brittle material is obtained,

$$d = \left(\frac{\sqrt{20}K_{Ic}}{\rho c \dot{\varepsilon}}\right)^{2/3} \tag{7.19}$$

where, $\dot{\varepsilon}$, is the linear strain rate, and, K_{Ic}, is the fracture toughness. Nominal fragment size data for oil shale have been obtained as a function of strain rate (Grady and Kipp, 1980). Material properties for the oil shale studied are, $K_{Ic} = 0.9\,\text{MN/m}^{3/2}$, $\rho = 2300\,\text{kg/m}^3$, and $c = 4000\,\text{m/s}$. The prediction of Equation (7.19) is shown to provide a reasonable description of the data in Figure 7.3. Further work (Grady, 1982a; 1982b; Grady and Benson, 1983; Costin and Grady, 1984) are showing that Equation (7.18) or (7.19) provide a useful predictive relation for numerous dynamic fragmentation applications and that energy rather than flaw considerations govern the fragmentation. It also suggests

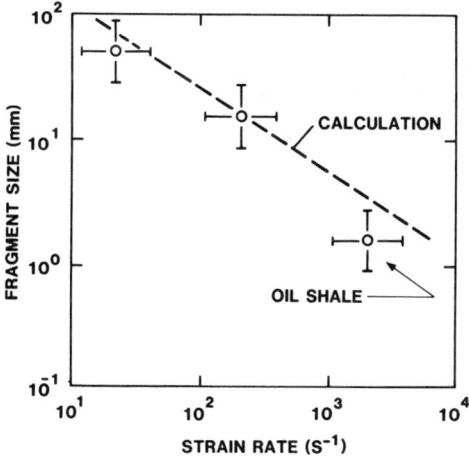

Figure 7.3 Strain-rate dependence of average fragment size in oil shale

that kinetic energy rather than strain energy is more important in many cases in determining fragment sizes.

7.3.3 Application to well shooting

An interesting and perhaps useful application of the fragment size expression derived from the model based on energy principles can be found in the process known as well shooting or the dynamic stimulation of wellbores through explosive or propellant loading (Warpinski et al, 1979; Schmidt et al., 1981). The multiple fracturing of wellbores holds promise for stimulating naturally fractured reservoirs through intersection with the pre-existing fracture network. The use of high-strength explosives has been found to have detrimental effects, however, through crushing and plastic flow near the borehole which can result in a residual compressive stress field and a tightly locked formation rather than the desired multiple fracture pattern and enhanced permeability (Smith et al., 1978). Extensive experiments by Schmidt et al. (1981) in Nevada Test Site tuff have shown that propellants with appropriate burning rates can produce stress loading behaviour which optimizes the dynamic multifracturing process without excessive crushing or plastic damage. They conclude that perhaps the most crucial parameter is the initial stress loading rate in the wellbore.

Certain aspects of this problem can be understood in terms of the energy expression for fragment size determined in the previous section. By assuming a geometry more appropriate to the radial cracking process occurring in the dynamically loaded wellbore, an expression similar to that of Equation (7.18) or (7.19),

$$N = \left(\frac{\rho c \dot{\varepsilon}}{6 K_{\text{Ic}}}\right)^{2/3} \quad (7.20)$$

can be obtained for the number of fractures per unit length occurring at a strain rate, $\dot{\varepsilon}$. Assuming a wellbore diameter, D, the number of cracks initiating from the wall is predicted to be,

$$N_D = \pi D \left(\frac{\rho c \dot{\varepsilon}}{6 K_{\text{Ic}}}\right)^{2/3} \quad (7.21)$$

This relation can be compared with the experiment of Schmidt et al. (1981) in NTS tuff in which an intermediate burning rate propellant provided a loading rate of 20 psi/μs, to a peak pressure of 13,800 psi. This translates to a circumferential strain rate at the wellbore wall of approximately, $\dot{\varepsilon} = 30/s$. Using the material properties for tuff of, $\rho = 1600 \text{ kg/m}^3$, $c = 2000 \text{ m/s}$, $K_{\text{Ic}} = 0.3 \text{ MN/m}^{3/2}$, and a wellbore diameter, $D = 0.15 \text{ m}$, Equation (7.21) provides a fracture number of, $N_D = 6.7$. This compares well with the seven major fractures and several minor fractures actually observed by mine back methods in this experiment.

The Mechanics of Fracture Under High-Rate Stress Loading 141

Although the agreement is encouraging and well shooting is an interesting application of the energy balance approach to predicting fracture spacing in dynamic fracture, this is only one aspect of a complex pulse tailoring problem. Extensive calculations have been performed by Swenson (1984) focused on optimizing all aspects of wellbore stimulation by multiple fracturing.

Considering the relatively gentle loading rates associated with wellbore stimulation, it is not presently clear why an energy balance considering only kinetic energy and fracture surface energy while ignoring the strain energy should be successful in predicting fracture spacing. It would also seem from dimensional considerations that the wellbore diameter might enter the relation in a more complicated way. The calculation clearly suggests, however, that the fracture process is not simply predicted by a description of inherent flaws in the neighbouring rock. A complete theory of the multiply fracturing event should also include the energy concepts identified here.

7.4 PARTICLE SIZE DISTRIBUTIONS OCCURRING IN DYNAMIC FRAGMENTATION

In the opening section of this report concepts relating to the transient strength and dynamic fracture stress of brittle solids subjected to impulsive stress-wave loading were considered. The following section provided a discussion of flaw structure and energy factors governing the new fracture surface area created in catastrophic fragmentation. It remains to consider a concept which is frequently the end concern in numerous impulsive fracture applications, namely, the distribution in sizes of the particles created in the event.

Numerous areas in science and everyday experience can be cited. One current application is concerned with recovery of oil from oil shale. Economic recovery requires effective fragmentation and distribution of void volume through explosive blasting methods (Boade *et al.*, 1981). More generally, aspects of fragmentation and resulting fragment distributions are important to a number of explosive or percussive rock breakage applications including deep drilling (Varnado, 1978), explosive or propellant stimulation of gas and oil wells (Warpinski *et al.*, 1979), and quarry, mining and construction blasting (Langefors and Kihlstrom, 1967). On a less technical chord, theories concerned with the accretion of planetary bodies have been proposed which depend on impact fragmentation and resulting fragment distributions (Matzui and Mitzutani, 1977). Distributions of iron particles on the ocean floor apparently result from the catastrophic ablation of meteorites entering the earth's atmosphere (Yamakashi *et al.*, 1981), and the natural occurrence of the explosive eruption of a volcano or the catastrophic impact of a major meteorite (O'Keefe and Ahrens, 1976) involve processes of fragmentation which can distribute debris through the earth's atmosphere.

Because of the continuing practical interest in material fragmentation, studies on various aspects of the problem can be traced through the literature for more than a century. The statistical nature of fragmentation was recognized early and stimulated efforts in identifying size distribution relations which correctly described the resulting fragmentation. Various standard distributions such as Poisson (Lienau, 1936; Bennett, 1936), binomial (Gaudin and Meloy, 1962), log normal (Kolmogorov, 1941), and Weibull (Rosin and Rammler, 1933), have been used and functions associated with the names of Rosin and Rammler (1953), Schuhmann (1940), and Gaudin and Meloy (1962) have acquired popularity in certain applications.

The strictly statistical approaches to fragmentation, however, tacitly ignored the dynamics of the fragmentation event and did not provide a means of correlating energy, or some other measure of the loading conditions, with the fragment distribution. These ideas appear to have first been explored in the studies of Mott (1947) and have been pursued more recently by Grady (1981a, 1981b). In this approach more significance is attributed to the dynamics of fracture activation and growth, including the nucleation process and the influence of material deformation properties. Mott (1947), in considering a restricted geometry, combined the spatial randomness of the fracture process with the growth of plastic tensile release waves and predicted fragment distributions dependent on both dynamic and material properties. More recent studies (Shockey et al., 1974; Margolin, 1983) have focused on developing physically founded laws governing the nucleation, growth and coalescence of fracture during one- and two-dimensional stress-wave propagation. A different approach to the statistics of fragmentation has been proposed by Griffith (1943), where particle fracture surface energy is related to the distribution through a unique application of classical Boltzmann statistics concepts.

7.4.1 The Weibull distribution

The statistical representations which have been used to describe fragment size distributions are almost as numerous as the fragmentation phenomena. Most were selected from classical statistical formulas. The Weibull representation is a flexible two-parameter analytic formula which has been found successful in describing a large body of fragmentation data. Application of the Weibull formula to fragment distributions appears to have been first suggested by Gates (1915). The most extensive comparisons of empirical data with this statistical representation was made by Rosin and Rammler (1953) in terms of mining and ore reduction applications.

Considerable effort has focused on supplying a theoretical framework for a Weibull description of fragmentation. Bennett (1936), Gilvarry (1961), and Kuznetsov and Faddeenkov (1975) have focused significant efforts in this direction, however, a clear theoretical basis has yet to be demonstrated. In fact, exact applicability of a Weibull representation to fragmentation has theoretical problems at the fine end of the distribution in terms of divergence of expressions for the fragment number and fragment surface area.

The purpose here is not to provide theoretical support for a Weibull description of fragmentation. A Weibull distribution has been found very successful in terms of describing fragmentation over the range of available data and is used here as a convenient means for examining systematic features of fragmentation phenomena.

According to the Weibull representation of fragmentation, the cumulative distribution of fragment volume fraction (or mass fraction) finer than size, x, is

$$V(x) = 1 - e^{-(x/\sigma)^n} \tag{7.22}$$

The volume density distribution is provided by the derivative of Equation (7.22),

$$v(x) = \frac{n}{\sigma}\left(\frac{x}{\sigma}\right)^{n-1} e^{-(x/\sigma)^n} \tag{7.23}$$

In Equation (7.22) or (7.23), σ, is the scale parameter and is closely related to the mean fragment size. The shape parameter, n, determines the variance and the skewness of the distribution. The Weibull representation has the flexibility of describing a very flat distribution with, $n \simeq 1$, to a strongly centred distribution for large values of n.

The exponential expression in Equation (7.22) and (7.23) provides the cutoff for the large particle upper end of the distribution. If attention is focused only on the fine portion of the distribution (i.e. x sufficiently smaller than σ), then simple power representations are obtained with,

$$V(x) = \left(\frac{x}{\sigma}\right)^n \tag{7.24}$$

and

$$v(x) = \frac{n}{\sigma}\left(\frac{x}{\sigma}\right)^{n-1} \tag{7.25}$$

This distribution is frequently convenient in application and has further convenience in the present paper where comparison with other distributions which are not Weibull in their upper end behaviour is desired.

Experimental evidence indicates that the shape parameter, n, can range from as low as 0.5 to an upper limit of about 5. A theoretical upper limit of $n = 6$ is suggested by one statistical approach (Grady and Kipp, 1984). Behaviour of the distribution shape over various methods of fragmentation is typified by the plot in Figure 7.4. A large body of fragmenting munitions data, exemplified by the work of Weimer and Rogers (1979) show values of n ranging from about 4 to 5. Direct impact fragmentation experiments show n from about 2 to 3 (Shockey et al., 1974) and in tension Hopkinson bar tests, where substantial shearing

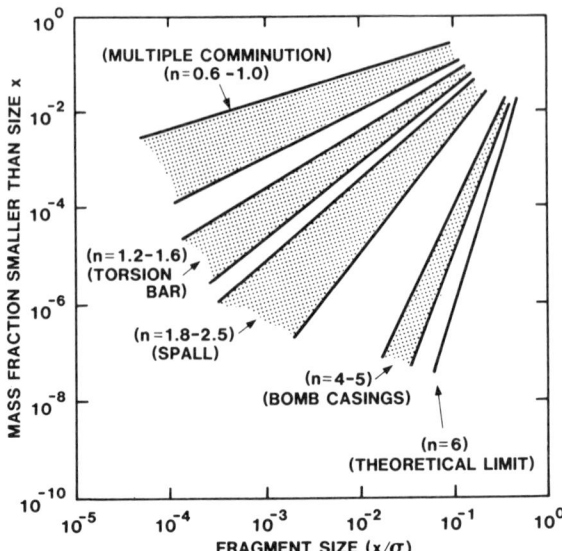

Figure 7.4 Behaviour of the fragment size distribution dispersion parameter over various fragmentation methods

fragmentation occurs, values from about 1.2 to 1.8 are typical (Grady et al., 1981; Costin and Grady, 1984). A large and diverse body of fragmentation data show distributions with n equal or close to unity. Ball milling comminution of minerals results in distributions with n near one (Rosin and Rammler, 1933). Explosive crushing experiments on glass spheres also provide values of n very close to unity (Bergstrom et al., 1962).

Several examples can be cited from the geophysical literature. Parkins and Tilles (1968) have suggested that interplanetary debris, presumably due to comet disintegration or to asteroid break-up is best described by, $n \simeq 1$, at least for the fine particles. While Matzui and Mizutani (1977) have found this particular distribution appropriate in calculations of planetesimal fragmentation. Yamakashi et al. (1981) have shown that magnetic spherules in deep sea sediments attributed to atmospheric ablation or to earlier fragmentation of iron meteorites, are also well described with a particle size distribution with n near unity.

The data available are far from complete, or systematic, but evidence is mounting which suggests that the rather broad range of the distribution parameter, n, may be related, at least in part, to the type or method of fragmentation. It appears that single tensile fragmentation leads to fragment size distributions with large values of n. In contrast, fragmentation with significant shearing and continued comminution leads to lower values, with indications that $n = 1$ may be a limiting value.

7.4.2 The single fracture (Poisson) limit

In fragmentation phenomena, events in which abrupt, single or initial one-time fragmentation occurs, as opposed to events in which continued reduction of already broken fragments occurs, can be identified. For instance, a rapidly expanding spherical shell of material will undergo an abrupt fragmentation process when the tensile strength is reached. Individual fragments will continue to fly outward on roughly radial trajectories and no further breakage will occur. The distribution in fragment size is determined by the physics and statistics of the process at the time of failure. In contrast, when a brittle solid is deformed in shear, the initial fracture event is followed by continued comminution as fragments roll and tumble and repeatedly impact one another. As another example of continued comminution, if the spherical shell in the first example were a ductile solid or liquid and if the initial fragments were ejected at a sufficiently high velocity into a finite atmosphere, further fragmentation of these particles would occur through aerodynamic break-up, ablation, or burning processes. The final size distribution of particles is observed to reflect this multiplicity of break-up processes.

The present section will focus on theoretical concepts leading to the fragment size distributions which result from the first example, namely, that of single fracture. The added effect of continued comminution will be considered in the following section.

With one notable exception, which will be addressed later (Griffith, 1943), theories focused on predicting fragment size distributions have started with the fundamental premise of a randomly cracked body. This immediately invokes concepts of Poisson statistics which were central to the developments of Bennett (1936), Lienau (1936), Gilvarry (1961), and Mott (1943). The approaches differ and that of Lienau and Mott, who apparently were not biased by a desire to arrive at a Weibull representation, appear to be the more correct application of the concepts. Extension of the statistical concepts to two and three dimensions proposed by Mott can be questioned, however, and an alternative method has been pursued by Grady and Kipp (1984). The concepts are most readily appreciated in one dimension, however, where fractures are considered to be points distributed randomly on an infinite line.

Consequently, consider, after the fracture event, an infinite one-dimensional line or rod along which fractures occur randomly with an expected or average size, x_0. Randomly distributed points on an infinite line obey Poisson statistics and the probability of finding, n, fractures in a length, x, is given by,

$$P(n, x) = \left(\frac{x}{x_0}\right)^n e^{-x/x_0}/n! \tag{7.26}$$

To determine the probability distribution in fragment lengths, we first obtain the probability of a length, x, having no fractures in it,

$$P(0, x) = e^{-x/x_0} \tag{7.27}$$

and then the probability of finding one fracture in a length, dx,

$$P(1, dx) = \frac{1}{x_0} dx \tag{7.28}$$

Then, the probability of finding a fragment of length, x, within a range of x to $x + dx$ is given by,

$$dP = P(0, x)P(1, dx) = \frac{1}{x_0} e^{-x/x_0} dx \tag{7.29}$$

which provides an exponential fragment size probability distribution of the form (Lienau, 1936),

$$p(x) = \frac{1}{x_0} e^{-x/x_0} \tag{7.30}$$

The dynamic ductile fracture of radially expanding thin metal rings which closely approximates the one-dimensional fragmentation process has been found to be well described by the predicted exponential distribution (Grady et al., 1984).

Mott (1943) recognized the applicability of the development of Lienau (1936) and used it as the basis for deriving the well-known Mott distribution for the fragmentation of thin shells. He reasoned that Equation (7.30) might still apply to two-dimensional fragmentation (and, by logical extension to three dimensions) and related the linear measure of the random variable to the fragment mass through, $x \sim m^{1/2}$. A change of variables then provides the probability distribution,

$$p(m) = \frac{1}{2\sqrt{\mu m}} e^{-\sqrt{m/\mu}} \tag{7.31}$$

where μ is the expected value of the fragment mass. Integration provides the familiar Mott cumulative fragment number distribution,

$$N(m) = N_0(1 - e^{-\sqrt{m/\mu}}) \tag{7.32}$$

against which a large body of exploding shell data have been compared.

Grady and Kipp (1984) have argued that, in the application of Poisson statistics to two- and three-dimensional fragmentation, a measure of area, or volume, respectively, should be selected as the random variable. For volume fragmentation an exponential probability density distribution in the fragment volume,

$$p(v) = \frac{1}{v_0} e^{-v/v_0} \tag{7.33}$$

is obtained.

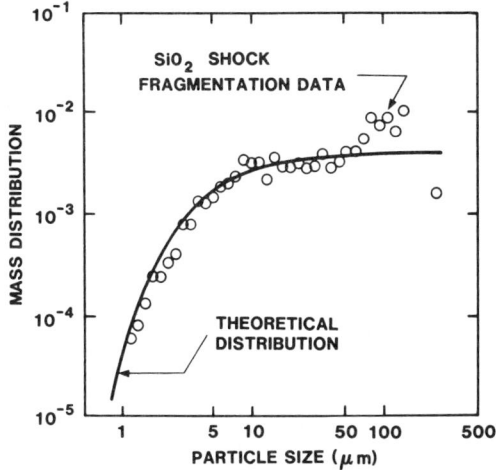

Figure 7.5   Particle size distribution from shock fragmentation of quartz

When related to the cumulative mass distribution shown in Figure 7.4 the Mott development leads to a shape parameter of $n = 4$ while that of Grady and Kipp provide $n = 6$. It is currently believed that an upper limit of $n = 6$ corresponds to truly random fragmentation. It is not yet clear, however, whether natural fragmentation is a strictly random statistical process, at least in the sense discussed here.

7.4.3 The multiple fracture (Boltzmann) limit

As pointed out earlier, experimental values for the shape parameter, seem to range from about 1 to 5, and although not yet well understood, there are indications that the more complex the fragmentation event in terms of opportunities for fragments to experience repeated breakage, the closer the value of n is to unity. For example, in the process of mineral reduction through ball milling (Rosin and Rammler, 1933), resulting distributions show values of n very close to one, and numerous further examples exist, some of which were cited earlier.

There are some immediate theoretical problems with a cumulative mass fraction distribution of the form $M(x) = x/\sigma$ (compare with Equation (7.24)). The mass fraction density distribution is $dM/dx = 1/\sigma$, while the fragment number and area distributions should behave as $dN/dx \sim x^{-3}/\sigma$ and $dA/dx \sim x^{-1}/\sigma$, respectively. Owing to the behaviour of the distribution at small particle size both the integrated fragment number and area are infinite. Consequently, a theory which explains the flat, $n = 1$, distribution should also account for a cutoff at small particle size.

A theoretical approach which holds some promise for explaining both features was suggested Griffith (1943). This study was not pursued and has lain dormant for a number of decades. Nevertheless its appeal to fundamental concepts is attractive and the ideas warrant further investigation. This approach is based on equipartition of energy concepts where the energy of interest is the new fracture surface energy created in the fragmentation event (Griffith, 1943). The basic premise is the same as that used in the development of classical Boltzmann statistics. Namely, the number of material elements in an energy state is determined by the most probable distribution of all the material elements over the available energy states consistent with the total energy. The most probable number of elements is found to be,

$$n_j = g_j e^{-\alpha - \beta \varepsilon_j} \quad (7.34)$$

where the parameters α and β are determined by the total number, N, and energy, E, of the system and g_j is the density of states corresponding to the energy level, ε_j. In the present application the energy, E, corresponds to the new fracture surface energy. It is not clear that recourse to temperature is either necessary or has meaning. Since the surface energy of a fragment will be of the order, γx^2, and the volume, of the order, x^3, a relation between a particle energy state and size, of proportionality, $1/x$, is arrived at. Griffith (1943) arrived at a distribution of the form,

$$\frac{dM}{dx} = B e^{-b/x}, \quad (7.35)$$

where B and b are constant parameters of the distribution. In that development a uniform density of states was assumed which is not obvious and is one feature of the theory which should be considered more deeply. Nevertheless, the resulting distribution leads to a flat, $n = 1$, behaviour at large particle size and has an exponential cutoff for small x.

Grady and Varga (1984) have carefully analysed the particle size distribution resulting from an explosively shocked quartz sample to a shock pressure of approximately 8 GPa. A photometric apparatus was used to accurately determine the particle size distribution down to less than 1 μm. The particle size distribution is compared with the theoretical distribution of Equation (7.35) in Figure 7.5. For values of $B = 0.004 \,\mu\text{m}^{-1}$ and $b = 4.5 \,\mu\text{m}$, the theoretical expression provides a remarkably good description of the data including the rollover between 1 and 10 μm, the relatively flat, $n = 1$, behaviour between about 10 and 200 μm, and the large particle cutoff at 250 μm. From the reasonably good agreement with the present fragment data, it would seen that the statistical energy theory of Griffith (1943) warrants more serious attention. The agreement may be fortuitous, however. The application is significantly different from the usual physical processes described by Boltzmann statistics and the treatment of the density of states is open to question.

The Mechanics of Fracture Under High-Rate Stress Loading

7.5 CONTINUUM MODELLING OF DYNAMIC FRACTURE

Practical application of the concepts of dynamic fracture mechanics typically involve complex loading situations where the stresses and stress rates vary in both space and time over the region of interest. Applications to consider which illustrate the complexity of problems of interest include: explosive blasting in single or multiple boreholes, with or without nearby free surfaces; impact or explosive cratering; and percussive or drag bit drilling, to name a few. Such applications involve unique stress-wave propagation geometries leading to the dynamic fracture process and are difficult to evaluate by strictly analytic methods.

With the advance of high-speed computer methods, efforts have been directed toward developing continuum descriptions of the fracture, fragmentation, and wave propagation to evaluate complex fracturing events. Davison and Stevens (1973) have established the fundamental concepts necessary in a thermodynamically consistent continuum description of dynamic fracture. Several groups have pursued models based on the activation, growth and coalescence of inherent distributions of fracture-producing flaws to predict crack and fragment size spectra in brittle fracture (Shockey et al., 1974; Grady and Kipp, 1980; Dienes et al., 1980; Margolin et al., 1982). Others workers have preferred to apply well developed concepts from plasticity to the problem of fracture (Butkovich, 1976; Johnson, 1978; Glenn. 1976), predicting regions and levels of damage, non-recoverable void volume, and tensile or shear fracture. The most appropriate approach has not yet been identified.

7.5.1 Directions in continuum modelling

The present discussion of continuum modelling of dynamic fracture is not intended to be an exhaustive review. Rather, it points out the variety of approaches which have been, and are still being, pursued to provide methods for calculating dynamic fracture phenomena. Such work is still quite active and considerable effort appears to remain before the best approaches emerge as mature calculational techniques.

One of the earliest, and simplest, methods for establishing a state of dynamic fracture within a stress-loaded body was to specify a tensile stress threshold (or spall threshold) at which point the corresponding elements were assumed to instantaneously fractured. The technique did illustrate certain of the stress-wave propagation features characteristic of fracturing bodies; however, it is now recognized that a fracture threshold independent of the size of the body and rate of loading is not a physical property.

Davison and Stevens (1973) studied the application of high intensity and short duration loads to brittle materials which results in intense local fracturing or spalling of the body. They introduced a concept of fracture damage in terms of a vector field describing the size and orientation of distributed penny-shaped cracks throughout the region of fracture. Damage was allowed to occur gradually according to some specified law of growth determined by the changing stress state at the point of fracture.

This general idea has been pursued by others. Curran and coworkers in a series of papers (Shockey et al., 1974; Seaman et al., 1976; Curran et al., 1977) have developed a theory of dynamic brittle fracture based on the nucleation and growth of penny-shaped crack fracture damage which evolves gradually to full coalescence of fragments. The inherent distribution of fracture-producing flaws is regarded as observable and petrographic methods are described in their work for determining such distributions. Laws based on the current stress state are specified to drive fracture nucleation and growth. The model has been implemented in two-dimensional stress wave codes and has been used extensively in several geo-engineering related applications.

A model of continuum fracturing devoted primarily to explosive fragmentation of oil shale was developed by Grady and Kipp (1980). The general framework followed that of Davison and Stevens (1973) in that fracture damage represented a scalar variable measure of crack-like defects which could grow under appropriate tensile stress loading. Physics of the activation and growth process, however, was guided by a dynamic application of Weibull statistical concepts which leads to the known size dependence of fracture stress observed in static testing (Jaeger and Cook, 1969). This approach allowed the fracture damage activation parameters to be determined directly from experimental measured fracture stress and fragment size dependence on strain rate (Grady and Kipp 1979a, b, 1980).

A 'bedded crack model' of dynamic fracture for brittle and quasi-brittle materials has been developed by Margolin and coworkers (Dienes and Margolin, 1980; Margolin and Adams, 1982; Margolin, 1983). Fracture damage is based on a microphysical description of fracture following the Griffith theory and considerable care in consistently relating damage and material integrity through an effective modulus theory was achieved. The model is amenable to computational simulation and dynamic fracture features such as rate dependent fracture strength appear naturally within the workings of the model.

A somewhat different approach to stress-wave induced fracture is represented in the work of Butkovich (1976) developed to calculate underground explosive fracture and induced permeability in coal. The method is more akin to conventional elastic–plastic calculations in that stress-space surfaces of yield or failure are established to determine onset of fracture. Fracture due to shearing is explicitly treated and two parameters are associated with fracture damage; a shearing related distortional strain and a tensile induced cracking or porosity which is related to the permeability.

A similar plasticity model of dynamic fracture has been described by Johnson (1978) and applied to explosive fracture in oil shale. A scalar fracture damage

The Mechanics of Fracture Under High-Rate Stress Loading 151

parameter is related to the damage-induced reduction in the unconfined yield stress of the material, although the parameter is a mathematical concept rather than a measured property. Damage growth is related to the over stress in excess of a pressure dependent yield surface, with no damage growth above a brittle–ductile transition point on the yield surface. Computer simulations of explosives placed in boreholes provided successful descriptions of extent and regions of fracture damage and dependence on explosive energy and geometrical features.

7.5.2 An application of continuum fracture modelling

Computer simulation of a complex dynamic fracture application can be illustrated by calculations performed in support of large-scale explosive fragmentation experiments conducted in the Colony Oil Shale Mine near Grand Junction, Colorado (Harper and Ray, 1980). The computer model used was that of Grady and Kipp (1980) and various extended applications have been considered by Boade et al. (1981) and Chen et al. (1983). In this model stress and strain are related through,

$$\sigma_{ij} = K(1-D)\varepsilon\delta_{ij} + 2G(1-D)(\varepsilon_{ij} - \tfrac{1}{3}\varepsilon\delta_{ij}) \tag{7.36}$$

where K and G are the intrinsic moduli of the oil shale and the time changing effective moduli are determined by a scalar measure of the fracture damage, D. The constitutive description in tension is provided by two rate equations for the fracture damage and fracture surface area,

$$\dot{D} = (36\pi)^{1/3}[n(\varepsilon)]^{1/3}C_g D^{2/3}(1-D), \tag{7.37}$$

$$\dot{A} = (48\pi^2)^{2/3}[n(\varepsilon)]^{2/3}C_g D^{1/3}(1-D), \tag{7.38}$$

where the specific forms are determined by physical considerations of internal cracks to high rate loading and the Weibull crack statistics concepts noted earlier (Grady and Kipp, 1979a, b). The parameter C_g is a crack propagation velocity and $n(\varepsilon)$ is a crack activation law specified by the Weibull fracture theory.

The complexity of the stress waves generated by explosive charges, and the appearance of relief waves that emanate from free surfaces or regions previously fractured, necessitate the use of wave-propagation codes to address realistic problems. The codes numerically integrate the conservation equations of mass, momentum, and energy, along with the constitutive equations for the material. The fracture model described above was incorporated into the Lagrangian two-dimensional wave code, TOODY-IV (Swegle, 1978).

A calculation is illustrated for one Colony Mines fragmentation experiment in Figure 7.6. This experiment involved the detonation of a 5.2 kg, 0.75 m long charge of AN-FO at the bottom of a 0.1 m diameter, 2.0 m deep borehole drilled into the floor of the mine. Profiles of the excavated crater measured at 90° intervals around the axis of the borehole are shown in the figure. Based on earlier

152   Mechanics of Geomaterials

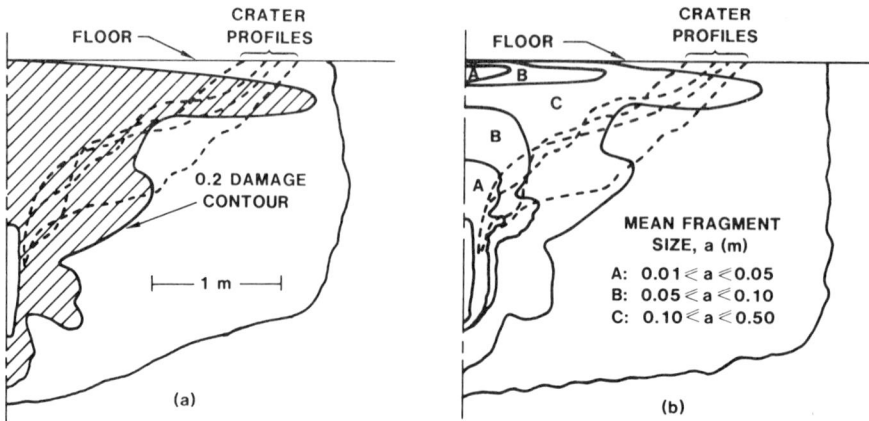

Figure 7.6   Computational calculation of (a) fracture damage and, (b) mean fragment size, about an explosively loaded borehole in oil shale

experiments (Grady et al., 1980), a damage level of 0.2 had been selected as a criterion for fracture damage levels sufficient for ejection. The numerical calculation established the damage contours and domains of fragment size shown in Figure 7.6 and are regarded as a reasonably good simulation of the cratering experiment. This type of calculation has been extended to evaluate concepts in explosive blasting (Boade et al., 1981). Multiple charges with variations in placement and timing have been accessed. Optimization of charge decking in the same borehole has also been performed by calculation.

7.6 SUMMARY

There appears to be a clear relation between the fracture threshold of a single crack subjected to high strain rate loading and the response of a crack system in determining the transient strength and fracture stress of a rapidly loaded body. Although certain conditions pertaining to the density and distribution of flaws are apparently required, studies on rocks over a variety of material conditions and strain rates seem to support this observation.

Perhaps one of the most interesting and important observations which is emerging from current studies on the bulk or continuum fracture properties of materials subjected to very rapid or impulsive loading is the importance of some measure of the fracture surface energy. A fracture toughness or strain energy release rate appears to be a necessary ingredient fundamental to both the dynamic strength of materials and in determining the number of cracks participating in the fracture process along with the final fragment size in the failed material.

This concept has been noticeably absent in the large body of continuum and computer modelling which has appeared in the literature to date. Considerably more emphasis has been focused on describing an initial condition of the body in terms of inherent flaws and weaknesses; perhaps more than is warranted. The ideas which are emerging suggest that both inherent flaw properties and energy conditions are probably important, but at the higher loading rates the latter effect becomes increasingly dominant.

Lastly, the description fragment size distributions resulting from fracture events and relating these distributions to the different geometries, material properties, and loading conditions is an interesting and complex problem which, at present, is very poorly understood. Distributions are observed to range from sharply centralized to broadly dispersed and there are indications that these differences may relate to the mode and multiplicity of fracture. Physically based statistical laws including those of Poisson and Boltzmann are being compared with fragmentation data; however, observations and conclusions at the present time are very tentative.

BIBLIOGRAPHY

Bennett, J. G. (1936) 'Broken coal', *J. Inst. Fuel*, **10**, 22–39.

Bergstrom, B. H., Sollenberger, C. L., and Mitchell, W., Jr. (1961) 'Energy aspects of single particle crushing', *Trans. AIME*, **220**, 267–372.

Bergstrom, B. H., Sollenberger, C. L., and Mitchell, W., Jr. (1962) 'Energy and size distribution aspects of single particle crushing', *Proc. U. S. Symp. Rock Mech. 5th*, pp. 155–172.

Birkimer, D. L. (1971) 'A possible fracture criterion for the dynamic tensile strength of rock', *Proc 12th Symp. on Rock Mech.* (G. B. Clark, Ed.), **573**.

Boade, R. R., Kipp, M. E., and Grady, D. E. (1981) 'A blasting concept for preparing vertical modified in situ oil shale retorts', Report SAND81-1255, Sandia National Laboratories.

Bond, F. C. (1952) 'Third theory of comminution', *Trans. AIME*, **193**, 484–494.

Butkovich, T. R. (1976) 'Calculation of fracture and permeability enhancement from underground explosions in coal', *Proc. Am. Soc. Mech. Eng.*, Mexico City, Sept. 19–23.

Charles, R. J. (1957) 'Energy-size reduction relationships in comminution', *Trans. AIME*, **208**, 80–88.

Chen, E. P., and Sih, G. C. (1977) *Elastodynamic Crack Problems* (G. C. Sih, Ed.), Noordhoff International Publishing.

Chen, E. P., Kipp, M. E., and Grady, D. E. (1983) 'A strain-rate sensitive rock fragmentation model', *Mechanics of Oil Shale*, K. T. Chan and J. W. Smith, (Eds), Applied Science Limited, Chapter 10.

Cohn, S. N., and Ahrens, T. J. (1981) 'Dynamic tensile strength of lunar rock types', *J. Geophys. Res.*, **86**, 1794–1802.

Costin, L. S., and Grady, D. E. (1984) 'Dynamic fragmentation of brittle materials using the torsional Kolsky bar', *Proc. Third Int. Conf. on Mech. Properties of Materials at High Rates of Strain*, Oxford, England, April, 1984.

Curran, D. R., Seaman, L., and Shockey, D. A. (1977) 'Dynamic failure in solids, *Phys. Today*, **30**, 46–55.

Davison, L., and Stevens, A. L. (1973) 'Thermomechanical constitution for spalling bodies', *J. Appl. Phys.*, **44**, 588–674.

Dienes, J. K., and Margolin, L. G. (1980) 'A computational approach to rock fragmentation', *Proc. U.S. Symposium on Rock Mech.* 21st, Rolla, Missouri, May, 1980.

Faddeenkov, N. N. (1975) 'Distribution of fragment sizes in block-type solid rock crushing by blasting', *Sov. Min. Sci.*, **11**, 660–664.

Forrestal, M. J., Grady, D. E., and Schuler, K. W. (1978) 'An experimental method to estimate the dynamic fracture strength of oil shale in the 10^3 to 10^4s^{-1} strain rate regime', *Int. J. Rock Mech. Min. Sci.*, **15**, 263–265.

Freund, L. B. (1973) 'Crack propagation in an elastic solid subject to general loading—III. Stress wave loading', *J. Mech. Phys. Solid*, **21**, 47–61.

Gates, A. O. (1915) *Trans. Am. Inst. Mining, Petrol, Engrs.*, **52**, 875.

Gaudin, A. M., and Meloy, T. P. (1962) 'Model and a comminution distribution equation for single fracture', *Trans. Soc. Min. Eng. AIME*, **223**, 40–50.

Gilvarry, J. J. (1961) 'Fracture of brittle solids, I, Distribution function for fragment size in single fracture (theoretical)', *J. Appl. Phys.*, **32**, 391–399.

Glenn, L. A. (1976) 'The fracture of a glass half-space by projectile impact', *J. Mech. Phys. Solids*, **24**, 93–106.

Grady, D. E. (1981a) 'Fragmentation of solids under impulsive stress loading', *J. Geophys. Res.*, **86**, 1047–1054.

Grady, D. E. (1981b) 'Application of survival statistics to the impulsive fragmentation of ductile rings', *Shock Waves and High-Strain-Rate Phenomena in Metals* (M. A. Meyers and L. E. Murr, Eds), 181–192.

Grady, D. E. (1982a) 'Local inertial effects in dynamic fragmentation', *J. Appl. Phys.*, **53**, 322–325.

Grady, D. E. (1982b) 'Analysis of prompt fragmentation in explosively loaded uranium cylindrical shells', SAND82-0140, Sandia National Laboratories, Albuquerque, New Mexico, February.

Grady, D. E., and Benson, D. A. (1983) 'Fragmentation of metal rings by electromagnetic loading', *Exp. Mech.*, **4**, 393–400.

Grady, D. E., and Hollenbach, R. E. (1979) 'Dynamic fracture strength of rock', *Geophys. Res. Letters*, **6**, 73–76.

Grady, D. E., and Kipp, M. E. (1979a) 'Oil shale fracture and fragmentation at higher rates of loading', *Proc. U.S. Symp. Rock Mech.*, 20th, Austic, TX, pp. 403–406.

Grady, D. E., and Kipp, M. E. (1979b) 'The micromechanics of the impact fracture of rock', *Int. J. Rock Mech. Min. Sci.*, **16**, 293–302.

Grady, D. E., and Kipp, M. E. (1980) 'Continuum modelling of explosive fracture in oil shale', *Int. J. Rock Mech. Min. Sci.*, **17**, 147–157.

Grady, D. E., and Kipp, M. E. (1984) 'Statistical size distribution in random single fracture', (in preparation).

Grady, D. E., and Lipkin, J. (1980) 'Criteria for impulsive rock fracture', *Geophys. Res. Letters*, **7**, 255–258.

Grady, D. E. and Varga, K. S. (1984) 'Shock comminution of crystalline quartz' (in preparation).

Grady, D. E., Lipkin, J., and Costin, L. S. (1981) 'Energy and particle size effects in the fragmentation of oil shale with a torsional split Hopkinson bar', *Proc. U.S. Symp. on Rock Mech.*, 22nd, MIT, Cambridge, MA.

Grady, D. E., Benson, D. A., and Kipp, M. E. (1984) 'Energy and statistical effects in the dynamic fragmentation of metal rings (to be published) *Third Int. Conf. on Mechanical Properties of Materials at High Rates of Strain*, Oxford, England, April, 1984.

Green, S. J., and Perkins, R. D. (1968) 'Uniaxial compression tests at varying strain rates on three geologic materials', *Symp. on Rock Mech.* (K. E. Gray, Ed.), Austin, Texas, pp. 35–54.

Griffith, A. A. (1920) 'The phenomena of rupture and flow in solids', *Phil. Trans. Royal Soc., A*, **34**, 137–154.
Griffith, L. (1943) 'A theory of size distribution of particles in a comminuted system', *Can. J. of Res.*, **21**, 57–64.
Harper, M. D., and Ray, J. M. (1980) 'Experimental design and crater profiles of intermediate scale experiments in oil shale', Los Alamos National Laboratory, Rpt. LA-8553-PR, June.
Ivanov, A. G., and Mineev, V. N. (1980) 'Scale effects in fracture', *Expl. Comb. and Shock Waves*, 617–638.
Johnson, J. N. (1978) 'Explosive produced fracture of oil shale', Los Alamos Scientific Laboratory, Rpt. LA-7357-PR, 38–45.
Kick, F. (1885) 'Das Gesetz der Proportionalem Widerstand and Seine Anwendung', Leipsig.
Kipp, M. E., Grady, D. E., and Chen, E. P. (1980) 'Strain-rate dependent fracture initiation', *Int. J. Fracture*, **16**, 471–478.
Kolmogorov, A. N. (1941) 'On the log-normal law of distribution of fragment dimensions in crushing' (in Russian), *Dokl. Akad. Nauk. SSSR*, **31**, 99.
Kumar, A. (1968) 'The effect of stress rate and temperature on the strength of basalt and granite', *Geophysics*, **33**, 501–510.
Kuznetsov, V. M., and Faddeenkov, N. N. (1975) 'Fragmentation schemes', *Fiz. Gor. Vzr.*, **11**, 637–645.
Langefors, U., and Kihlstrom, B. (1967) *The Modern Technique of Rock Blasting*, 2nd edn., John Wiley, New York, 1967.
Lankford, J. (1976) 'Dynamic strength of oil shale', *Soc. Pet. Eng. J.*, **16**, 17–22.
Lienau, C. C. (1936) 'Random fracture of a brittle solid', *J. Franklin Inst.*, **221**, 485–494, 674–686, 769–783.
Lindholm, U.S., Yeakley, L. M., and Nagy, A. (1974) 'The dynamic strength and fracture properties of dresser basalt, *Int. J. Rock Mech. Min. Sci.*, **11**, 181–191.
Lipkin, J., Grady, D. E., and Cambell, J. D. (1977) 'Dynamic flow and fracture of rock in pure shear', 18*th U.S. Symp. on Rock Mech.*, Keystone, Colorado, 3B2-1, June, 1977.
Lundberg, B. (1976) 'A split Hopkinson bar study of energy absorption in dynamic rock fragmentation, *Int. J. Rock Mech. Min. Sci.*, **13**, 187–197.
Margolin, L. G. (1983) 'Numerical simulation of fracture', *Proc. Int. Conf. on Constitutive Laws for Engineering Matrials* (C. S. Desai and R. H. Gallagher, Eds), Tucson, Arizona, p. 567.
Margolin, L. G., and Adams, T. F. (1982) 'Numerical simulation of fracture', *Proc. U.S. Symp. Rock Mech.*, 23rd, Berkeley, CA, August, 1982.
Matzui, T., and Mitzutani, W. (1977) 'Why is a minor planet minor', *Nature*, **270**, 506–507.
Mott, N. F. (1947) 'Fragmentation of shell cases', *Proc. R. Soc. London*, **300**, 300–308.
Mott, N. F., and Linfoot, E. H. (1943) 'A Theory of fragmentation', Extra-Mural Research No. F72/80, Ministry of Supply (England), January.
O'Keefe, J. D., and Ahrens, T. J. (1976) 'Impact ejecta on the moon', *Proc. Lunar Sci. Conf.*, 7*th*, 1976, pp. 3007–3025.
Oucherlony, F. (1980) 'Review of fracture toughness testing of rock', Report DS 1980:15, Swedish Detonics Research Foundation, Stockholm, Sweden.
Parkins, D. W., and Tilles, D. (1968) 'Influx measurements of extra-terrestrial material', *Science*, **159**, 936–946.
Rinehart, J. S. (1965) 'Dynamic fracture strengths of rocks', *Proc. Seventh Symp. Rock Mech.*, Pennsylvania State University, 1965.
Rosin, P., and Rammler, E. (1933) 'The laws governing the fineness of powdered coal', *J. Inst. Fuel*, **7**, 29–36.

Schmidt, R. A., Warpinski, N. R., Finley, S. J., and Shear, R. C. (1981) 'Multi-frac test series final report', SAND81-1239, Sandia National Laboratories, Albuquerque, New Mexico, November, 1981.

Schuhmann, R. (1940) 'Principles of comminution, in size distribution and surface calculations', Mining Technology, Rpt. AIME TP 1189, pp. 1–11, Am. Inst. Min., Metall., and Pet. Eng., New York, July, 1940.

Seaman, L., Curran, D. R., and Shockey, D. A. (1976) 'Computational models for ductile and brittle fracture', *J. Appl. Phys.*, **47**, 4814–4826.

Shockey, D. A., Curran, D. R., Seaman, L., Rosenberg, J. T., and Petersen, C. F. (1974) 'Fragmentation of rock under dynamic loads', *Int. J. Rock Mech. Min Sci.*, **11**, 303–317.

Smith, C. W., Bass, R. C., and Tyler, L. D. (178) 'Puff'n tuff, a residual stress gas fracturing Experiment', *Proc. U.S. Sym. Rock Mech.*, 19th, Stateline, Nevada, 1978.

Steverding, B., and Lehnigk, S. H. (1976) 'The fracture penetration depth of stress pulses', *Int. J. Rock Mech. Min Sci.*, **13**, 75–80.

Swegle, J. W. (1978) 'TOODY IV—a computer program for two-dimensional wave-propagation', SAND78-0522, Sandia National Laboratories.

Swenson, D. V., and Taylor, L. M. (1984) 'A finite element model for the analysis of tailored pulse stimulation of boreholes', *Num. Anal. Methods in Geomech*, (to be published).

Tuler, F. R., and Butcher, B. M. (1968) 'A criterion for the time dependence of dynamic fracture', *Int. J. Frac. Mech.*, **4**, 431–437.

Varnado, S. (1978) 'Geothermal drilling and completion technology development', Report 78-0670C, Sandia National Laboratories, July.

Von Rittinger, P. R. (1867) *Lehrbuch der Aufbereitungskunde*, Ernst and Korn, Berlin, p. 19.

Warpinski, N. R., Schmidt, R. A., and Cooper, P. W. (1979) 'High-energy gas frac: multiple fracturing in a wellbore', 20th *Proc. U.S. Symp. Rock Mech.*, pp. 143–152.

Weimer, R. J., and Rogers, H. C. (1979) 'Dynamic fracture phenomena in high-strength steels', *J. Appl. Phys.*, **50**, 8025–8030.

Yamakashi, K., Nogami, K., and Shimamura, T. (1981) 'Size distribution of siderophile element concentrations in black magnetic spherules from deep-sea sediments', *J. Geophys. Res.*, **86**, 3129–3132.

Mechanics of Geomaterials
Edited by Z. Bažant
© 1985 John Wiley & Sons Ltd

Chapter 8

Behaviour of Solids with a System of Cracks: Discusser's Report

A. Ehrlacher

8.1 INTRODUCTION

The two preceding chapters dealing with 'the behaviour of solids with a system of cracks' cover the main fields of the subject. The first one, 'Inelastic properties of solids with random cracks' of Professor Zaitsev is concerned with the quasi-static problems. The second one, 'The mechanics of fracture under high-rate stress loading' of Professor Grady is concerned with the dynamic problems.

Zaitsev insists on the randomness of the distribution of cracks and inhomogeneities in the solids. A distinction is drawn between rocks and hardened cement paste which are quasi-homogeneous at the scale of the problem, and concrete which is significantly inhomogeneous.

Grady deals with four problems. In the first part he investigates the strain rate dependence of the critical traction stress of a solid. In the second part, he shows how kinetic energy considerations can predict the size of particles produced in dynamic fragmentation without reference to the initial distribution of flaws. In the third part, he looks at the dynamic fragmentation from a statistical point of view, and in the last part, he uses a damage parameter D with an appropriate rate equation to produce a continuum modelling of dynamic fracture.

First of all, let us emphasize the fact that, concerning solids with a system of cracks, the questions we may have to answer are significantly different if we consider a static loading or a dynamic loading.

Consider the stress intensity factor K_I at the crack tip of a crack with a characteristic dimension a. If we apply a static loading, K_I is proportional to $\sqrt{a}$, while if we apply a dynamic loading with a constant strain rate, for example, the stress intensity factor is independent of the crack size and proportional to $t^{3/2}$ for small time t in comparison with a/c_s (c_s is the shear wave velocity).

Let K_{IC} be the toughness of our material. In the static case the K_I of the greatest cracks reaches K_{IC} at first and only those cracks propagate. In the dynamic case, with a constant strain rate $\dot{\varepsilon}_0$, all the cracks with a characteristic size a greater than a certain value a_0 will have the same stress intensity factor K_I, and then a lot of cracks will reach the critical value K_{1C} and propagate at the same time, giving a fragmentation of the solid. The value a_0 is decreasing with the strain rate as $\dot{\varepsilon}_0^{-2/3}$.

Because of these qualitatively different behaviours of the system of cracks, we shall discuss them separately. In section 8.2, we shall discuss the quasi-static problems. The dynamic problems will be discussed in section 8.3.

8.2 QUASI-STATIC PROBLEMS

Among, the quasi-static problems dealing with micro-cracked solids, at least three questions arise naturally. The first one is about the influence of the presence of micro-cracks on the macroscopic behaviour. The second one is about the consequences on the macroscopic behaviour of the propagation of these micro-cracks. The third one is about the coalescence of micro-cracks which produces a macro-crack and leads to the failure of the solid.

The numerous references given in the Zaitsev's chapter show that there are many methods to try to answer these questions, at least in the quasi-homogeneous case. Before we discuss the Monte Carlo method proposed by Zaitsev, we shall comment the macroscopic thermodynamical approach and the homogenization technique, which are among the more classical ways of tackling that sort of question.

8.2.1 Macroscopic thermodynamical approach

It is possible to take into account the macroscopic experimental results directly by the use of a set of thermodynamic internal state variables, with some of them characterizing the microstructure of the material. Then a thermodynamic potential (the free energy for example) and a dissipative pseudo-potential or rate equation satisfying thermodynamic restrictions (see, for example, Germain (1973) or Nguyen (1981)) can define the macroscopic behaviour of the material.

The difficulty of this approach is to postulate a good choice for the internal state variables and the potentials. To do that, one generally adopts a simplified physical model for the microstructure. The crossing from the microlevel to the macrolevel is often done using the homogenization technique.

In the case of micro-cracked bodies, damage parameters are used extensively as internal variables summarizing the effects of micro-cracks on the macroscopic behaviour. Kachanov (1958), Lemaitre and Chaboche (1978), Lemaitre et al. (1979), or Mazars and Lemaitre (1982) use for example a scalar damage parameter $D(t)$ to explain the decreasing of the elastic modulus (for experimental results see, for example, Terrien (1980)).

$$E(t) = (1 - D(t))E_0 \qquad (8.1)$$

The damage parameter, the value of which increases from zero to one, is physically bounded by the fraction of flaws' surface in a cross section of the solid.

Behaviour of Solids with a System of Cracks

A certain solution for brutal damage in elastic brittle solids where parameter D can take only two values, $D = 0$ for the same material, $D = 1$ for the totally damaged material, has been given by Bui et al. (1980), (1981), (1983). Shah and Suaris (1983a, 1983b) give a thermodynamic approach with a vectorial representation for damage to account for the planar nature of micro-cracks.

If we want to take into account the non-isotropic natures of the microcracking process, it is possible to use a tensorial representation for damage $\mathbf{D}(t)$, in order to change the initial elastic matric, $\mathbf{E}_0$, to the actual one, $\mathbf{E}(t)$ (Bui, Dang Van, and Stolz, (1982)).

$$\mathbf{E}(t) = (\mathbf{I} - \mathbf{D}(t))\mathbf{E}_0 \qquad (8.2)$$

where $\mathbf{I}$ is the identity tensor.

The thermodynamic force associated with the rate of damage $\dot{D}$ is the elastic strain energy W (Lemaitre and Chaboche, 1978), and then generally $\dot{D}$ is postulated as a function of W. The approach with damage parameters is especially suitable for tension. When dealing with tension and compression, the formulation is far more complex because of the micro-crack closure which can restore the initial elasticity matrix in compression.

It is then possible to use instead of damage parameters other internal state variables such as a fracturing strain (Bažant, 1980). That choice can also take into account the sliding of the two opposite surfaces of closed micro-cracks.

8.2.2 Homogenization technique

The second very classical way of looking at quasi-homogeneous solids is to consider that each point at the macroscale represents at the microlevel a small solid which would be in an uniform stress state if it included no flaw or other inhomogeneity. The size and position of flaws or inhomogeneities are characterized by a finite number of parameters at the macroscale. The value of the fields at each point at the macroscale is the mean value of the same field at the microlevel in the volume of the microelement, these fields being characterized by a finite number of parameters.

For consistency, one needs an additional condition, the localization condition, which may be the classical Hill-Mandel condition (Mandel, 1966; Hill, 1967), or a periodicity hypothesis, or other conditions (see, for example, Suquet, 1982).

Generally the use of this homogenization technique needs some hypothesis on the microstructure. For example, should we consider only one crack in the microelement, or two interacting cracks, or more...? Once we have chosen our microstructure and our localization condition, we have to solve the problem of finite loading parameters on the microelement. If we have such a solution, the mean value of the fields and the rate of the characteristic parameters give us the macroscopic behaviour.

A very good application of this technique for micro-cracked bodies has been recently done by Andrieux (1983), who looked at a single cracked microstructure in plane strain and used a solution depending on two parameters characterizing the discontinuities of displacements on the crack surfaces. He deduced then the macroscopic behaviour of a micro-cracked body which seems to well approximate the experimental results in compression and in tension for concrete. From the solution of two interacting cracks, it is possible to apply the same technique to a more sophisticated microstructure.

This homogenization technique is often the first stage of a macroscopic thermodynamic approach. It permits us to construct a continuum model with a set of internal state variables at the macrolevel and their rate equations from a simple model of the microstructure, the behaviour of which is known. Then from a given state of the solid, that is to say from a given initial field of internal variables, we can calculate the evolution of the fields for a given loading. When no solution exists any more, the failure occurs.

The important point we must underline in view of Zaitsev's chapter is that we must know the initial field of internal variables to know the further evolution. In certain problems, such as plasticity problems, the initial values of internal variables are, to a certain extent, of little importance, and can be given the value of zero.

On the other hand, when dealing with micro-cracked bodies, it is generally not possible to hypothesize that the initial distribution of flaws is zero, and that initial distribution is of the greatest importance since the damage concentrates in the weakest zone. We must then account for the initial random distribution of flaws or internal variables in the case of a continuum modelling.

8.2.3 The Monte Carlo method

Zaitsev chooses to investigate the influence of the initial random distribution of flaws in a rather interesting manner. He numerically simulates the evolution of a system of microstructures, the behaviour of which is known.

Dealing with different problems, he chooses different microstructures with different behaviour, as in his figures (6.7, 6.17, 6.23). Each microstructure being characterized by a finite number of parameters, he chooses certain statistical laws for the initial values of these parameters, and he generates a random macrostructure with the help of a computer. The behaviour of each microstructure being well known, he can simulate a loading and compare the results for different initial distributions of flaws and experimental data.

Zaitsev's first simulation (Zaitsev, 1980; Zaitsev and Wittmann, 1981) is done with cracks, the centre, the length, and the orientation of which is statistically uniformly distributed, and he considers the interaction between two neighbouring cracks. His second simulation (Zaitsev, 1981; Wittmann and Zaitsev, 1981) has been done with a microstructure made with pores and cracks. Each pore has two coplanar pre-existing cracks. The position, length and orientation of the microstructure is statistically uniformly distributed.

Then he makes simulations with circular inclusions with one pre-existing interfacial crack and pre-existing cracks in the matrix, another simulation for concrete with polygonal inclusions with one pre-existing bond crack, and for lightweight concrete he makes a simulation with round inclusions and small pores. For each simulation the calculation gives the global σ–ε curve and the macro-crack loading to the failure of the sample. Note that he does not use a continuum modelling of his micro-cracked sample because he generates directly some microstructure, but there is no essential difference from a random generation of initial internal variable values by means of the finite element method for continuum modelling.

In both cases the Monte Carlo method provides an efficient tool to study the qualitative response of a sample to a certain loading, the influence of the parameters of the random distribution, and a global statistical characterization of the behaviour of the sample.

However, before answering these questions we must determine how many microstructure or finite elements are needed to obtain significant results (Zaitsev chooses 50 microelements), and how many numerical simulations are sufficient to deduce statistically satisfactory results. Zaitsev makes twenty realizations of the Monte Carlo method to deduce the results of his Figure 6.19.

Then the influence of the parameters of the random distribution may thus be quite easily simulated, while a statistical characterization of the global behaviour is not a simple question. We have a rather similar problem when we want to deduce the behaviour of the material from experimental data. Nevertheless, this method is very efficient since it can be adapted to a number of problems simply by changing the microelement or the interaction between microelements, as it can be seen in Zaitsev's chapter. Especially, this method can be used to give a probabilistic response to the problem of determining the risk of failure of a complex structure submitted to a complex quasi-static loading.

8.3 DYNAMIC PROBLEMS

We have already emphasized the fact that while for static loading only one or a few macro-cracks appear in the solid, for dynamic loading all the micro-cracks the length of which is greater than a certain size a_0 can propagate. This leads to dynamic fragmentation of the solid (a_0 is decreasing with the strain rate as $\varepsilon_0^{-2/3}$).

8.3.1 Dynamic stress intensity factor

The dynamic stress intensity factor for a single crack has been extensively studied in many papers among which we can mention de Hoop (1958), Craggs (1963), Broberg (1960), Baker (1962), Cotterel (1964), Eshelby (1969), Achenbach (1970),

Achenbach and Nuismer (1971), Glennie and Willis (1971), Chen and Sih (1973), Bui (1978), and especially Kostrov (1966, 1975) and Freund (1972a, 1972b, 1973, 1974). Atkinson and Eshelby (1968), Kostrov and Nikitin (1970), Freund (1972), and Bui, Ehrlacher, and Nguyen (1980) have studied the problems bounded with the energy release rate at the crack tip.

From these works we can deduce, for the single case of a non-propagating crack and for a short time in comparison with a/c_s (where a is the crack length and c_s the shear wave velocity), that for loading at a constant strain rate the stress intensity factor at the crack tip is independent of the crack size and depends on time as $t^{3/2}$ Kipp et al., 1980). If we hypothesize that the crack tip will propagate as soon as the stress intensity factor reaches the critical value K_{IC}, we can deduce the fracture time and the strain rate dependent fracture stress as $\dot{\varepsilon}_0^{1/3}$.

Grady hypothesizes that for a system of cracks loaded at constant strain rate all the cracks whose lengths are sufficiently high behave in the same way, and he deduces the general strain rate dependence of the fracture stress as $\dot{\varepsilon}_0^{1/3}$ for the sufficiently flawed material (that seems not to be the case for Solenhofen Limestone (Grady and Lipkin, 1980). Then he demonstrates that other classical dynamic fracture criteria such as those of Tuler and Butcher, Birkimer, Steverding and Lehnigk, or even that of von Rittinger, give similar results.

8.3.2 The fragment size

In view of applications, the most important question is to determine the fragment size occurring in the dynamic fracture process. There are three ways to approach the average fragment size d. The simplest one is a relation between the fragment size and the number of activated cracks. If we denote as N the number of activated cracks per unit volume, the fragment size d is of order $N^{-1/3}$. If we recall the order of the characteristic size of the activated cracks given by Grady,

$$a_0 \sim (c_s K_{IC}/E\dot{\varepsilon}_0)^{2/3} \tag{8.3}$$

and if we hypothesize that the number of cracks per unit volume the size of which is greater than a_0 is of the order of

$$N(a_0) \sim \alpha/a_0^3 \tag{8.4}$$

where α is a non-dimensional parameter, we find that the fragment size d would be of the order of:

$$d \sim \alpha^{-1/3} a_0 \sim \alpha^{-1/3} (c_s K_{IC}/E\dot{\varepsilon}_0)^{2/3} \tag{8.5}$$

The second way to approach the fragment size is to consider the strain energy stored in the solid before the fragmentation occurs. When the cracks propagate, this strain energy is transformed into surface energy for one part, and into kinetic energy for the other part. It is well known that the more rapid is the crack propagation, the more important is the latter part. For example, consider a statically loaded body with a single crack. If the crack does not propagate, there is a static energy release rate G^{stat} which characterizes the fields near the crack tip. If suddenly the crack propagates at speed $\dot{a}$, then the energy release rate is

$$G = g_I(\dot{a}) \cdot G^{stat} \qquad (8.6)$$

where $g_I(\dot{a})$ is a universal function of the crack tip velocity decreasing from one to zero when $\dot{a}$ increases from zero to the Rayleigh wave velocity C_R. (Atkinson and Eshelby, 1968; Kostrov and Nikitin, 1970; Freund, 1972; Ehrlacher, 1980.)

The function $g_I(\dot{a})$ can be analytically written in a complicated manner, but is very well approximated by the linear function

$$g_I(\dot{a}) \approx 1 - \dot{a}/C_R \qquad (8.7)$$

Then, $\dot{a}g_I(\dot{a})G^{stat}$ is the rate of the energy dissipated through the crack tip propagation, and $\dot{a}(1 - g_I(\dot{a}))G^{stat}$ is the rate of transformation of the strain energy into kinetic energy.

Because in general, during the fragmentation process, the velocity of the crack tips is of the order of the Rayleigh wave velocity ($\dot{a}/C_R \sim 0.6$–0.7), the most important part of the strain energy is transformed into kinetic energy. Then the fragment size is not of the order of $(K_{IC}/\sigma_c)^2$, as an energy balance neglecting the kinetic energy would have suggested, but it is one or more orders of magnitude greater. This point is reinforced by the fact that the energy needed to create a new surface is not constant but seems to be an increasing function of the crack tip velocity because the dissipative mechanisms are more complex for rapid cracks.

By contrast, kinetic energy considerations instead of strain energy considerations give rather interesting and applicable results (see Grady's chapter). The order of the fragment size which can be deduced (Equation 7.19 of Grady's chapter) shows the same dependence) on toughness K_{IC} and strain rate $\dot{\varepsilon}_0$ as that in the flaw distribution approach. As it can be seen from Figure 7.3 of Grady's chapter, the experimental data seem to be well predicted by the strain rate dependence $\dot{\varepsilon}_0^{-2/3}$.

8.3.3 Continuum modelling approach

From the work of Davison and Stevens (1973), the continuum modelling approach for dynamic fragmentation is of interest. Grady's chapter shows very well the variety of the continuum modelling approaches through numerous references. The model proposed by Grady and Kipp uses a classical scalar damage parameter, D, with an appropriate rate equation deduced from physical considerations on statistical concepts. The comparison of his Figures (7.6(a)) and (7.6(b)) shows very well how effective such a direct continuum approach can be with the use of a wave propagation code.

These types of continuum modelling give a good approximation of the shape of the boundary between the sane zone and the fragmentation zone, and a good agreement between the value of the calculated damage parameters in the fracture zone and the experimental measurements of the fragment size. Perhaps we can hope that a continuum modelling will describe, through a good choice of internal state variables and rate equations, the statistical nature of the fragment size at each point on the macroscale.

8.4 CONCLUSION

The quasi-static and the dynamic behaviours of a micro-cracked solid are quite different, since in the static case only the more susceptible micro-cracks can propagate and coalesce to induce the failure of the solid in two or a few pieces, while in the dynamic case many micro-cracks propagate at the same time, inducing fragmentation of the solid.

In the quasi-static case the behaviour of the solid is very sensitive to the initial distribution of flaws. Then the Monte Carlo method proposed by Zaitsev is a very powerful tool to compare the experimental results to numerical simulations, and to predict the risk of failure of a complex structure under complex loading.

When dealing with dynamic problems, the initial distribution and size of flaws may be important for predicting the fragment size in the failed material. However, Grady's approach from the kinetic energy point of view can give good predictive results without referring to this initial distribution.

For both static and dynamic loadings, it seems that a continuum modelling with the help of damage parameters and appropriate rate equations can provide a relatively simple and rather efficient tool for applications.

REFERENCES

Achenbach, J. D. (1970) *Z.A.M.P.*, **21**, 887–900.
Achenbach, J. D. and Nuismer, R. (1971) *Int. J. Fract. Mech.*, **23**, 6, 969–976.
Andrieux, S. (1983) Thesis, Ecole Nationale des Ponts et Chaussées, Paris.
Atkinson, C., and Eshelby, J. D. (1968) *Int. J. Fract. Mech.*, **4**, 3–8.
Baker, B. R. (1962) *J. Appl. Mech.*, **29**, 449–458.
Bažant, Z. P. (1980) *J. of Eng. Mech. Div., Proc. of the ASCE*, **106**.
Broberg, K. B. (1960) *Arkiv för Fisik*, **18**, 159–192.
Bui, H. D. (1978) *Mécanique de la Rupture Fragile*, Masson, Paris.
Bui, H. D., Dang Van K., and Stolz C. (1981) *C.R.A.S.*, **292**, série II, p. 251.
Bui, H. D., Dang Van K., and Stolz C. (1982) *C.R.A.S.*, **294**, série II, p. 1155.
Bui, H. D., and Ehrlacher, A. (1980) *Int. Conf. on Fracture*, ICF 5, Cannes, vol. 2, pp. 533–553.
Bui, H. D., Ehrlacher, A., and Nguyen, Q. S. (1980) *J. de Mécanique*, **19**, 697–723.
Bui, H. D., Ehrlacher, A., Renard, C. (1983) *Cong. Franç. de Méca.*, Lyon.
Chen, E. P. and Sih, G. C. (1973) *Int. J. Solids Struc.*, **9**, 897–919.
Cotterel, B. (1964) *J. Appl. Mech.*, **31**, 177–199.
Craggs, J. W. (1983) *Fracture of Solids* (D. C. Drucker and J. Gilman (Eds), Wiley, New York.
Davison, L., and Stevens, A. (1973) *J. Appl. Phys.*, **44**, 588–672.

de Hoop, A. I. (1958) *Doctoral dissertation*, Technische Hogeschool, Delft.
Ehrlacher, A. (1980) *Int. Conf. on Fracture, ICF* 5, Cannes.
Eshelby, J. D. (1969) *J.M.P.S.*, **17**, 177–199.
Freund, L. B. (1972) *J. of Elasticity*, **2**, 341–349.
Freund, L. B. (1972a) *J.M.P.S.*, **20**, 129–140.
Freund, L. B. (1972b) *J.M.P.S.*, **20**, 141–152.
Freund, L. B. (1973) *J.M.P.S.*, **21**, 47–61.
Freund, L. B. (1974) *J.M.P.S.*, **22**, 137–146.
Germain, P. (1973) *Cours de Mécanique des Milieux Continus*, Masson, Paris.
Glennie, E. B., and Willis, J. R. (1971) *J.M.Ø.S.*, **19**, 11–30.
Grady, D. E., and Lipkin, J. (1980) *Geophys. Res. Letters*, **7**, 255–258.
Hill, R. (1967) *J.M.P.S.*, **15**, 79.
Kachanov, L. M. (1958) *Inv. Ak. N.S.S.R.*, **8**, 26.
Kipp, M. E., Grady, D. E., and Chen, E. P., (1980) *Int. J. Fract.*, **16**, 471–478.
Kostrov, B. V. (1966) *P.M.M.*, **30**, 6, 1042–1049.
Kostrov, B. V. (1975) *Int. J. Fract.*, **11**, 47–56.
Kostrov, B. V., and Nikitin, L. V. (1970) *Arch. Mech. Stos.*, **22**, 749–776.
Lemaitre, J., and Chaboche, J. L. (1978) *J. Mécan. Appl.*, **2**, 3, 317.
Lemaitre, J., Cordebois, J. P., and Dufailly, J. (1979) *C.R.A.S.*, **288**, série B, 391.
Mandel, J. (1966) *Proc. 11th Int. Cong. Appl. Mech.*, Munich, p. 502.
Mazors, J., and Lemaitre, J. (1982) *Annales I.T.B.T.P.*, no. 401.
Nguyen, Q. S. (1981) *IUTAM Symp. on '3-dim. Const. Rel. and Ductile Fracture'*, North-Holland, Amsterdam, pp. 315–330.
Shah, S. P., and Suaris, W. (1983a) *A.S.C.E.-EMD, Spec. Conf.* Purdue Univ.
Shah, S. P., and Suaris, W. (1983b) *SMIRT 7 Conf.*, Chicago.
Suquet, P. (1982) *Thesis*, Université P. et M. Curie, Paris.
Terrien, M. (1980) *Bul. de Liaison L.C.P.C.*, **105**.
Wittmann, F. H., and Zaitsev, Y. V. (1981) *Int. Conf. on Fract., ICF* 5, Cannes.
Zaitsev, Y. V. (1980) *Proc. 7th Int. Conf. on Chemistry of Cement*, IV, 176–180.
Zaitsev, Y. V. (1981) *Int. Conf. on Fract., ICF* 5, Cannes.
Zaitsev, Y. V., and Wittmann, F. H. (1981) *Matériaux et Constructions*, **14**, 357–365.

PART IV

SHEAR LOCALIZATION, FAULTING, AND FRICTIONAL SLIP

Mechanics of Geomaterials
Edited by Z. Bažant
© 1985 John Wiley & Sons Ltd

Chapter 9
Constitutive Relations for Frictional Slip

A. L. Ruina

9.1 INTRODUCTION

What is the appropriate boundary condition for modelling slip, or possible slip, on the boundary between two solids? Assuming no hydrodynamic lubrication, this is the problem of determining a constitutive relation for frictional slip. Two general questions motivate the use of frictional boundary conditions. The first is the question of the strength of a structure that contains joints on which slip may occur. The second is the question of stability of motion once slip deformation does occur.

For many purposes the question of strength of a rock joint is well answered by Coulomb's law: the shear traction required to cause slip is proportional to the normal traction. Coulomb's law, which is analogous to perfect plasticity for solid deformation, states a condition for slip but does not consider the fact that the friction force varies once slip does occur. This variation is neglected for two reasons: (1) it is often a small fraction of the total shear traction required to cause slip, and (2) for purposes of simple structural strength analysis, all that is of interest is the peak friction force (or traction) required to cause slip, without regarding that slip may occur both before and after this peak. This peak friction is often called the 'strength' of the interface. Coulomb's law states that the strength of an interface is proportional to the normal force across the interface. The proportionality constant is called the coefficient of friction μ. Some variants of Coulomb's law take into account a slightly more complex dependence on normal stress (e.g. 'Byerlee's law' (Byerlee, 1978)), and also try to relate the normal stress dependence to surface geometry and bulk material properties (Barton, 1976).

If one is concerned with the stability of slip, then the variation of the friction force as slip progresses, even if small, is of primary importance. In particular, slip instabilities in elastic systems depend, except in special cases, on the friction coefficient decreasing as slip progresses. The special cases that predict instability without a decrease in the friction coefficient during slip involve either (1) systems in which the normal stress decreases as slip progresses, or (2) systems in which the imposed load increases with increased slip (e.g. a spring with a negative incremental modulus). In most cases, however, a decrease in friction coefficient with ongoing slip is required for slip instabilities (and the resulting release of elastic or gravitational potential energy).

The characterization of friction force variation during slip, as needed for stability analysis, is the subject of this presentation.

Organization of the presentation is as follows: (1) statement of the scope of the frictional constitutive relations. (2) description of some classical constitutive relations and their deficiencies, and (3) description of the state variable friction laws which resolve several deficiencies in the classical friction laws. Consequences of the constitutive relations for determination of slip stability, although the ultimate goal, will only be mentioned as needed to motivate the discussion. For the quickest view of the subject the reader should skip to Section 9.4.2., titled 'Basic phenomenology—step changes in slip rate'.

The state variable friction laws have been discussed in some detail in previous papers (Dieterich, 1978, 1979a, 1980, 1981; Ruina, 1980, 1983; Rice and Ruina, 1983). These laws have been used to discuss slip instabilities in the above papers as well as in Dieterich (1979b), Kosloff and Liu (1980), Mavko (1980, 1984) and Gu et al. (1984). The aim here is to introduce the state variable constitutive relations in the context of constitutive relations for slip without undue attention to some of the technical details presented in earlier papers.

9.2 PROBLEM STATEMENT

9.2.1 Continuum 'point' of slip surface

Figure 9.1 shows a region of a solid that includes a frictional interface that may undergo slip. The length L drawn is meant to represent the smallest size scale over which one can distinguish the deformation from a rigid displacement. The

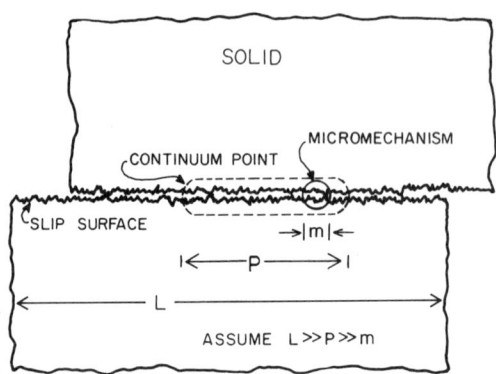

Figure 9.1 Two solids in frictional contact. The scale over which deformations are observed is L, the size of a continuum point is P, and the continuum point consists of micromechanisms of size m

existence of a constitutive relation requires the concept of a continuum 'point'. A continuum 'point' is a region of the interface between the solids and may include a small layer of solid as well. The continuum 'point' has some size P that must be negligible compared to L. The 'point' is presumed to include many microscopic features. The microscopic features may include microscopic asperities that fracture, deform or slip and may include chemical diffusion and local heating processes as well. These micromechanical elements are assumed to have characteristic size m such that the continuum 'point' may include many of these mechanisms. Thus for the continuum description to be sensible a hierarchy of size scales is implicit such that $L \gg P \gg m$. Further, over a wide range, the constitutive relation must be independent of the size of the continuum point.

Some experiments (e.g. Barton, 1981) show that the shear strength of a rock joint depends on the size of the joint. Data from these experiments can thus not be viewed as descriptive of a continuum 'point' with size P in the range of the sample size. It is possible, however, that a much larger sample could represent a continuum 'point'. On the other extreme it is possible that the size effects mentioned can be explained, as Barton explains them, as the consequence of frictional constitutive relations applied to various smaller-scale surface points within his samples.

A peculiar ambiguity that occurs frequently in the discussion of friction is that the separation between models of friction and models using friction is not always clear. This is exemplified by various uses of the word asperity. In the discussion above the word asperity has been used to describe part of the micromechanics of friction. Understanding of the micromechanics requires that the asperities be modelled. What boundary conditions would one use for describing the asperities? 'Frictional' boundary conditions could be used, but then friction would be an ingredient of the friction model. This approach is not necessarily circular since the micro-asperities may in fact govern the slip over larger-scale inhomogeneities that in turn are averaged to make up the macroscopic friction description.

On the other hand the word asperity is used in connection with earthquake modelling to mean a macroscopic region on a fault surface with higher strength than its surroundings. In this case one would imagine that frictional boundary conditions are quite likely to apply but that different boundary conditions (or different frictional strengths) apply in different surface regions.

In summary, interest is in describing the mechanical behaviour of a continuum point or small region of slipping surface. This description is for use in modelling at a larger scale. The results of such modelling may or may not be useful for modelling at a still larger scale. At every scale, however, a frictional boundary condition of some kind is needed.

9.2.2 Macroscopic mechanical variables in the constitutive relations

The primary mechanical variables of interest are shown in the picture of a continuum point in Figure 9.2. The material constraint on the relation between

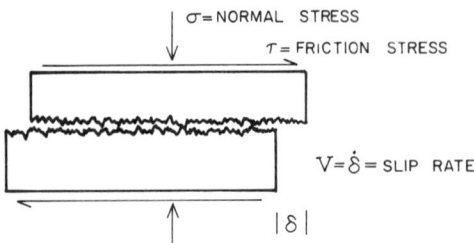

Figure 9.2 The macroscopic variables of interest for the continuum point: the normal stress σ, the shear (friction) stress τ, the slip displacement δ, and the slip rate $V = d\delta/dt$

the normal traction σ, the shear traction τ, and the slip displacement δ is of interest. The slip displacement δ can be precisely defined as the excess relative shear displacement between two solid points on opposing sides of the slip surface over that which would be predicted by the laws of bulk deformation of the solid. Experiments indicate that this definition is close to the approximation that the slip surface has no macroscopic elastic deformation. That is, δ is (approximately) totally inelastic. The slip rate is $V = d\delta/dt$ (where t is time).

A full mechanical description could conceivably include other deformation variables. Most obvious is the surface separation displacement. Experiments that elucidate the primary features to be discussed have neither measured nor controlled this variable. Future work will include surface separation in the description.

Also missing is slip in more than one direction. There is no reason to believe that an isotropic surface remains isotropic once slip has occurred. Thus if one is interested in a surface that may slip in more than one direction the effects of induced anisotropy would have to be considered. Such effects would require a second slip variable for description. Few relevant results are known.

Further deformation variables that one may want to include for generality are neglected because their effects are unknown, of little use and likely to have little effect on the primary slip and traction variables of interest. Such deformation variables are the relative rotation of the two surfaces and the in plane strain in the two solids.

The slip surface is actively coupled to the adjoining solid in several other ways that are not discussed here. For example, effects of thermal coupling between the slip surface and the solid are neglected and the slip process is then idealized to be, at least at the macroscopic level, without temperature fluctuations. Similarly neglected are fluid pore pressure and chemical potential effects.

Overall then, the constitutive relation is imagined to be for a given material in a given environment. The constitutive relation may be different for different environments. The micromechanics of the slip may include thermal fluctuations,

Constitutive Relations for Frictional Slip

pore fluid pressures and chemical potential gradients but these effects are assumed to be coupled to the adjoining solid only through their effects on the macroscopic slip and shear traction.

One last simplification, which is required by lack of data from all but the most recent and still incomplete experiments, is that the normal traction σ remain constant during slip.

9.2.3 Idealization of a testing machine

Unfortunately, much of what is observed in friction experiments is due to the imperfect nature of testing machines. A perfect testing machine could prescribe a precise value for the slip displacement δ at every instant in time. One might at first think of a perfect machine as one that prescribes a controlled value of the friction traction or force) τ at every instant in time. However, such a machine would, for many surfaces, lead to unstable slip and would thus conceal the constitutive relations presented here.

A real testing machine might be idealized by the spring-block model in Figure 9.3 (assuming accelerations are low enough so that inertia can be neglected). A part of the machine, represented by the screw in Figure 9.3, moves at a controlled and possibly variable rate. In a real testing machine this load point would be a motor, a hydraulic valve, or possibly just an electronic signal. In any case, the load point is mechanically (and possibly electrically) removed from the slip surface. The idealization in Figure 9.3 is that the transmitted friction traction τ is proportional to the difference between the load point displacement and the slip displacement. The stretch of the spring, with stiffness k, represents all of the elastic deformation of the machine and the sample. A real machine may also have rate effects that could be represented, say, by additional spring and viscous 'dashpot' elements in series and parallel with the spring in Figure 9.3. This would not alter the conclusions discussed below so long as the machine has an elastic, rather than viscous, response for sudden deformations. In the discussion that follows one must take into account that real machines are not rigid and are most simply idealized as having a finite elastic stiffness.

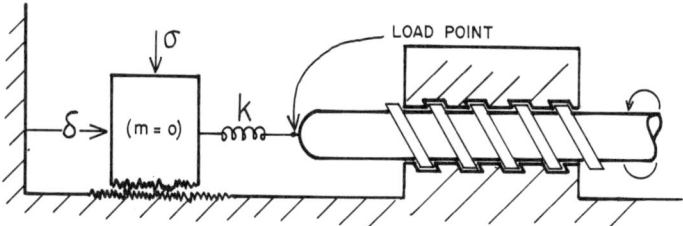

Figure 9.3 The idealized testing machine. A massless block is dragged on a slip surface by a spring with stiffness k. The spring is stretched by a load point represented by the screw in the figure

174   *Mechanics of Geomaterials*

The coupling between normal stress and slip in 'triaxial' and asymmetric biaxial testing machines is, for simplicity, excluded here. In order to include such machines in this discussion one would have to make assumptions about the effects of changing normal stress on the friction stress.

9.2.4 Overall relation between shear traction and slip

Figure 9.4 shows an idealized qualitative relation between shear traction τ and measured displacement 'δ' for a surface undergoing first slip in a testing machine with some finite but large stiffness (the quotes on 'δ' indicate that the measured displacement includes some elastic deformation). First, slip is imposed at a roughly constant rate by controlled motion of the load point. The shear stress rises rapidly at first and the displacement measure 'δ' shows only elastic deformation (the slope of τ vs 'δ' is high if the displacement transducer is close to the slip surface). As slip begins, the friction force increases less rapidly and then may or may not pass through a peak value. The dashed lines indicate these two possibilities. Eventually, as slip progresses, the friction force approaches a constant value. Next, machine motion is stopped for some amount of time and then resumed again at a roughly constant rate. When the machine stops the friction force drops. When the machine resumes motion the friction force rises, goes over a peak and then relaxes to a constant level over the distance D (this last result can only be observed in a testing machine that is stiff enough to inhibit stick–slip instabilities). If the machine is then stopped and started again then the transient response in friction stress will be repeated.

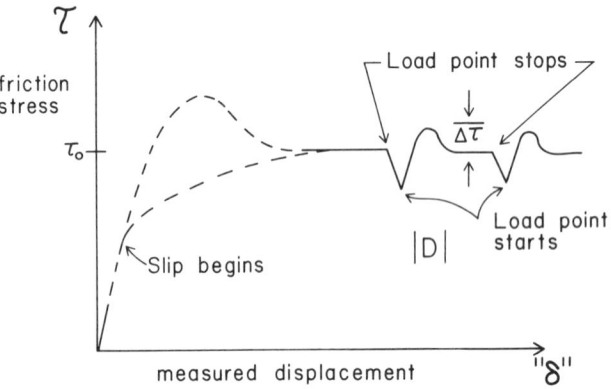

Figure 9.4 Overall qualitative friction response of a surface at constant σ in a stiff but not rigid machine. First sliding may follow one of the dashed curves before τ becomes constant. When the testing machine load point is stopped the friction force drops. It rises and goes over a peak when load point motion is resumed. The transient change in friction stress is $\Delta\tau$ and occurs over the distance D. Stopping the load point again and then resuming slip reproduces the transient change in τ

Constitutive Relations for Frictional Slip

Typically the friction traction τ_0 varies from about 0.4σ to 0.7σ and the transient peak $\Delta\tau$ may be up to about 0.1σ (depending on the sample and the experiment). The transient distance D may be in the range of $1\,\mu m$ to $1\,mm$ (also depending on sample and somewhat on machine stiffness). Describing the variations in the friction force during transient responses associated with resumed sliding is of major interest in the analysis of slip instabilities.

9.3 CLASSICAL CONSTITUTIVE RELATIONS

Three useful classical constitutive relations that describe some features of the transient responses in Figure 9.4 are: (1) 'static–dynamic' friction, (2) slip rate dependent friction, and (3) slip displacement dependent friction. These relations are discussed below.

9.3.1 'Static–dynamic' friction

The static–dynamic friction law is illustrated in Figure 9.5. Until τ/σ reaches a level μ_s no sliding occurs. The transient $\Delta\tau$ in Figure 9.4 is approximated as occurring instantaneously. The instant slip begins, the friction coefficient drops from μ_s to μ_d. μ_s is called alternately the 'static', 'stationary' or 'starting' coefficient of friction. μ_d is called the 'dynamic', 'kinetic', or 'sliding' coefficient of friction. As will be seen, this friction law is a limiting form of the two other classical friction laws. Slight variants of this law are that the static friction depends on (1) the time of stationary contact (Dieterich, 1972) or (2) the rate at which τ increased just previous to slip (Johannes *et al.* 1973). The problems mentioned below are not resolved by these variants.

One prediction of the static–dynamic friction law is that slip initiation is always unstable no matter how stiff the testing machine. This prediction is

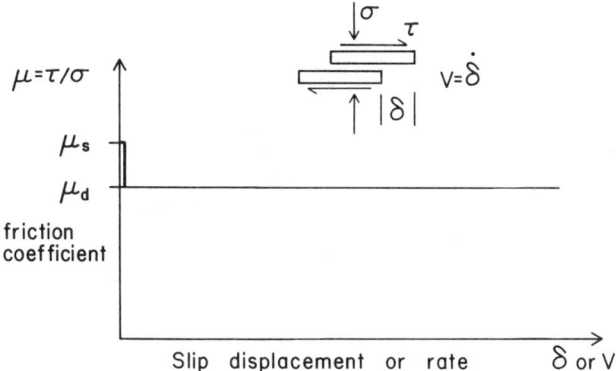

Figure 9.5 'Static–dynamic' friction. The friction coefficient drops from μ_s to μ_d the instant that δ or V exceeds zero

contradicted by the experiments of Dieterich (1978) in which the presence of stick–slip instability was found to depend on sufficiently small machine stiffness k.

Another prediction is discussed with reference to Figure 9.6. Two linear elastic solids have a sliding contact surface and a remote loading τ_{applied}. On part of the surface with size L imagine that slip has already begun so $\tau = \mu_d \sigma$. The remaining surface has not slipped. If the applied traction τ_{applied} is greater than $\mu_d \sigma$ there is a stress concentration at the boundary between the slipped and unslipped regions. In fact, according to linear elasticity, the stress τ is unbounded at the point adjacent to the slipped region. This violates the assumption of no slip (since then $\tau > \mu_s \sigma$). Thus, no point can be adjacent to a slipping region. If any region slips and if $\tau_{\text{applied}} > \mu_d \sigma$ then the whole body slips. Since at least some regions of slip (imperfections) are expected on the boundary of contacting solids, the static–dynamic friction law paradoxically predicts that macroscopically the friction stress τ can never exceed the dynamic friction $\mu_d \sigma$. In other words, this friction law, if correct could never be observed for slip between elastic solids.

In the language of linear elastic fracture mechanics, as applied to shear fractures (with $\mu_d \sigma$ subtracted by superposition) (Palmer and Rice, 1973), this friction law corresponds to $K_{\text{crit}} = G_{\text{crit}} = 0$ (K_{crit} and G_{crit} are the critical stress intensity factor and energy release rate, respectively).

Both of the problems named above depend on the fact that the distance D in Figure 9.4 was neglected since it is small. In some applications, however, the relative magnitude of D to some other length is important. In the case of the spring block model D can only be approximated as zero if it is small compared to the spring relaxation distance for the relevant force drop. On the boundary of the elastic solid D can be taken as zero if it is small compared to the elastic displacement $L\Delta\tau/G$, where G is the elastic shear modulus and L is the length of the slipped region. (This is equivalent to the fracture process zone being a small

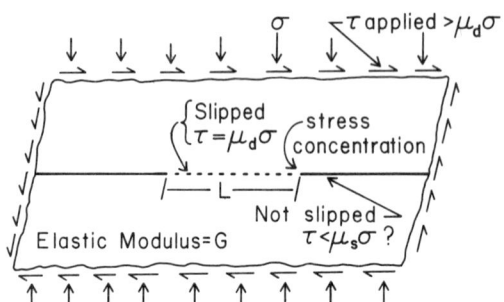

Figure 9.6 The stress concentration at the edge of a slipped region in an elastic solid. The far field applied stress is τ_{applied}. The stress concentration at the boundary of the slipped region with length L contradicts, with 'static–dynamic' friction, the assumption that τ is finite

fraction of L. If one wants to take $D = 0$ strictly, without compensation by inclusion of an equivalent K_{crit}, then the statement $K_{\text{crit}} = 0$ implies the further restriction $D \ll (\tau_{\text{applied}} - \sigma\mu_d)^2 L/G\Delta\tau$.) As noted, however, treating D as zero here predicts no macroscopic static friction.

Thus, the static–dynamic friction model might only be sensible in spring-block systems if D is small compared to block motions but large compared to block deformations.

Even if the above condition is met two other shortcomings are possible: (1) the static friction may not just depend on time of stationary contact or load rate but on the history of load during the nominally stationary contact, and (2) dynamic friction is, in general, not exactly a constant but varies with slip rate.

9.3.2 Slip rate dependent friction

Assuming that friction is a function of slip rate alone, $\tau = f(V)$ (with f roughly proportional to σ), one can predict some aspects of the transient response in Figure 9.4. Figure 9.7 illustrates such a friction relation. In this view the static friction coefficient μ_s is $f(0)/\sigma$. At the start of sliding, transient changes in stress τ like those in Figure 9.4 can be predicted. As the slip rate accelerates from zero the friction force drops (assuming $df/dV < 0$). If the major part of this drop occurs at very low slip rates the static–dynamic approximation may be valid.

However, a mechanical stability analysis predicts that if $df/dV < 0$, steady sliding is unstable in an ideal testing machine and thus is not possible. This prediction contradicts the common observation of steady sliding between surfaces for which the friction traction is a decreasing function of steady slip rate (e.g. Dieterich, 1979a).

Thus, again we reach a paradox. If friction is a decreasing function of rate alone it cannot be observed in steady sliding.

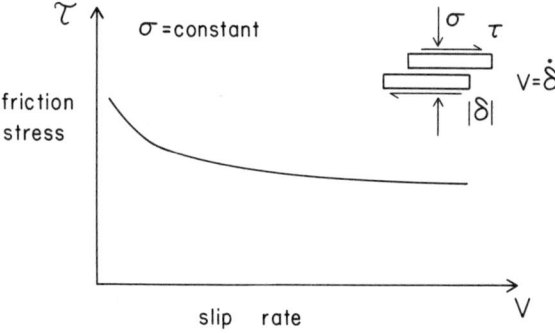

Figure 9.7 Slip rate dependent friction

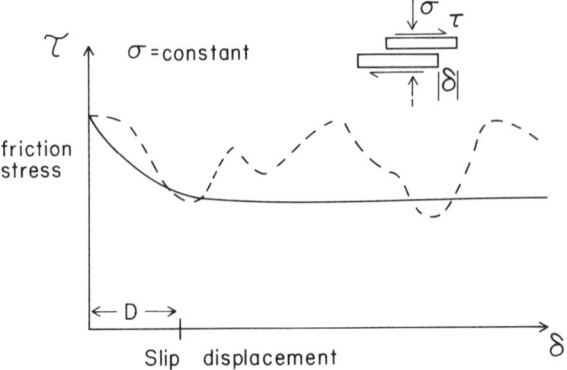

Figure 9.8 Slip displacement dependent friction. The solid line shows τ monotonically decreasing with characteristic slip displacement D. The dashed line shows that τ may vary from point to point on the slip surface

9.3.3 Slip displacement dependent friction

The solid curve in Figure 9.8 shows friction as a function of displacement starting from stationary (still) contact, $\tau = f(\delta)$. This function could be used as a constitutive relation. Unlike the previous two laws this friction law is not self-contradictory. It leads to the prediction that first slip is conditionally stable depending on the relative magnitude of the slope $-df/d\delta$ to the machine stiffness.

A slip displacement constitutive relation like that pictured in Figure 9.8 is deficient in that (1) it does not include observed time or rate effects, and (2) in order to produce more than one transient response it must be artificially reset when slip stops.

Another slip displacement dependent friction law is that $\tau = f(\delta)$, rather than monotonically decaying from first slip, varies from point to point (Byerlee, 1970) as in the dashed line in Figure 9.8. This law can predict instabilities but not steady sliding.

9.3.4 General combinations of classical constitutive relations

The friction relations just discussed can be combined to resolve some of the difficulties. For example, one could construct a friction law with a load history dependent static friction that decays to a rate dependent dynamic level after slip has progressed a distance D. When slip stops the relation is started again. The constitutive law just mentioned would have to be somewhat arbitrary if expressed in detail. Further, it still does not resolve the paradox that steady slip is observed in stiff machines for surfaces in which the friction traction is decreased by an increase in the steady sliding rate.

Constitutive Relations for Frictional Slip

The source of the problems with the classical laws is the central role of a particular variable, special to an experiment or plot, in the constitutive relations. All of the classical relations assume a form $\tau = f(x)$ where x is an experimental parameter (time, load rate, slip rate, or slip displacement). A more general approach, begun in the description of metal friction (e.g. Rabinowicz, 1958, 1959), is to assume that the friction traction depends on a variable θ that evolves with slip or time in such a way that the effects in different particular experiments can be predicted. The development of the idea of an internal state variable (or variables) is the subject of the remainder of this presentation.

9.4 STATE VARIABLE CONSTITUTIVE LAWS

The idea of characterizing rock friction by an evolving variable θ, where θ is a measure of the quality of surface contact, is due to Dieterich (1978, 1979a) and has been generalized in Ruina (1980, 1983) and Rice and Ruina (1983). The development of the state variable constitutive relations began as an attempt to unite the three classical laws just discussed. The original idea expressed by Dieterich (1978) which is quite similar to ideas of Rabinowicz (1958) was the following: slip at the steady speed d_c/t is equivalent to still contact (no slip) for time t. In this equivalence d_c is a typical micro-asperity size and the equivalence is attributed to the fact that in both cases the contact time of the asperity is the same. Dieterich defined a variable θ ('average contact time') which is equal to time when no slip occurs and to d_c/V during steady slip. The friction constitutive relation was then written $\tau = f(\theta)$. Later Dieterich (1979a) noted that the friction traction depended on slip rate as well as on the internal variable θ. He also created an evolution law for the determination of θ during slip histories that involved varying slip rates. The evolution law was, however, technically defective in that the internal variable θ it calculates is not equal to time for stationary contact (Ruina, 1980, 1983). Discovery of this error partially motivated study of the general nature of internal variable constitutive laws for friction. The idea has now been abstracted so that the quality of contact variable θ can be defined without regard to a particular experiment or microscopic interpretation. The state variable constitutive relations are phenomenological and at present only have tentative qualitative microscopic justification.

9.4.1 Steady sliding and steady state

In the state variable constitutive relations that have been proposed by Dieterich and Ruina the concept of steady sliding has a central role. An arbitrary slip history is viewed as a perturbation, possibly a very large perturbation, from steady sliding. In practice the surface may never undergo steady slip, but the constitutive relation is based on the possibility of steady sliding if the slip surfaces were in a mechanically stiff environment. Associated with steady slip at a

180   *Mechanics of Geomaterials*

constant rate is a constant surface state (which is different for different rates). In the development below, the state variable constitutive relations are motivated by experiments where sliding is steady or perturbed from being steady. Application of these relations to non-steady slip then can return features that the classical descriptions aimed to explain, but without self-contradiction.

9.4.2 Basic phenomenology—step changes in slip rate

Experiments of Dieterich (1978, 1979a, 1980, 1981), Ruina (1980), Teufel (1980) and Johnson (1981) are simplified and idealized in Figure 9.9. Effects associated with first sliding (the dashed lines in Figure 9.4) are neglected. These experiments were conducted at slip rates between about 0.01 and 100 μm/s. An infinitely stiff testing machine with no inertia is imagined in this idealization. The features pictured are closely approximated in stiff servo-controlled testing machines.

Figure 9.9 shows the surface response to step changes in slip rate. After steady slip at rate V_1, the slip rate is suddenly increased to V_2, held constant, and then suddenly returned to V_1. Figure 9.9(a) shows the imposed slip rate as a function of slip displacement. Figure 9.9(b) shows the consequent friction traction as a function of slip displacement. Figure 9.9(c) shows the friction traction plotted against slip rate. Different parts of the experiment are marked with the letters A to G. The solid curves represent steady sliding, the dashed curves the immediate response to step changes in slip rate, and the dotted curves the relaxation to the new steady sliding.

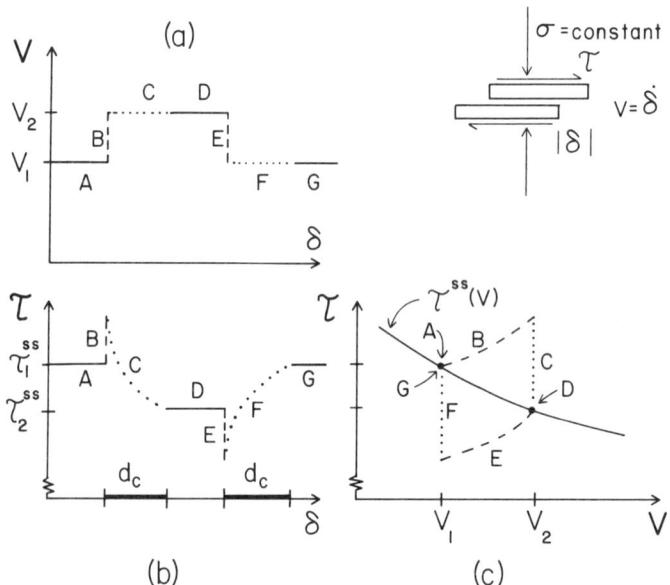

Figure 9.9   Idealized response to step changes in slip rate

At A the slip rate is constant at V_1 as it has been for some time. Consequently the friction traction is the constant τ_1^{ss}. At D the slip rate has been constant at the new rate V_2 for some time and the friction traction has the steady value τ_2^{ss}. At G the slip rate is again V_1 and the slip is again τ_1^{ss}. One could imagine steady slip at a range of rates. At each rate V there is a corresponding $\tau^{ss}(V)$. Points A, D, and G lie on the curve $\tau^{ss}(V)$, represented as a solid line in Figure 9.9(c). The curve $\tau^{ss}(V)$ is not an adequate constitutive relation, however. It would predict that steady sliding could not be observed if, as in the figure, $d\tau^{ss}(V)/dV < 0$ (as mentioned earlier for $\tau = f(V)$). Also, the curve $\tau^{ss}(V)$ does not describe the transition from one steady slip rate to another.

The dashed curve B shows the instantaneous response to a sudden increase in slip rate. Ideally, the curve is traversed in zero time. The dotted curve C shows the relaxation to the new steady state. This relaxation occurs with characteristic distance d_c. In this illustration the relaxation is greater than the initial instantaneous response. The relaxation being greater than the instantaneous response produces overall slip rate weakening ($d\tau^{ss}(V)/dV < 0$). This is not a general feature. The dashed curve E shows the response to a sudden decrease in slip rate. The subsequent relaxation to steady state, the dotted curve F, again occurs with roughly the same characteristic distance d_c.

Some further approximate features of the response to step changes in slip rate, that are not clearly shown in Figure 9.9 are that: (1) the transient changes in τ are on the order of $0.01\sigma \ln(V_2/V_1)$, and (2) the relaxation of τ to steady state seems well approximated by an exponential (or sum of two exponentials) function of slip displacement.

The data in Figure 9.9 (or equivalent experiments) shows the variation of friction stress for step changes in slip rate. The question remains as to how to use this data to determine the relationship between friction stress τ and slip displacement δ for more complex slip histories. In the remaining presentation curves like those in Figure 9.9 are used to relate τ and δ during arbitrary slip histories: (1) qualitatively, (2) exactly for nearly steady sliding, (3) approximately using one state variable, (4) more accurately using two state variables, and (5) exactly using an impracticable general functional relation.

9.4.3 Qualitative description

Figure 9.9 shows that the friction traction τ is governed by two competing effects. The dashed curves B and E show that, for instantaneous changes in slip rate, τ has a positive viscous-like dependence on slip rate. This instantaneous positive rate dependence competes with the relaxation shown in the dotted relaxation curves C and F. The relaxation may be viewed as a non-instantaneous negative dependence on slip rate. The stress jumps up at B due to the instantaneous positive rate dependence and then slowly drops as the non-instantaneous negative rate dependence takes effect.

One may infer that these two effects are due to a general qualitative microscopic model. The various microscopic deformation and failure mechanisms have a positive stress dependence on rate. The inherent surface strength, or overall 'quality' of surface contact, depends negatively on the rate of sustained slip. For instantaneous changes in slip rate the surface 'quality' is unchanged and the positive rate dependence of the deformation and failure mechanisms is visible. As slip progresses the 'quality' evolves towards a value that is less for fast slip than for slow slip.

More specifically, Dieterich (1978, 1979a) has proposed that the 'quality' be measured by the total area of asperity contacts. This area, in turn, can be connected directly with average time of asperity contact since, in Dieterich's view, time dependent asperity creep causes the area increase. Tolstoi (1967) has indirectly asserted that surface separation displacement is a measure of the 'quality' of surface contact. The 'quality' of surface contact might also be reduced by surface contaminants adsorbed between contacting asperities, where the amount of trapped contaminant may depend on the past macroscopic slip rates. Detailed microscopic models are lacking.

In any case, it is evident from Figure 9.9 that sudden changes to slow slip rates cause the surface 'quality' to increase. This increase corresponds roughly to a healing process that may, in some limit, be equivalent to time dependent static friction.

In order to infer a quantitative constitutive law from data of the type in Figure 9.9 further assumptions are required. The approach in the next section is to only consider small deviations from constant rate sliding. The approach in the subsequent section is to make assumptions about the quantification of the surface 'quality' mentioned above as a state variable.

9.4.4 Nearly steady sliding

In general, a linear system can be characterized by its response to special input functions. If attention is limited to nearly steady sliding the frictional constitutive relation can be viewed as a linear system. The 'input' function is the variation in the slip rate from a given constant rate V_1. The 'output' function is the variation in the friction stress from the steady value associated with V_1, $\tau^{ss}(V_1)$. If, in Figure 9.9, V_2 is only slightly greater than V_1 the curves A to D in Figure 9.9(b) represent the response of the friction surface to a step function input.

Any function can be written as a superposition of step functions. Similarly the response of a linear system to an arbitrary function can be written as a superposition of step responses. Thus, for small variations from steady sliding, curves A to D in Figure 9.9(b) completely describe the friction law. The result just named is stated more precisely by the superposition integral in Rice and Ruina (1983, see Figure 1 and Equations (6) and (7) there in).

Rice and Ruina (1983) make use of this linearized description of the frictional

Constitutive Relations for Frictional Slip

constitutive relation to analyse the stability of steady slip in a quite general way. One important result of these calculations (generalized from Ruina (1980, 1983)) is that, if the jump B in Figure 9.9(b) is positive, steady slip is possible in a sufficiently stiff machine. The positive instantaneous rate dependence is consistent with all of the previously cited experiments except Dieterich (1978), in which the experimental resolution was relatively low. The results are only applicable to nearly steady sliding, however. Thus the linearized description just mentioned, although quite general for nearly steady sliding, can only suggest the nature of extremely unsteady and thus nonlinear sliding (as in stick–slip instabilities and fault rupture).

9.4.5 The general one state variable constitutive relation

When the sliding is not close to steady the friction law is not linear and cannot be rigorously characterized by the response to a step change in slip rate. The state variable approach assumes that the state of the surface can be characterized by a number θ (or numbers θ_i more generally). The state variable θ evolves with slip depending on the current state of the surface and the slip rate. One then tries to find a state variable law that describes the data. All of the general features in Figure 9.9 can be reproduced by a constitutive law of the following form (Ruina, 1980, 1983), which includes all of the forms proposed by Dieterich (1978, 1979a, 1980, 1981) as special cases:

$$\tau = \sigma F(\theta, V) \tag{9.1a}$$

$$d\theta/dt = G(\theta, V) \tag{9.1b}$$

In this general form, the functions F and G need to be specified to make the law specific. The variable θ is the measure of 'quality' of contact and is *defined* by the evolution law (9.1b).

Equation (9.1a) states that the friction stress depends on the current value of θ and on the slip rate. Certain restrictions on the functions F and G are required to maintain consistency with Figure 9.9. In order for the jumps (dashed curves B and E in Figure 9.9) in τ to have the right sign the function F must be an increasing function of V. In order for the steady state to be approached, the function G must be such that, at constant V, θ approaches a value $\theta^{ss}(V)$ (Ruina, 1980, 1983).

Figure 9.9 can now be interpreted in terms of Equation (9.1). At a sliding is steady. V and τ are constant and θ must be constant also. Thus at point A, $d\theta/dt = G(\theta, V) = 0$. At B, when the slip rate is suddenly increased, τ jumps because of Equation (9.1a). θ does not jump since it is assumed that the function G is always finite. Thus the dashed curve in Figure 9.9(c) is a curve of constant θ. Similarly, the dashed curve E in Figure 9.9(c) is a curve of constant θ. One can imagine sudden jumps in slip rate starting at many different slip rates. Corresponding to each of these jumps a dashed curve in Figure 9.9(c) could be drawn. Each curve would be a plot of $\tau = F(\theta, V)$ for a fixed θ.

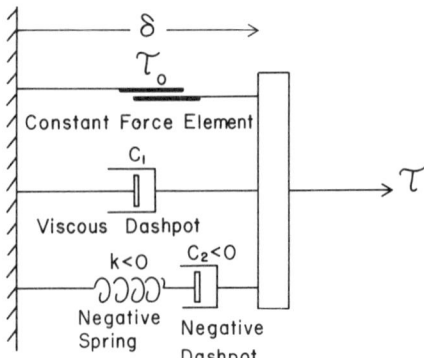

Figure 9.10 Phenomenological model of the one state variable friction law for nearly constant rate sliding. The model has a perfectly plastic element, a positive dashpot, a negative dashpot, and a negative spring

The dotted curve C is traversed at constant V. Thus, on curve C the change in τ is due entirely to change in θ. The dotted curve C shows θ changing in accordance with the differential Equation (9.1b). θ is tending towards its steady state value $\theta^{ss}(V)$. The characteristic distance d_c in Figure 9.9(b) is thus a property of the function $G(\theta, V)$.

Curves E and F can be explained similarly.

The form (9.1) is consistent with Figure 9.9, even if the step changes in slip rate are large. Equation (9.1) can be applied when the slip rate V is an arbitrary function of time. Thus Equation (9.1) serves as a constitutive relation for slip. It is not the most general form, however, and it is possible to show that experimental data consistant with Figure 9.9 may be impossible to describe with the form (9.1).

For nearly steady sliding, the friction law in Equation (9.1) can be described with the phenomenological model shown in Figure 9.10. The three branches of the model are: (1) the constant τ_0 (a perfectly plastic element), (2) a viscous dashpot, and (3) a negatively viscous dashpot (a dashpot that aids rather than damps motion) in series with a spring with a negative stiffness. The second branch causes the instantaneous rate dependence and the third branch causes the slower acting, but negative rate dependence.

9.4.6 Simplified Dieterich law—simple one state variable relation

Some greater insight may be provided by examining an example of Equation (9.1) with the functions $F(\theta, V)$ and $G(\theta, V)$ specified exactly. The following relations are proposed in Ruina (1980, 1983) to approximate some of the laws proposed by

Dieterich. They are consistent with all of the comments made above with regard to Figure 9.9.

$$\tau = \sigma(\mu_0 + \theta + A\ln(V/V_c)) \tag{9.2a}$$

$$d\theta/dt = -(V/d_c)(\theta + B\ln(V/V_c)) \tag{9.2b}$$

In these equations μ_0, A, B, d_c and V_c are material constants. The constant μ_0 is roughly 0.5 and gives the approximate coefficient of friction, A and B determine the rate sensitivity and are of the order of 0.01 in fits to experiments, d_c is the characteristic distance in the relaxation to steady state and ranges from 1 μm to 1 mm in experimental fits, and V_c is an arbitrary constant that is included for dimensional consistency. If the adjoining solid has linear deformation behaviour, the constant μ_0 has no effect on slip stability. Thus in this simple friction relation there are three relevant constants: A, B, and d_c.

Equation (9.2) can be explained as follows: the friction coefficient depends additively on a constant μ_0, the quality of contact variable θ, and an increasing function of the slip rate. Thus, for a given value of θ (a fixed surface state) μ depends only on $\ln(V/V_c)$. Use of the natural log function obviously is not precise since it predicts negative friction at sufficiently slow rates. However, since the constant A is of the order of 0.01 this paradox only appears at slip rates well below those that have been checked by experiment (1 angstrom per 10,000 years). The variable θ evolves as governed by Equation (9.2b). If θ is greater than $-B\ln(V/V_c)$ then $d\theta/dt$ is negative (note that $B\ln(V/V_c)$ can be negative). If θ is less than $-B\ln(V/V_c)$ then $d\theta/dt$ is positive. Thus θ is always chasing the value $-B\ln(V/V_c)$. At constant slip rate Equation (9.2b) predicts that θ exponentially approaches $-B\ln(V/V_c)$ with the characteristic distance d_c.

For steady sliding at speed V Equation (9.2) becomes

$$\tau^{ss}(V) = \sigma(\mu_0 + (A-B)\ln(V/V_c)) \tag{9.3}$$

For steady sliding to be a decreasing function of slip rate $B > A$. Equation (9.3) again appears objectionable because it predicts negative friction at high slip rates. Again, this occurs well beyond experimental observations (one million times the speed of light) and is not a fundamental difficulty.

Equation (9.2b) predicts that the state variable θ does not evolve in the limit of zero slip speed. However, an effect approximating time dependent static friction is calculated in computer simulations using Equation (9.2) since θ increases for small but non-zero slip rates.

Equation (9.2) is only intended as a simple model that incorporates approximately many of the experimentally observed frictional responses. In differs from experimental observations in two possibly important ways: (1) experiments of both Dieterich (1979a) and Teufel (1980) show that $\tau^{ss}(V)$ does not continue to decrease with V at high V, and (2) Ruina (1980, 1983) showed that at least two state variables were needed to fit his experiments.

9.4.7 Two state variables

Ruina (1980, 1983) was required to use two state variables to describe his experimental data well. The constitutive relation he used was Equation (9.2) with an additional state variable and corresponding evolution law. The use of two state variables instead of one has the interesting consequence for spring-block modelling that apparently aperiodic motions can be predicted (Gu et al., 1984; Ruina 1983).

9.4.8 General functional form of the constitutive relations

One may imagine a list of constitutive relations ranging from the general and inapplicable to the specific and oversimplified. In rough order one may consider the following possible constitutive relations:

$$\tau(t) = \hat{\mathcal{F}}[\sigma(\text{all past } t), V(\text{all past } t)] \tag{9.4a}$$

$$\tau(t) = \sigma \mathcal{F}[V(\text{all past } t)] \tag{9.4b}$$

$$\tau(t) = \sigma F[V(t), \text{state}] \tag{9.4c}$$

$$\tau(t) = \sigma F(V(t), \theta_1, \theta_2 \ldots) \tag{9.4d}$$

where

$$d\theta_i/dt = G_i(V(t), \theta_1, \theta_2 \ldots) \quad i = 1, 2 \ldots$$

$$\tau(t) = \sigma F(\theta, V) \quad \text{where } d\theta/dt = G(\theta, V) \tag{9.4e}$$

$$\tau(t) = \text{classical } t, V, \text{ or } \delta \text{ dependence} \tag{9.4f}$$

$$\tau = \mu\sigma \quad \text{where } \mu = \text{constant} \tag{9.4g}$$

Equation (9.4a) shows the general functional dependence of τ on all past V and σ. Equation (9.4b) limits attention to constant normal stress and is the form presented in Rice and Ruina (1983). Equation (9.4c) incorporates the history of slip in the state which is described by state variables in (9.4d). Equation (9.4d) also includes evolution laws for the change of the state variables with time. Equation (9.4e) uses only one state variable. Equation (9.4f) refers to the classical constitutive laws discussed earlier and (9.4g) is Coulomb's law.

In general one should stay as close to the bottom of this list as application allows. However, this presentation has attempted to demonstrate that there may be good reasons to move up towards the middle of the list in order to obtain better understanding of some instability phenomena.

9.5 SUMMARY

The subject of constitutive relations for frictional slip has been presented with particular emphasis on relations useful for stability analysis. In particular, the state variable laws first introduced by Dieterich have been described. For more detailed treatment of the development of these laws the reader should consult the references.

Many open questions remain. The description presented here still does not describe many observations (unmentioned in this presentation) and is not well understood microscopically. Nonetheless, it is hoped that relations of the general type presented here will, by means of mechanical models, contribute further to the understanding of slip instabilities.

ACKNOWLEDGEMENTS

Jim Rice and Turon Onat for informative discussions. Diana Cox, Frank Horowitz, Ebby Askariaman, Les Schaffer and Claude Froidevaux for editorial comments; Cheryl Nickels and Eniko Farkas for drafting; Zdenek Bažant and Mary Hill for patience; the Department of Interior USGS Earthquake Prediction Program for funding.

REFERENCES

Barton, Nick (1976) 'The shear strength of rock and rock joints', *Int. J. Mech. Min. Sci. β Geomech. Abstr.*, **13**, 255–279.

Barton, Nick (1981) 'Some size dependent properties of joints and faults', *Geophysical Research Letters*, **8**, 667–670.

Byerlee, J. D. (1970) 'The mechanics of stick-slip', *Tectonophysics*, **9**, 475–486.

Byerlee, J. (1978) 'Friction of Rocks', *Pure Appl. Geoph.*, **116**, 615–626.

Dieterich, J. H. (1972) 'Time-dependent friction in rocks', *J. Geophys. Res.*, **77**, 3690–3697.

Dieterich, J. H. (1978) 'Time-dependent friction and the mechanics of stick slip', *Pure. Appl. Geophys.*, **116**, 790–806.

Dieterich, J. H. (1979a) 'Modeling of rock friction, 1, Experimental results and constitutive equations', *J. Geophys. Res.*, **84**, 2161–2168.

Dieterich, J. H. (1979b) 'Modeling of rock friction, 2, Simulation of preseismic slip', *J. Geophys. Res.*, **84**, 2169–2175.

Dieterich, J. H. (1980) 'Experimental and model study of fault constitutive properties', in *Solid Earth Geophysics and Geotechnology* (edited by S. Nemet-Nasser), American Society of Mechanical Engineers, New York, pp. 21–30.

Dieterich, J. H. (1981) 'Constitutive properties of faults with simulated gouge', in *Mechanical Behavior of Crystal Rocks, Geophys. Monogr.* 24 (edited by N. L. Carter, M. Friedman, J. M. Logan, and D. W. Stearns), AGU, Washington, D.C., pp. 103–120.

Gu, Ji-Chang, Rice, J. R., Ruina, A. L., and Tse, S. T. (1984) 'Slip motion and stability of a single degree of freedom elastic system with rate and state dependent friction, *J. Mech. Phys. Sol.*, **32**, 167–196, 1984.

Johannes, V. I., Green, M. A., and Brockely, C. A. (1973) 'The role of the rate of application of the tangential force in determining the static friction coefficient', *Wear*, **24**, 381–385.

Johnson, T. (1981) 'Time dependent friction of granite: Implications for precursory slip on faults', *J. Geophys. Res.*, **86**, 6017–6028.

Kosloff, D. D., and Liu, H.-P. (1980) 'Reformulation and discussion of mechanical behavior of the velocity-dependent friction law proposed by Dieterich', *Geophys. Res. Lett.*, **7**, 913–916.

Mavko, G. M. (1980) 'Simulation of creep events and earthquakes on a spatially variable model', *Eos Trans. AGU*, **61**, 1120.

Mavko, G. M. (1984) 'Large-scale earthquakes from a laboratory friction law', *J. Geophys. Res.* in press.

Palmer, A. C., and Rice, J. R. (1973) 'The growth of slip surfaces in the progressive failure of over-consolidated clay', *Proc. Roy. Soc. Lond. A.*, **332**, 527–548.

Rabinowicz, E. (1958) 'The intrinsic variables affecting the stick–slip process', *Proc. Phys. Soc. London*, **71**, 668–675.

Rabinowicz, E. (1959) 'A study of the stick–slip process', in *Friction and Wear*, (edited by Davies), Elsevier, New York, pp. 149–164.

Rice, J. R., and Ruina, A. L. (1983) 'Stability of steady frictional sliding', *J. Appl. Mech.*, **50**, 343–349.

Ruina, A. L. (1980) 'Friction laws and instabilities: A quasistatic analysis of some dry frictional behavior', PhD thesis, Brown University, Providence, RI.

Ruina, A. L. (1983) 'Slip instability and state variable friction laws', *J. Geoph. Res.*, **88**, B12, 10359–10370.

Teufel, L. W. (1981) 'Frictional instabilities in rock: Effect of stiffness, normal stress, sliding velocity and rock type', paper presented at the 18th Annual Meeting, Soc. for Eng. Sci., Brown University, Providence, RI, 1981 (also in review for *Geoph. Res. Lett.*)

Tolstoi, D. M. (1967) 'Significance of the normal degree of freedom and natural normal vibrations in contact friction', *Wear*, **10**, 199–213.

Mechanics of Geomaterials
Edited by Z. Bažant
© 1985 John Wiley & Sons Ltd

Chapter 10

Shear Localization in Rocks Induced by Tectonic Deformation

B. Evans and T. -F. Wong

10.1 INTRODUCTION

Shear localization is observed in geologic materials on scales ranging from the size of thin sections up to zones that are hundreds of kilometres long and perhaps 30 km or so in width. Geologic deformation occurs over a broad spectrum of temperature, pressure, and time scales. At one end of the spectrum, localization results in shear bands commonly referred to as 'faults', in which case the onset of localization occurs with relatively little inelastic strain (typically not more than a few per cent) and insignificant dislocation activity. On the other hand, 'shear zones' are quite common in geologic settings for which deformation is believed to have occurred at temperature and pressure high enough for crystal plasticity and possibly steady state creep processes to be operative over the time scale involved.

It is well known that predictions from localization analyses are very sensitive to the details of the constitutive equations used. In general the inelastic behaviour of rock is expected to be pressure-, temperature-, and rate-sensitive with varying degrees of strain softening or hardening under the pressure and temperature range of geophysical interest (Figure 10.1). Therefore, relevant instability analysis can be carried out only for situations for which input on mechanical behaviour from detailed laboratory and field studies is available.

Since geologic materials are so rich in examples of strain localization, we will not attempt to present a comprehensive review of the whole subject. The transition from the brittle to the ductile field is extremely complex (Carter and Kirby, 1978; Paterson, 1979). While a localization analysis in this context is well worth study, our current understanding of rock rheology in the transitional regime is too limited for developing a meaningful theoretical framework. We will therefore limit our scope and focus on two end members: brittle faulting where the deformation is highly pressure-sensitive and relatively rate-insensitive, and shear zone formation where the deformation is highly rate-sensitive and basically pressure-insensitive. Following current rock mechanics usage, the latter mode of deformation will be referred to as 'fully plastic.'

Recent developments in theoretical, field, and laboratory work have allowed

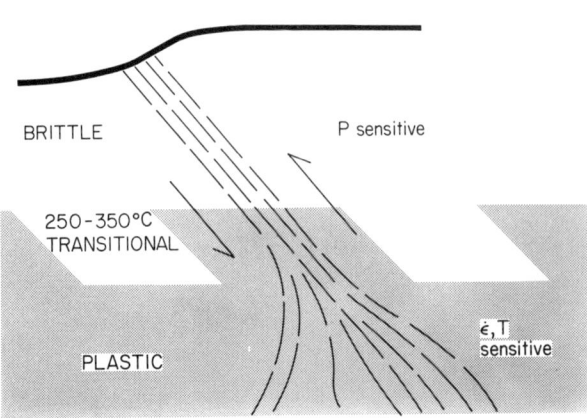

Figure 10.1 Schematic drawing of a two-layer model of a large fault zone (after Sibson)

new insights into the brittle faulting process. An active research area in continuum mechanics is the analysis of shear localization as a bifurcation phenomenon. Such an approach, for a rate-insensitive, but pressure-sensitive and dilatant material, has provided a unified framework for the continuum description of faulting (Rudnicki and Rice, 1975). On a microscopic scale, systematic electron microscopy studies (Wong, 1982) illustrate the complexity of the micromechanical processes and provide a comparison with the continuum description. At the same time, detailed mapping of geologic faults (Segall and Pollard, 1983; Aydin and Johnson, 1983) and seismological study of mining-induced earthquakes (McGarr et al., 1979) highlights some difficulties regarding the scaling question. The presence of water has significant effects on the brittle behaviour of rocks. This subject is discussed in another presentation (Rudnicki, this volume, Ch. 15); we will focus our discussion on brittle faulting in dry rocks.

An analysis of flow instability has to take into account the strain-rate sensitivity, activation enthalpy, strain hardening (or softening), and adiabatic heating (Argon, 1973; Jonas, et al., 1976). One major advance in rock rheology study in the past decade has been in the careful determination of the flow laws for steady state creep of materials thought to be constituents of the upper mantle and lower crust (Goetze, 1978; Kirby, 1980). Such experimentally determined constitutive relations have provided the essential input parameters for basically all the theoretical analyses of ductile shear zone formation. Mechanisms proposed include 'thermal runaway' (Yuen et al., 1978; Fleitout and Froidevaux, 1980), 'plastic instability' (Poirier et al., 1979), and 'chemical weakening' (Sorensen, 1983).

Increased understanding has been gained through the extensive use of transmission electron microscopy and chemical microanalysis techniques in the study of deformation-induced microstructures. Systematic sampling in the field and comparison of the microstructures in field samples and laboratory specimens have provided important constraints on the operative mechanisms (Christie and Ord, 1980; Kohlstedt and Weathers, 1980; White et al., 1980).

In a certain sense, the analysis of shear localization for geologic materials proceeds along lines similar to earlier developments in applied mechanics for metal, polymer, and metallic glass. This parallel development does not appear to be broadly appreciated nor fully exploited. We will try to emphasize such parallel developments where appropriate in the following discussion.

10.2 LOCALIZATION ANALYSIS FOR FRICTIONAL, DILATANT MATERIALS

When loaded under compressive stresses in the brittle field, a rock fails by development of a localized shear band. This process, commonly called 'faulting' in a geologic context has been extensively studied, and a comprehensive review was recently given by Paterson (1978). We will first summarize recent theoretical results from bifurcation analyses of constitutive relations proposed for pressure-sensitive, dilatant materials and then compare them with experimental observations.

The theory of localization of plastic deformation was reviewed by Rice (1976), and results pertinent to geologic materials such as rock and soil were summarized by Cleary and Rudnicki (1976). Since then, there have been several papers proposing constitutive relations for rock-like materials (e.g. Bažant and Kim, 1979; Nemat-Nasser and Shokooh, 1980). Most of these involve pressure-dependent yielding and dilatancy but otherwise are similar in spirit to the flow theory of metal plasticity. We will focus on the following relation proposed by Rudnicki and Rice (1975) for brittle rock since it has the essential features of this class of constitutive relations with the least complication:

$$D_{ij} = \frac{1}{2G}\left(\overset{\triangledown}{\sigma}_{ij} - \frac{v}{1-v}\overset{\triangledown}{\sigma}_{kk}\delta_{ij}\right) + \frac{1}{h}P_{ij}Q_{kl}\overset{\triangledown}{\sigma}_{kl} \qquad (10.1)$$

with

$$P_{ij} = \sigma'_{ij}/2\tau + \beta/3\delta_{ij}$$

$$Q_{ij} = \sigma'_{ij}/2\tau + \mu/3\delta_{ij}$$

where D_{ij} is the symmetric part of the rate of deformation, and $\overset{\triangledown}{\sigma}_{ij}$ is the spin-invariant Jaumann rate of Cauchy stress. The first term in the bracket represents the usual Hookean behaviour, whereas the second part is the inelastic contribution. σ'_{ij} is deviatoric part of the stress tensor, and τ is the square root of the second stress invariant J_2. Note that the inelastic deformation is described by three parameters: a hardening modulus h, a dilatancy parameter β, and a frictional parameter μ.

The deformation is therefore rate-insensitive, and the relation reduces to the classical Prandtl–Reuss relation when the pressure dependence of yielding and dilatancy are insignificant. Rudnicki and Rice (1975) argued on physical grounds that subsequent yield surfaces should have a vertex structure. Non-normality coupled with yield surface vertices render the analysis more involved than that in classical metal plasticity. Recently, Needleman (1979) presented a detailed plane-strain analysis of the incompressible case. Rudnicki and Rice (1975) neglected terms of the order stress divided by shear modulus in their final computation. Vardoulakis (1980) and Anand and Spitzig (1982) recently suggested that such terms can be important in certain situations, and compare predictions from the more involved calculation with experimental data on granular sand.

Fitting stress–strain data for brittle rock, Rudnicki and Rice (1975) estimated μ to range from 0.4 to 0.9 and β from 0.2 to 0.4. For such materials, the bifurcation analysis predicts a localized zone inclined at about $30°$ to σ_1 (the maximum principal compressive stress) for a sample loaded to fracture under axisymmetric compression. This is in general agreement with experimental observations (Paterson, 1978). For a fixed μ and β, the critical value of h at onset of localization is strongly dependent on the loading configuration. In general, localization under plane-strain deformation has to occur at the hardening stage (Anand and Spitzig, 1982). For a wide range of μ and β, the critical h for axisymmetric compression is generally predicted to be very negative (well into strain softening). The result is modified if the influence of yield surface vertex (Rudnicki and Rice, 1975) and stress-induced anisotropy (Rudnicki, 1977) are included.

10.3 MICROMECHANICS OF BRITTLE FAULTING

Earlier experiments with strain gauges (Hadley, 1975), acoustic emission (Lockner and Byerlee, 1980), and holography (Soga et al., 1978) indicate pre-failure localization of strain at a hardening stage. Direct microscopic observation requires samples deformed stably through the post-failure region with displacement control. Using a testing machine of high stiffness and special loading–unloading technique (Wawersik and Brace, 1971), Wong (1982) obtained complete suites of pre- and post-failure samples of Westerly granite for scanning electron microscope (SEM) study.

The SEM observation points out the complexity of the localization process in this relatively isotropic, low porosity (about 1 per cent) rock. Inception of faulting in the sense of localized deformation extending over two or more grains is quite evident in samples stressed to just beyond peak stress. The micromechanical processes leading to the formation of a through going fault include a number of mechanisms dependent on mineralogy and grain orientation. These include three types of geometric instability well known to the material scientists: microbuckling of slender columns in grains segmented by micro-crack arrays (Figure 10.2), with characteristic dimension of a grain size (Evans and Adler, 1978), kinking in biotite grains (Figure 10.3 which have reached the plastic yield stress (Frank and

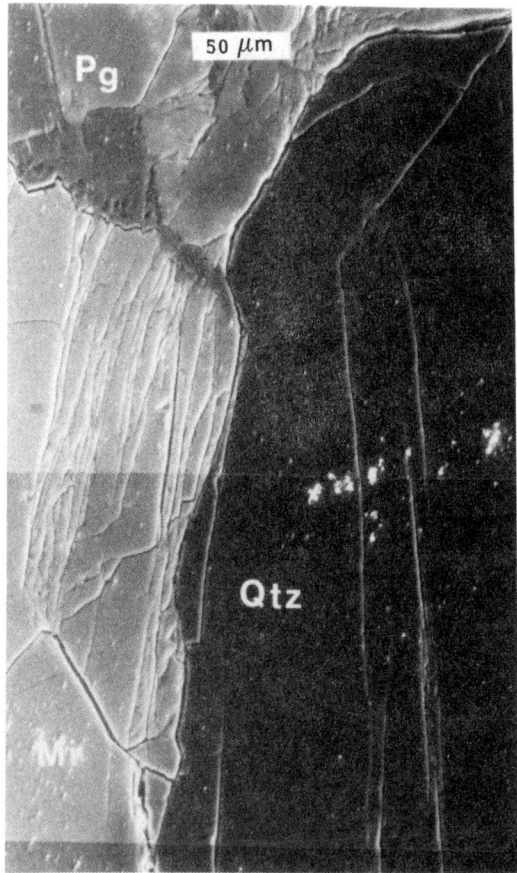

Figure 10.2 Axial crack arrays in microcline (mi) and quartz (qtz) observed in a sample retrieved just after loaded to the peak stress ($\sigma_1 - \sigma_3 = 1.21$ GPa). The confining pressure (σ_3) was 250 MPa and temperature was 150°C. The microcline grain has segmented into slender columns and incipient microbuckling is evident. Mineral at the top is plagioclase (pg). Maximum compression (σ_1) was vertical

Stroh, 1952); rotation and crushing of 'joint blocks' formed by pore-emanated cracks in plagioclase (Figure 10.4), with characteristic dimension dictated by the pore spacing (Goodman, 1976). In addition, shear slip along cracks at high angle to σ_1, favourable for frictional displacement is evident in the post-failure samples (Figure 10.5).

A continuum description such as (10.1) of the brittle behaviour of a complicated, polycrystalline material such as rock is probably adequate over a continuum element large enough for the effects of grain scale inhomogeneity and

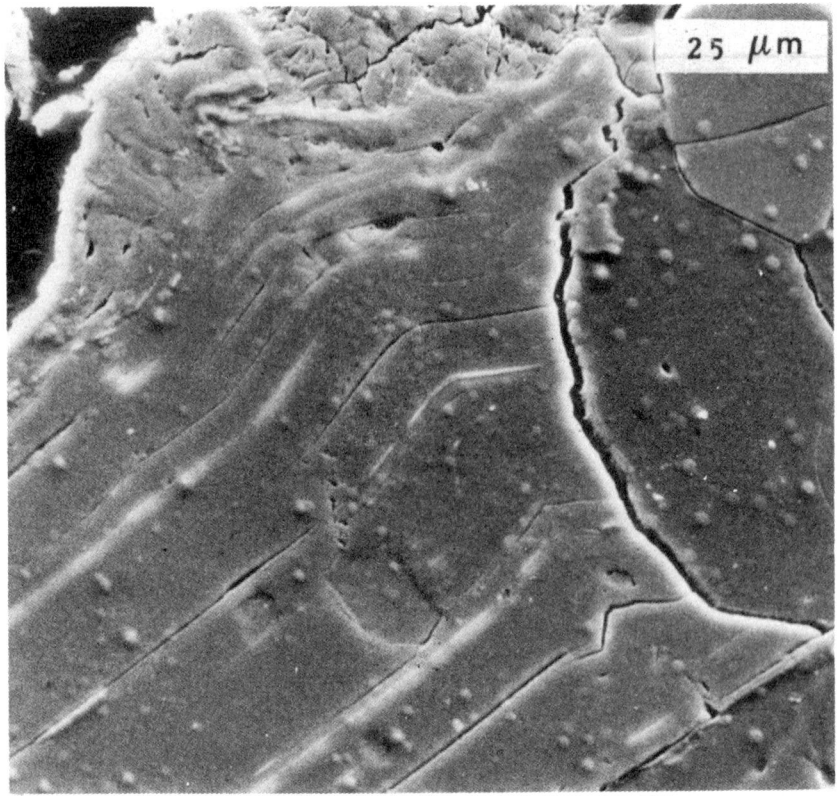

Figure 10.3 Kinking in a biotite grain located right next to the shear band in a post-failure sample deformed at 250 MPa and 150°. Note the cracks along the easy slip planes. Maximum compression was vertical

anisotropy to average out. The SEM observation shows that the mineral quartz, comprising about one-third by volume of the granite, has limited participation in the localization process in the initial post-failure stage. In other words, localized deformation extending over a continuum element with grains of all major mineral types is not observed until the sample has been deformed well into the post-failure stage. In this limiting sense, the SEM observation agrees with the theoretical prediction discussed above.

10.4 PHYSICAL THEORY OF DILATANCY IN BRITTLE ROCK

A number of physical theories have been developed aiming to determine the macroscopic stress–strain behaviour of a brittle rock in terms of the micromechanical processes. A group of models, commonly referred to as 'sliding crack models' postulate that frictional sliding along grain boundaries or cracks inclined

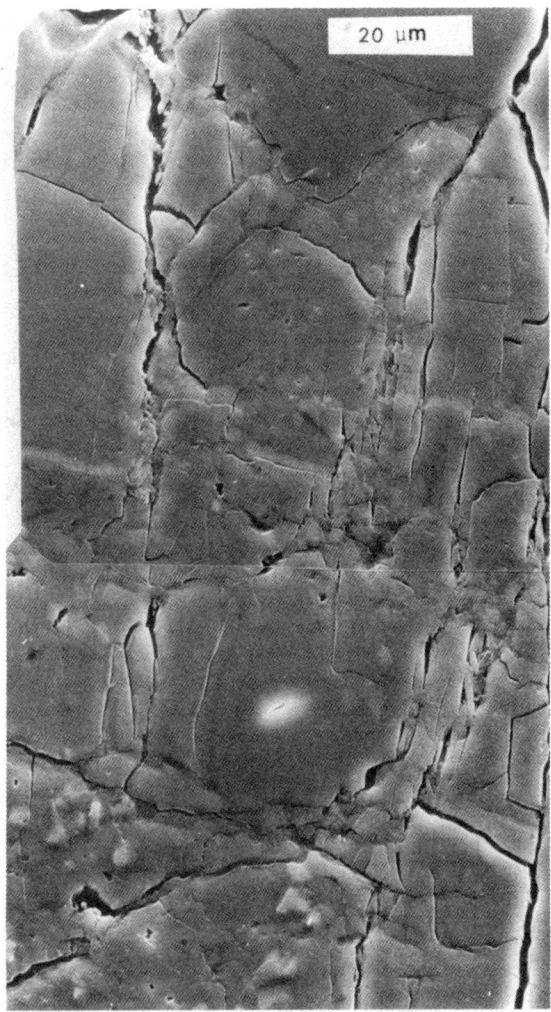

Figure 10.4 Pore emanated cracks in a plagioclase grain in a post-failure sample. The spacing of the pores seems to dictate the characteristic dimension of the 'block' structures which undergo rotation and crushing to accommodate strain localization. Maximum compression was vertical

at high angle to σ_1 pull open other tensile cracks, causing the latter to extend parallel to σ_1 and giving rise to dilatancy (Kachanov, 1982; Moss and Gupta, 1982). Kachanov's approach is in a sense similar to the slip theory of Batdorf and Budiansky (1949). Kachanov shows that there is limited path-independence for stress histories which do not depart too much from proportional loading, and he elaborates on the development of yield surface corners.

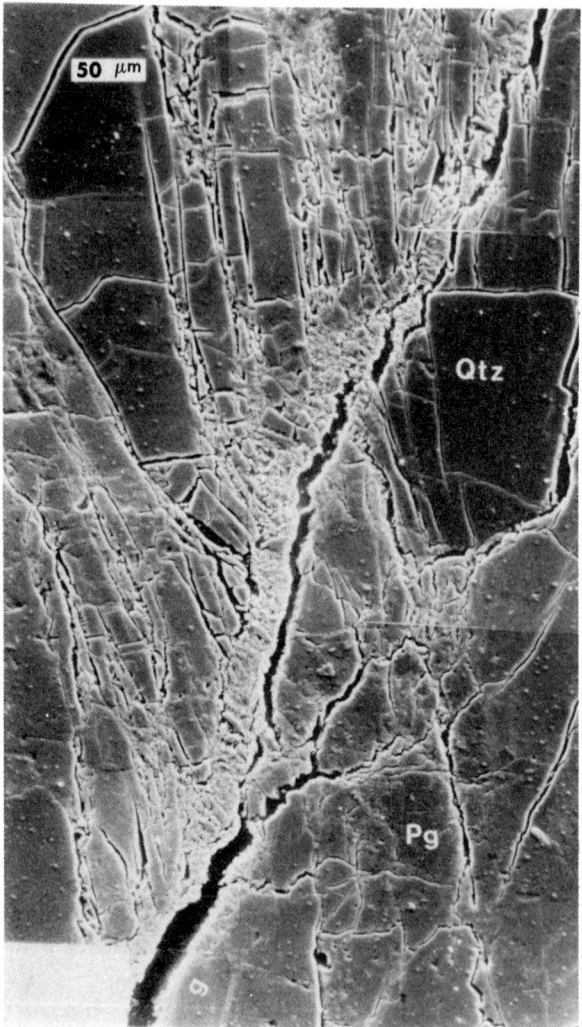

Figure 10.5 Coalescence of a crack array in quartz (qtz) and a crack network in plagioclase (pg) with a grain boundary (g) to form a through going fault. The long axial cracks in quartz outline slender columns that were continuous and have buckled during the instability process, whereas not many well-defined elongated columns can be traced in the plagioclase grain. At high magnification, the longer cracks in the latter can be seen to be joining up with pores. Shear slip along the grain boundary is evident. Maximum compression was vertical

On the other hand, Stevens and Holcomb (1980) argued on the basis of SEM observations and their own study of hysteresis in stress cycling tests that the 'sliding crack' is unrealistic. SEM studies of Tapponier and Brace (1976) concluded that dilatancy is primarily a consequence of two types of cracking: (1) widening and extension of pre-existing discontinuities, such as grain boundaries, cracks, and pores; (2) initiation and propagation of cracks at localities with high contrast in elastic moduli, such as at transverse grain boundaries between different minerals. Recently, Kranz (1979) and Batzle et al. (1980) suggested that geometric irregularities such as asperities at grain boundaries may be important in the initiation of stress-induced cracks.

A conclusion common to all the SEM studies is that the stress-induced cracks are in general 'mode I' tensile cracks. The crack orientation has a highly anisotropic distribution, and is mostly at low angles (say, less than 15°) to σ_1. Neither Tapponier and Brace (1976) nor Kranz (1979) found a significant number of inclined cracks. Although Wong (1982) did observe a number of cracks inclined at high angles to σ_1 in granite samples deformed under pressure and temperature higher than previous work, no appreciable mode II or III deformation along such cracks are evident in the pre-failure samples.

A number of mechanisms have been proposed as alternatives to the 'sliding crack'. Stevens and Holcomb (1980) suggested the 'reversible Griffith crack', and Janach and Guex (1980) proposed the 'shear bubble' model which hinges on the behaviour of interfacial cracks with different elastic properties on either side. The implications of such alternative micromechanical models for phenomenological descriptions have not been quantitatively explored in detail.

There are, however, some recent observations in agreement with the 'sliding crack model'. Focal mechanism study of acoustic emission from brittle rock undergoing dilatancy concludes that many of the emission sources cannot be adequately represented by a tensile crack (Sondergeld and Estey, 1982), strongly suggesting that shear slip has to be involved. Tensile opening of micro-cracks is easily resolved under the SEM. Whereas recognition of appropriate 'strain markers' helps one to identify shear displacement discontinuity in the field, no unambiguous approach is available for identifying shear cracks under the SEM.

10.5 THE SCALING PROBLEM

The size effect is of particular importance in rock mechanics because of the large span in dimension between laboratory samples and rock mass involved in engineering practice or tectonic processes. Most of the data available are in centimetre-size cylindrical specimens deformed in a conventional triaxial configuration. Several laboratories are interested in developing large-scale testing facilties, and the progress was reported in a recent workshop (Cook and Heard, 1981).

The size effect for uniaxial compressive strength has been investigated to an extent. The ratio of laboratory to field strengths for several relatively weak rocks can be as high as 10, and it seems that there is a critical size of about 1 m such that larger specimens suffer no further decrease in strength (Singh, 1981; Brace, 1981). It is fair to say that more thorough studies, both theoretical and experimental, are necessary before more definitive conclusions can be drawn.

A question naturally arises as to the relevance of laboratory results to the understanding of large-scale faulting. There are not many comprehensive field studies of the evolution of brittle faulting. Two recent studies on different scales concluded that the mechanical processes in the field are very similar to the laboratory observation. McGarr et al. (1979) reported that localized deformation due to mining-induced earthquakes possesses features quite similar to those seen by Hallbauer et al. (1973) in laboratory specimens of the country rock stressed to failure. A careful mapping of the Navajo and Entrada sandstones in Utah was carried out by Aydin and Johnson (1983), who concluded that many of their observations can be adequately interpreted by Rudnicki and Rice's model.

However, a recent field study by Segall and Pollard (1983) concluded that laboratory observation may not be universally applicable in the field. Their observation of faulting in granodiorite in the central Sierra Nevada indicates that faults are nucleated on pre-existing joints (large-scale tensile cracks). The joints subsequently act collectively as a weak zone and undergo significant shear motion probably induced by a stress field rotated over time. More field studies of this detail are needed to determine whether such a localization process is general.

10.6 LOCALIZATION OF FLOW IN THE PLASTIC REGIME

Lapworth (1885) was the first to describe a fine-grained, well-laminated rock along the Moine Thrust zone in Scotland and named it mylonite. Since then, geologists have recognized that mylonites indicate shear zones of localized deformation which may extend to continental dimensions (Figure 10.6). Mylonites in such shear zones usually show grain size reduction, and some times show evidence of melting. The deformation mechanism is inferred to be rate-sensitive plastic flow in many instances.

A convenient (but not necessarily exact) criterion commonly used for the localization analysis of rate-sensitive materials states that the onset of localization occurs if the load-bearing capacity decreases with strain (Backofen, 1972; Argon, 1973). Recently, there have been some studies (e.g. Bai, 1982; Steif et al., 1982) which explicitly consider the development of runaway instability of strain associated with an initial imperfection.

There are two important differences between the usual material science and the geologic analysis of strain localization in this context. First, the change of area with loading is an important consideration, e.g. the development of necking in metals. Although there are extensional tectonic settings for which necking seems to be important (Tapponier and Francheteau, 1978) most shear zones occur in

Figure 10.6 Aerial photograph of the Nordre Stromfjord shear zone in Greenland. The rocks above the fjord are largely undeformed while those below the fjord have suffered strains about 6.0. The total offset along the 120 km long feature is about 15 km. The width of the fjord in the photograph is about 3 km (Reproduced by permission (A. 509/84) of Geodætisk Institute, Denmark.)

200   Mechanics of Geomaterials

geologic settings for which loading can be approximated as simple shear and therefore the area variation needs not be considered.

Second, although thermal softening induced by adiabatic heating is of importance to metals only for dynamic loading (Culver, 1973; Costin et al., 1979) it can be significant for geologic materials with much lower thermal diffusivity. As a matter of fact, most of the theoretical analyses have aimed to assess the possible contribution of thermal softening. Poirier (1980) recently presented a comprehensive analysis with the geologic problem in mind. He considered a constitutive relation for simple shear of the following form:

$$\sigma = M^{(1+n+m)}\tau_0 \, \varepsilon^n \, \dot{\varepsilon}^m \exp\left(\frac{mQ}{RT}\right) \tag{10.2}$$

where M is the Taylor factor (Kochs, 1958), τ_0 is a constant possibly weakly dependent on temperature, ε is the strain, $\dot{\varepsilon}$ is the strain rate, n is the strain hardening coefficient ($n = \partial\sigma/\partial\varepsilon|\dot{\varepsilon}, T$), m is the strain rate sensitivity ($\partial\sigma/\partial\dot{\varepsilon}|\varepsilon, T$), and Q is the activation energy.

He took for convenience the criterion for onset of localization that:

$$\frac{\partial \ln \sigma}{\partial \varepsilon} = (1+n+m)\frac{d\ln M}{d\varepsilon} + \frac{d\ln \tau_0}{d\varepsilon} + \frac{n}{\varepsilon} - m\frac{d\ln \dot{\varepsilon}}{d\varepsilon} - \frac{mQ}{RT^2}\frac{dT}{d\varepsilon} = 0 \tag{10.3}$$

From left to right, the contributions from the different terms are referred to as 'geometric softening', 'structural softening', 'strain softening', 'strain rate softening', and 'thermal softening', respectively. Of course, they will be referred to as 'hardening' with the signs reversed.

It is quite evident from field evidence that metamorphism and deformation are often concurrent (Beach, 1980). Chemical reactions between minerals, influx or egress of water, and thermal perturbation can radically alter the coefficients above. Although we discuss separate softening mechanisms below, it is important to recognize that the decoupling of processes is strictly pedagogical.

10.7 THERMAL SOFTENING

Among the mechanisms proposed for shear zone formation, thermal softening is probably the most thoroughly analysed from a theoretical point of view. Most of the geologic studies follow the approach of Gruntfest (1963) who showed that thermal runaway occurs under constant stress (σ_0) boundary conditions if the parameter:

$$Gu = \frac{a\sigma_0^2 \, 1^2}{k\eta_0} \tag{10.4}$$

achieves a critical value. Here, η_0 is the viscosity at room temperature T_0, 1 is a characteristic dimension, and k is the thermal conductivity. The viscosity η at temperature T is given by $\eta = \eta_0 \exp(-a(T - T_0))$. Gruntfest's formalism has been adapted to analyze the instability of magma flow (Fujii and Uyeda, 1974) and glaciers (Clark et al., 1977).

Although an instability condition such as (10.3) can in principle be applied to materials with a wide range of rheology, our current understanding of flow behaviour of rock is limited. A fair number of experimental studies have been performed on steady state creep of rocks (Goetze, 1978; Kirby, 1980) but, quantitative results on the flow behaviour in the brittle–ductile transition are too limited to be of much use for such an analysis (see Tullis and Yund, 1977; Caristan, 1982 for recent reviews).

Yuen et al. (1978) solved the one-dimensional problem of coupled heat transport and deformation of two half-spaces with an instantaneous stepwise increase in slip at the interface. The rheology was taken to be Newtonian viscous, with the effective viscosity adjusted to agree with experimentally determined power law creep results. An important conclusion of this study is that thermal runaway probably is uncommon with constant velocity boundary conditions (which are considered to be the more realistic situation for geologic application). In fact, the shear zone broadens out to a dimension that scales as the square root of time.

If a characteristic dimension is introduced into the problem by specifying a width for a pre-existing weak zone (e.g. induced by a thermal anomaly), a transient stage is possible with strain gradually localized towards the centre of the zone (Fleitout and Foridevaux, 1980). As expected, the subsequent long-term behaviour will be similar to that considered by Yuen et al. (1978).

Poirier et al. (1979) considered a more realistic rheology by including transient creep behaviour extrapolated from Goetze's (1971) experimental data for Westerly granite. An initial imperfection with a flow stress lower than the surroundings (at the same temperature and strain rate) is introduced. No runaway instability is possible with constant velocity boundary condition even for total adiabaticity. As a matter of fact, the long-term response of the weakened zone is to deform uniformly at a strain rate which scales as the initial strength defect and the inverse of strain rate sensitivity.

A conclusion common to all these studies is that the thermal anomaly characteristically extends to a dimension wider than that for strain concentration, a fact which would be significant in the field comparison of mineral assemblages and strain data. However, the degree of strain concentration predicted by the model calculations is not as drastic as some of the field observations. The results may be modified if one performs the computation for two- or even three-dimensional geometry. Necking, for example, is sensitive to the geometry and loading configuration (Backofen, 1972). However, the major weakness of the current models is probably the inadequacy of the constitutive relations used. We will discuss below the field and laboratory observations which all point to the complexity of softening mechanisms. A quantitative description of these mechanisms is urgently needed before more realistic considerations of shear zone formation are possible.

10.8 GEOMETRIC, STRUCTURAL, AND STRAIN SOFTENING

In addition to the inference that plastic flow has occurred, three elements seem to be intrinsic to the definition of a mylonite: (1) occurrence of a planar zone, narrow with respect to the surrounding undeformed rock, (2) reduction of grainsize from protolith to mylonite zone, (3) enhanced foliation or lineation (Tullis et al., 1982). The first of these elements is, of course, the evidence for localization of shear (Ramsay, 1980). However, it is important to remember that while the field evidence certainly indicates at least transient strain localization, the steady state shear zone structure is not known. In fact, Sorenson (1983) has argued, on the basis of field evidence, that strain softening of the rocks may indeed be followed by strain hardening.

It is generally believed that the grain size reduction in shear zones occurs by recrystallization during deformation, i.e. dynamic recrystallization (Bell and Etheridge, 1973; White, 1977; Christie and Ord, 1980; Kohlstedt and Weathers, 1980) although small grain sizes could also result from the formation of new minerals (Robin, in Tullis et al., 1982). Experiments on dynamically recrystallizing metals document weakening of 0.20 of the yield stress or more at the onset of recrystallization (Glover and Sellars, 1973; Sellars, 1978; Ion et al., 1982). Micrographs of these experiments show newly formed grains along previously existing grain boundaries. There is a striking similarity between the textures in the dynamically recrystallized metals and in materials deformed in the superplastic regime on the one hand, and the progression of textures in a mylonitic shear zone on the other hand. For example, compare micrographs from Glover and Sellars (1973) or Ion et al., (1982) with Figure 7 (from Kohlstedt and Weathers, 1980). Thus it has been suggested that the strain localization in the shear zone results from strength differences between the coarse grained protolith and the fine-grained mylonite. The difference may result from geometric weakening due to preferred orientation of the dynamically recrystallized grains, from structural weakening induced by increased recovery due to recrystallization from enhanced grain boundary sliding (see White, 1977, for a review) or from a transition to diffusional superplastic flow (Boullier and Gueguen, 1975). An additional source of structural weakening may derive from the strength differences due to rearrangement of the geometric distribution of the phases in polymineralic rocks (Poirier, 1980).

Although comparisons between dynamically recrystallized metals and textures in geologic shear zones are intuitively and aesthetically appealing, recent experiments on dynamically recrystallizing rocks are equivocal. Post (1977) has observed localized zones of finely recrystallized grains in a series of mechanical tests on Mt Burnett dunite. The formation of recrystallized zones was sometimes

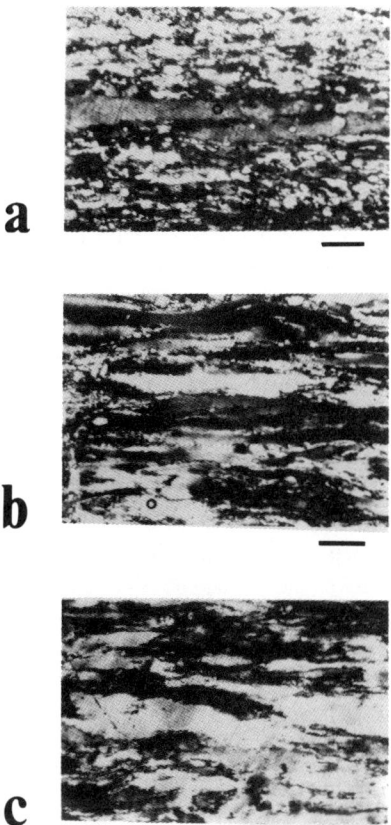

Figure 10.7 Transmitted light photomicrographs of the development of microstructure in a quartzite from the Idaho Springs–Ralston shear zone showing progressively larger amounts of dynamic recrystallization towards the centre of the shear zone. The similarity of these textures to dynamically recrystallized metals is striking. The scale bar is 0.1 mm. (From Kohlstedt and Weathers, 1980, by permission of the American Geophysical Union)

associated with stress drops of 0.3–0.6 of the peak stress. Those stress drops have been ascribed to either the onset of grain-size sensitive flow (Twiss, 1976; Goetze, 1978) or transient creep effects (Post, 1977; Zeuch, 1982, 1983). However, other experiments in peridotite (Chopra and Paterson, 1981), quartzite, (Christie et al., 1983), and limestone (Schmid, Paterson, and Boland, 1980) deformed to large strains do not show strain weakening during recrystallization.

Recent experiments on peridotites (Chopra and Paterson, 1981) and halite

(Guillope and Poirier, 1979) may provide important keys to reconciling these results: Chopra and Paterson (1981) showed that the presence of water promotes recrystallization and have suggested that the decreased strength of 'wet' samples is due to stress relaxation along grain boundaries. Guillope and Poirier (1979) have demonstrated that dynamic recrystallization may occur either by progressive subgrain rotation or grain boundary migration. Poirier (1980) has suggested that no weakening would be expected during the operation of the former mechanism. Thus, on the basis of present experimental data, the presence of finely recrystallized material cannot necessarily be presumed to be a sufficient condition for strain weakening. It is possible that both the production of fine grains and the operation of strain weakening may depend critically on the introduction of water or some other chemically weakening agent (Etheridge and Wilkie, 1979; Sorenson, 1983).

10.9 METAMORPHIC REACTIONS AND STRAIN SOFTENING

It has long been recognized that deformation in shear zones commonly is accompanied by metamorphic reactions (Teall, 1885; Beach, 1980). Oftentimes, petrologic examination reveals that the shear zone was an open chemical system and that minerals in the zone have undergone extensive hydration (Kerrich et al., 1977; Beach, 1980). However, this need not always be true since, in at least some cases, localized strain may develop in rocks undergoing prograde reactions (e.g. Brodie, 1980, 1981).

Mechanical tests on metals, ceramics and minerals alike show that both brittle and plastic strength may depend on the chemical environment. For instance, deformation by pressure solution is strongly dependent on the chemical properties of an intergranular fluid (Robin, 1978). Deformation of quartz by dislocation and diffusional mechanisms is also dramatically affected by the presence of water (see a review by Paterson and Kekulawala, 1979) and perhaps by the activity of other chemical species as well (Hobbs, 1981). Furthermore exothermic metamorphic reactions (e.g. retrograde hydration reactions) will provide an additional heat source.

The metamorphism may also cause structural weakening since the reaction products form a new mineral assemblage which may be weaker than the unreacted rock (White and Knipe, 1978); this is particularly so in the case of hydration of feldspar to form phyllosilicates. In addition, new minerals dispersed throughout the matrix may inhibit grain growth allowing grain-size sensitive mechanisms including diffusional flow, pressure solution or grain boundary sliding to operate with enhanced strain rates (Etheridge and Wilkie, 1979).

It is clear that chemical reactions may profoundly affect the mechanical strength of rocks; apparently the converse is equally true. Reduction of the grain size of a potential reactant will increase the grain boundary area available as a reactant site (Mitra, 1978; Wintsch, 1975) and can potentially greatly increase the metamorphic reaction rate. Deformation may also increase the rate of introduction of water, either through dilatancy enhanced permeability (Sibson, et al., 1975) or by plastic strain. It has been shown that the temperature of the α–β phase transition in quartz is affected by non-hydrostatic stress (Coe and Paterson, 1969) and it is probable that other mineral equilibria will also be affected (Yund and Tullis, 1980; Wintsch, 1975, 1981). Finally, chemical potential gradients due to non-hydrostatic stresses may well couple with gradients between reactants and products to cause increased diffusion rates (Brodie, 1980) or to provide favourable nucleation sites for new minerals (Knipe, 1979). Thus concurrent metamorphic and deformation processes are tightly coupled, and to understand the mechanical behaviour of reacting mineral assemblages will require the consideration of both sets of phenomena.

10.10 CONCLUSION

We reviewed here the study of shear localization in geologic materials in two contexts. An interesting observation is that our understanding of the two phenomena, brittle faulting and shear zone formation, seem to have followed quite different lines of development.

Laboratory studies of dilatancy and fracture in brittle rock have been very numerous in the past twenty years or so. The quantitative evaluation of the stress–strain relationship as well as microscopy study of stress-induced microstructures have motivated seminal work on the formulation of constitutive equations incorporating dilatancy and pressure-sensitivity. The subsequent bifurcation analyses provide insight on the importance of both loading configurations and physical parameters on the onset of localization.

In contrast, only limited field study with a quantitative approach has been made, although the phenomenon is widely observed in tectonic settings. As a result, a number of important questions concerning the scaling problem remain unanswered.

On the other hand, the tectonic study of shear zones has been an active area of research. Field evidence frequently indicates that fluids have entered sheared regions and that metamorphism and deformation are concurrent and, quite probably, intricately coupled processes. Geometric, structural, and strain softening may result from spatial redistribution of a weak phase, refinement of grain size, and subsequent transition to grain-size sensitive creep, reduction of the Taylor factor due to the production of preferred orientation, or increased recovery due to transient creep effects.

In contrast to the study of brittle phenomena, quantitative laboratory studies of localization during plastic flow are very limited, partly due to the experimental difficulty and partly due to the complexity of the process. Although a generalized framework can be adapted from materials science, advances in theoretical interpretation are hampered by the absence of appropriate constitutive equations for strain softening as specialized to geologic materials. Thermal softening has been analysed theoretically to some detail, and thermal runaway appears to be unlikely unless imperfections with drastic strength reduction are present or unless the boundary conditions are approximated by constant stress conditions.

ACKNOWLEDEGMENT

This work was supported by the National Science Foundation through contracts EAR-8008284, -8108560 (B.E.) and -8218379 (T.F.W.) Kai Sorensen and Maura Weathers kindly provided us with Figures 10.6 and 10.7 respectively.

REFERENCES

Anand, L., and Spitzig, W. A. (1982) 'Shear-band orientations in plane strain', *Acta Metall.*, **30**, 553–561.
Argon, A. S. (1973) Stability of plastic deformation, in *The Inhomogeneity of Plastic Deformation, ASM*, pp. 161–189.
Aydin, A., and Johnson, A. M. (1983) 'Analysis of faulting in porous sandstones', *J. Structural Geol.*, **5**, 19–31.
Backofen, W. A. (1972) *Deformation Processing*, Addison-Wesley, Reading Mass.
Bay, Y. L. (1982) 'Thermo-plastic instability in simple shear', *J. Mech. Phys. Solids*, **30**, 195–207.
Batdorf, S. B., Budiansky, B. (1949) 'A mathematical theory of plasticity based on the concept of slip', Technical note 1871, Nat. Adv. Committee on Aeronautics.
Batzle, M., Simmons, G., and Siegfried, R. W. (1980) 'Microcrack closure in rocks under stress', *J. Geophys. Res.*, **85**, 7072–7090.
Bažant, Z. P., and Kim, S. S. (1979) 'Plastic-fracturing theory for concrete', *J. Eng. Mech. Div. ASCE*, **105**, 407–428.
Beach, A. (1980) 'Retrogressive metamorphic processes in shear zones', *J. Struct. Geol.*, **2**, 257–263.
Bell, T. H., and Etheridge, M. A. (1973) 'Microstructure of mylonites and their descriptive terminology', *Lithos*, **6**, 337–348.
Boullier, A. M., and Gueguen, Y. (1975) 'SP-mylonites: origin of some mylonites by superplastic flow', *Contrib. Mineral Petrol.*, **50**, 93–104.
Brace, W. F. (1981) 'The effect of size on mechanical properties of rock', *Geophys. Res. Lett.*, **8**, 651–652.
Brodie, K. H. (1980) 'Variations in mineral chemistry across a phlogopite peridotite shear zone', *J. Struct. Geol.*, **2**, 265–272.
Brodie, K. H. (1981) 'Variation in amphibole and plagioclase composition with deformation', *Tectonophysics*, **78**, 385–402.
Caristan, Y. (1982) 'The transition from high temperature creep to fracture in Maryland diabase', *J. Geophys. Res.*, **87**, 6781–6790.
Carter, N. L., and Kirby, S. H. (1978) 'Transient creep and semibrittle behavior of crystalline rocks', *Pageoph*, **116**, 807–839.
Chopra, P. N., and Paterson, M. S. (1981) 'The experimental deformation of dunite', *Tectonophysics*, **78**, 453–473.
Christie, J. M., Ord, A. (1980) 'Flow Stress from Microstructures of Mylonites: Example and Current Assessment,' *J. Geophys. Res.*, **85**, 6253–6262.

Christie, J. M., Koch, P. S., Ord, A., and George, R. P. (1983) 'The effect of water on the rheology and microstructures of experimentally deformed quartzite', *J. Geophys. Res.*, in press.

Clarke, G. K. C., Nitzan, U., and Paterson, W. S. B. (1977) 'Strain heating and creep instability in glaciers and ice sheets', *Rev. Geophys. Space Physics*, **15**, 235–247.

Cleary, M., and Rudnicki, J. (1976) 'On the initiation and propagation of dilatant rupture zones in geological materials', in *The Effect of Voids on Material Deformation*, ASME.

Coe, R. S., and Paterson, M. S. (1969) 'The α–β inversion in quartz: a coherent phase transition under non-hydrostatic stress', *J. Geophys. Res.*, **74**, 4921–4948.

Cook, N. G. W., and Heard, H. C. (1981) 'NSF–UC Berkely workshop on large scale laboratory testing in Geomechanics', *Geophys. Res. Lett.*, **8**, 645–723.

Costin, L. S., Crisman, E. E., Hawley, R. H., and Duffy, J. (1979) 'On the localization of plastic flow in mild steel tubes under dynamic torsional loading', in *Mechanical Properties at High Rates of Strain* (J. Harding, Ed.), The Institute of Physics, Bristol, pp. 90–100.

Culver, R. S. (1973) 'Thermal instability strain in dynamic plastic deformation', in *Metallurgical Effects at High Strain Rates*, (R. W. Rhode et al., Eds.), Plenum, N. Y., pp. 519–530.

Etheridge, M. A., and Wilkie, J. C (1979) 'Grain size reduction, grain boundary sliding and the flow strength of mylonites', *Tectonophysics*, **58**, 159–178.

Evans, A. G., and Adler, W. F. (1978) 'Kinking as a mode of structural degradation in carbon fiber composites', *Acta Metall.*, **26**, 725–738.

Fleitout, L., and Froidevaux, C. (1980) 'Thermal and mechanical evolution of shear zones', *J. Struct. Geol.*, **2**, 159–164.

Frank, F. C., and Stroh, A. N. (1952) 'On the theory of kinking', *Proc. Phys. Soc.*, **B65**, 811–821.

Fujii, N., and Uyeda, S. (1974) 'Thermal instabilities during flow of magma in volcanic conduits', *J. Geophys. Res.*, **79**, 3367–3370.

Glover, G., and Sellars, C. M. (1973), 'Recovery and recrystallization during high temperature deformation of α-iron', *Met. Trans.*, **4**, 765–775.

Goetze, C. (1971) 'High temperature rheology of Westerly granite', *J. Geophys. Res.*, **76**, 1223–1230.

Goetze, C. (1978) 'The mechanics of creep in olivine', *Phil. Trans. Roy. Soc. London*, **A228**, 99–119.

Goodman, R. E. (1976) *Methods of Geological Engineering*, West, St. Paul, 472 pp.

Gruntfest, I. J. (1963) 'Thermal feedback in liquid flow, plane shear at constant stress', *Trans. Soc. Rheol.*, **7**, 195–207.

Guillope, M., and Poirier, J. P. (1979) 'Dynamic recrystallization during creep of single crystalline halite: an experimental study', *J. Geophys. Res.*, **84**, 5557–5567.

Hadley, K. (1975) 'Azimuthal variation of dilatancy', *J. Geophys. Res.*, **80**, 4845–4850.

Hallbauer, D. K., Wagner, H., and Cook, N. G. W. (1973) 'Some observations concerning the microscopic and mechanical behavior of quartzite specimens in stiff, triaxial compression tests', *Int. Journ. Rock Mech. Min. Sci.*, **10**, 713–726.

Hobbs, B. E. (1981) 'The influence of metamorphic environment upon the deformation of minerals', *Tectonophysics*, **78**, 335–383.

Ion, S. E., Humphreys, F. J., and White, S. H. (1982) 'Dynamic recrystallization and the development of microstructure during the high temperature deformation of magnesium', *Acta. Metall.*, **30**, 1909–1919.

Janach, W., and Guex, L. H. (1980) 'In-plane propagation of shear microcracks in brittle rocks under triaxial compression', *J. Geophys. Res.*, **85**, 2543–2553.

Jonas, J. J., Holt, R. A., and Coleman, C. E. (1976) 'Plastic stability in tension and compression', *Acta Metall.*, **24**, 911–918.
Kachanov, M. L. (1982) 'A microcrack model of rock inelasticity', *Mech. Mat.*, **1**, 19–41.
Kerrich, R., Fyfe, W. S., Gorman, B. E., and Allison, I. (1977) 'Local modification of rock chemistry by deformation', *Contrib. Mineral Petrol.*, **65**, 183–190.
Kirby, S. H. (1980) 'Tectonic stresses in the lithosphere: constraints provided by the experimental deformation of rocks', *J. Geophys. Res.*, **85**, 6353–6363.
Knipe, R. J. (1979) 'Chemical changes during slaty cleavage development', *Bull. Mineral.*, **102**, 206–209.
Kochs, U. F. (1958) 'Polyslip in polycrystals', *Acta Metall.*, **6**, 85–94.
Kohlstedt, D. L., and Weathers, M. S. (1980) 'Deformation-induced microstructures, paleopiezometers and differential stresses in deeply eroded fault zones', *J. Geophys. Res.*, **85**, 6269–6285.
Kranz, R. L. (1979) 'Crack growth and development during creep in Westerly granite', *Int. J. Rock Mech. Min. Sci.*, **16**, 23–36.
Lapworth, C. (1885) 'The Highland controversy in British geology: its causes, course and consequences', *Nature*, **32**, 558–559.
Lockner, D. A., and Byerlee, J. D. (1980) 'Development of fracture planes during creep in granite', in *Proc. of 2nd Conf. on AE/Microseismic Activity in Geological Structures, and Materials* (H. R. Hardy, Jr., and F. W. Leighton, Eds.), Trans. Tech. Pub., Clausthal, Germany.
McGarr, A., Spottiswoode, S. M., Gay, N. G., and Ortlepp, W. D. (1979) 'Observations relevant to seismic driving stress, stress drop, and efficiency', *J. Geophys. Res.*, **84**, 2251–2261.
Mitra, G. (1978) 'Ductile deformation zones and mylonites: the mechanical processes involved in the deformation of crystalline basement rocks', *A. J. Sci.*, 1057–1084.
Moss, W. C., and Gupta, Y. M. (1982) 'A constitutive model describing dilatancy and failure in brittle rock', *J. Geophys. Res.*, **87**, 2985–2990.
Needleman, A. (1979) 'Non-normality and bifurcation in plane strain tension and compression', *J. Mech. Phys. Solids*, **27**, 231–254.
Nemat-Nasser, S., and Shokooh, A. (1980) 'On finite plastic flows of compressible materials with internal friction', *Int. J. Solids Struct.*, **16**, 495–514.
Paterson, M. S. (1978) *Experimental Rock Deformation—The Brittle Field*, Springer Verlag, New York, 254 pp.
Paterson, M. S. (1979) 'The mechanical behaviour of rock under crustal and mantle conditions', in *The Earth: Its Origin, Structure and Evolution* (M. W. McElhinny, Ed.), Academic Press, pp. 469–489.
Paterson, M. S., and Kekulawala, K. R. S. S. (1979) 'The role of water in quartz deformation', *Bull. Mineral.*, **102**, 92–98.
Poirier, J. P. (1980) 'Shear localization and shear instability in materials in the ductile field', *J. Struct. Geol.*, **2**, 135–142.
Poirier, J. P., Bouchez, J. L., and Jonas, J. J. (1979) 'A dynamic model for aseismic ductile shear zones', *Earth and Planet. Sci. Lett.*, **43**, 441–453.
Post, R. L. (1977) 'High temperature creep of Mt Burnett dunite', *Tectonophysics*, **42**, 75–110.
Ramsay, J. G. (1980) 'Shear zone geometry: a review', *J. Struct. Geol.*, **2**, 83–99.
Rice, J. R. (1976) 'The localization of plastic deformation', in *Theoretical and Applied Mechanics* (W. T. Koiter, Ed.), North-Holland, Amsterdam.
Robin, P. Y. F. (1978) 'Pressure solution at grain-to-grain contacts', *Geochim. Cosmochim. Acta*, **42**, 1383–1389.

Rudnicki, J. W. (1977) 'The effect of stress-induced anisotropy on a model of brittle rock failure as localization of deformation', *Proc. 18th U.S. Symposium on Rock Mechanics*.
Rudnicki, J. W., and Rice, J. R. (1975) 'Conditions for the localization of deformation in pressure-sensitive dilatant materials', *J. Mech. Phys. Solids*, **23**, 371–394.
Schmid, S. M., Paterson, M. S., and Boland, J. N. (1980) 'High temperature flow and dynamic recrystallization in Carrara marble', *Tectonophysics*, **65**, 245–280.
Segall, P., and Pollard, D. D. (1983) 'Nucleation and growth of strike slip faults in granite', *J. Geophys. Res.*, **88**, 555–568.
Sellars, C. M. (1978) 'Recrystallization of metals during hot deformation', *Phil. Trans. R. Soc.*, **135**, 513–516.
Sibson, R. H., Moore, J. and Rankin, A. H. (1975) 'Seismic pumping—a hydrothermal fluid transport mechanism', *J. Geol. Soc. Lond.*, **131**, 653–659.
Singh, M. M. (1981) 'Strength of rock', in *Physical properties of Rocks and Minerals* (Y. S. Toulonkian and C. Y. Ho, Eds.), McGraw-Hill, pp. 83–121.
Soga, N., Mizutani, H., Spetzler, H. and Martin, R. J., III (1978) 'The effect of dilatancy on velocity anisotropy in Westerly granite', *J. Geophys. Res.*, **83**, 4451–4458.
Sondergeld, C. H., and Estey, L. H. (1982) 'Source mechanisms and microfracturing during uniaxial cycling of rock', *Pageogh*, **120**, 151–166.
Sorensen, K. (1983) 'Growth and dynamics of the Nordre Stromfjord shear zone', *J. Geophys. Res.*, **88**, 3419–3438.
Steif, P. S., Spaepen, F., and Hutchinson, J. W. (1982) 'Strain localization in amorphous metals', *Acta Metall.*, **30**, 447–455.
Stevens, J. L., and Holcomb, D. J. (1980) 'A theoretical investigation of the sliding crack model for dilatancy', *J. Geophys. Res.*, **85**, 7091–7100.
Tapponier, P., and Brace, W. F. (1976) 'Development of stress-induced microcracks in Westerly granite', *Int. Journ. Rock Mech. Min. Sci.*, **13**, 103–112.
Tapponier, P., and Francheteau, J. (1978) 'Necking of the lithosphere and the mechanics of accreting plate boundaries', *J. Geophys. Res.*, **83**, 3955–3970.
Teall, J. J. H. (1885) 'The metamorphism of dolerite into hornblende schist', *Q. Jour. Geol. Soc.*, **41**, 133–145.
Tullis, J., Snoke, A. W., and Todd, V. R. (1982) 'Significance and petrogenesis of mylonitic rocks', *Geology*, **10**, 227–230.
Tullis, J., and Yund, R. A. (1977) 'Experimental deformation of dry Westerly granite', *J. Geophys. Res.*, **82**, 5705–5718.
Vardoulakis, I. (1980) 'Shear band inclination and shear modulus of sand in biaxial tests', *Int. J. Num. Analy. Meth. Geomechanics*, **4**, 103–119.
Wawersik, W. R., and Brace, W. F. (1971) 'Post failure behavior of a granite and a diabase', *Rock Mech.*, **3**, 61–85.
White, S. A., Burrows, S. E., Carreras, J., Shaw, N. D., and Humphreys, F. J. (1980) 'On mylonites in ductile shear zones', *J. Struct. Geol.*, **2**, 175–187.
White, S. (1977) 'Geological significance and recrystallization processes in quartz', *Tectonophysics*, **39**, 143–170.
White, S. H., and Knipe, R. J. (1978) 'Transformation and reaction enhanced ductility in rocks', *J. Geol. Soc., London*, **135**, 513–516.
Wintsch, R. P. (1975) 'Feldspathization as a result of deformation', *Bull. Geol. Soc. Amer.*, **86**, 35–38.
Wintsch, R. P. (1981) 'Syntectonic oxidation', *Am. J. Sci.*, **281**, 1223–1239.
Wong, T. -F. (1982) 'Micromechanics of faulting in Westerly granite', *Int. J. Rock. Mech. Min. Sci.*, **19**, 49–64.
Yuen, D. A., Fleitout, L., Schubert, G., and Froidevaux, C. (1978) 'Shear deformation

zones alone major transform faults and subducting slabs', *Geophys. J. Royal Astro. Soc.*, **54**, 93–119.

Yund, R. A., and Tullis, J. (1980) 'The effect of water, pressure, and strain on Al/S: order disorder kinetics in feldspar', *Contin. Min. and Pet.*, **72**, 297–302.

Zeuch, D. H. (1982) 'Ductile faulting, dynamic recrystallization, and grain size sensitive flow of olivine', *Tectonophysics*, **83**, 293–308.

Zeuch, D. H. (1983) 'On the inter-relationship between grain size sensitive creep and dynamic recrystallization of olivine', *Tectonophysics*, **93**, 151–168.

Mechanics of Geomaterials
Edited by Z. Bažant
© 1985 John Wiley & Sons Ltd

Chapter 11

Shear Localization, Faulting, and Frictional Slip: Discusser's Report

J. R. Rice

Compressive deformation of brittle geomaterials in the post-peak softening range, in the presence of confining pressure, is normally accompanied by the localization of deformation into a single (or a few) discrete fault zone(s) which traverse a test specimen. Evans and Wong have reviewed deformation micromechanisms in rocks deformed under such conditions, giving emphasis in the oral presentation to the brittle field and to the micro-crack-based mechanisms of inelastic deformation that occur then. They discuss how local instabilities, such as buckling of micro-crack columns, lead to zones of concentrated shear and crushing at the level of a few grains, and how these zones are ultimately joined together in forming the throughgoing fault. They also review the continuum mechanical background for localization. For rate-independent solids localization corresponds to the loss of ellipticity for the equations governing quasi-static displacement increments $\delta \mathbf{u}$, that is, for the system of equations that express continuing stress field equilibrium in the form

$$(L_{ijrs}\delta u_{r,s})_{,i} = 0 \qquad (11.1)$$

where $\mathbf{L}$ is the incremental tangent modulus for the particular cone (e.g. plastic loading versus unloading), in a space of incremental displacement gradients, in which the considered deformations proceed. See Rice (1976) for review of the underlying theory. As Evans and Wong emphasize, constitutive equations developed to show the near-peak and post-peak response of frictional geomaterials generally lead to predictions that localizations will occur. This happens within the theory when the matrix (in indices j and r) $n_i L_{ijrs} n_s$, which is nonsingular for all directions $\mathbf{n}$ during deformation in the prelocalized state, becomes singular for some particular set of directions $\mathbf{n}$ (marking the orientations of the characteristic segments which then emerge).

In a certain technical sense, the meeting of the condition just cited corresponds to the first occurrence of an instantaneously non-hardening state, marking the transition from hardening to softening type deformation. The response is non-hardening at the critical condition in the sense that it corresponds to a situation for which the traction increment $n_i L_{ijrs} \delta u_{r,s}$ on a plane with normal **n** is unaltered by a change of incremental displacement gradient of form $\delta g_r n_s$, corresponding to shear and extension relative to that plane. This is indeed a rather technical characterization of a non-hardening state, and will not in general coincide with the conventional understanding in which a true or Cauchy measure of stress passes through a maximum along a non-bifurcated deformation path. (Hardening is used in this latter sense by Evans and Wong.)

Despite the qualitative accordance between observed faulting in rock and predictions of localization theory, as cited by Evans and Wong, it seems accurate to state that no convincing quantitative understanding has yet been achieved in this case. Perhaps this is due largely to the present inability to represent reliably, in constitutive relations, behaviour revealed by the electron microscopy studies of micro-cracking, internal buckling, fragmentation and crushing presented so well in their figures. The situation is far better for ductile metallic single crystals, in which case the micromechanics of deformation by dislocation slip is well understood and can be translated with little ambiguity into macroscopic constitutive relations, based on shearing along the discrete slip orientations of the crystal. In that case recent work (Peirce, Asaro, and Needleman, 1982; Asaro, 1983) has led to remarkable advances in the explanation of flow inhomogeneities, including macroscopic shear bands, kinks, and patchy slip, on the basis of localization and related instability predictions for the macroscopic inelastic constitutive relations. Thus there is some reason for confidence in the continuum approach to localization, although it can serve for useful predictions only when the constitutive relations can be made to correspond adequately to actual behaviour. Since the test for localization involves response moduli for deformation increments that are, in general, directed differently from those of the prelocalized state, the requirements for constitutive description are rather subtle and not easily approached experimentally.

Some notable features of shear localization, revealed for the various material models that have been studied thus far, include the apparent sensitivity of predictions to overall deformation mode (plane strain compression more conducive to localization than axisymmetric), to vertex yielding (as for multi-slip in crystals and as expected for frictional slip on micro-cracks in rocks), and to imperfections or local non-uniformities of material properties. Such local non-uniformities are found, in theoretical studies, to lead often to concentration of deformation in a mode such that the local zone is driven to the localization condition well before the surrounding material is so deformed. Apparently, the local non-uniformity can itself result from necking, bulging, or surface rumpling bifurcation modes which occur earlier. The work reported by Evans and Wong also suggests the importance of non-uniformities developed at the few-grain scale in triggering localization.

Among the general issues to be raised for discussion, we might consider that of how one formulates and solves boundary value problems in the post-elliptic regime. There is an obvious need for such a development for modelling failure processes throughout geomechanics. The question is probably not separable from that of how one models a localized zone when no further continuum solution exists along it. Presumably, for a severely localized zone, one must then go over a traction-displacement (versus stress–strain) description, and most of the remainder of this discussion is devoted to just that topic. The issues of non-existence of continuum solutions and transition to traction-displacement relations tend to be blurred, perhaps usefully at the present level of understanding, by the finite grid or element size used in numerical studies. Numerical solutions in the non-hardening and, especially, softening regimes must be expected to have an essential dependence on element size, and there would seem to be nothing within rate-independent constitutive models as conventionally formulated which could lead to a thickness for these localized zones. One wonders if there is any physically based foundation for 'strain gradient' theories, inclusive of a characteristic length scale, which could address the thickness issue meaningfully.

Evans and Wong have also reviewed the evidence for shear localization in the ductile creep regime for rock polycrystals at high compressive stress. Here the possibilities are raised that temperature and deformation induced metamorphism, together with thermal softening from shear heating, contribute to the process.

While the focus of Evans and Wong was on the inception of localization, Ruina outlined recent advances on understanding deformation in the fully localized state, describable as slip along a fault or friction surface.

One portion of such studies relates to interpretation of the post-peak phase of the confined triaxial test, in a sufficiently stiff (or servocontrolled) apparatus, in terms of slip along a throughgoing fault—at least in circumstances that are simple enough that a single fault forms. In that case an interpretation of the post-peak deformation, advanced by Palmer and Rice (1973) to model the instability of overconsolidated clay slopes and applied recently (e.g. Wong, 1982) to characterize rock faulting at high confining pressure, is to regard the portions of the specimen outside the fault zone as unloading elastically and those within as contributing a concentrated slip δ across a mathematical surface that represents the fault. The 'slip weakening' relation between shear stress τ and slip δ on the surface may then be inferred from the post peak test and, it may be argued, such a τ versus δ relation is the fundamental information to be extracted from this type of test.

The perspective that arises on slope instability, traceable to earlier descriptions of progressive failure by Skempton and Bishop, is that localized deformation is induced at a site of stress concentration and the remaining course of the progressive failure consists of the crack-like spread of a planar rupture zone, subject to a slip-weakening relation between τ and δ on its surface. Similar concepts are in use for large-scale earthquake shear faults (e.g. Andrews, 1976;

Day, 1982; Rice, 1980; Stuart, 1979) and in model studies of the dynamic spread for frictional slip zones along the interface of large laboratory friction specimens (Okubo and Dieterich, 1981). The approach to shear rupture just described is based on cohesive zone models of tensile fracture, evolved through contributions of Barenblatt, of Dugdale, of Bilby, Cottrell, and Swinden, and of others. Indeed, the presentations at this symposium by Hillerborg, Bažant, Ballarini et al. and Reinhardt show the productive use of such cohesive models for the tensile fracture of concrete, where the requirement that aggregate particles be pulled out against friction from the adjoining fracture faces causes a gradual degradation of tensile stress σ with opening w across the fracture.

The result, whether in modelling shear or tensile fracture, is a smeared-out zone of gradual strength degradation with ongoing relative displacement across the faces of the advancing fracture near its tip. If the size of the zone of strength degradation (which can be estimated for shear ruptures following Palmer and Rice, 1973, and Rice, 1980) is small compared to the overall size of the slipping region along a fault, then the resulting description of the fracture becomes equivalent to that of elastic-brittle crack mechanics for a crack which sustains residual friction τ_f on its surfaces and which can extend when the singularity at its tip releases the fracture energy

$$G = \int_0^{\delta*} [\tau - \tau_f] \, d\delta \qquad (11.2)$$

per unit area; here the integral refers to the area between the slip weakening τ versus δ relation and its residual asymptote, $\tau = \tau_f$, considered to be achieved at some slip δ_*.

The elastic brittle crack mode is one of highly non-uniform slip. By contrast, it may be the case for a small through-faulted laboratory specimen that failure occurs by essentially uniform slip everywhere on the rupture surface. Shear rupture in this uniform slip mode may be described by a single degree of freedom model, represented as the classical block which is slid along a surface by imposition of some prescribed motion to one end of a spring whose other end is attached to the block. As Ruina commented, the condition for stability of the slip-weakening process is then that the unloading stiffness of the spring be greater in magnitude than that of the (negative) slope of the slip-weakening relation between friction force and block displacement.

The focus of most descriptions of failure via the (tacitly) rate-insensitive slip weakening model is on a single failure episode, for a system that has never previously slipped or that has somehow recovered from a previous episode and begins life anew. Ruina, however, has outlined a far broader constitutive description appropriate for faults that have sustained arbitrary numbers of previous slips and that could, depending on loading rate, stiffness in elastic interaction with their surroundings, and constitutive parameters, sustain repeated further slip episodes of either stable or unstable type. He shows how experimental results reveal that inevitably τ increases with slip rate $V(=d\delta/dt)$ for sudden impositions of changes in V, accomplished such that they can be considered to occur at a fixed 'state' of the slipping surface. That is, the friction process is characterized by a positive instantaneous viscosity. However, the further changes of τ in some sustained programme of slip are such that one says that the 'state' of the surface is itself not fixed but, rather, it evolves with ongoing slip. The result of this evolution is such that the 'steady state' value τ^{ss}, which τ approaches with sustained slip at fixed V, may in different circumstances of normal stress, temperature, and choice of material exhibit either positive or negative viscosity. The latter is, of course, an indication that the surface is prone to stick–slip instability. See Ruina (1983) and Gu et al. (1984) for constitutive descriptions, rooted in work by Dieterich (1979, 1981), for which one or, more accurately, two evolving state variables are employed.

The resulting constitutive structure has the following form, when state is characterized by the variables $\theta_1, \theta_2, \ldots, \theta_n$ and, for simplicity, one holds normal stress constant:

$$\tau = F(V, \theta_1, \theta_2, \ldots, \theta_n); \quad d\theta_i/dt = G_i(V, \theta_1, \theta_2, \ldots, \theta_n), \quad i = 1, 2, \ldots, n \quad (11.3)$$

Evidently, $\partial F/\partial V > 0$ and, from what has been said, it can be assumed that the functions G_i are such that for slip at fixed V, each θ_i evolves stably towards a steady state value $\theta_i^{ss} = \theta_i^{ss}(V)$ satisfying $G_i = 0$ for $i = 1, 2, \ldots, n$, so that the steady state strength is

$$\tau^{ss} = \tau^{ss}(V) = F[V, \theta_1^{ss}(v), \theta_2^{ss}(V), \ldots, \theta_n^{ss}(V)] \quad (11.4)$$

Actually, for variations of V within factors of a few powers of 10, the resulting rate and evolving state based fluctuations of τ are a rather small part of the total friction strength. Nevertheless, it is precisely this small part which can in certain circumstances correspond to weakening with ongoing slip, giving the possibility of frictional slip instability, and in other circumstances can correspond to state changes that could be identified as restrengthening in nearly stationary contact.

Accordingly, the analysis of fault slip and its stability within the scope of this rate and state dependent framework has become the focus of much recent work. Unfortunately, the subject is still sufficiently in its infancy that almost all progress to date has been made only for fault slip as modelled by the single degree of freedom spring-loaded sliding block system (e.g. Dieterich, 1980; Rice and Ruina, 1983; Gu et al., 1984). This misses out much of the interesting mechanics and physics of faulting, related to non-uniform crack like spread of slip zones, and such is an important area for future progress.

REFERENCES

Andrews, D. J. (1976) 'Rupture velocity of plane-strain shear cracks', *J. Geophys. Res.*, **81**, 5679–5687.

Asaro, R. J. (1983) 'Micromechanics of crystals and polycrystals', in *Advances in Applied Mechanics*, Vol. 23, Academic Press, pp. 1–115.
Day, S. M. (1982) 'Three-dimensional simulation of spontaneous rupture: the effect of nonuniform prestress', *Bull. Seism. Soc. Am.*, **72**, 6, 1881–1902.
Dieterich, J. H. (1979) 'Modelling of rock friction', *J. Geophys. Res.*, **84**, 2161–2175.
Dieterich, J. H. (1980) 'Experimental and model study of fault constitutive properties', in *Solid Earth Geophysics and Geotechnology* (Ed. S. Nemat-Nasser), Appl. Mech. Div., Vol. 42, Amer. Soc. Mech. Engrs., New York, pp. 21–30.
Dieterich, J. H. (1981) 'Constitutive properties of faults with simulated gouge', in *Mechanical Behavior of Crystal Rocks* (Eds, N. L. Carter, M. Friedman, J. M. Logan, and D. W. Stearns), Geophys. Monogr. Ser. No. 24, Amer. Geophys. Union. pp. 103–120.
Gu, J. -C., Rice, J. R., Ruina, A. L., and Tse, S. T. (1984) 'Slip motion and stability of a single degree of freedom elastic system with rate and state dependent friction', *J. Mech. Phys. Solids*, **32**, 167–196.
Okubo, P. G., and Dieterich, J. H. (1981) 'Fracture energy of stick–slip events in a large-scale biaxial experiment', *Geophys. Res. Letters*, **8**, 8, 887–890.
Palmer, A. C., and Rice, J. R. (1973) 'The growth of slip surfaces in the progressive failure of overconsolidated clay slopes', *Proc. Roy. Soc. Lond.*, Ser. A332, 527.
Peirce, D., Asaro, R. J., and Needleman, A. (1982) 'An analysis of nonuniform and localized deformation in ductile single crystals', *Acta Metallurgica*, **30**, 1087–1119.
Rice, J. R. (1976) 'The localization of plastic deformation', in *Theoretical and Applied Mechanics* (Proc. 14th Int. Congr.) (Ed. W. T. Koiter), North-Holland, Amsterdam, p. 207.
Rice, J. R. (1980) 'The mechanics of earthquake rupture, in *Physics of the Earth's Interior* (Proceedings of the International School of Physics 'Enrico Fermi', Course 78, 1979) (Ed. A. M. Dziewonski and E. Boschi), Italian Physical Society (printed by North-Holland, Amsterdam), pp. 555–649.
Rice, J. R. and Ruina, A. L. (1983) 'Stability of steady frictional slipping', *Trans. ASME. J. Appl. Mech.*, **50**, 343–349.
Ruina, A. L. (1983) 'Slip instability and state variable friction laws', *J. Geophys. Res.*, **88**, 10359–10370.
Stuart, W. D. (1979) 'Strain softening prior to two-dimensional strike slip earthquakes', *J. Geophys. Res.*, **84**, 1063–1070.
Wong, T. -F. (1982) 'Shear fracture energy of Westerly granite from post-failure behavior', *J. Geophys. Res.*, **87**, B2, 990–1000.

PART V

FRACTURE PROPAGATION AND FRACTURE ENERGY

Mechanics of Geomaterials
Edited by Z. Bažant
© 1985 John Wiley & Sons Ltd

Chapter 12

Fracture Propagation in Rock

A. R. Ingraffea

12.1 INTRODUCTION

Suppose one wants to predict numerically the stability and trajectory of a set of discrete cracks in rock or a rock mass. Further, suppose one wants to apply a rigorous, fracture-mechanics-based approach to this situation. Two problems immediately present themselves. First, how does one measure, accurately and inexpensively, the fracture toughness of the rock? Second, how does one formulate a numerical model which captures all the physics required by fracture mechanics as applied to rock?

The intent of this paper is to answer both of these questions. A survey of actual and potential applications of fracture mechanics principles to the problem of crack propagation in rock is unnecessary. One has only to scan the Proceedings of the United States National Symposia on Rock Mechanics over the past decade to understand that such applications range over virtually all time and distance scales of interest to geotechnical, structural, and materials engineers. An excellent starting point for such a perusal is the excellent keynote paper by Fairhurst and Cornet (1981).

Again, rather than summarizing all developments in fracture toughness testing of rock, this paper will detail a single method which is likely to become a practical standard. The reader is referred to excellent summaries by Barton (1983) and Ouchterlony (1980a) for range and background on this topic.

While other, more exotic numerical methods are being brought to bear on the modelling problem, approaches based on standard finite and boundary element formulations are detailed here. The reader will surely find counterpoint to these methods in the accompanying Discussion to this paper.

Finally, what of the synergism between computer and numerical method in the area of rock fracture modelling? State-of-the-art interactive graphic techniques can, as will also be shown here, greatly facilitate the solution to this difficult class of problem.

12.2 FRACTURE TOUGHNESS TESTING

As part of his activity with the ASTM Subcommittee E24.07 on Fracture of Brittle Non-metallic Materials, the writer conducted an international survey of fracture toughness testing techniques for rock. Results of that survey, conducted in 1979, indicated that no fewer than ten different specimen geometries had been employed for this purpose. Since then, interest has continued to grow rapidly as evidenced by the recent compilations by Ouchterlony and Barton mentioned above. Together, these summaries of test results indicate that at least 70 rock types have been tested using no fewer than two dozen different geometries.

12.2.1 Evolution of practical fracture toughness testing specimens

Selection of the most appropriate geometry for practical application, as opposed to basic research, testing can be approached by applying the following constraints. The specimen must:

(1) Be easily and inexpensively prepared from core with minimal wastage. It is anticipated that a large number of tests would be required to accommodate in a statistically meaningful way variation in test parameters and lithologies. This constraint in itself effectively eliminates from consideration geometries requiring many and accurate machining operations. In the writer's best judgement, the only candidate geometries remaining after application of this constraint are core-based; these are the hollow pressurized cylinder (PC) (Abou-Sayed and Simonson, 1977), the short-rod (SR) (Barker, 1977a), the single-edge-cracked (or chevron-edge-notched) round-bar-in-bending (SECRBB, CENRBB) (Ouchterlony, 1980b), and the disc-shaped compact specimen (DC(T)) (Newman, 1979). Figure 12.1 shows how it would be possible, using the last three geometries, to use a single length of core to

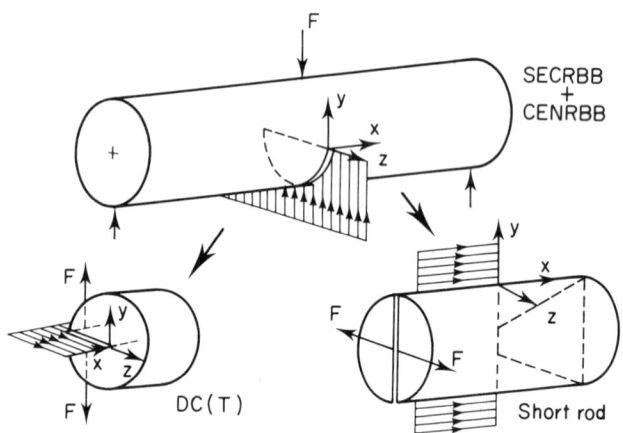

Figure 12.1 Three core-based test specimens from one piece of core. From Ouchterlony and Zongqi (1983)

measure toughness in three mutually perpendicular directions (Ouchterlony and Zongqi, 1983). It is apparent that if there is not a large anisotropy in toughness, the more-difficult-to-test DC(T) specimen would be superfluous.
(2) Be able to produce valid fracture toughnesses with one dimension as small as two inches. Since only preliminary testing can indicate minimum dimension requirements for valid results, it is difficult to say whether this constraint would, *a priori*, eliminate any of the candidates listed above with the common NX core size. It should be noted, however, that only one of those geometries, the short-rod, offers the possibility of testing cracks longer than the core diameter.
(3) Be easily instrumented and loaded, even at high temperatures. This constraint effectively eliminates the PC from consideration because of the effect of high temperature on the pressurizing fluid and pressure seals. Loading and instrumentation on the SECRBB/CENRBB and SR are straightforward at room temperature, but require some modification for high temperature testing.
(4) Possess a firm analytical basis in the form of compliance and stress intensity factor calibration. The SR (Beech and Ingraffea, 1982; Bubsey *et al.*, 1982; Ingraffea *et al.*, 1983b), the SECRBB/CENRBB (Ouchterlony, 1980b, 1981a) and the DC(T) (Newman, 1979) meet this constraint.
(5) Have a proven track record of use on rock. The writer (Ingraffea *et al.*, 1982, 1983a), Barker (1977), and Atkinson (1984) have together tested as least 30 different rock types using the SR. The SECRBB/CENRBB has also seen extensive use on rock (Ouchterlony, 1980b, 1981b).

Other constraints may well arise. For example, if available core is subject to discing, the SECRBB/CENRBB may not be suitable in that, with its crack propagating in the direction of the discing planes, it might produce biased toughness measurements.

Application of these constraints and observations indicates that only two candidate specimens currently exist, the SECRBB/CENRBB and the SR. Actually, of the two RBB specimens, only the CENRBB should receive further considerations since, like the SR, it has the distinct advantages of a chevron notch design. These considerable advantages are:

(1) No fatigue pre-cracking is required since a natural crack is produced during the stable growth phase of loading.
(2) Neither crack length nor displacement measurements are required if the specimen conforms to LEFM size restrictions. Even if small-scale inelasticity is permitted, that is, the specimen is used for J_{1C} or K_{Ic}^p (defined below) measurement, crack length is still not required.
(3) To evaluate K_{1c} on specimens meeting size restrictions, only the maximum load applied by a soft testing machine is required.

In a comparison between the SR and CENRBB specimens, the CENRBB is easier to prepare, while for the SR substantially less core length is required, load-line-displacement is easier to measure and, as noted above, much longer natural crack lengths are available.

The author has used the SR extensively in fracture toughness testing of a wide variety of rocks at Cornell University (Ingraffea et al., 1982, 1983a). In the next section experience with the SR is detailed with descriptions of specimen calibration, testing technique, and result interpretation.

12.2.2 The short-rod testing system

Considerable effort has been expended in establishing complete testing systems built around the short-rod (Barker, 1979; Ingraffea et al., 1983a). The starting point for such a system is accurate compliance and stress-intensity factor calibration of the specimen. An entire symposium (Ingraffea et al., 1983a; Raju and Newman, 1983) was recently held on this topic. A proposed standard short-rod geometry, Figure 12.2, very similar to that used in all testing at Cornell, was analysed by a number of workers using both finite and boundary element techniques. The results of these analyses can be expressed as follows. For critical stress-intensity factor,

$$K_{Ic} = \frac{P_{max} \bar{Y}_{min}}{B\sqrt{W}} \quad (12.1)$$

where

P_{max} = maximum applied load
B = specimen diameter
W = specimen length

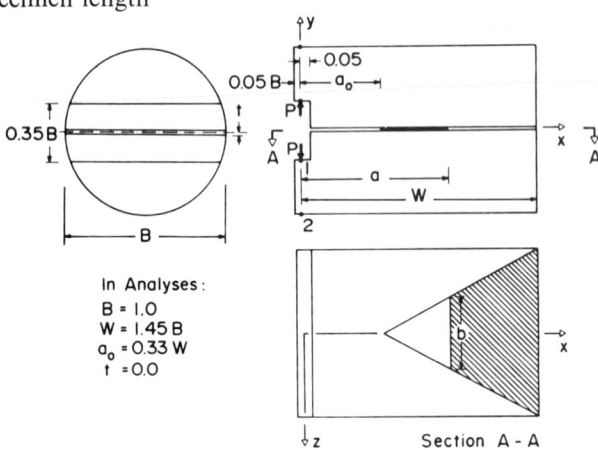

Figure 12.2 Preferred short-rod specimen geometry and nomenclature. From Ingraffea et al. (1983b)

and $\bar{Y}_{min}$ is the minimum value of the average, normalized stress intensity factor given by

$$\bar{Y} = \frac{K_1 B \sqrt{W}}{P} = \sqrt{W} \left[\frac{1}{2b} \frac{d(CEB)}{d(a/B)} \right]^{1/2} \qquad (12.2)$$

here

b = length of the crack front
a = crack length
C = compliance at point 1 in Figure 12.2.

The $\bar{Y}_{min}$ value found in the author's analyses is 28.3 and occurs at a critical crack length, a_c, of $0.83B$. This value is an excellent agreement with those obtained analytically and experimentally by other workers (Bubsey et al., 1982; Raju and Newman, 1983).

A least-squares fit to computed compliances produced the following express:

$$CEB = -3049 + 20124.7(a/B) - 52052.3(a/B)^2$$
$$+ 67368.7(a/B)^3 - 43334.7(a/B)^4$$
$$+ 11192.9(a/B)^5 \qquad (12.3)$$

which is valid for $0.65 < a/B < 1.1$. Equations (12.1) to (12.3) are valid if LEFM restrictions are met by the specimen size in use. The short-rod can also be used, however, to produce J_{Ic} or J_{Ic}-like measures of toughness through simple calculations described later.

A practical testing system should not involve time-consuming, expensive specimen preparation procedures. Three simple operations are used to prepare a short-rod specimen directly from rock cores. First, the specimen is cut to a nominal length of $1.45B$ using a standard, water-cooled, rock cut-off saw. The ends are made parallel by proper adjustment of the core guide prior to cutting. Second, the diameteral cuts are made. The specimen is held at the proper angle for each of two cuts necessary to produce the chevron notch by a simple fixture which prohibits core rotation about its axis between cuts. Lastly, metal end plates are epoxied to the top surfaces to act as loading lines for the splitting force. The use of plates as opposed to the groove shown in Figure 12.2 was chosen for load transfer because the latter method would require a grinding operation. Preparation time is such that a technician with minimal training can prepare 20–30 specimens a day from rock cores. A prepared specimen is shown in Figure 12.3.

The method of testing of a short-rod is also straightforward and involves application of an opening load to the mouth of the specimen. As the load is increased, a crack initiates at the point of the chevron slot and advances longitudinally in a stable manner, tending to split the specimen in half (Figure 12.3, right). If micro-cracking and plasticity effects are negligible, the opening load reaches a maximum when the crack reaches a critical location; thereafter, the crack-advancing load decreases with further crack growth. The maximum

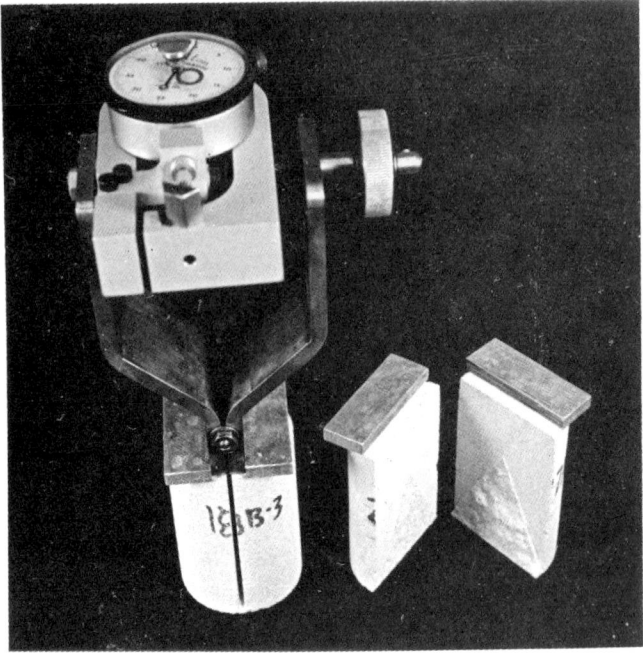

Figure 12.3 NX size short-rod specimens of Indiana limestone. Prepared for test, left. After test, right

load is linearly related to the fracture toughness, K_{Ic}, from fundamental principles of linear elastic fracture mechanics by way of Equation (12.1).

In tests at Cornell the splitting force is applied with a simple mechanical testing apparatus. The loading device is hand-held and -operated and is shown in Figure 12.3. Turning the actuating knob creates the splitting force which is read directly on an integral force gauge. General test procedure consists of inserting the jaws of the loading device between the end plates, turning the knob, and recording the failure load which is given by a following needle on the gauge.

Testing of a range of rock types showed that application of the splitting force sometimes caused horizontal shearing failure in specimens with weak bedding planes when these planes were close to perpendicular to the expected fracture plane (Figure 12.4, lower right). The method used to eliminate this phenomenon was an application of an axial pressure prior to testing. This pressure serves to give the bedding planes greater shearing resistance. The clamps used to apply the pressure were calibrated. Accordingly, a known torque on the clamps produced a known pressure. The modification to the general testing procedure for axial loading involved applying and tightening the clamps to give the desired axial pressure prior to insertion of the jaws of the loading device. As before, the failure load is recorded and is substituted into a relationship like Equation (12.1) which

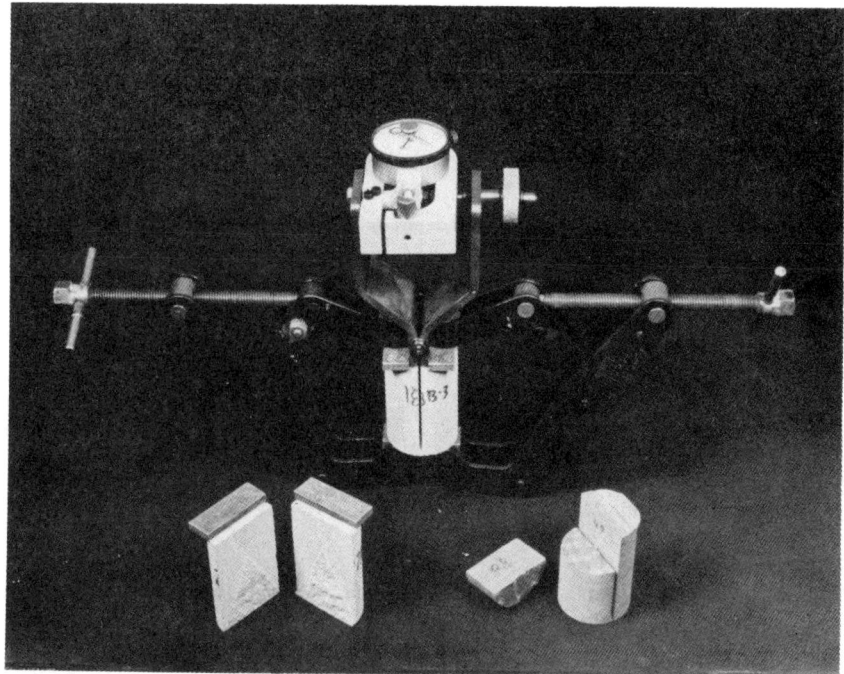

Figure 12.4 Premature failure of specimen, lower right. Calibrated clamps on specimen, rear

has been corrected for axial pressure (Ingraffea et al., 1983a). Figure 12.4 shows the clamps attached to a specimen ready for testing.

If the crack tip process zone, which might be experiencing micro-cracking and/or plasticity, is too large relative to the crack length to be negligible, but still sufficiently small compared to other specimen dimensions, the measured critical stress intensity assumes the role of K_Q. An inelasticity correction factor can then be applied to K_Q to produce a K_{Ic}^p value. Hereafter K_{Ic}^p is used to denote the 'elasto-plastic' K_{Ic}, whose meaning and measurement were first described by Barker (1979). K_{Ic}^p is similar to J_{Ic} in that it is a value of toughness which would be equal to a K_{Ic} obtained from a specimen large enough to meet LEFM restrictions. There are substantial differences between the two and also certain advantages in using the K_{Ic}^p approach. These are discussed by Barker (1979) who used a K_{Ic}^p approach to obtain valid K_{Ic} values for Indiana limestone from specimens which were too small to produce K_{Ic} from LEFM calculations (Barker, 1977b).

If the specimen is sub-size, a record of load versus any displacement proportional to the load-line displacement is required to measure K_{Ic}^p. The equation proposed by Barker (1979) for computing K_{Ic}^p is,

$$K_{Ic}^p = \left(\frac{1+p}{1-p}\right)^{1/2} K_Q \qquad (12.4)$$

where

K_Q = critical stress intensity as computed from Equation (12.1), where, for this case, P_{max} is replaced by the load corresponding to the critical crack length, $a_c = 0.83B$.
p = inelasticity correction index.

The correction index is computed from a load–displacement plot, such as the hypothetical example shown in Figure 12.5, by way of,

$$p \equiv \frac{\Delta x_0}{\Delta x} \qquad (12.5)$$

The reader should study (Barker, 1979, 1977b) References 19 and 20 for justification of Equations (12.4) and (12.5).

Figure 12.6 is a plot of apparent toughness, K_Q, versus crack length for over 100 tests on Indiana limestone. Four investigators and five specimen geometries were used over a period of ten years to generate these results. Of special interest here are the short-rod results from Cornell, labelled as present study, and by Barker. Results from 34 tests at Cornell (slightly corrected for deviation from

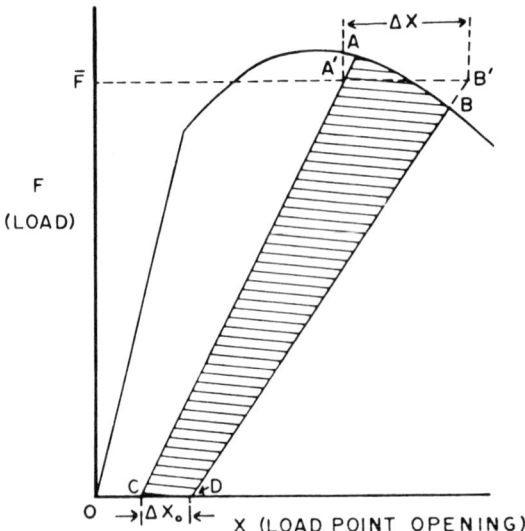

Figure 12.5 Idealized load versus load point opening loading cycle for short-rod specimen with small-scale inelasticity. From Barker (1979)

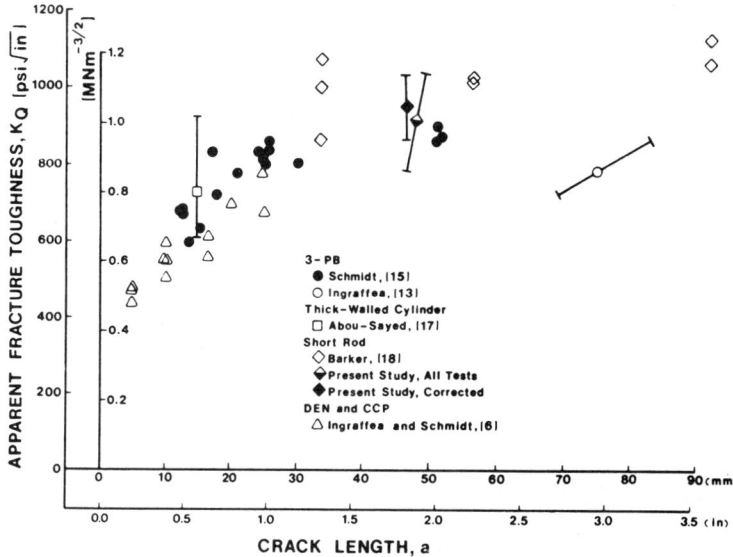

Figure 12.6 Summary of K_Q versus crack length test results for Indiana limestone. References cited and figure itself are in Ingraffea et al. (1983a)

nominal dimensions) plot well within the K_Q vs a trend established by other geometries. Barker's results are actually values of K_{Ic}^p and indicate that, for Indiana limestone, the NX core size used for the Cornell tests is not quite large enough for LEFM use.

It is certainly safe to say that, with the introduction of the short-rod testing system, toughness measurements, even when limited by available core size, are now practicable. With information controlling the behaviour of a discrete crack available from testing, it follows that techniques for employing such information in an analytical model should be addressed.

12.3 NUMERICAL MODELLING OF DISCRETE CRACK PROPAGATION

Why should one study fracture propagation? Is not prediction of fracture initiation the object of fracture mechanics? In his papers on rupture under tensile and compressive loading, Griffith (1921, 1924) proposed conditions for fracture initiation which he presumed to be coincident with structural instability. The vast majority of the fracture mechanics research since Griffith has addressed the problem of predicting structural failure as the immediate consequence of fracture initiation. Yes, considerable attention has been focused on sub-critical crack growth as in fatigue and ductile fracture. However, the amount of propagation before fracture initiation in these cases is typically small compared to that which potentially occurs after. Why, then, should one be interested in modelling propagation: where a crack goes, what it does along the way, and how much energy it takes to get there?

There is nothing in the rules of rock mechanics (or fracture mechanics, either) which says that a fracture, once initiated, is always unstable. It may stop. Where? Why? What was its trajectory? What must be done to get it going again? Moreover, these questions must be addressed for each crack since in rock mechanics study of the propagation of a single crack is not often the case. So, already one has:

Observation # 1: Fracture propagation can be stable in the load control sense. Assuming linear elastic fracture mechanics (LEFM) conditions (unless otherwise noted), this implies that stress-intensity is decreasing with increasing crack length. As shall be seen, this situation is generally due to the preponderance of compressive *in situ* stresses and loadings in geotechnical engineering. Obviously, it will not be sufficient just to compute stress-intensity factors for an initial crack configuration. A good numerical model should be able to update stress-intensity factors as crack length changes.

Observation # 2: Propagation of multiple cracks is common in realistic problems of rock mechanics. A numerical model should be versatile enough to accommodate propagation of many cracks.

This latter observation leads to another difficulty in modelling of fracture propagation. With each growth increment of a given crack, a new boundary value problem is generated. Displacement and traction boundary conditions may change, stress trajectories are altered, even loading may change in direction and intensity. As a consequence, propagation of one crack may cause initiation of another, and cracks may influence each other's stability and trajectory. This is clearly the case in dynamic fracture because of stress wave reflections. But without a periodic 'look' at the full stress field during quasi-static propagation, one might overlook:

Observation # 3: Each increment of fracture changes the structure. One should be able to predict the effects of this change on the stress field and on other cracks.

With the possibility of multiple cracks propagating quasi-statically one must dispense with another simplification often applied in fracture mechanics, that of self-similar propagation. Curvilinear (mixed-mode, to some people) crack propagation is common in rock mechanics. Therefore, one must have:

Observation # 4: Cracks curve during propagation in response to a change the stress field. A numerical model should be able to predict a changing crack trajectory.

Moreover, if one accepts that mixed-mode stress intensity factors, for example K_I and K_{II}, can be present along a crack front, then three follows:

Observation #5: Theoretically, mixed-mode fracture initiation can occur when $K_I \leqslant K_{Ic}$. Consequently, a numerical model must incorporate an interaction theory which accurately predicts the critical mixture of stress intensity factors which will cause the next increment of propagation.

Of course, fracture can be foe as well as friend to the geotechnical engineer. Mine, tunnel, dam abutment, and rock foundation instability problems are often the result of unpredicted fracture propagation. A numerical model which can predict the likelihood of a roof fall or rock burst, and can suggest a method of fracture stabilization or an alternate form of energy release can be an invaluable design tool.

Clearly, the problems of crack propagation modelling, even with the simplifying assumptions of LEFM, are manifold. The problem begins with fracture initiation, so one has to go a bit further than simple stress-intensity factor solutions. This is the purpose of the present section. Within the context of modern techniques of stress analysis, the finite and boundary element methods (FEM and BEM), the following topics are addressed:

(1) A method for efficient, accurate calculation of stress intensity factors for substitution into.
(2) Mixed-mode fracture initiation theories for critical mixture and angle change predictions.
(3) Methods for crack increment length prediction for a given load change, or, conversely, the prediction of the load required to drive a crack a given distance.
(4) Algorithms for incorporating the above into efficient computer programs.
(5) Techniques for efficient remeshing to accommodate discrete crack growth.
(6) The use of interactive computer graphics in the highly adaptive, nonlinear field of fracture propagation modelling.

This will be accompanied with example problems whenever possible. The first of these will be presented in the next section to provide a physical basis for the observations particular to rock fracture just presented.

12.3.1 The nature of fracture propagation in rock

The following example problem will serve a number of purposes. It will lend physical insight into characteristics particular to fracture propagation in rock. It

230   Mechanics of Geomaterials

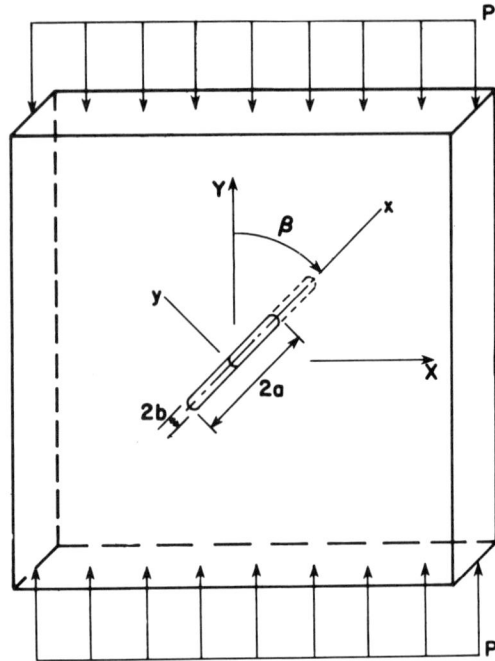

Figure 12.7 Angle-notched plate loaded in uniaxial compression. From Ingraffea (1977)

will clarify some misconceptions and their implications regarding the theoretical fracture resistance of rock structures. Finally, it will serve as a basis for development and comparison on the techniques required for modelling fracture propagation.

Example #1: Observations on fracture propagation under compression
The problem is shown in Figure 12.7 and is recognized to be that addressed by Griffith in his second paper (1924). Rock plates like those shown in Figure 12.7 were tested by the author (Ingraffea, 1977; Ingraffea and Ko, 1981; Ingraffea and Heuze, 1980) with the following results:

(1) As predicted by Griffith (1924), first crack growth occurs from points initially under tensile stress concentration on the notch (see Figure 12.8). This set of two, symmetrically placed cracks was labelled primary.
(2) Primary crack trajectory was curvilinear.
(3) In contrast to what Griffith (1924) expected, propagation of primary cracks was observed to be stable: an ever-increasing load was required to increase crack length.
(4) After considerable primary crack propagation, a second set of two, symmetri-

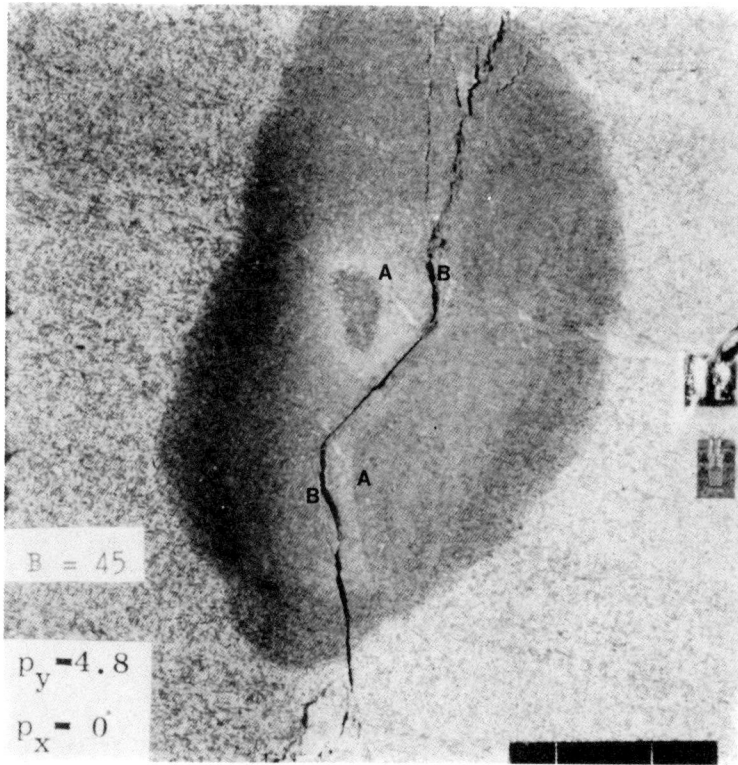

Figure 12.8 Results of test on Indiana limestone plate. Primary cracks to points A. Secondary cracks nucleated near points B. From Ingraffea (1977)

cally placed, cracks appeared. These were labelled secondary and originated in the interior of the plate in a newly formed tensile stress zone (see Figure 12.9).

(5) Failure of the plate, defined as a throughgoing rupture, was a result of unstable secondary crack propagation, Figure 12.8, at a load level in the range of 3 to 5 times the primary crack initiation load.

These observations were typical for plates of Indiana limestone and St Cloud charcoal granodiorite, with $30° \leq \beta \leq 90°$.

Observations 2 and 3 differentiate the observed fracture response of this configuration from that usually observed in tension-loaded structures. Stable primary crack propagation indicates that, within the assumption of LEFM, the associated energy release rate, G, decreases with increasing crack length for a constant load. The curvilinear nature of the primary crack path is a result of a variable, mixed-mode stress intensity being applied to the incrementally advancing crack tip.

232 *Mechanics of Geomaterials*

Figure 12.9 Composite micrograph of granodiorite fractures. Notch tip, A; primary crack tip, B; nucleating secondary crack, C and D. From Ingraffea (1977)

Items (3) to (5), above, deserve special attention. The often-quoted, and quite incorrect, theoretical ratio of compressive to tensile strength of rock is based on the supposition that the *initiation* of what are here called primary cracks is synonymous with rupture. That such is not the case has been observed by many experimentalists (e.g. Cotterell, 1972; Bombalakis, 1968; Hoek and Bienawski, 1965) and digested by few of anybody else: sub-critical crack growth can occur under ideal LEFM conditions and monotonically increasing load.

Observations 4 and 5 are particular to rock. In tests on glass, polymethylmethacrilate (PMMA) and CR39 in the same configuration (Cotterell, 1972; Bombalakis, 1968; Hoek and Bienawski, 1965) only primary cracking was evident and rupture did not occur. This phenomenon is shown in Figure 12.10 which

Figure 12.10　Results of test on PMMA plate. Primary cracking only; no rupture occurs. From Ingraffea (1977)

depicts the primary crack behaviour of a PMMA plate loaded to near its compressive yield stress. Results 4 and 5 therefore indicate a fundamental difference in the fracture response of rock structures as compared to glass, plastic, and metals. As we shall see later, the high (though not theoretically predictable!) compressive to tensile strength ratio of rock compared to those materials leads to what the author has called the strength ratio effect (Ingraffea, 1979). This effect explains the initiation of what are called secondary cracks in the present problem, and is the proximate cause of rupture in this as well as many other problems in rock fracture. The strength-ratio effect is actually a corollary to Observation 3 mentioned earlier, but bears individual emphasis:

Observation #6: Owing to fracture propagation, new regions of tensile stress can be generated. Although the magnitude of those tensile stresses may be low compared to an applied compressive stress, the relatively low tensile strength of rock makes such regions potential sites for nucleation of additional cracks. A model for fracture propagation should be capable of predicting formation of such sites.

With these experimental observations, and with a list of requirements for modelling of fracture propagation in hand, one can begin model formulation.

12.3.2 Stress intensity factor computation

The prediction of load level, angle change, and length corresponding to each increment of fracture propagation requires accurate computation of mixed-mode stress intensity factors. Their efficient computation is also desirable since, as noted above, many analyses may need to be performed in a single problem. Virtually all of the numerical methods applicable to elastostatics have been used for stress intensity factor computation. A particularly simple yet accurate method which does not depart from standard FEM and BEM approaches is available. This method is described in detail elsewhere (Barsoum, 1976, 1977; Freese and Tracey, 1977) however, it will be briefly summarized here.

Consider the typical crack-tip region shown in Figure 12.11. Assume that the region has been discretized with standard, isoparametric, linear-strain triangles and that two of these are arrayed as shown. It has been shown that these elements will reproduce the LEFM-predicted displacement and stress fields if, as shown in Figure 12.11, the mid-side nodes of all element sides emanating from the crack tip are moved to the quarter-point position. Shih et al. (1976) have further shown that the displacements computed at the labelled nodes in Figure 12.11 can be correlated with the theoretical displacements at their positions with the stress intensity factors as weight functions. Solving for the stress intensity factors leads directly to,

$$K_I = \sqrt{\frac{2\pi}{L}} \frac{G}{\kappa + 1} [4(v'_B - v'_D) + v'_E - v'_C]$$

$$K_{II} = \sqrt{\frac{2\pi}{L}} \frac{G}{\kappa + 1} [4(u'_B - u'_D) + u'_E - u'_C]$$

(12.6)

in which

$L=$ length of singularity element side along the ray
$v' =$ crack-opening nodal displacements
$u' =$ crack-sliding nodal displacements
$\kappa = (3 - 4v)$ for plane strain.

The primes indicate that the global-coordinate nodal displacements have been transformed to the crack-tip coordinate system defined in Figure 12.11. The above procedure has been generalized to the three-dimensional case by Ingraffea and Manu (1980). This approach is also available, with only slight modification, for BEM codes which use isoparametric element formulations (Cruse and Wilson, 1977; Blandford et al., 1982). For studies on the accuracy of the method the reader should see Barsoum (1976, 1977) and Ingraffea and Manu (1980).

Algorithmically, the displacements and coordinates of the crack face nodes

Fracture Propagation in Rock 235

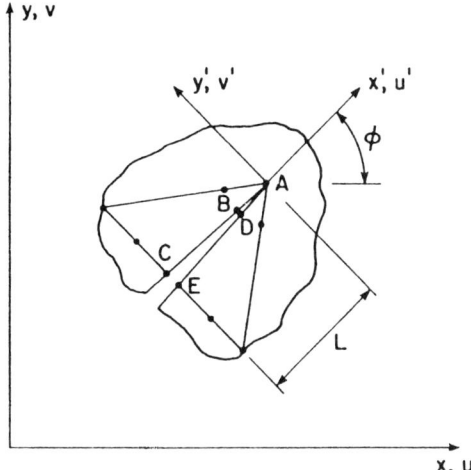

Figure 12.11 Nodal lettering for stress intensity factor computation using Equation (12.6)

belonging to the quarter-point FEM or BEM elements need to be flagged and retrieved for each crack increment solution. These are then transferred to a simple subroutine which codes Equations (12.6). In a crack propagation analysis, this computation would be done for each crack tip after each load or growth increment. The computed stress intensity factors would determine the stability and angle change of each crack tip according to one of the theories outlined in the next section.

12.3.3 Theories of mixed-mode fracture

The determination of fracture initiation from an existing flaw in combined Mode I and Mode II requires knowledge of the stress intensity factors, determined analytically and functions of geometry and load, and the appropriate fracture toughness, a material state property, determined experimentally. These parameters combine in a theoretical mixed-mode fracture initiation function analogous to a multi-axial stress state yield function of plasticity. A number of theories are available which produce such a function (Erdogan and Sih, 1963; Sih, 1973; Hussain et al., 1974). Only one of these theories will be briefly summarized here.

The $\sigma(\theta)_{\max}$ theory was first formulated by Erdogan and Sih (1963). The parameter governing fracture initiation in their theory is the maximum circumferential tensile stress, $\sigma(\theta)_{\max}$, near the crack tip.

Given a crack under mixed-mode conditions, the stress state near its tip can be expressed in polar coordinates as,

$$\sigma_r = \frac{1}{\sqrt{2\pi r}} \cos\frac{\theta}{2} \left[K_1\left(1 + \sin^2\frac{\theta}{2}\right) + \tfrac{3}{2} K_{11} \sin\theta - 2 K_{11} \tan\frac{\theta}{2} \right] + \cdots$$

$$\sigma_\theta = \frac{1}{\sqrt{2\pi r}} \cos\frac{\theta}{2} \left[K_1 \cos^2\frac{\theta}{2} - \frac{3}{2} K_{11} \sin\theta \right] + \cdots \quad (12.7)$$

$$\tau_{r\theta} = \frac{1}{\sqrt{2\pi r}} \cos\frac{\theta}{2} [K_1 \sin\theta + K_{11}(3\cos\theta - 1)] + \cdots$$

The $\sigma(\theta)_{\max}$ theory states that:

(1) Crack extension starts at the crack tip and in a radial direction.
(2) Crack extension starts in a plane normal to the direction of greatest tension, i.e. at θ_0 such that $\tau_{r\theta} = 0$.
(3) Crack extension begins when $\sigma_{\theta_{\max}}$ reaches a critical, material constant value.

The theory is stated mathematically using Equation (12.7),

$$\sigma_\theta \sqrt{2\pi r} = \text{constant} = \cos\frac{\theta_0}{2} \left[K_1 \cos^2\frac{\theta_0}{2} - \frac{3}{2} K_{11} \sin\theta_0 \right] = K_{1c} \quad (12.8)$$

or

$$1 = \cos\frac{\theta_0}{2} \left[\frac{K_1}{K_{1c}} \cos^2\frac{\theta_0}{2} - \frac{3}{2}\frac{K_{11}}{K_{1c}} \sin\theta_0 \right] \quad (12.9)$$

and

$$\tau_{r\theta} = 0 \Rightarrow \cos\frac{\theta_0}{2} [K_1 \sin\theta_0 + K_{11}(3\cos\theta_0 - 1)] = 0 \quad (12.10)$$

Equations (12.9) and (12.10) are the parametric equations of a general fracture initiation locus in the K_1–K_{11} plane, shown in Figure 12.12. Also, the direction of the initial fracture increment, θ_0, can be found from Equation (12.10) which gives,

$$\begin{aligned} &\theta_0 = \pm \pi \text{(trivial)} \\ &K_1 \sin\theta_0 + K_{11}(3\cos\theta_0 - 1) = 0. \end{aligned} \quad (12.11)$$

In summary, the governing equations of the $\sigma(\theta)_{\max}$ theory are (12.19) and (12.10). Algorithmically, the stress intensity factors for a given crack tip location and loading are first substituted into Equation (12.10) to obtain the new angle of propagation, θ_0. The stress intensity factors and the angle θ_0 are then substituted into Equation (12.9). If it is not satisfied, the stress intensity factor pair plots either within or outside the fracture locus shown in Figure 12.12. If within, then that crack cannot propagate without a sufficient increase in stress intensity factors. If

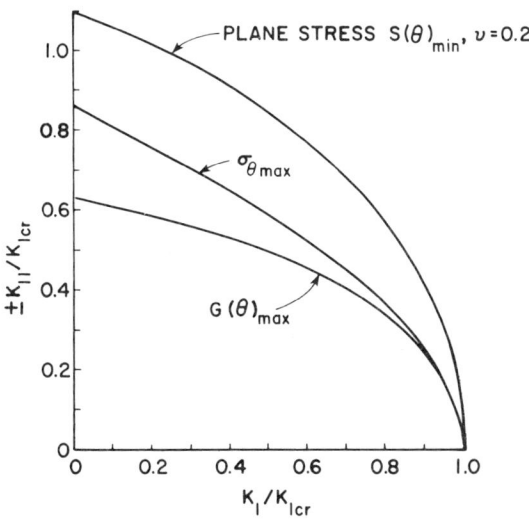

Figure 12.12 Fracture initiation loci for mixed-mode theories

outside, then the crack is unstable and can continue to propagate until it reaches a free surface or until the stress intensity factor pair returns to within the locus.

In a quasi-static fracture propagation analysis the governing equations for one of the theories would be applied at the end of each growth step or load step. Recall Observation 3: it may not be necessary to increase loads to bring the stress intensity factors of previously stable crack tip onto the fracture locus. The propagation of another crack may cause the same effect. Algorithmically, this implies that the interaction factor for each crack tip, the right-hand side of Equation (12.9), be updated in memory after each crack or load increment. As shall be seen later, depending on the mode of interaction between the program and the user, the former or the latter will use the interaction factors to decide which one or more of the crack tips should be propagated in a given fracture step.

12.3.4 Comparison of mixed-mode fracture theories

Figure 12.12 compares the interaction effects predicted by the $\sigma(\theta)_{max}$, $S(\theta)_{min}$ (Sih, 1973), and $G(\theta)_{max}$ (Hussain et al., 1974) theories. It can be seen that $S(\theta)_{min}$ theory is the least conservative of the three shown. It also predicts that the Mode II fracture toughness of most rocks ($v < 0.3$) is larger than K_{Ic}, while the other theories predict a smaller value.

How does theory compare to experiment? Ingraffea (1977) and Kordisch and Sommer (1978) contain much data and lengthy discussion relevant to this question, but the comparisons are all based on materials other than rock. The author has performed wide spectrum, mixed-mode fracture initiation tests on

Indiana limestone and Westerly granite (Ingraffea, 1981). The results are shown in Figure 12.13(a) and 12.13(b) for the limestone and granite, respectively. It appears, based on this somewhat limited data, that the $S(\theta)_{min}$ theory is the most accurate of the three theories used for comparison.

It should be emphasized, however, that a crack finding itself under substantial

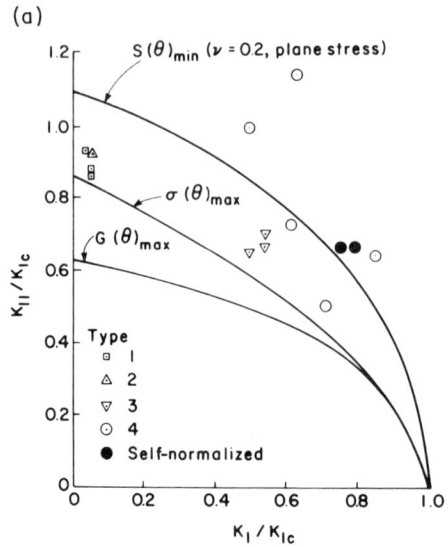

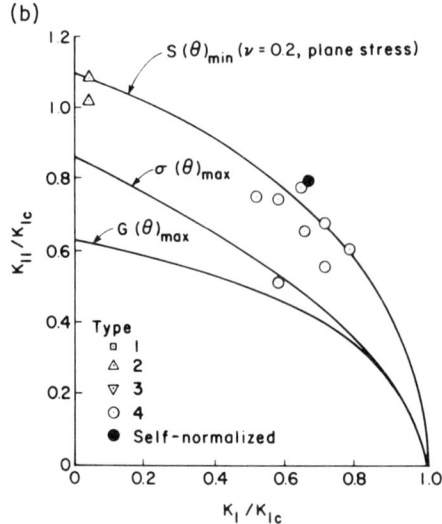

Figure 12.13 Results of mixed-mode fracture initiation tests. (a) Indiana limestone; (b) Westerly granite. From Ingraffea (1981)

Fracture Propagation in Rock

Mode II loading does not long remain in the high K_{II}/K_I domain of interaction. Such a crack quickly changes trajectory to minimize or eliminate the K_{II} component. Consequently, the life of a crack propagating quasi-statically is spent in the high K_I/K_{II} region of the interaction plane where the differences among the theories are minimal. It is the author's opinion that, except for fracture increments under high K_{II}, use of any of the referenced theories would result in substantially the same trajectory and load history. This will be seen in example problems to follow.

12.3.5 Predicting crack increment length

Previous sections showed how to compute stress intensities, and how to use them to predict local stability and angle change. However, Observation 1 is a remainder that to complete a fracture propagation model one must also be able to predict either, (a) the length of a fracture increment for a given load change, or, (b) the load change required to derive a crack a specified length. These predictions are relatively simple and straightforward.

The fundamental principle here is that a fracture, once initiated, will continue to propagate as long as there is sufficient energy or, equivalently, effective stress intensity, available. Effective stress intensity, K^*, here refers to a mixed-mode case and is the combination of Mode I, II, and III stress intensity factors required by the particular mixed-mode theory in use. The right-hand side of Equation (12.9) can, therefore, be viewed as a normalized effective stress intensity factor.

One must consider a number of possible stability cases in creating an algorithm for predicting fracture increment length. These possibilities will be addressed through Figures 12.14 and 12.15. For simplicity, assume that LEFM applies for all crack lengths. Further assume that one is investigating propagation along some predicted direction θ_0.

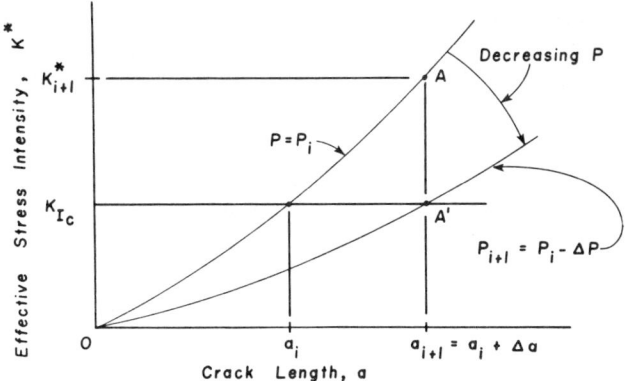

Figure 12.14 Stress intensity factor variation for Case 1

240   Mechanics of Geomaterials

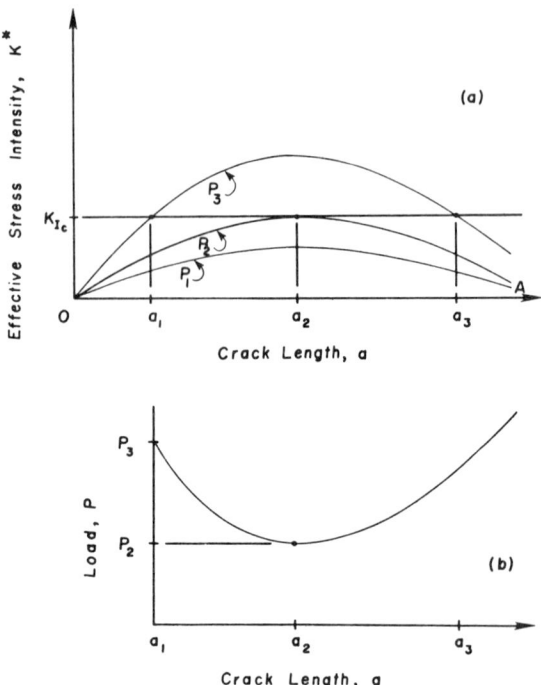

Figure 12.15   Stress intensity factor variation for Case 2; (b) load variation for Case 2

Case 1: Effective stress intensity increases monotonically with crack length, curve OA in Figure 12.14. If the initial flaw size is less than a_i, no propagation occurs. For $a = a_i$, propagation can occur and it will continue at $P = P_i$; that is, a condition for local instability has been met. Of course, an algorithm could be written which would place such a scenario in displacement or crack-length control: A crack increment, Δa, could be specified and the load decrement required to just bring the crack tip to a $a_i + \Delta a$ could be computed. This situation is depicted by curve OA' in Figure 12.14. To compute P_{i+1} recall that LEFM specifies that at instability,

$$K_{Ic} = \alpha_i P_i \sqrt{a_i} = \alpha_{i+1} P_{i+1} \sqrt{a_{i+1}} \qquad (12.12)$$

where α = factor depending on geometry and interaction theory.

Therefore,

$$P_{i+1} = \frac{\alpha_i}{\alpha_{i+1}} P_i \sqrt{\frac{a_i}{a_{i+1}}} \qquad (12.13)$$

Equation (12.13) is only directly useful, however, if the α_{i+1} coefficient is known at Step *i*. For arbitrary problems, this is certainly not the case. An alternative is to

Fracture Propagation in Rock   241

propagate the fracture an amount Δa in the direction θ_0 and compute K^*_{i+1} at load level P_i. The new load level is then,

$$P_{i+1} = \left(\frac{K_{Ic}}{K^*_{i+1}}\right) P_i \qquad (12.14)$$

as can be seen in Figure 12.14. Behaviour described in this case is typical of many of the Mode I fracture specimens used to measure K_{Ic} of rock. It can also occur in a variety of circumstances in practical rock fracture problems.

Case 2: Effective stress intensity increases, reaches a maximum value, and then decreases with increasing crack length, curve OA in Figure 12.15(a). For the value of K_{Ic} shown and at load level P_1, no crack propagation is possible. At load level P_2, propagation is possible only at crack length $a = a_2$, but the corresponding, theoretical fracture increment length is $\Delta a = 0$. At load level P_3, propagation can occur for a crack of length a_1, and it would be unstable in load control. Again, as in Case 1, above, using a crack length or displacement control algorithm the crack of initial length a_1 could be propagated stably to length a_2 by decreasing the load incrementally from level P_3 to level P_2, as shown in Figure 12.15(b).

For crack lengths longer than a_2, fracture propagation is stable in the load control sense. An effective stress intensity monotonically decreasing with increasing crack length implies that a monotonically increasing load is required for continued propagation. In Figure 12.15, it can be seen that if the load is again increased to P_3 propagation to crack length a_3 is possible.

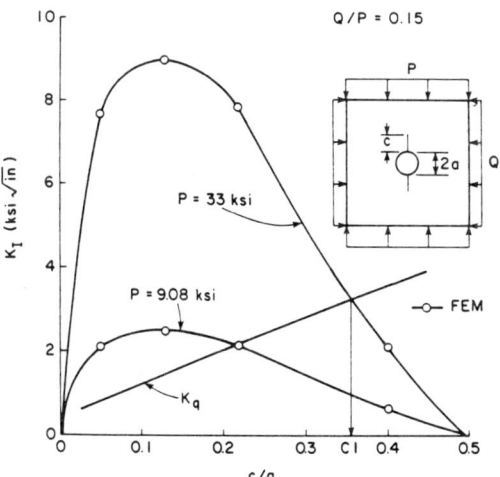

Figure 12.16 Stress intensity factor variation for crack propagating from circular hole (1 ksi = MPa, 1 ksi $\sqrt{\text{in}}$ = 1.1 MPa $\sqrt{\text{m}}$). From Ingraffea (1979)

An example of this behaviour for a pure Mode I case is shown in Figure 12.16 which is taken from a study on fracture propagation around underground openings (Ingraffea, 1979). Cracks are induced at crown and invert of a circular opening in a plate under the indicated biaxial compression. The computed K^* (here $K_I^* = K_I$) versus crack length relationship at two load levels is shown. Although it was assumed that the effective toughness, K_Q, increased with crack length, the propagation scenario is the same as described above. Propagation is unstable at first, but ceases each time the K_I and K_Q curves intersect. This type of propagation behaviour is very common in rock mechanics. It occurs, for example, in hydrofracturing.

If one is starting with crack length a_1, and load level P_3 (Figure 12.15), the prediction technique is the same as described under Case 1: propagate the fracture an amount Δa in the direction θ_0 at load P_3, compute the effective stress intensity for the new crack length, and apply Equation (12.14).

Suppose, however, that one is at load level P_2 and crack length a_2. One can still use the same algorithm: the only difference is that the quantity in parentheses in Equation (12.14) will now always be less than one.

The reverse of Case 2 is also possible: effective stress intensity can at first decrease and then begin to increase with increasing crack length. (See, for example, Saouma et al. (1982) and Beech and Ingraffea (1982).) This implies nothing new algorithmically, however, since the implications of this situation are handled by the techniques described in Cases 1 and 2.

A number of alternative numerical techniques for fracture increment length prediction are available (Ingraffea, 1977; Blandford, 1981; Saouma, 1981). Some are based on energy balance, some are more approximate than others. The simple technique described here is theoretically exact for pure Mode I, colinear propagation. Recalling Observation 4, however, we can see that any technique which employs finite, straight fracture increments will be approximate. One is updating effective stress intensity incrementally, rather than continuously. A curvilinear trajectory is being modelled piecewise by straight segments. Stress intensity factors and angle changes will be somewhat in error. The error depends on the specified length of the fracture increment.

The analogy here is with dynamic analysis where the time step controls accuracy and stability of the solution. It is the author's experience with his codes that predicted trajectories sometimes oscillate about an average path. This is a manifestation of error in K_{II} simulation which is a result of 'kinking' the crack path rather than allowing it to continuously curve. Spuriously high K_{II} values are computed which, alternating in sign with each increment, zigzag the crack. However, it is quite possible that if too large an increment is used divergence of predicted trajectory could occur.

All the theoretical ingredients for fracture propagation modelling under mixed-mode, LEFM assumptions have now been presented. These ingredients have been combined in computer programs developed by the author and his students at Cornell University. In the next section, some general observations concerning these programs will be presented.

12.4 FRACTURE PROPAGATION PROGRAMS

Research and application thrusts into fracture propagation modelling at Cornell University can be divided into areas of numerical method and user–computer interface.

12.4.1 Numerical methods

Incremental fracture propagation codes have been developed using both the finite and boundary element methods. As will be shown, each of these methods has characteristics which make it the appropriate choice for given structure, dimensionality of model, or interface hardware.

In general, the boundary element method as used here is suited to elastic, homogeneous structures containing few propagating cracks. The boundary element method is superior in efficiency and accuracy to the finite element method for modelling of three-dimensional crack propagation problems. Since only the boundaries of the structure, including the crack faces, need to be discretized, the data base for a boundary element method analysis is much smaller than that of a finite element method analysis. Also, since perspective views of three-dimensional meshes are not encumbered with all the interior nodes and elements of a finite element mesh, effective user–computer interface can be obtained with low-level computer graphics equipment.

A two-dimensional code, Boundary Element Fracture Analysis Program, BEFAP (Blandford, 1981; Ingraffea et al. 1983c), is operational, and an example of its use is described later. The three-dimensional version is under development in connection with the Cornell Program for Computer Graphics for use with high-level computer graphics hardware (see next section for description of high- and low-level computer graphics hardware).

For problems involving inhomogeneities, interfaces, or many cracks, BEFAP currently is not suitable. For problems of this type, as well as those involving material nonlinearity, the Finite Element Fracture Analysis Program, FEFAP (Saouma, 1981; Saouma and Ingraffea, 1981), has been developed. Again, the two-dimensional version of FEFAP is operational, and the three-dimensional version is being implemented in a high-level interactive computer graphics environment. Examples of problems solved using FEFAP will also be presented later.

12.4.2 User–computer interface

Three levels of user–computer interfacing are available for operation the BEFAP/FEFAP group:

(1) Interactive without graphics: the standard keyboard entry of data, editing, of files, and spooling of output to a printer.
(2) Low-level interactive computer graphics: storage tube graphic display devices. Display of initial mesh, deformed mesh, principal stress vectors, crack increment trajectories, G curves, and load displacement curves. Interactive programming capability, meaning that the user participates in the real time solution of the problem by, for examples, editing each mesh update, or selecting the length of a crack increment or magnitude of a load increment.
(3) High-level interactive computer graphics: vector refresh graphic display devices. With high-level graphics, all of the capabilities of low-level exist but the display is continually updated. This means that selected regions of the display can be changed nearly instantaneously without the necessity for redraw of the entire display. The mesh can be 'zoomed' or panned to highlight detail, and three-dimensional objects can be translated and rotated to enhance the user's perception of a complex object and its mesh.

The operational version of BEFAP currently operates only in the interactive mode, while FEFAP operates in interactive, low-level, and high-level interactive graphic modes.

In the example problems to follow, one will notice an evolution in the user–computer interface towards increasing use of interactive computer graphics. This evolution is still under way but the original objectives are the same. These are:

(1) *Minimize manual generation of input data.* This applies both to the total amount of data necessary to define the problem and the physical act of transferring this information to the computer. The user should communicate geometrical information to his code by way of interactive graphics, e.g. a digitizing tablet and pen in conjunction with a vector refresh display terminal or a cursor and key system such as on some storage tube terminals. Only the geometrical information absolutely required for automatic mesh generation should be input in this manner.
(2) *Make the programs interactive and adapative.* On request, the user should be informed of real time progress of the analysis by way of graphic displays. Moreover, he should be given the freedom to modify the course of the analysis by changing the data base while it is in progress.
(3) *Results should be displayed in a simple and effective way.* The user should be able to see graphic display of intermediate and final stress and displacement fields, load histories, crack patterns.

12.4.3 Automatic remeshing

All the programs developed at Cornell University employ automatic remeshing algorithms. This means that the programs, after computing stress intensities, interaction, angle change, and change in load or fracture length, automatically relocate each crack tip and remesh accordingly. The remeshing algorithms developed by Blandford (1981) for BEFAP and Saouma (1981) for FEFAP are very versatile. They accommodate mixtures of element types and allow a wide range of crack configurations to be modelled. This versatility will be evident in the example to follow.

12.4.4 Example solutions

Example # 2: Simulation of the tests described in example 1
Numerical Method: Finite element and Boundary element Solutions
User–Computer Interface: Punch Cards and Manual Remeshing for Finite Element Solution; Interactive without Graphics for Boundary Element Solution

The author used finite elements to simulate (Ingraffea, 1977; Ingraffea and Heuze, 1980) the behaviour shown in Figure 12.8 for the first, and most rudimentary, example of fracture propagation modelling using the techniques described in this chapter. Each fracture increment required manual remeshing and a job resubmission. Dozens of manhours were required to produce the results of ten primary crack increments.

More recently the same problem was analysed by Blandford, 1981 and Saouma, 1981 using BEFAP. A typical mesh is shown in Figure 12.17. Initial data preparation required about 3 man-hours. The analysis itself, involving seven primary crack increments and computation of domain stresses, required about 10

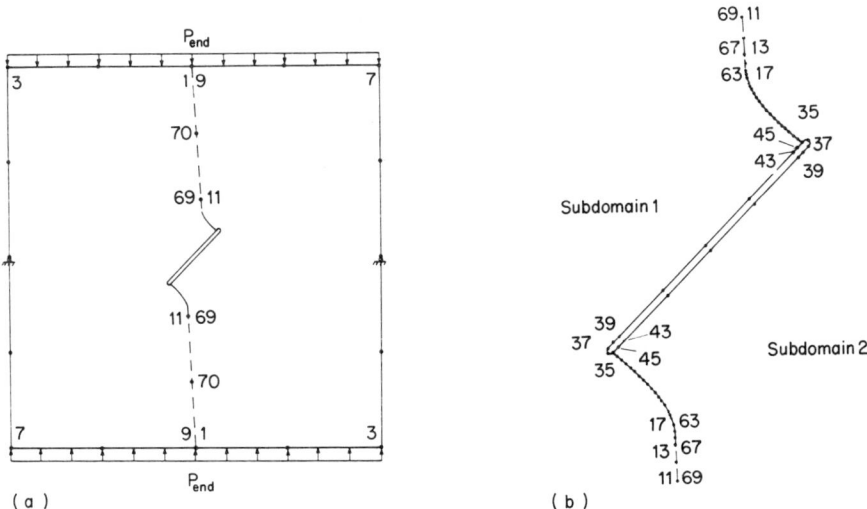

Figure 12.17 (a) Typical boundary element mesh for problem # 1; (b) typical notch tip detail. From Ingraffea *et al.* (1983c)

246 Mechanics of Geomaterials

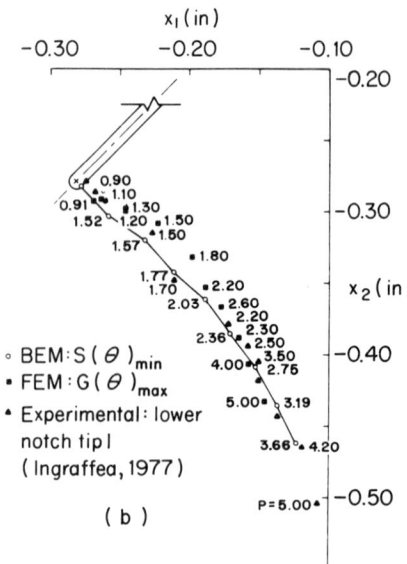

(b)

Figure 12.18 Comparison of predicted and observed primary crack trajectories for problem of Example 1. P in ksi (ksi = 6.9 MPa, 1 in = 25.4 mm). From Ingraffea et al. (1983c)

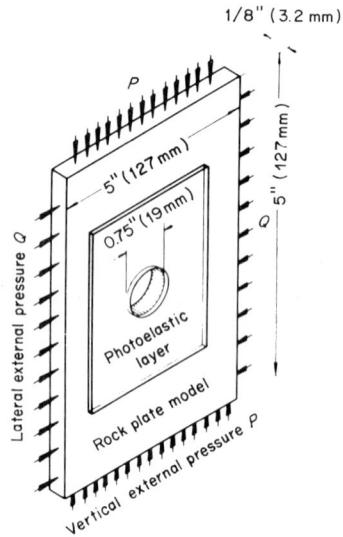

Figure 12.19 Rock plate model used in experimental study of collopse of underground cavity. From Hoek (1964)

Fracture Propagation in Rock 247

CPU minutes on an IBM 370/168. Still, however, the computed stress field had to be manually plotted to mark the area of secondly crack nucleation. A comparison of typical finite element, boundary element, and experimental results is shown in Figure 12.18.

Example #3: Collapse of underground cavity
Numerical Method: Finite Element
User–Computer Inerface: High-level Interactive Computer Graphics

The structure shown in Figure 12.19 represents a cross-section through a deep underground cavity loaded by overburden and horizontal stresses. The model shown was tested by Hoek (1964). An analysis of this structure was performed using the high-level interactive computer graphics facilities at the Cornell Program for Computer Graphics. An interactive graphics preprocessor (Haber et al., 1978) was used to generate the initial finite element mesh, shown in Figure 20(a), and its attributes. Generation of all initial data required about 10 man-minutes.

The fracture response of this structure is similar to that of the previous example. Primary cracks initiate at crown and invert and propagate in the manner described previously with respect to Figure 12.16.

The initial major principal stress field is depicted in Figure 12.20(b). Shown is a photograph taken from a colour postprocessor display (Schulman, 1981). Regions in which the stress exceeds the postulated tensile strength are shown here in black. Such postprocessing can be performed at the end of each fracture increment; fields of principal, normal, and shear stress, strain energy density, and displacement can be quickly displayed. Moreover, no additional man-effort is required to generate an image since the postprocessor data base is common with that of the preprocessor.

Next, secondary cracks nucleate in the plate interior in a tension zone developed in response to primary crack propagation. This zone can be seen as the blackened area shown in the postprocessor image of Figure 12.21(a). A 'zoomed' detail of the final mesh showing the predicted secondary crack path is shown in Figure 12.21(b). Experimental results for a similar problem (Lajtai and Lajtai, 1975) are shown in Figure 12.22.

Example #4: Fracture propagation under indentation loading
Numerical Method: Finite Element
User–Computer Interface: Low-level Interactive Computer Graphics

The mechanisms of fracture propagation under a tunnel boring machine roller cutter have been much studied but, in the author's opinion, are not yet completely understood. Paul (Paul and Gangal, 1969) proposed that at the point of contact with the rock surface high bearing stresses would generate a bulb of very high hydrostatic pressure. He used finite element analysis to prove that such a

248 *Mechanics of Geomaterials*

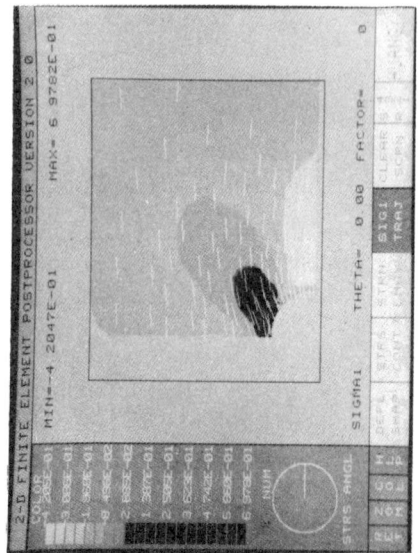

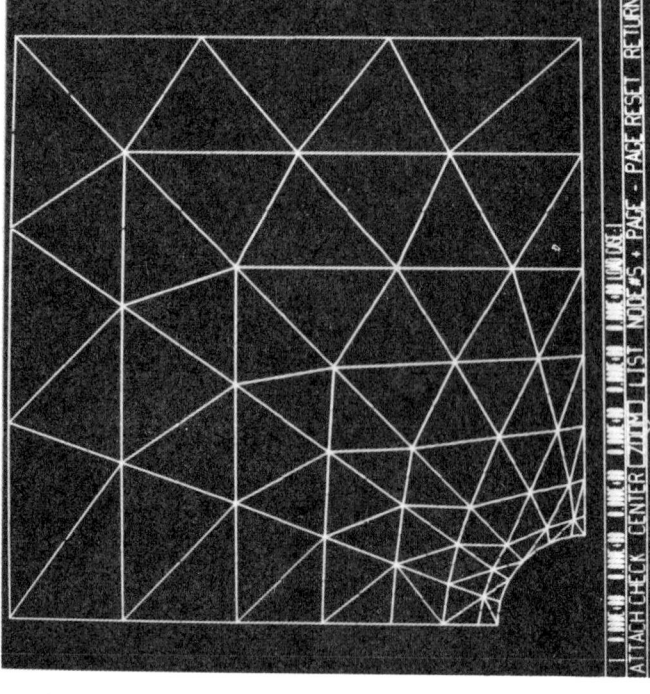

Figure 12.20 (a) Initial mesh for example #3; (b) initial major principal field. From Ingraffea (1979)

Fracture Propagation in Rock

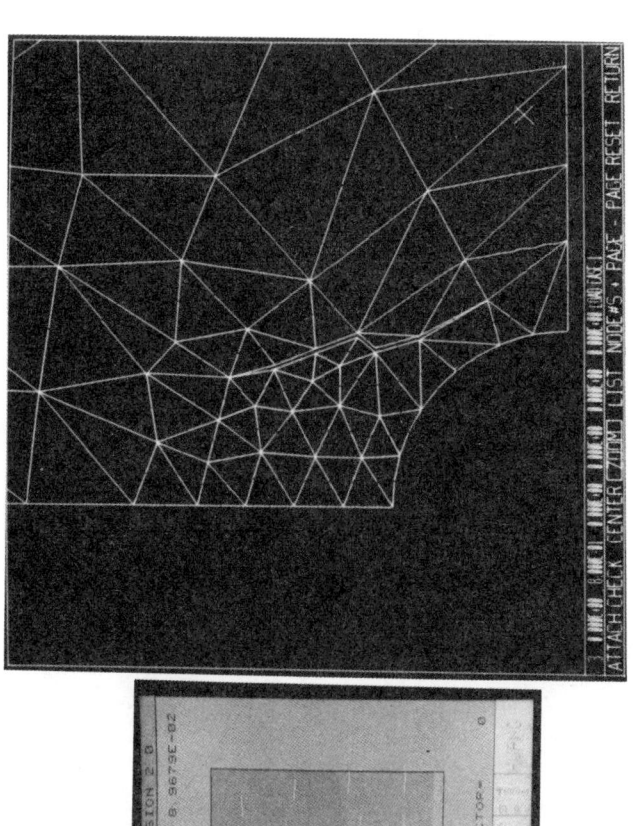

Figure 12.21 (a) Major principal stress field after stabilization of primary crack at point A; (b) detail of final mesh. From Ingraffea (1979)

250 Mechanics of Geomaterials

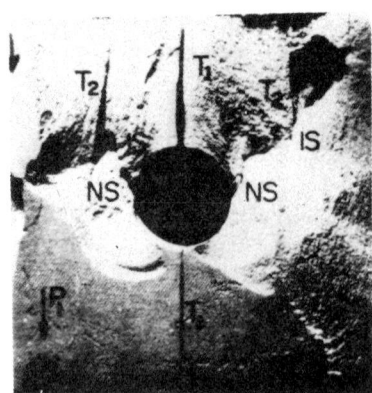

Figure 12.22 Fracture patterns observed in tests on plaster models similar to the structure in Figure 12.19. From Lajtai and Lajtai (1975)

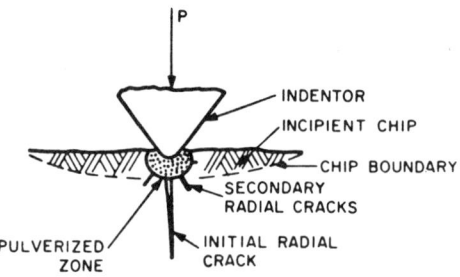

Figure 12.23 Postulated fracture pattern under indentor loading. From Paul and Gangal (1969)

condition would produce a primary radial crack as shown in Figure 12.23. FEFAP was used to model this occurrence as well as to predict the trajectories of the secondary radial cracks which Paul (Paul and Gangal, 1969) surmised would occur after primary crack stabilization.

The initial and final meshes are shown in Figure 12.24. The predicted fracture pattern closely resembles those observed after TBM roller cutter passage over granite (Wang et al., 1978), Figure 12.25.

FEFAP, designed by Saouma and the author (Saouma, 1981; Saouma and Ingraffea, 1981), is highly interactive and adaptive. The user receives information graphically after each analysis step, and is put in control of each subsequent step. For example, Figure 12.24(a) shows a typical fracture increment control page and its question/response dialogue. The automatic mesh modification algorithm in FEFAP can accommodate multiple cracks, a mixture of Q8, LST, and quarter-

Fracture Propagation in Rock

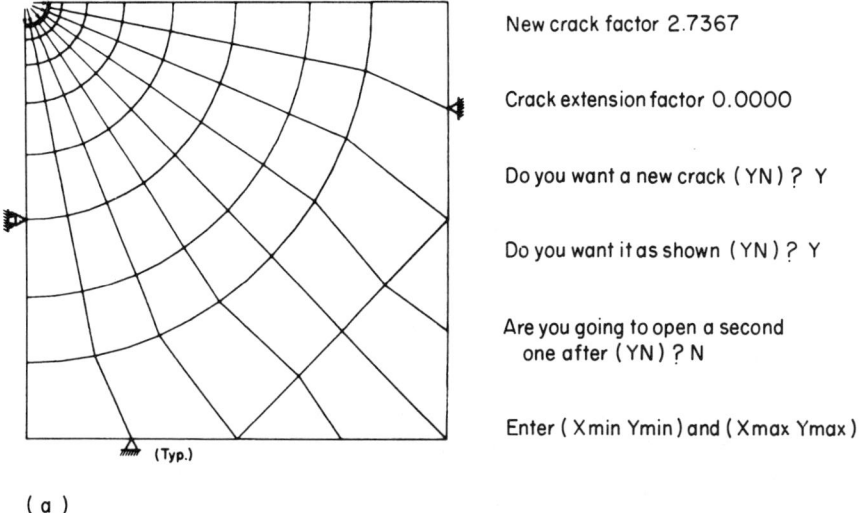

New crack factor 2.7367

Crack extension factor 0.0000

Do you want a new crack (YN) ? Y

Do you want it as shown (YN) ? Y

Are you going to open a second one after (YN) ? N

Enter (Xmin Ymin) and (Xmax Ymax)

(a)

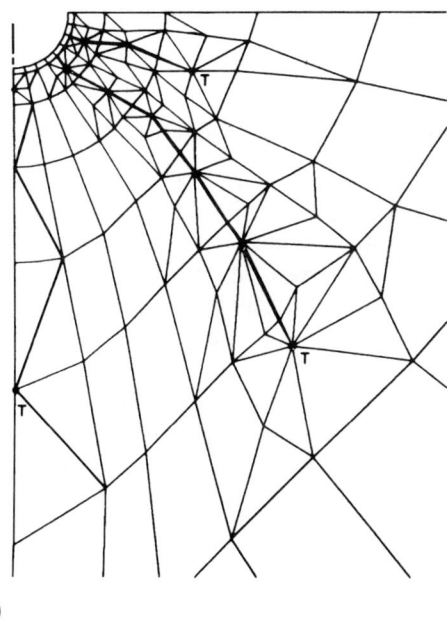

(b)

Figure 12.24 Initial, (a), and final, (b), meshes for Example 4. Crack tips are at points T

252 Mechanics of Geomaterials

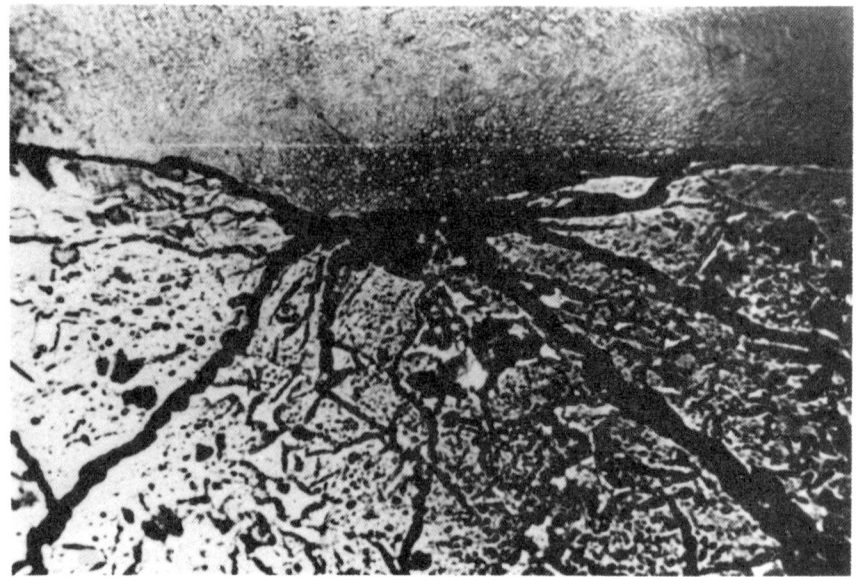

Figure 12.25 Fracture on cross-section of granite plate after TBM cutter pass. From Wang et al. (1978)

point singular elements, and interior or symmetry line cracks. The user is given the option of interactively modifying a generated mesh.

12.5 FRACTURE PROPAGATION MODELLING—THE FUTURE

The techniques described in this paper are certainly not the only ones available for modelling of fracture propagation. Alternative approaches can be based on other numerical methods, theories, and algorithms. However, it is the author's strongly held opinion that, regardless of which model is pursued, interactive computer graphics will play a decisive role in determining the viability of any programme in the marketplace of 'real world' problems. The continuing, rapid revolution in graphics hardware capability and software development and the ever-increasing cost-effectiveness of fast, virtual-memory, minicomputers are the driving forces in the evolution of sophisticated fracture propagation programs.

Nowhere will this be more evident than in the area of fully three-dimensional modelling. The present high cost of performing analysis of three-dimensional structures is due largely to the human effort required to define the crack geometrical data, element topology, boundary conditions, and material properties. In fact, the complexity of error detection, or even slight modification, with three-dimensional meshes can substantially reduce the cost effectiveness of a program. The user falls back onto a two-dimensional or axisymmetric model that

Fracture Propagation in Rock

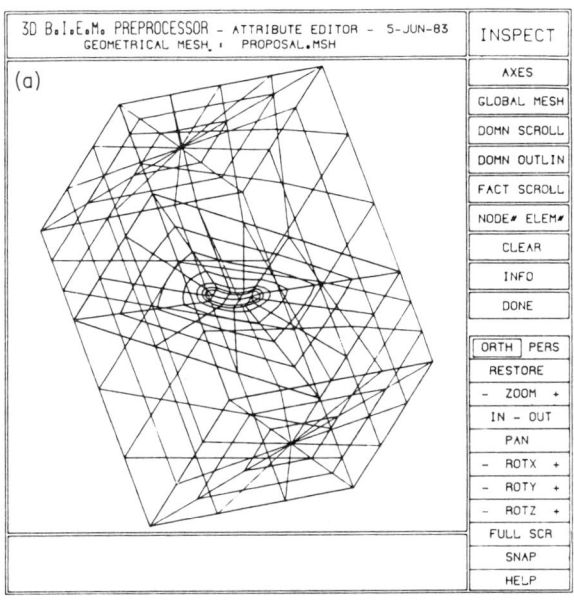

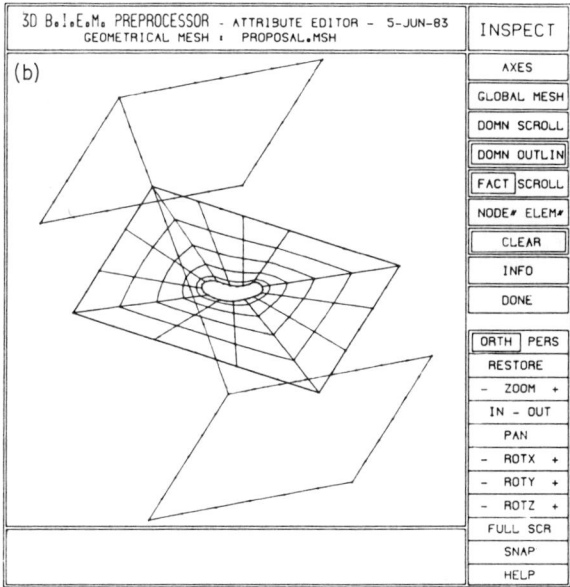

254 *Mechanics of Geomaterials*

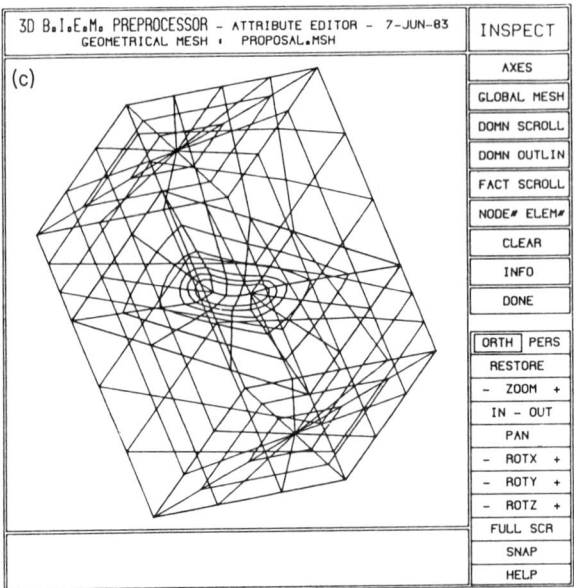

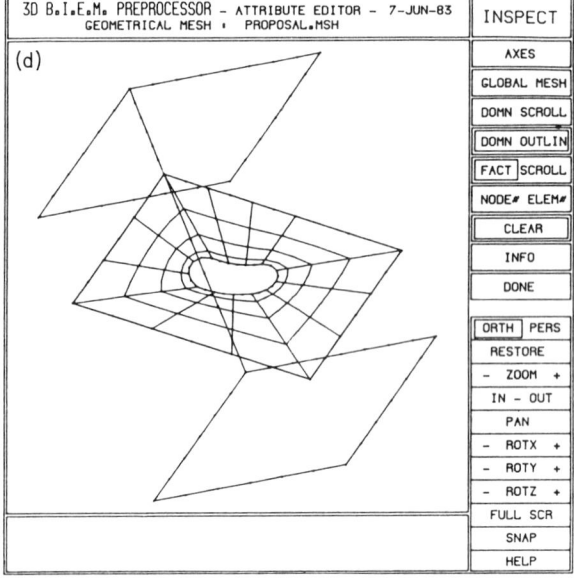

Figure 12.26 Views of three-dimensional boundary element mesh for simulated hydro-fracture; (a) complete mesh, (b) crack plane with fracture surface removed for clarity, (c) complete mesh after increment of cracking, (d) same as (b) above, after increment

Fracture Propagation in Rock

is 'good enough', sacrificing the realism of the three-dimensional problem in the face of the reality of tremendous labour cost.

However, interactive/adaptive preprocessing can eliminate a large percentage of such cost while simultaneously placing the engineer back in control of computer analysis. For example Perucchio and the author (Perucchio *et al.*, 1982; Perucchio and Ingraffea, 1983) have developed three-dimensional finite and boundary element preprocessors for use with fracture propagation codes in a high-level computer graphics environment. An example use of the boundary element preprocessor is the simulated hydrofracture propagation sequence shown in Figure 12.26.

Figure 12.26(a) shows the initial-stage BEM mesh, while Figure 12.26(b) shows only the fracture plane with the fracture surface removed. Figures 12.26(c) and 12.26(d) are analogous images after a fracture propagation step.

The complete boundary element mesh of Figure 12.26(a) can be rotated, translated, zoomed, and depth-cued. The user need not sacrifice to the CPU his engineering insight into the physics of the problem. Figure 12.26(b) shows that the mesh can be taken apart to study cross-sections, sub-domains, or even elements of interest. This capability will be invaluable during the course of a fracture propagation analysis.

Given the changes in the speed and cost of computers and the increasing use of interactive graphics, the author sees the use of truly three-dimensional fracture propagation codes, with the versatility of existing two-dimensional programs, as a certainty in only a few years.

ACKNOWLEDGEMENTS

The author would like to thank the following former and present students for making the developments reported here possible: Professor Victor Saouma, Professor George Blandford, Dr Corneliu Manu, Dr Renato Perucchio, Mr Walter Gerstle, Dr Jay Beech, Professor Priscilla Nelson, and Mr Kirk Gunsallus.

The author would also like to thank Professor Donald Greenberg, Director of the Cornell Program for Computer Graphics, for the continued use of his facility.

Research grants for the development of the techniques and programs described here came from the National Science Foundation.

REFERENCES

Abou-Sayed, A. S., and Simonson, E. R. (1977) 'Fracture toughness, K_{1c}, of triaxially-loaded Indiana limestone', *Proceedings, 18th US Symposium on Rock Mechanics*, 2A3-1, 2A3-8, 1977.

Atkinson, B. (1984) 'Subcritical crack growth in geological materials', to appear in *Journal of Geophysical Research*.

Barker, L. M. (1977a) 'A simplified method for measuring plane strain fracture toughness', *Engineering Fracture Mechanics*, **9**, 361.
Barker, L. M. (1977b) 'K_{Ic} measurements using short rod specimens—the elastic–plastic case', Terra Tek *Report TR 77-91R*. Nov.
Barker, L. M. (1979) 'Theory for determining K_{Ic} from small, non-LEFM specimens supported by experiments on aluminum', *International Journal of Fracture*, **15**, 6, 515–536.
Barker, L. M. (1981) 'Short-rod and short-bar fracture toughness specimen geometries and test methods for metallic materials', *ASTM STP 743*, pp. 456–475.
Barsoum, R. S. (1976) 'On the use of isoparametric finite elements in linear fracture mechanics', *International Journal of Numerical Methods in Engineering*, **10**, 25.
Barsoum, R. S. (1977) 'Triangular quarter-point elements as elastic and perfectly-plastic crack tip elements', *International Journal of Numerical Methods in Engineering*, **11**, 85.
Barton, C. (1983) 'Systematic jointing in the cardium sandstones along the Bow River, Alberta, Canada', Ph.D. dissertation, Yale University.
Beech, J., and Ingraffea, A. R. (1982a) 'Three-dimensional finite element stress intensity factor calibration of the short rod specimen', *International Journal of Fracture*, **18**, 3, 217–229.
Beech, J., and Ingraffea, A. R. (1982b) 'Three-dimensional finite element stress intensity factor calibration of the short rod specimen', *International Journal of Fracture*, **18**, 3, 127.
Blandford, G. E. (1981) 'Automatic two-dimensional quasi-static and fatigue crack propagation using the boundary element method, Ph.D. dissertation, Cornell University.
Blandford, G., Ingraffea, A. R., and Liggett, J. A. (1982) 'Two-dimensional stress intensity factor calculations using boundary elements methods', *International Journal for Numerical Methods in Engineering*, **17**, 387.
Bombalakis, E. G. (1968) 'Photoelastic study of initial stages of brittle fracture in compression', *Tectonophysics*, **6**, 461.
Bubsey, R. T., Munz, D., Pierce, W. S., and Shannon, J. L., Jr. (1982) 'Compliance calibration of the short rod chevron-notch for fracture toughness testing of brittle materials', *International Journal of Fracture*, **18**, 2, 125–133.
Cotterell, B. (1972) 'Brittle fracture in compression', *International Journal of Interactive Mechanics*, **8**, 195.
Cruse, T. A., and Wilson, R. B. (1977) *Boundary-Integral Equation Method for Elastic Fracture Mechanics*, AFSOR-TR-0355.
Erdogan, F., and Sih, G. C. (1963) 'On the crack extension in plates under plane loading and transverse shear', *ASME Journal of Basic Engineering*, **85**, 519.
Fairhurst, C., and Cornet, F. (1981) 'Rock fracture and fragmentation', *Proceedings of the 22nd U.S. Symposium on Rock Mechanics*, MIT, June 28–July 2, 21–46, 1981.
Freese, C. E., and Tracey, D. M. (1977) 'The natural isoparametric triangle versus collapsed quadrilateral for elastic crack analysis', *International Journal of Fracture*, **12**, 767.
Griffith, A. A. (1921) 'The phenomenon of rupture and flow in solids', *Philosophical Transactions of the Royal Society of London*, Ser. **A221**, 163
Griffith, A. A. (1924) 'Theory of rupture', *Proceedings of the First International Congress of Applied Mechanics*, Delft, pp. 55–63.
Haber, R. B., Shephard, M. S., Gallagher, R. H., and Greenberg, D. P. (1978) 'A generalized graphic preprocessor for two-dimensional finite element analysis', *Computer Graphics*, A Quarterly Report of SIGGRAPH-ACM, **12**, 3, 323.
Hoek, E. (1964) 'Rock fracture around mining excavations', *Proceedings 4th International*

Conference on Strata Control and Rock Mechanics, Columbia University, New York, p. 334.
Hoek, E., and Bieniawski, Z. T. (1965) 'Brittle fracture propagation in rock under compression', International Journal of Fracture Mechanics, 1, 139.
Hussain, M. A., Pu, S. L., and Underwood, J. (1974) 'Strain energy release rate for a crack under combined Mode I and Mode II, frac. analysis', ASTM STP, **560**, 2.
Ingraffea, A. R. (1977) 'Discrete fracture propagation in rock: laboratory tests and finite element analysis, Ph.D. dissertation, University of Colorado, 347 pp.
Ingraffea, A. R. (1981) 'Mixed-mode fracture initiation in Indian limestone and Westerly granite, Proceedings 22nd US Symposium on Rock Mechanics, Cambridge, MA, p. 186.
Ingraffea, A. R. (1979) 'The strength-ratio effect in rock fracture', Proceedings 20th US Symposium On Rock Mechanics, Austin, TX, p. 153.
Ingraffea, A. R., and Heuze, F. E. (1980) 'Finite element models for rock fracture mechanics', International Journal of Numerical and Analytical Methods in Geomechanics, **4**, 25.
Ingraffea, A. R., and Ko. H. -Y. (1981) 'Determination of fracture parameters for rock', Proceedings of First USA-Greece Symposium Mixed Mode Crack Propagation (Sih, G. C. and Theocaris, P. S., Eds.), Sijithoff & Noordhoff, Alphen aan den Rijn, Netherlands, p. 349.
Ingraffea, A. R., and Manu, C. (1980) 'Stress-intensity factor computation in three dimensions with quarter-point elements', International Journal of Numerical Methods in Engineering, **15**, 10, 1427.
Ingraffea, A. R., Gunsallus, K. L., Beech, J. F., and Nelson, P. (1982) 'A fracture toughness testing system for prediction of tunnel boring machine performance', Proceedings of the 23rd US Symposium on Rock Mechanics, Berkeley, California, August 1982, pp. 463–470.
Ingraffea, A. R., Gunsallus, K. L., Beech, J. F., and Nelson, P. P. (1983a) 'A short-rod based system for fracture toughness testing of rock', ASTM STP 855 Chevron-Notched Specimens: Testing and Stress Analysis, Louisville, p. 152.
Ingraffea, A. R., Perucchio, R., Han, T. -Y., Gerstle, W. H., and Huang, Y. P. (1983b) 'Three-dimensional finite and boundary element calibration of the short-rod specimen', ASTM STP 855 Chevron-Notched Specimens: Testing and Stress Analysis, Louisville, p. 49.
Ingraffea, A. R., Blandford, G., and Liggett, J. A. (1983c) 'Automatic modelling of mixed-mode fatigue and quasi-static crack propagation using the boundary element method', Proceedings 14th National Symposium on Fracture Mechanics, ASTM STP 791, I-407.
Kordisch, H., and Sommer, E. (1978) 'Bruchkriterien bei uberlagenden normal und scherbeanspruchung von rissen', Report W6.77, Fraunhofer-Gesellschaft, Institut fur Festkorpermechanik, Freiburg.
Lajtai, E. A., and Lajtai, V. N. (1975) 'The collapse of cavaties', International Journal of Rock Mechanics and Mining Science, and Geomechanical Abstracts, **12**, 81.
Newman, J. C. (1979) 'Stress intensity factors and crack opening displacements for round compact specimens', NASA TM 80174, Langley Research Center, Oct.
Ouchterlony, F. (1980a) 'Review of fracture toughness testing of rock', DS1980:15, Swedish Detonic Research Foundation, Stockholm.
Ouchterlony, F. (1980b) 'A new core specimen for the fracture toughness testing of rock', DS 1980:17, Swedish Detonic Research Foundation (SveDeFo), Stockholm, Sweden, 18.
Ouchterlony, F. (1981a) 'Extension of the compliance and stress intensity formulas for the single edge crack round bar in bending', Fracture Mechanics for Ceramics, Rocks, and Concrete (Eds. S. W. Freiman and E. R. Fuller Jr.), STP 745, ASTM, Philadelphia, pp. 237–257.

Ouchterlony, F. (1981b) 'A simple R-curve approach to fracture toughness testing of rock with sub-size SECRBB specimens', DS 1981:18, SveDeFo, Stockholm, Sweden, 41 pp.

Ouchterlony, F., and Zongqi, S. (1983) 'New methods of measuring fracture toughness on rock cores', *Proc. First. Int. Symp. Rock Frag. Blasting*, Lulea, Sweden, Aug. 22–26, 1983 (in press).

Paul, B., and Gangal, M. D. (1969) 'Why compressive loads on drill bits produce tensile splitting in rock', SPE 2392, *Proceedings 4th Conference on Drilling and Rock Mechanics*, University of Texas at Austin, p. 109.

Perucchio, R., Ingraffea, A. R., and Abel, J. F. (1982) 'Interactive computer graphic preprocessing for three-dimensional finite element analysis', *International Journal of Numerical Methods in Engineering*, **18**, 909.

Perucchio, R., and Ingraffea, A. R. (1983) 'Interactive computer graphic preprocessing for three-dimensional boundary integral element analysis', *Journal of Computers and Structures*, **16**, 153.

Raju, I. S., and Newman, J. C. (1983) 'Three-dimensional finite element analysis of the chevron-notched fracture specimen', ASTM STP from *Symposium on Chevron-Notched Specimens: Testing and Analysis*, Louisville, Kentucky, April 21 1983 (in press).

Saouma, V. E. (1981) Interactive finite element analysis of reinfored concrete: a fracture mechanics approach, Ph.D. dissertation, Cornell University.

Saouma, V. E., and Ingraffea, A. R. (1981) 'Fracture mechanics analysis of discrete cracking', *Proceedings, IABSE Colloquium on Advanced Mechanics of Reinforced Concrete*, Delft, p. 393.

Saouma, V., Ingraffea, A. R., and Catalano, D. (1982) 'Fracture toughness of concrete: K_{Ic} revisited', *ASCE, Journal of the Engineering Mechanics Division*, **108**, 1152.

Schulman, M. A., The interactive display of parameters of two- and three-dimensional surfaces, M.S. thesis, Cornell University.

Shih, C. F., de Lorenzi, H. G., and German, M. D. (1976) 'Crack extension modeling with singular quadratic isoparametric elements', *International Journal of Fracture*, **12**, 647.

Sih, G. C. (1973) 'Some basic problems in fracture mechanics and new concepts', *Engineering Fracture Mechanics*, **5**, 365.

Wang F. -D., Ozdemir, L., and Snyder, L. (1978) 'Prediction and verification of tunnel boring machine performance', Paper presented at Euro-Tunnel, Basel, Switzerland.

Mechanics of Geomaterials
Edited by Z. Bažant
© 1985 John Wiley & Sons Ltd

Chapter 13
Fracture in Concrete and Reinforced Concrete

Z. P. Bažant

13.1 INTRODUCTION

Although cracking represents a salient feature of the behaviour of concrete structures, not only under ultimate loads but also at service states, fracture mechanics has not been used in practical analysis of structures. Structural engineers had a good reason; the linear fracture mechanics was found to be inapplicable to typical concrete structures, and the premises of ductile fracture mechanics did not match material behaviour. However, in various recent investigations, particularly those at the Technical University of Lund, Northwestern University, and Politecnico di Milano, it has been shown that fracture mechanics can be applied to concrete structures provided that one takes into account the effect of a large micro-cracking zone or fracture process zone that always exists at the fracture front.

The objective of the present paper is to review the results of the investigations at Northwestern University, many of them carried out under a cooperative agreement with Politecnico di Milano (as part of the Italy–US Science Cooperation Program), and also to present some new results on a continuum model for strain-softening and on R-curve analysis. It is not possible to include a comprehensive review of all the work on fracture of concrete; other work may be consulted for that (Wittmann, 1983; Mindess, 1983; Ingraffea, 1984; Shah, 1984; ASCE, 1982; Bažant, 1984).

13.2 BLUNT CRACK BAND MODEL

The simplest way to model cracking in a finite element program is to assume that the cracks are continuously distributed over the area of the finite element and manifest themselves by a reduction of the elastic modulus in the direction normal to the cracks. Complete cracking corresponds to a reduction of the elastic modulus to zero. In this description, introduced by Rashid (1968), the crack band front obviously cannot be narrower than the width of the frontal finite element.

It has not been generally recognized, however, that, normally, the width of the crack band front also cannot be wider than a single element. Of course, one could enforce the crack front to be of a multiple-element width, however, that would not be justified mechanically since localization of strain into a single-element width generally leads to a release of elastic energy. There is a further reason why a multiple-element width at the crack band front is not a correct model; if we make the loading step sufficiently small, then only one element cracks during the loading step, and this relieves the stresses in the finite element that is on the side of the element that has just cracked, thus preventing an increase of the crack front width, except if a uniform strain distribution is enforced by heavy reinforcement. Even if two finite elements at the crack front had exactly the same stress values, it would be unrealistic to assume that they both crack simultaneously since the statistical scatter of material properties will always cause one of these elements to crack before the other does. Thus, one may adopt the blunt crack band model with a single-element wide front as a realistic and numerically convenient model for cracking in concrete (Bažant, 1982; Bažant and Cedolin, 1979, 1980, 1983; Bažant and Oh, 1983a; Cedolin and Bažant, 1980; Bažant et al., 1983b). A similar approach can be applied to rock (Bažant, 1982; Bažant and Oh, 1982).

Regardless of whether the zone of micro-cracking at the fracture front in

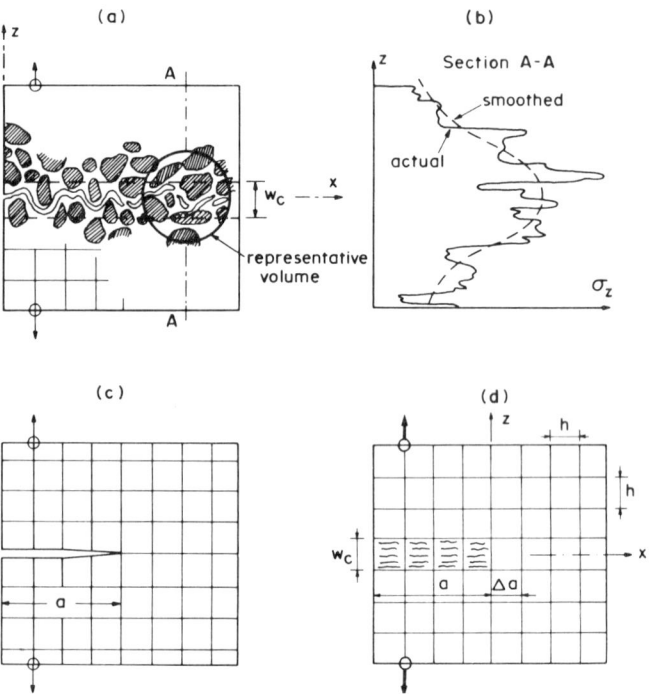

Figure 13.1 Heterogeneous microstructure, random scatter of stresses, and finite element models (after Bažant and Oh, 1982)

concrete is very wide or not, two elementary justifications may be offered for the blunt crack band model. One of them is the heterogeneity of the material. We treat the material as a smoothed, homogeneous continuum in the macroscopic sense. In this treatment, the macroscopic stresses and strains represent the averages of the actual (microscopic) stresses and strains over a certain so-called representative volume of the material whose size must be at least several times the maximum aggregate size (Figure 13.1). Obviously, the rapid and scattered variation of stresses and strains over smaller distances cannot be described by a continuum approach. Therefore, using finite elements of sizes less than several times the aggregate size would not allow any improvement in the description of the actual stress and strain fields within concrete. Even if one wishes to treat a continuous sharp crack in concrete, the blunt crack band model does not represent the reality any worse than a sharp inter-element crack model because the actual crack path is not straight but highly tortuous.

As another justification of the blunt crack band model for describing sharp fractures in concrete, one may cite the recently documented fact that a sharp inter-element crack and a blunt crack band of single-element width yield approximately the same results if the mesh is not too crude (roughly when there are at least fifteen finite elements in a square mesh across the cross-section). Both models give energy release rates that differ not more than a few per cent from the exact elasticity solution. To illustrate it, Figure 13.2 exhibits some of the numerical results from Bažant and Cedolin (1979). In these calculations, the normal stress in the direction perpendicular to the cracks was assumed to drop suddenly to zero when the energy criterion for crack band propagation became satisfied. The finite element mesh in Figure 13.2 covers a cut-out of an infinite elastic medium loaded at infinity by uniform normal stress $\bar{\sigma}$ perpendicular to a line crack of length $2a$. The nodal loads applied at the mesh boundary are calculated as the resultants of the exact stresses in the infinite medium, based on Westergaard's exact solution which is shown as the solid curve. The data points in Figure 13.2 show numerical results for the square mesh shown (mesh A), as well as for finer meshes B and C (not shown) for which the element size was 1/2 and 3/8 of the element size shown, respectively. A similar equivalence of results for the sharp inter-element crack and the blunt crack band can be demonstrated when the stress is considered to drop gradually rather than suddenly to zero (Bažant and Oh, 1983a).

Aside from the foregoing justifications, the blunt crack band model appears to be more convenient for finite element analysis. When a sharp inter-element crack gets extended through a certain node, the node must be split into two nodes. This increases the total number of nodes and changes the topological connectivity of the mesh. Unless the nodes are renumbered, the band structure of the structural stiffness matrix is lost. Moreover, if the direction in which an inter-element crack should extend is not known in advance, one needs to make trial calculations for various possible locations of the node ahead of the crack front through which the

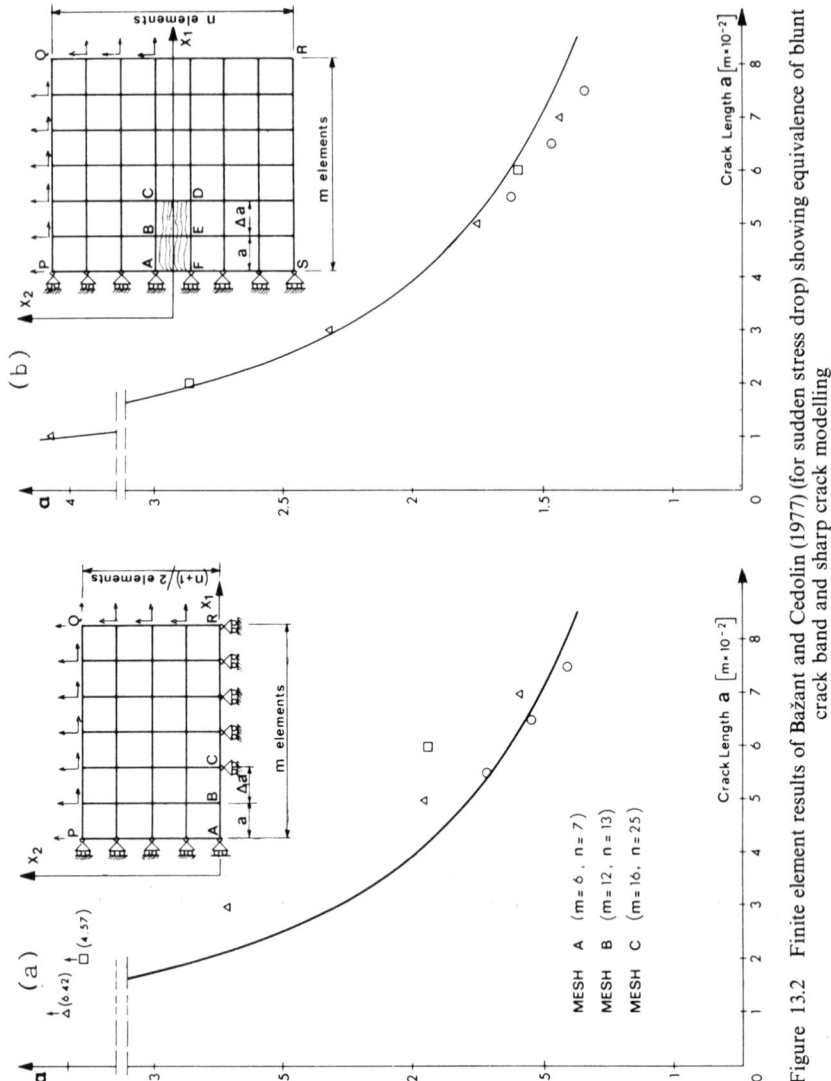

Figure 13.2 Finite element results of Bažant and Cedolin (1977) (for sudden stress drop) showing equivalence of blunt crack band and sharp crack modelling

Fracture in Concrete and Reinforced Concrete 263

crack should pass, in order to determine the correct direction of crack propagation. On the other hand, in the blunt crack band model, a fracture propagating in any direction through the mesh can be modelled as a zigzag crack band with any direction of the cracks relative to mesh lines. All that needs to be done to model an oblique crack direction is to reduce the elastic stiffness in the direction normal to the cracks.

Recently, various attempts to observe the distribution of micro-cracks ahead of the fracture front in concrete have been made (Mindess and Diamond, 1980; Cedolin et al., 1983a, 1983b). From strain measurements by Moiré interferometry (Cedolin et al., 1983a, 1983b), it appears that the width of the micro-crack zone at the fracture front is about one aggregate size. Within this width, there is a crack concentration. However, the line along which the densest micro-cracks are scattered is not straight but rather tortuous (Figure 13.1), which would not be modelled by a straight inter-element crack any better than by a crack band. Correlation of the crack band model to such microscopic observation is, of course, difficult since the micro-crack density varies while in the crack band it is assumed to be uniform. The question then is at which micro-crack size to draw the distinction. Thus, the width of the microscopically observed crack band front depends on the definition of the width of the micro-cracks that are counted within the crack band.

One significant difference from ductile fracture of metals consists in the size of the fracture process zone, defined as the zone in which the material undergoes strain-softening, i.e. the maximum principal stress decreases at increasing strain. This zone is large for concrete but relatively small for metals, even in the case of ductile fracture. In the latter case, there is a large yielding zone, but the material does not undergo strain-softening in this zone (Figure 13.3).

The stress–strain relation with strain-softening for the fracture process zone may be replaced by a strain–displacement relation if the displacement represents the integrated value of the strains across the width of the crack front. In this sense, the present blunt crack band model is equivalent to the previous line crack models with softening stress–displacement relations, introduced by Barenblatt, 1959; Dugdale, 1960; Kfouri and Miller, 1974; Kfouri and Rice, 1977; Knauss,

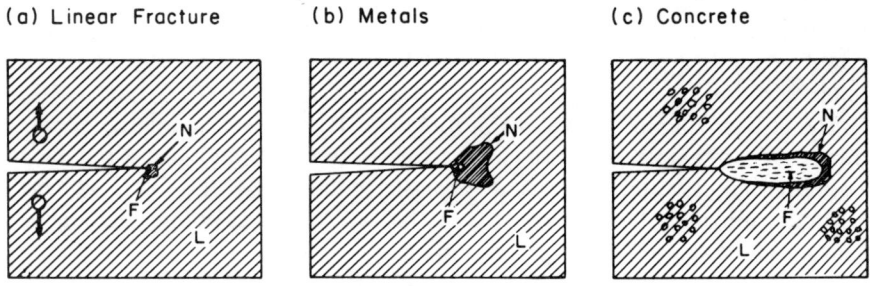

Figure 13.3 Nonlinear zone N and fracture process zone F for various materials

1974; Wnuk, 1974; Hillerborg et al., 1976. For concrete this approach was pioneered by Hillerborg, Modéer, and Petersson (1976) and Petersson (1980) in their model of a fictitious sharp inter-element crack.

Let us now outline one possible form of the softening stress–strain relation for the fracture process zone. Let the virgin crack-free material be described by the elastic stress–strain relation

$$\varepsilon = \mathbf{C}\sigma \tag{13.1}$$

Here, σ, ε are the column matrices of the cartesian normal components of strain and stress, in cartesian coordinates $x_1 = x$, $x_2 = y$, and $x_3 = z$. **C** is a 3 × 3 square compliance elastic matrix of the virgin material, with components C_{11}, $C_{12}, \ldots, C_{33}$. For the sake of simplicity, we may now assume that all microcracks spread over the finite element are normal to axis z. Appearance of such cracks has no effect on the lateral strains ε_x and ε_y, and the only effect is an increase in the averaged normal strain ε_z in the direction perpendicular to the cracks. This may be described by cracking parameter μ introduced only in one diagonal term of the compliance matrix (Bažant, 1982; Bažant and Oh, 1983a)

$$\mathbf{C}(\mu) = \begin{bmatrix} C_{11} & C_{12} & C_{13} \\ & C_{22} & C_{23} \\ & & C_{33}\mu^{-1} \end{bmatrix} \tag{13.2}$$

The cracking parameter μ is 1 for the initial crack-free state, and approaches zero for the final fully cracked state. It has been shown (Bažant and Oh, 1983a), that the limit of the inverse of the compliance matrix $C(\mu)$ as $\mu \to 0$ is, exactly, the well-known stiffness matrix for a fully cracked elastic material, $\mathbf{D}^{fr}$. This matrix is identical to the elastic stiffness matrix for the plane state of stress, which exists in the material between the cracks.

The cracking parameter may be calibrated so as to yield the desired tensile stress–strain relation with strain-softening $\sigma_z = EF(\varepsilon_z)$, in which $E = 1/C_{33} =$ Young's modulus. Then one has $\mu = F(\varepsilon_z)/\varepsilon_z$. Function $F(\varepsilon_z)$ may be given as a bilinear stress–strain diagram (Figure 13.4), characterized by tensile strength f'_t, softening modulus E_t (negative), and limit strain ε_0 for which full cracks are formed. For computer analysis, the foregoing stress–strain relation is differentiated to obtain an incremental form to be used in a program with step-by-step loading.

The strength limit, f'_t, needs adjustment to take into account the effect of multiaxial stress state. In particular, the tensile strength limit is decreased due to normal compressive stresses σ_x and σ_y parallel to the crack plane. Correction may be done according to the well-known biaxial failure envelope for concrete (Bažant and Oh, 1983a) (Figure 13.4(d)).

The use of cracking parameter μ resembles the so-called continuous damage mechanics, in which damage is characterized by parameter ω which corresponds

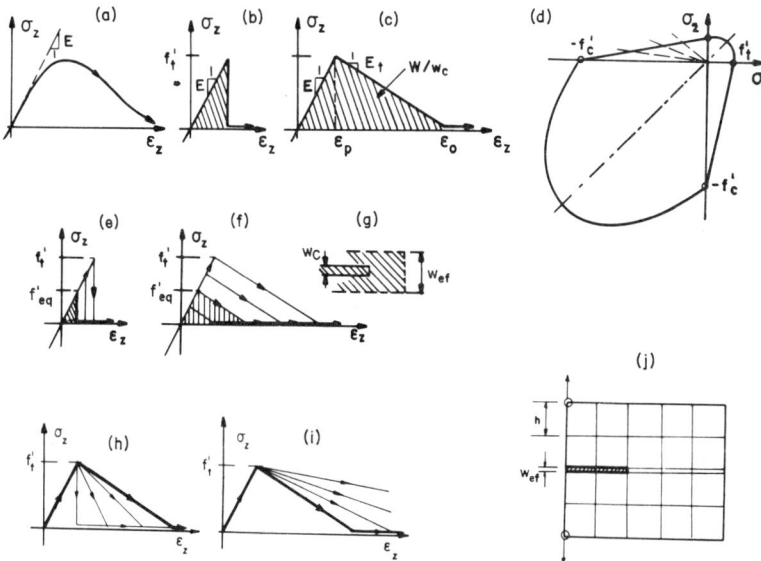

Figure 13.4 Tensile stress–strain diagrams assumed for fracture analysis

to $1 - \mu$. There is, however, a fundamental difference in that the damage due to micro-cracking is considered to be inseparable from a zone of a certain characteristic width that is a material property, as we will explain later.

The energy consumed by crack formation per unit area of the crack plane, i.e.

$$G_f = W_f w_c, \quad W_f = \int_0^{\varepsilon_0} \sigma_z \, d\varepsilon_z \tag{13.3}$$

represents the fracture energy; w_c = width of the crack band front (fracture process zone), and W_f = work of maximum principal tensile stress per unit volume = area under the uniaxial tensile stress–strain curve (Figure 13.4).

The magnitude of w_c is obviously an important factor. If the stress–strain relation, including its strain-softening range, is considered to be a material property, which seems logical, then the larger is w_c the larger is the fracture energy G_f. It has been previously demonstrated (Bažant and Cedolin, 1979, 1980), however, that finite element calculations yield results independent of the choice of the element size (except for a negligible numerical error converging to zero) only if the fracture energy G_f is considered as a material constant. Equation (13.3) then indicates that the width w_c of the crack band front must also be a material constant, to be determined by tests. Indeed, if the value of w_c is changed without adjusting the strength limit f'_t or the strain-softening modulus E_t, the predicted values of loads needed for further crack propagation may change drastically (Bažant and Cedolin, 1979, 1980). For the bilinear tensile stress–strain relation

(Figure 13.4), energy balance requires that

$$W_f = \tfrac{1}{2}(C_{33} - C^t_{33})f'^2_t w_c = \tfrac{1}{2}f'_t \varepsilon_0 w_c$$

or

$$w_c = \frac{2G_f}{f'^2_t} \frac{1}{C_{33} - C^t_{33}} \qquad (13.4)$$

in which $C^t_{33} = 1/E_t$ (negative) and $C_{33} = 1/E$. Thus, the effective width of the crack band front may be determined by measuring the tensile strength, the fracture energy, and the softening modulus E_t.

Note that Equation (13.4) is similar to the well-known Irwin's expression for the size of the yielding zone. It should be also noted that determination of w_c from mechanical measurements depends on the knowledge of the strain-softening slope E_t. If this slope is changed, a different value of w_c is obtained, and fracture test data may still be fitted equally well, within a certain range of w_c. In fitting test data for concrete fracture from the literature, it has been noted that good fits could be obtained for w_c ranging from $2d_a$ to $4d_a$ where d_a is the maximum aggregate size. The front width

$$w_c = 3d_a \qquad (13.5)$$

was nearly optimum, and at the same time was consistent with the softening modulus E_t as observed in the direct tension tests of Evans and Marathe (1968), Heilmann et al., (1969), Rüsch and Hilsdorf (1963), Petersson (1981), and Reinhardt and Cornelissen (1984).

Most of the important test data from the literature (Brown, 1972; Carpinteri, 1980–1981; Entov and Yagust, 1975; Gjørv et al., 1977; Huang, 1981; Kaplan, 1961; Kesler et al., 1971; Mindess et al., 1977; Naus, 1971; Shah and McGarry, 1971; Shah, 1984; Sok et al., 1979; Schwartz et al., 1981; Walsh, 1972; Wecharatana and Shah, 1980; Petersson, 1981), have been fitted with good success using the present nonlinear fracture model (Bažant and Oh, 1983a). Some of the fits obtained in Bažant and Oh (1983a) by finite element analysis using square meshes are shown in Figures, 13.5 and 13.6, in which P_{max}, representing the maximum measured load, is plotted as a function of either the crack length (flaw depth) or the specimen size. The optimum fits obtainable with linear elastic fracture mechanics are shown for comparison in these figures as the dashed lines. In calculating these results, the loading point was displaced in small steps and the reaction, representing load P, was evaluated at each loading step by finite elements. The same bilinear stress–strain relation (Equation (2)) was assumed to hold for all elements.

Note that the crack band fracture theory models well not only the test results for notched fracture specimens, but also the results for unnotched beams, in which the nominal bending stress at failure decreases as the beam depth increases (Figure 13.6h). This phenomenon is due to the fact that the large fracture process zone (strain-softening zone) cannot be fully accommodated in a small

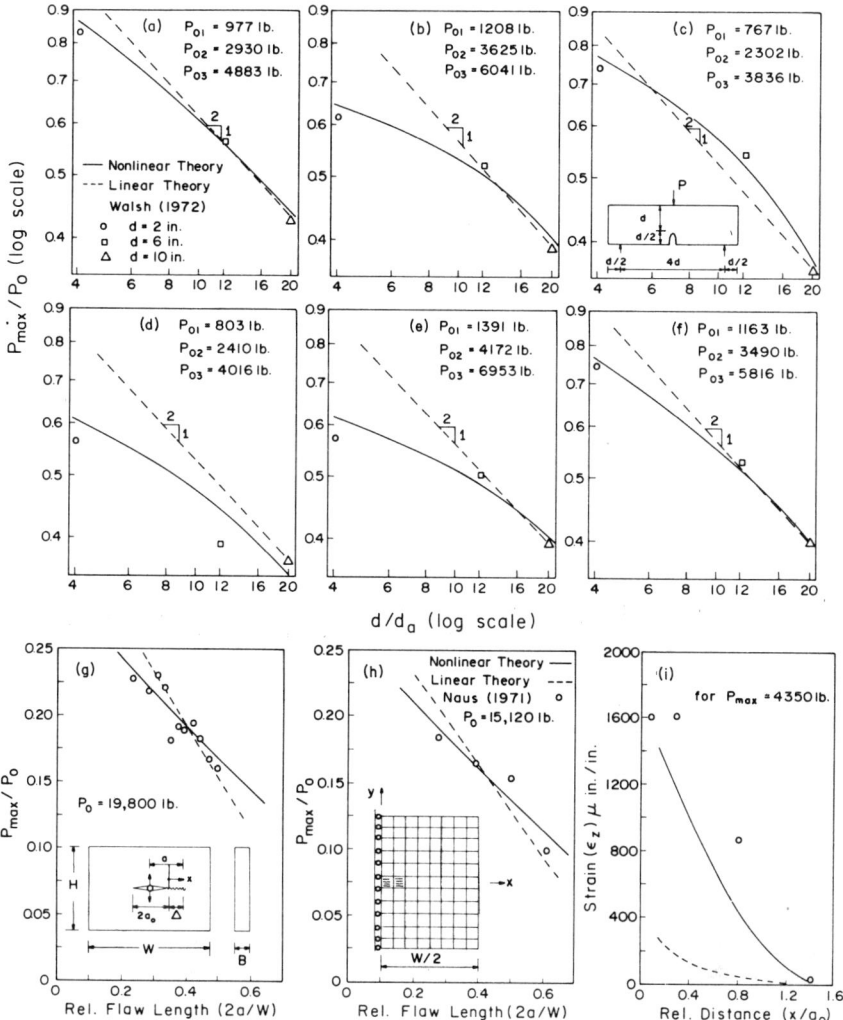

Figure 13.5 Results of crack band analysis compared with maximum load test data from the literature (after Bažant and Oh, 1983)

beam. The same phenomenon was previously modelled as a statistical size effect; however, explanation in terms of fracture mechanics, previously proposed by Hillerborg, appears to be more correct.

For metals, deviations from linear fracture mechanics have been described by the so-called R-curves (resistance curves), which represent the variation of apparent fracture energy as a function of the crack extension from a notch. Based on an idea of Irwin and Krafft et al. (1961), the R-curve may be considered for most situations as a fixed material property, although in reality this can be

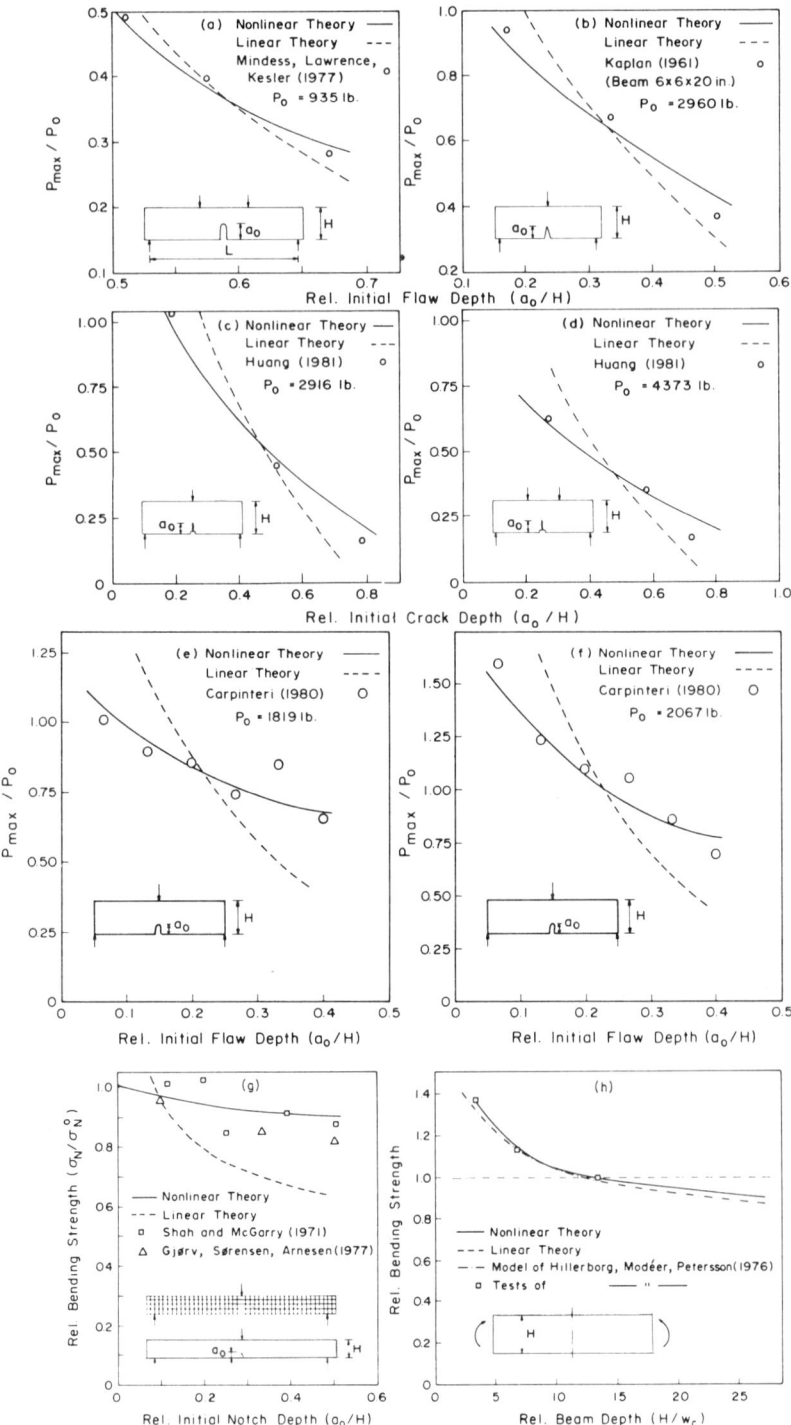

Figure 13.6 Crack Band Calculation Results Compared with Further Maximum Load Test Data From the Literature (after Bažant and Oh, 1983)

exactly true only asymptotically, for infinitely small crack extensions from a notch (for longer extensions, the R-curve should, in theory, also depend on the boundary geometry, location of the loads, crack path, etc.). It is noteworthy that the present theory achieves a good fit of test data without introducing any variation of fracture energy G_f, i.e. G_f is a constant. In fact, the present theory

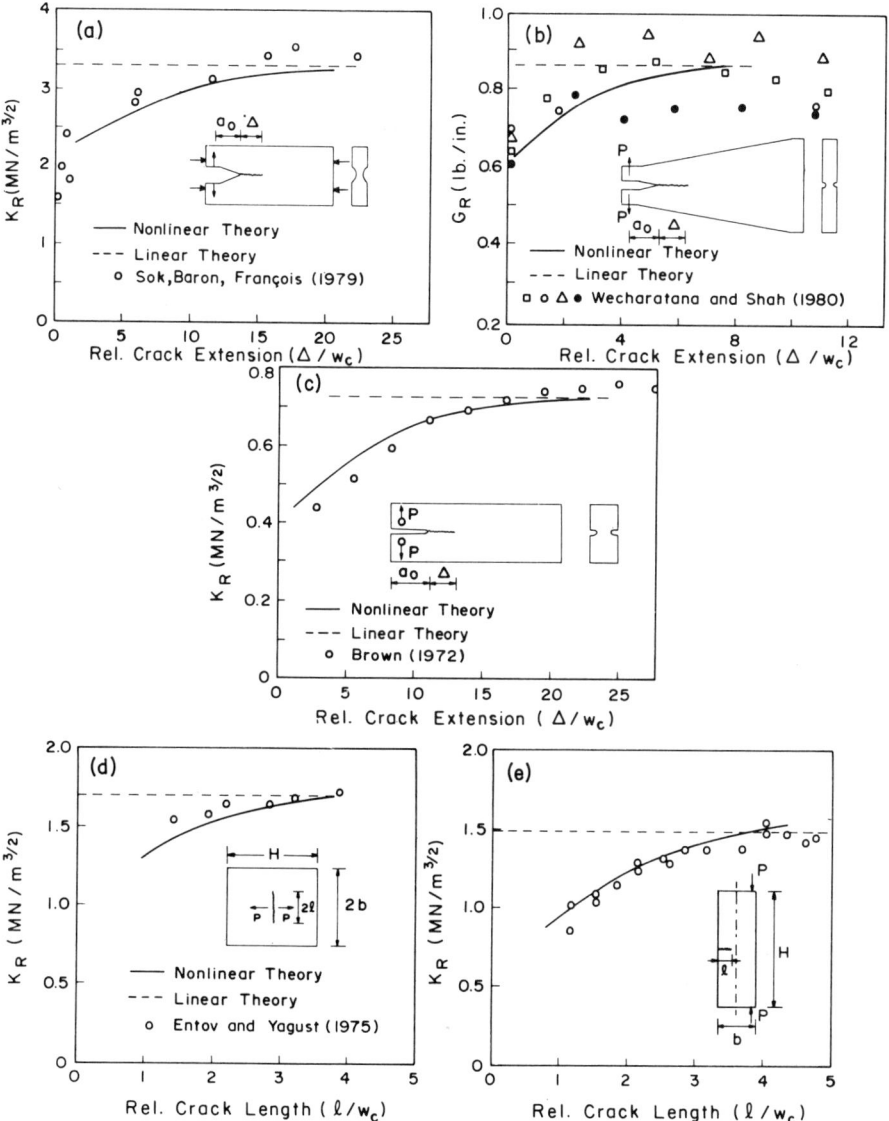

Figure 13.7 Crack band calculation results compared with measured R-curves from the literature (after Bažant and Oh, 1983)

allows calculating the R-curves. For this purpose one needs to evaluate the work of the nodal forces acting at the crack front element during a small crack band extension. In this manner, the R-curves have been calculated using the same fracture parameters as in the previous fitting of maximum load data. These calculations have led to good fits of R-curve data reported in the literature (Bažant and Oh, 1983a); see Figure 13.7, using test data from Brown, 1972; Entov and Yagust, 1975; Sok et al., 1979; Wecharatana and Shah, 1980. For the details of analysis, see Bažant and Oh, 1983a. It is worth noting that the present theory has been also used with equal success to fit the test data for various rocks (Bažant, 1982; Bažant and Oh, 1982).

Statistical analysis of the test data available in the literature revealed that the crack band theory allows a great reduction of the coefficient of variation ω_0 of the deviations of test data from the theory, as compared to previous theories. In the case of maximum load data, $\omega_0 = 0.066$ for the present theory, while for the best fits with linear elastic fracture mechanics, $\omega_0 = 0.267$. For the strength criterion, $\omega_0 = 0.650$ (see Bažant and Oh, 1983a). In the case of R-curve data, the present crack band theory yields for the deviations of test data from the theory the standard deviation $s = 0.083$, while linear fracture mechanics with constant fracture energy yields $s = 0.317$; see Bažant and Oh (1983a). These are significant improvements in the error statistics, and the accuracy of the present crack band theory is seen to be sufficient for practical purposes.

The analysis of test data from the literature also has resulted into an approximate empirical formula for the prediction of fracture energy on the basis of tensile strength f'_t, maximum aggregate size d_a, and Young's modulus E;

$$\tilde{G}_f \simeq 0.0214(f'_t + 127)f'^2_t d_a/E \qquad (13.6)$$

in which f'_t must be in psi (psi = 6895 Pa), and $\tilde{G}_f$ is in lb/in. Exploiting the relation $G_f \simeq 3d_a f'^2_t (E^{-1} - E_t^{-1})/2$, one can further obtain a prediction formula for the softening modulus

$$E_t \simeq \frac{-69.9E}{f'_t + 56.7} \qquad (13.7)$$

13.3 APPLICATION IN FINITE ELEMENT PROGRAMS

Finite elements of size $h = w_c = 3d_a$ may be too small for many practical applications. However, we cannot simply increase the element size because according to Equation (13.3) the energy consumed by fracture would increase proportionally with h, other parameters remaining unchanged. Obviously, in order to maintain the same energy consumption by fracture, the area under the tensile stress–strain diagram must be changed in inverse proportion to the element width h. This may be done most conveniently by adjusting the strength limit from the actual strength f'_t to a certain equivalent tensile strength f'_{eq}; Figure 13.4(f). If we use the bilinear stress–strain diagram and keep the softening modulus E_t constant, we obtain the following expression for the equivalent strength in a square mesh in which the fracture propagates parallel to the mesh line;

$$f'_{eq} = c_f \left(1 + \frac{E}{-E_t}\right)^{-1/2} \left(\frac{2G_f E'}{hr_f}\right)^{1/2} \qquad (13.8)$$

in which c_f is a calibration factor close to 1, depending on the type of finite element, and r_f is a correction for the compressive normal stress σ_3 parallel to the crack plane; $r_f \simeq 1 - v'\sigma_3/\sigma_1$ where $v' = v/(1-v)$, $v =$ Poisson ratio. We see from Equation (13.8) that the tensile strength limit must be reduced in inverse proportion to the square root of element size h.

Alternatively, one may make the declining slope steeper while keeping the same peak stress (Figure 13.4(h)). The slope is again chosen so as to ensure correct energy dissipation (Bažant and Oh, 1983a). The maximum element size that can be used with this approach corresponds to the vertical stress drop (Figure 13.4(e), (h)), and if still larger elements need to be used, then the strength limit needs to be reduced using a vertical stress drop (Figure 13.4(e)).

If the fracture path is known, one may also enlarge the finite elements outside the crack band while keeping the elements of the crack band at the correct width w_c (Figure 13.4(j)) and using the actual stress–strain diagram. As still another alternative, a transition to the line crack model of Hillerborg et al., may be obtained by reducing the band width below w_c, in which case the declining slope of the stress–strain diagram must be made milder (Figure 13.4(i)) so as to preserve the correct energy dissipation.

If the tensile strength limit (or the declining slope) is not changed when the element size changes, fracture analysis is unobjective in that the results may strongly depend on the analyst's choice of the element size. The first example of this was presented in Bažant and Cedolin (1983a) in which a rectangular panel, either plane or reinforced, was analysed for propagation of a symmetric central crack band. It was demonstrated that by changing the element size four times, the calculated value of the load needed for further crack band propagation changed by a factor of 2 (i.e. by 100 per cent). This implies a gross error in the prediction of energy absorption in the structure.

It should be realized, however, that keeping the strength limit and the declining slope the same regardless of the element size does not always lead to wrong results. In fact, for many situations finite element analyses with a constant tensile strength yielded good results, in agreement with tests. The reason why this happens is that many structures are fracture-insensitive, i.e. their failure depends primarily on other phenomena such as plastic yielding of steel rather than on cracking of concrete. The flexural failure of reinforced beams is a good example. To decide whether the problem is fracture-insensitive, the analyst needs to run the finite element calculations twice: once with the actual tensile strength f'_t, and once with a zero tensile strength. If the results do not differ significantly, one may forget about fracture mechanics.

For structures much larger than the aggregate size, the size of the fracture process zone may become negligible compared to the cross-section dimensions (this is true, for example, for gravity dams). If the finite elements are not very small, a small fracture process zone can be obtained by considering a vertical stress drop (Figure 13.4(b), (e)) instead of gradual strain-softening. A small fracture process zone is a prerequisite for the validity of linear elastic fracture mechanics, and indeed it is found (Bažant, 1982; Bažant and Oh, 1983a) that the use of a sudden stress drop after the tensile strength limit has been reached leads to results that are very close to the exact solutions of linear elastic fracture mechanics (Bažant and Cedolin, 1979). The energy release rates obtained with the present crack band model are just as close to the exact solution as those obtained with the sharp inter-element crack model (Bažant, 1982).

When a vertical stress drop is assumed, the energy criterion of fracture mechanics can be more closely approximated by a direct calculation of the energy release due to crack band extension, rather than by the use of a certain tensile stress–strain diagram with equivalent strength. A formula for the change in potential energy due to crack band extension was given in Bažant (1982) and Bažant and Oh (1983a). In this formula, one calculates the work of the nodal forces acting upon the frontal finite element during the fracture formation. One must also consider the differences between the initial and final strain energy within the cracked frontal finite element, as well as the work of distributed forces transmitted from reinforcement to concrete.

Instead of directly calculating the work of nodal forces on the frontal finite element, one may also obtain the exact energy release through the use of the J-integral. This method of analysis was developed by Pan and coworkers (1983).

An important advantage of the blunt crack band model is that the direction of mesh lines need not be changed if the fracture runs in a skew direction. The crack band propagation criterion then requires some adjustment in order to give results that do not depend on mesh inclination. We consider a rectangular mesh of mesh sizes Δx and Δy (Figure 13.8). An inclined crack band is represented as a zigzag crack band of overall orientation angle α_F. Let α_M be the orientation angle of the mesh lines x, and α_C be the direction of the cracks within the finite element (Figure 13.8). We seek the effective width w_{ef} of a smooth crack band that is equivalent to the zigzag band. Consider one cycle, of length l, on the line connecting the centroids of the elements in the zigzag band. The number of elements per cycle l in the x-direction is $N_x = l\cos\alpha/\Delta x$, and the number of those in the y-direction is $N_y = l\sin\alpha/\Delta y$ in which $\alpha = |\alpha_F - \alpha_M|$ provided that $0 \leq \alpha \leq 90°$. The area of the zigzag band per cycle l is $(N_x\Delta y)\Delta x + (N_y\Delta_x)\Delta y$. This area must equal the area

lw_{ef} of the equivalent smooth crack band, in order to assure the same energy content at the same stresses. This condition yields the effective width

$$w_{ef} = \Delta x \sin \alpha + \Delta y \cos \alpha \quad (0 \leqslant \alpha \leqslant 90°) \tag{13.9}$$

A somewhat different equation, giving similar results for $\alpha \leqslant 45°$, has been used in previous work (Bažant and Cedolin, 1980, 1983a; Cedolin and Bažant, 1980).

A different adjustment is needed when one considers a sudden stress drop and determines crack propagation directly from the energy change ΔU caused by extending the crack band into the next element. The propagation condition then is $\Delta U/\Delta a = -G_f$ where Δa is the length of extension of the crack band, which is equal to the mesh size h if the crack band propagates parallel to the mesh line (Bažant and Cedolin, 1979, 1980; Bažant, 1984). In the case of a zigzag band, Δa must be replaced by an effective crack band extension Δa_{ef} in the direction of the equivalent smooth crack band. We may assume Δa_{ef} to be the same for each crack band advance within the cycle l in Figure 13.8, whether this advance is in the x- or y-direction. Then $\Delta a_{ef} = l/N$ where $N = N_x + N_y =$ number of elements per cycle l. This condition yields

$$\Delta a_{ef} = \left(\frac{\cos \alpha}{\Delta x} + \frac{\sin \alpha}{\Delta y}\right)^{-1} \quad (0 < \alpha < 90°) \tag{13.10}$$

It has been demonstrated that the calculation results are objective not only with regard to the choice of element size but also with regard to the choice of mesh inclination relative to the fracture direction. Meshes of various inclination have been used to calculate the load versus crack-length diagram for the rectangular panel considered before; they have been found to yield essentially the same results, except that the scatter (numerical errors) are somewhat larger for the zigzag crack bands than for a smooth band; see Bažant and Cedolin (1983a).

It is one problem when the fracture direction is known and the zigzag crack band

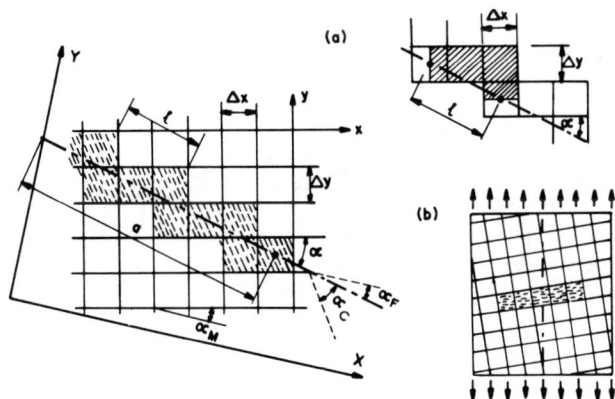

Figure 13.8 Zigzag crack band propagation through a square mesh

is placed so as to conform to the average fracture direction, and another problem when the fracture direction is unknown in advance and a choice of the next element to crack must be made. The latter problem is obviously more difficult. It has been found that any finite element mesh, including a square mesh, is not entirely free of a directional bias. This bias is the strongest when the angle of fracture direction with the mesh line is small. For example, if a square mesh in the centre-cracked rectangular panel is only moderately slanted (Figure 13.8(b)), then the equivalent strength criterion with the effective width given by Equation (13.8) indicates the crack band to extend straight along the mesh line, i.e. in the inclined direction, while correctly there should be side jumps so that the zigzag band would, on the average, conform to a horizontal direction. It appears rather difficult to avoid this type of bias. Various methods to avoid it are being studied (Marchertas et al., 1982; Pan and Marchertas, private communication; Bažant and Pfeiffer, in preparation; Bažant et al., 1983b), and various search routines to determine which element is the next to crack (i.e. the element straight ahead or the element on the left or on the right) are being investigated.

When concrete is reinforced, attention must also be paid to the question of bond slip of reinforcing bars embedded in concrete. It has been shown (Bažant and Cedolin, 1980), that neglect of the bond slip is impossible, leading to unobjective results strongly dependent on the mesh size and converging to a physically incorrect solution. If no slip is considered to occur at the nodes between the bars and concrete, and if the element size is varied, then the stiffness of the connection between the opposite sides across the crack band changes with the mesh size and would approach infinity for a vainishing mesh size, thus preventing any crack propagation at all. In reality, due to a limit on the bond stress that can be transmitted at the surface of steel bars, there is a certain bond slip length L_s on each side of a crack band (Figure 13.9). This length depends on

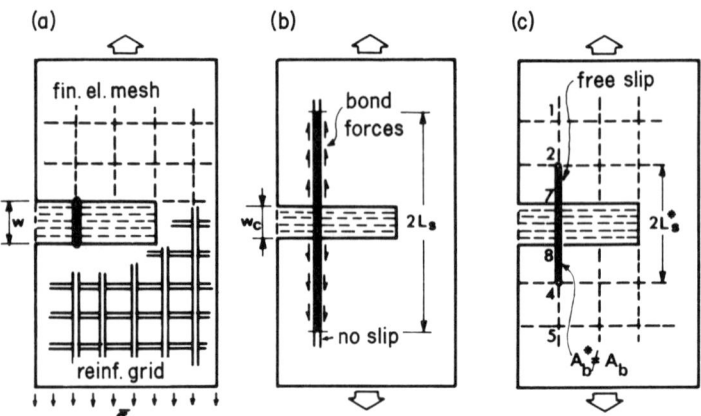

Figure 13.9 Illustrations of bond slip and equivalent free bond slip length (after Bažant and Cedolin, 1980)

the bar cross-section A_b, ultimate bond force U'_b, ratio n' of the elastic moduli of steel and concrete, stress σ_s in the bars at the point of crack band crossing, and the reinforcement ratio p (the bars are assumed to be spaced regularly and densely). The following approximate expression was derived (Bažant and Cedolin, 1980),

$$L_s = \frac{A_s}{U'_b}(\sigma_s - \sigma'_s) \simeq \frac{A_b}{U'_b}\frac{1-p}{1-p+n'p}\sigma_s \qquad (13.11)$$

For convenience of programming it is further possible to replace this actual bond slip length with an equivalent free bond-slip length L_s^* which coincides with a distance between certain two nodes of the mesh, and which permits neglecting the bond shear stresses, which are difficult to model in a finite element code. The cross-section area A_b must also be adjusted to a value A_b^*. The values of L_s^* and A_b^* are then determined from the condition that the extension of the steel bar over the length L_s, with bond shear stresses present, would be the same as the extension of a bar of cross-section area A_b^* and length L_s^* with zero bond stresses; (see Bažant and Cedolin, 1980).

When reinforcement is used, the expression for the equivalent strength must also be modified. The following formula has been derived from energy release considerations (Bažant, 1984; Bažant and Cedolin, 1983a).

$$f'_{eq} = c_f \left(\frac{2G_f E_c}{w_{ef} r_f}\right)^{1/2} \left(1 + c_p n' \frac{p}{L_s^*} \cos \alpha_s \right) \qquad (13.12)$$

in which α_s = angle of the reinforcing bars with the normal to the crack band, and c_p is an empirical coefficient to be found by numerical calibration, i.e. by comparisons of results for different mesh sizes (Bažant and Cedolin, 1983a).

It might be more realistic to treat reinforcement and bond slip by introducing two overlaid continua, one representing the plain concrete, and one representing the reinforcement mesh. These continua would be allowed to displace relative to each other and would transmit distributed volume forces from one to another, depending on the relative slip. This approach would be, however, much more complicated.

13.4 STRUCTURAL SIZE EFFECT

The dispersed and progressive nature of cracking at the fracture front may be taken into account by introducing the following hypothesis (Bažant, 1983a): The total potential energy release W caused by fracture in a given structure depends on both:

(1) the length a of the fracture, and
(2) the area of the cracked zone, $nd_a a$

in which n is a material constant characterizing the width of the cracking zone at

276 Mechanics of Geomaterials

the fracture front (Bažant and Oh, 1983a), $n \simeq 3$. The dependence of W upon crack length a describes that part of energy release that flows into the fracture front from the surrounding uncracked regions of the structure.

Parameters a and $nd_a a$ are not nondimensional. They are permitted to appear only in a nondimensional form, which is given by the following nondimensional parameters

$$\alpha_1 = \frac{a}{d}, \quad \alpha_2 = \frac{nd_a a}{d^2} \qquad (13.13)$$

They represent the nondimensional fracture length and the nondimensional area of the cracked zone. Furthermore, W must be proportional to volume $d^2 b$ of the structure, b denoting the thickness, and to the characteristic energy density $\sigma_N^2/2E_c$ in which $\sigma_N = P/bd$ = nominal stress at failure, P = given applied load, and d = characteristic dimension of the structure. Consequently we must have

$$W = \frac{1}{2E_c}\left(\frac{P}{bd}\right)^2 bd^2 f(\alpha_1, \alpha_2, \xi_i) \qquad (13.14)$$

in which f is a certain continuous and continuously differentiable positive function, and ξ_i represent ratios of the structure dimensions characterizing the shape of the structure.

To illustrate the structural size effect, we now consider structures of different sizes but geometrically similar shape, including the same ratio of fracture length to the characteristic dimension of the structure, and the same reinforcement ratio. Under this assumption, the shape parameters ξ_i are constant. Using the energy criterion for crack band propagation, $\partial W/\partial a = G_f b$, in which G_f is the fracture energy, we obtain (for constant ξ_i) $\partial f/\partial a = f_1(\partial \alpha_1/\partial a) + f_2(\partial \alpha_2/\partial a)$ where we introduce the notations $f_1 = \partial f/\partial \alpha_1, f_2 = \partial f/\partial \alpha_2$. Substituting this and Equation (13.14) into $\partial W/\partial a = G_f b$, we then get

$$\left(\frac{f_1}{d} + \frac{f_2 nd_a}{d^2}\right)\frac{P^2}{2bE_c} = G_f b \qquad (13.15)$$

Here the fracture energy may be expressed as the area under the tensile stress–strain curve, i.e. $G_f = nd_a(1 - E_c/E_t)f_t'^2/2E_c$, in which E_c is the initial Young's elastic modulus of concrete, E_t is the mean strain-softening modulus, which is negative, and f_t' is the direct tensile strength of concrete. Substituting this expression for G_f together with the relation $P = \sigma_N bd$ into Equation (13.15), we finally obtain the following simple formula

$$\sigma_N = Bf_t'\left(1 + \frac{d}{\lambda_0 d_a}\right)^{-1/2} \qquad (13.16)$$

in which σ_N = nominal stress at failure, $B = [(1 - E_c/E_t)/f_2]^{1/2}$, and $\lambda_0 = mf_2/f_1$.

Fracture in Concrete and Reinforced Concrete 277

B and λ_0 are constants when geometrically similar structures of different sizes are considered.

In the plot of log σ_N versus $\log(d/d_a)$ where d/d_a is the relative structure size, Equation (13.16) is represented by the curve shown in Figure 13.10. If the structure is very small, then the second term in the parenthesis in Equation (13.15) is negligible compared to 1, and so $\sigma_N = Bf'_t$ is the condition characterizing failure; it represents the strength criterion, which in Figure 13.10 corresponds to a horizontal line. This special case is obtained if W depends only on the crack-zone area but not on the fracture length.

If the structure is very large, then 1 is negligible compared to the second term in the parenthesis of Equation (13.10). Then $\sigma_N = \text{const.}/\sqrt{d}$. This is the type of size effect known to apply for linear elastic fracture mechanics. Thus, linear elastic fracture mechanics must always apply for a sufficiently large concrete structure, which is no doubt the case for a dam. It is interesting to note also that the preceding nondimensional analysis yields this limiting case when the starting

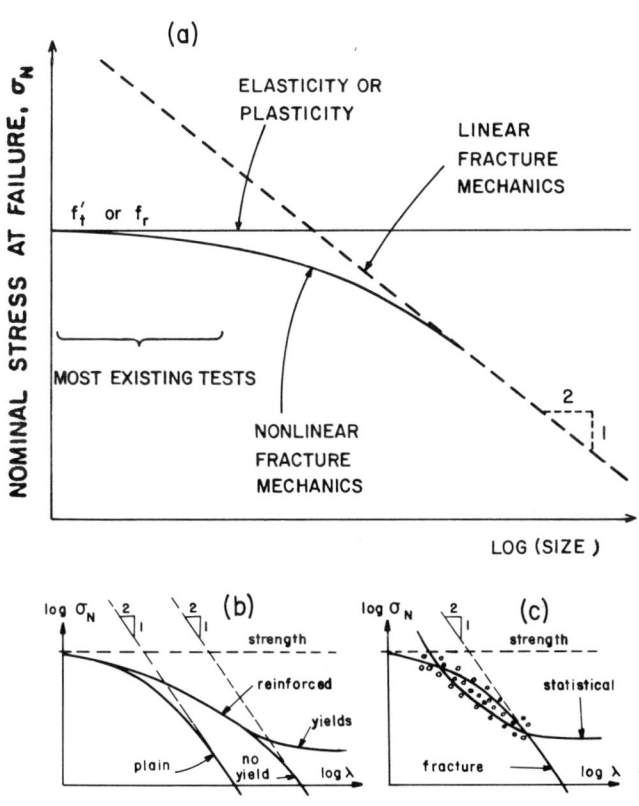

Figure 13.10 Various theories for structural size effect

hypothesis includes only the dependence of W on the fracture length but not on the cracked zone area. In Figure 13.10 the limiting case of linear fracture mechanics is represented by the straight line of downward slope $-1/2$.

The size effect in concrete structures failing due to cracking of concrete represents, as we have shown, a gradual transition from the strength criterion to the energy criterion of linear elastic fracture mechanics. Unfortunately, among the numerous test data on fracture of plain concrete as well as on brittle failures of concrete structures, as reported in the literature, only a very small fraction of the data involves specimens of sufficiently different sizes to check our preceding conclusion on the size effect. Among the available data on plain concrete, the size effect may be checked from the test data of Walsh (1972) (Figure 13.5). A very good agreement with Equation (13.16) is found from these data. As for brittle failure of concrete structures, a check can be made using certain data for the diagonal shear failures of concrete beams with longitudinal reinforcement but without stirrups. Results of such tests are shown in Figure 13.11 for the test data from Kani (1966), Leonhardt and Walther (1961–63), Bhal (1968), Walraven (1978), Taylor (1972), Rüsch et al. (1962), and Swamy and Qureshi (1971). In spite of the large scatter, which is due to comparing test data from different laboratories for different concretes, the declining trend is obvious. A horizontal line, corresponding to the strength criterion (as well as to the present ACI or CEB-FIP codes), is contradicted by the test data. At the same time, however, one can clearly see a substantial deviation from the straight line representing linear elastic fracture mechanics. For more detail, see Bažant and Kim (1984).

In the preceding analysis we have not paid any particular attention to reinforcement. If a densely and regularly distributed reinforcement is present, one finds that the size effect is again governed by Equation (13.16), however, with different constants, provided that the reinforcement remains elastic. Compared to plain concrete, the asymptotic straight line for linear elastic fracture mechanics is pushed in the plot of Figure 13.10 toward the right, i.e. there exists a greater range of sizes for which the strength criterion applies. Nevertheless, for sufficiently large structures, a transition to the size effect of linear fracture mechanics does occur. This conclusion, however, is true only as long as the reinforcement does not yield. If it does, then there is another transition in Figure 13.10 to a horizontal asymptote (see Bažant, 1983a).

The decrease of nominal stress at failure with the structure size has been explained in the past as a statistical phenomenon. The strength of concrete is randomly variable, and in a larger structure there is a greater chance of encountering a smaller strength, which could explain the size effect. However, since the random variations occur only within certain representative volumes of a fixed small size, the statistical size effect must lead to a horizontal asymptote. Thus, the asymptotic behaviour is quite different from that we obtained for fracture mechanics. Needless to say, the fracture-type size effect appears to be more realistic.

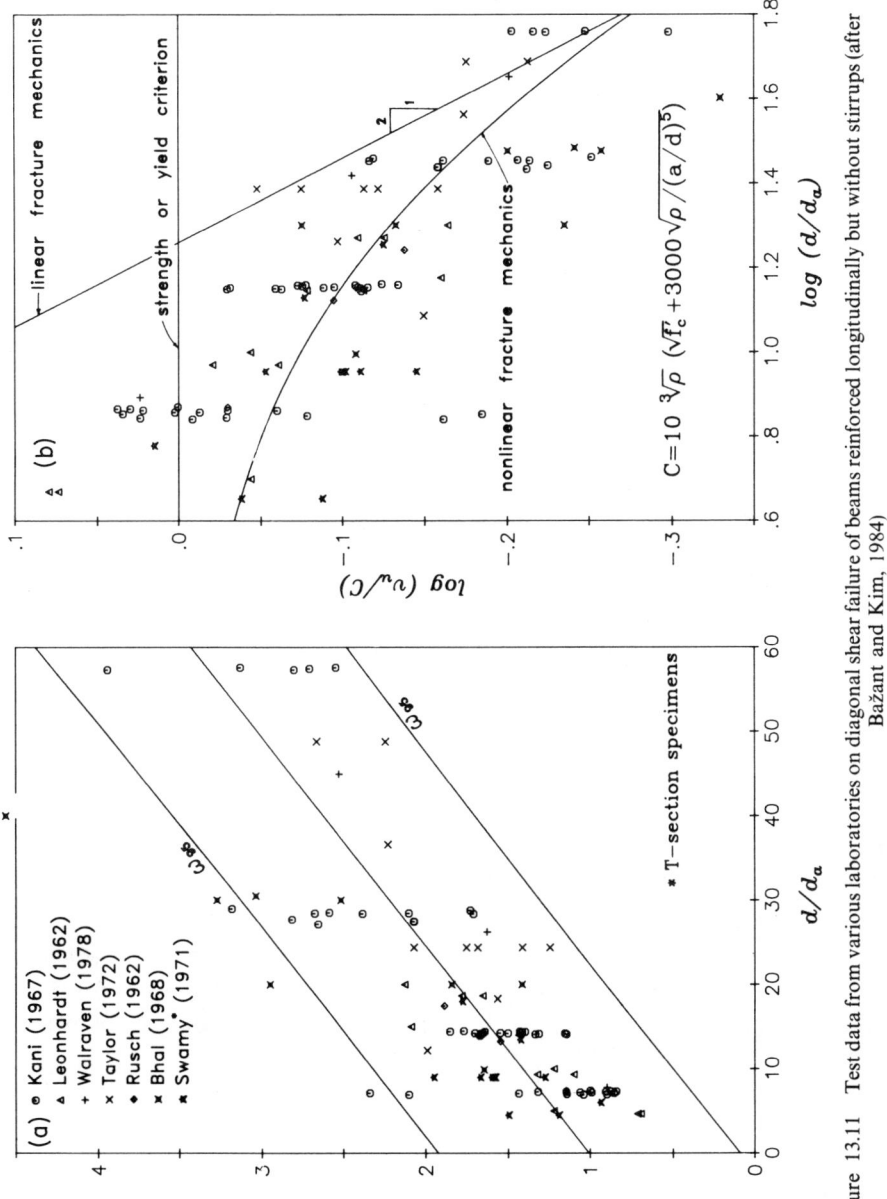

Figure 13.11 Test data from various laboratories on diagonal shear failure of beams reinforced longitudinally but without stirrups (after Bažant and Kim, 1984)

13.5 IMBRICATE CONTINUUM AND CONVERGENCE AT MESH REFINEMENT

A limiting and awkward feature of the crack-band theory is the fact that the finite element is not allowed to be smaller than a certain characteristic size w_c (unless, of course, the strain-softening slope is altered, but this is an artifice). This makes it impossible to obtain resolution of the displacement, strain, and stress fields on smaller scales, and particularly near the fracture front. Even though the actual stresses and strains are randomly scattered on such a small scale, their mean description is of interest for the overall response. Moreover, it is of interest from the mathematical point of view. A discrete model is not a mathematically rigorous concept if the discretization cannot be refined and a convergence to some continuum solution achieved. We will now describe a new type of a continuum which extends the crack-band theory to arbitrarily small meshes and achieves a convergent behaviour.

For statistically heterogeneous materials, the overall, averaged behaviour is defined for the so-called representative volume, which represents the smallest sample of the material for which the statistical properties of the microstructure are roughly the same regardless of the location from which the sample is taken. The microscopic, smoothed-out stress σ represents the force resultant over the cross-section of the representative volume divided by the cross-section area, and the microscopic, smoothed-out strain, which may be called the mean strain, may be defined as the relative change of length of line segment $\overline{12}$ in Figure 13.12 and may be expressed as

$$\bar{\varepsilon}(x) = \frac{1}{l}\left[u\left(x+\frac{l}{2}\right) - u\left(x-\frac{l}{2}\right)\right] \quad (13.17)$$

Here x = coordinate, u = displacement (macroscopic, smoothed-out), and l = characteristic length = size of the representative volume. Equation (13.17) means that $\bar{\varepsilon}$, as well as associated stress σ, refers to the cross-section $\overline{PQ}$ in the middle of segment $\overline{12}$ in Figure 13.12. Note that the stress and strain at cross-section $\overline{PQ}$ does not depend on the change of length of line segment $\overline{34}$ (Figure 13.12); rather, the length change of this line segment determines the stress at cross-section $\overline{RS}$. This fact may be expressed by means of the following hypothesis.

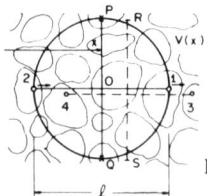

Figure 13.12   Illustrations for representative volume of a heterogeneous aggregate material

Fracture in Concrete and Reinforced Concrete

Hypothesis I. The stress $\sigma(x)$ at any point x (except in a boundary layer of thickness l) depends on the change of distance between points $x + l/2$ and $x - l/2$, but does not depend on the change of distance between any other two points lying a finite distance apart.

According to this hypothesis, we may construct the following finite element system for the special case of one-dimension. We consider a bar of unit cross-section (Figure 13.13(a)) and subdivide the length coordinate x by nodes $k = 1, 2, 3, \ldots$ into equal segments of length h, subjected to the condition that $h = l/n$ where n is an integer. According to the hypothesis, the stress at point $x_k + l/2$, denoted as σ_k, depends on the difference of displacements at points $x_k + l$ and x_k, i.e.

$$\sigma_k = \bar{E}(\bar{\varepsilon}_h)\varepsilon_k, \quad \bar{\varepsilon}_k = \tfrac{1}{2}(u_{k+n} - u_k) \tag{13.18}$$

in which the subscripts refer to nodal numbers ($k = 1, 2, \ldots$), and $\bar{E}$ is the secant modulus depending on the mean strain. It is now evident that if $h < l$ (i.e. $n > 1$), the mean strain $\bar{\varepsilon}_k$ represents the strain in a finite element which spans from node k to node $k + n$ and *skips* the nodes lying in between, as illustrated in Figure 13.13(b). Since there must be an element of length l attached to every node, the finite elements must obviously overlap, i.e. be imbricated; see the finite elements illustrated by circles in Figure 13.13(b). Due to this imbricate structure, we will call the limiting continuum for $h \to 0$ (or $n \to \infty$) the *imbricate* continuum. A two-dimensional generalization is illustrated in Figure 13.14(a).

It must now be observed, however, that the system of imbricate elements alone (i.e. the circular elements in Figure 13.13(b)) is not stable. One can easily check that the displacement distribution $u_k = A \sin(2\pi x_k/l)$ results in no stress in any imbricate element, i.e. an unresisted deformation can happen, which is in-

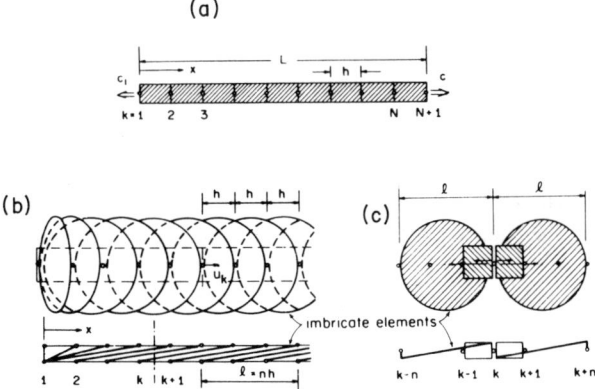

Figure 13.13 (a) Discretization of a one-dimensional bar. (b), (c) Arrangement of imbricate elements

admissible. Therefore, the discrete model must be stabilized by introducing additional finite elements of the smoothed-out macroscopic continuum which have length h and span from one node to the next (Figure 13.13(a)–(c)). These are the usual finite elements, and they will be called the local elements. The sum of the cross-section areas of these local elements and of the imbricate elements must equal 1. Denoting as c the cross-section (or thickness) of the local elements ($0 < c \leqslant 1$), we must assure that the combined cross-section area of the imbricate elements equals $1 - c$. Because n imbricate elements intersect any cross-section between two nodes (Figure 13.13(b)), the cross-section area or thickness of each imbricate element must be taken as

$$b_e = \frac{1-c}{n} \quad (13.19)$$

As the mesh is refined ($n \to \infty$), the thickness of each imbricate element approaches zero.

Even though n imbricate elements intersect each cross-section, there are only two imbricate elements attached to one node k (Figure 13.13(c)). In addition, there are two local elements attached to node k. Thus, the nodal equilibrium equation reads $b_e(\sigma_k - \sigma_{k-n}) + c(\tau_k - \tau_{k-1}) + p(x_k)h = 0$, in which p is the distributed load per unit length, and τ_k is the stress in the local element, which may be expressed as

$$\tau_k = E(\varepsilon_k)\varepsilon_k, \quad \varepsilon_k = \frac{1}{h}(u_{k+1} - u_k) \quad (13.20)$$

Here E is the local elastic modulus, interpreted as a secant modulus which depends on the local strain ε_k.

Substituting, according to D'Alembert principle, $p(x_k) = \rho \ddot{u}_k$ where superior dots denote time derivatives and ρ is the mass density, and rearranging algebraically the foregoing node equilibrium equation, we obtain the following equation of motion

$$\frac{1-c}{l^2}[\bar{E}(\bar{\varepsilon}_k)(u_{k+n} - u_k) - \bar{E}(\bar{\varepsilon}_{k-n})(u_k - u_{k-n})] + \frac{c}{h^2}[E(\varepsilon_k)(u_{k+1} - u_k) - E(\varepsilon_{k-1})(u_k - u_{k-1})] = \rho \ddot{u}_k \quad (13.21)$$

Although elastic materials are not the purpose of this formulation, it is interesting to note that for the special case of constant $\bar{E}$ and E, Equation (13.21) reduces to

$$(1-c)\bar{E}\frac{u_{k+n} - 2u_k + u_{k-n}}{l^2} + cE\frac{u_{k+1} - 2u_k + u_{k-1}}{h^2} = \rho \ddot{u}_k \quad (13.22)$$

The total stress, representing the sum of all forces within the cross-section of the bar, may be expressed as

$$S_k = \frac{1-c}{n}\sum_{j=1}^{n}\sigma_{k+1-j} + c\tau_k \quad (13.23)$$

Fracture in Concrete and Reinforced Concrete

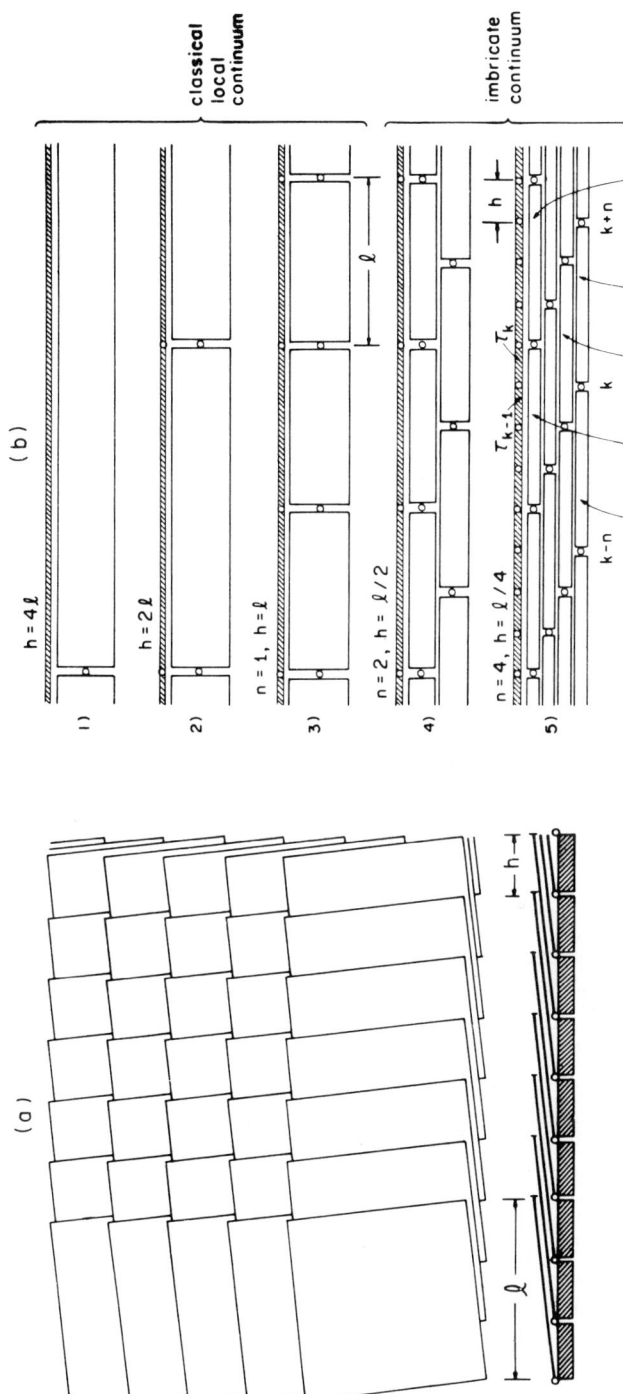

Figure 13.14 (a) Two-dimensional arrangement of square imbricate elements (the element sides should be perfectly horizontal and vertical but are misaligned for the purpose of illustration). (b) One-dimensional element arrangements for finer and finer subdivisions (the nodes on top of each other have the same displacement)

It is now obvious that the finite difference expressions in Equation (13.22) represent the well-known finite difference approximations to the second derivative $\partial^2 u/\partial x^2$. Similarly, finite difference approximations of the first derivatives may be detected in Equation (13.21). Thus, the continuum limit of Equation (13.21) ($h \to 0$ and $n \to \infty$ at constant l) has the form

$$(1 - c)D_x\sigma(x) + c\frac{\partial}{\partial x}\tau(x) = \rho\frac{\partial^2 u(x)}{\partial t^2} \quad \left(\text{for } \frac{l}{2} \leq x \leq L - \frac{l}{2}\right) \quad (13.24)$$

in which

$$\sigma(x) = \bar{E}D_x u(x), \quad \tau(x) = E\frac{\partial}{\partial x}u(x) \quad (13.25)$$

with D_x denoting an operator defined by the relation

$$D_x u(x) = \frac{1}{l}\left[u\left(x + \frac{l}{2}\right) - u\left(x - \frac{l}{2}\right)\right] = \bar{\varepsilon}(x) \quad (13.26)$$

Here $t =$ time, σ is the limit of the stress in the imbricate elements which may be called the broad-range stress (note that it does not represent a mean of some stresses, unlike $\bar{\varepsilon}$), and τ is the limit of the stresses in the local elements which may be called the local stress. Equations (13.24)–(13.26) are the continuum relations characterizing the imbricate continuum. The total stress in the continuum is obtained as the limit of Equation (13.23),

$$S(x) = \frac{1 - c}{l}\int_{-l/2}^{l/2} \sigma(x + s)ds + c\tau(x) = (1 - c)\bar{\sigma}(x) + c\tau(x) \quad (13.27)$$

in which $\bar{\sigma}$ represents the mean broad-range stress. Note that Equation (13.24) may be also written as $\partial S(x)/\partial x = \rho\ddot{u}(x)$.

The question now is whether the imbricate continuum can, indeed, describe strain-softening regions of finite size, and whether it can do so in a stable manner. It appears that, in order to achieve stability, the local stress–strain relation must not exhibit strain-softening, i.e. the tangent modulus based on $E(\varepsilon)$ must be positive, while the broad stress–strain relation may exhibit strain-softening, i.e. the tangent modulus $\bar{E}_t = \bar{E} + (\partial\bar{E}/\partial\bar{\varepsilon})\bar{\varepsilon}$, may become negative. The question of stability is easy to investigate by considering the broad stress–strain relation $\sigma = \bar{E}_t(\bar{\varepsilon} - \varepsilon_f)$ is which E_t and ε_f are constants. Then, substituting $u = A \exp[i\omega(x - vt)]$ where $i^2 = -1$, and ω, v, A are constants, one finds that the field equation in Equations (13.24)–(13.26) for $E = \bar{E} =$ constant reduces to an algebraic equation for the wave velocity v. As is well-known, the continuum is stable if v is not imaginary, i.e. $v^2 > 0$. It can be shown that this leads to the condition $c > 0$.

Computer calculations (Bažant et al., 1983b) verified that $c = 0.1$ yields a stable and well behaved solution. This is true even for strain-softening with tangent modulus $\bar{E}_t = -0.2E$ where E is the elastic modulus of the material. In these calculations, carried out by T. P. Chang (Bažant et al., 1983b), two broad-range stress–strain diagrams were considered: one which dips below the strain axis so as to achieve full strain-softening for the total stress S, and one which terminates by a horizontal segment when it reaches the strain axis, in which case the total stress S does not reach full strain-softening (Figure 13.15). In both cases, for strains exceeding strain ε_f at the end of strain-softening, a horizontal plateau is considered both for the broad-range and the local stress–strain diagram (Figure 13.15). Of these, the case where the strain-softening does not dip below the strain axis appears physically more reasonable, but both approaches appear to work numerically (Bažant, Chang and Belytschko, 1983b).

The field equations in Equations (13.24)–(13.26) apply only at distances more than l from the boundary. Within boundary segments of length l, the continuum equations are much more complicated, however, the discrete modelling is quite simple. One needs to first lay out all imbricate elements attached to every node, and then those imbricate elements which stick out beyond the boundary may be chopped and attached to the boundary node (see Figure 13.13(b)).

For the stress–strain diagrams in Figure 13.15, extensive computer studies of convergence at diminishing mesh size h have been carried out (Bažant et al., 1983b). As a practical example, it was considered that constant outward velocities c_1 in the outward direction are prescribed at the boundary nodes; Figure 13.13(a). For the case of an elastic bar, this produces step waves of strain with magnitudes c_1/v; as these two waves meet at mid-length, the strain is doubled. For the case of a strain-softening continuum (Figure 13.15), the velocity c_1 is selected such that the initial inward step-waves are just below the elastic strain limit. As these two waves meet at mid-length, strain rapidly increases into the strain-softening range. Examples of some results (Bažant et al., 1983b), showing strain distributions for various times obtained for various local element sizes h, are exhibited in Figure 13.16(a). These figures demonstrate that a finite-size strain-softening region is obtained, and that the response converges as h is diminished. Moreover, it has

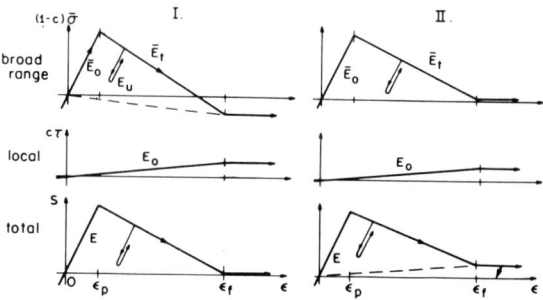

Figure 13.15 Broad-range, local and total stress–strain relations used in numerical example

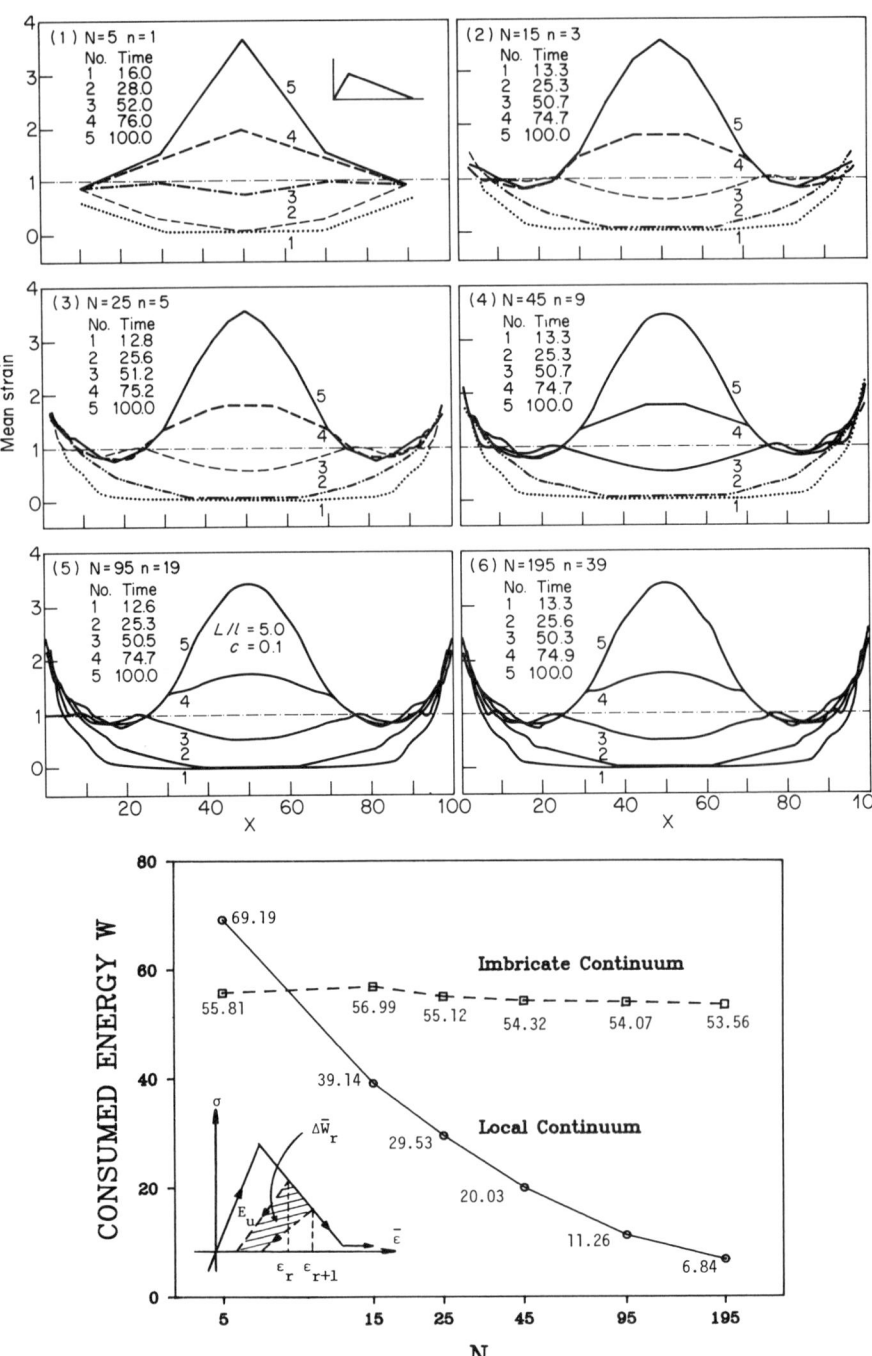

Figure 13.16 Mean strain distributions at various times obtained by T. P. Chang (Bažant et al., 1983b) for various numbers N of imbricate continuum subdivisions and energy dissipated by strain-softening as a function of N.

been checked that for the case of the usual, local continuum ($l = h$ at diminishing h), the solution converges to an exact solution for strain-softening in this bar, obtained in Bažant and Belytschko (1983). These results have been obtained by an explicit time-step algorithm (Bažant, Chang and Belytschko, 1983b).

Particularly interesting is a comparison of the total energy W dissipated in the bar due to strain-softening. Figure 13.16(b) shows a plot of W as a function of the number N of subdivisions; W is obtained from $\bar{W}_r = [(\sigma_r^2/E'_r) - (\sigma_{r+1}^2/E'_{r+1}) + (\sigma_r + \sigma_{r+1})\Delta\varepsilon]/2 =$ energy dissipated per step and per unit length, $E'_r =$ unloading modulus at time t_r. Note that for the imbricate continuum W converges to a finite value while for the classical local continuum it converges to zero, which is unrealistic.

A more rigorous derivation of the imbricate continuum may be based on the principle of virtual work. It follows directly from Hypothesis I that the work variation for a one-dimensional bar of length l may be expressed as

$$\delta W = \int_{l/2}^{L-l/2} \left\{ \frac{\sigma(x)}{l}\left[\delta u\left(x+\frac{l}{2}\right) - \delta u\left(x-\frac{l}{2}\right)\right] + \tau(x)\frac{\partial}{\partial x}\delta u(x) - p(x)\delta u(x) \right\} dx + \delta W_b = 0 \qquad (13.28)$$

in which δu is any kinematically admissible variation of displacements, and δW_b is the virtual work done within the boundary segments of length l. It may be shown by variational calculus that the condition $\delta W = 0$ for any δu leads to the continuum equations in Equations (13.24)–(13.26) (Bažant, 1983b).

The variational approach permits deriving the field equations for the imbricate continuum in three-dimensions. For this purpose, the operator D_s may be generalized as

$$D_i u_j\left(\mathbf{x}+\frac{l}{2}\mathbf{i}\right) = \frac{1}{l}\left[u_j\left(\mathbf{x}+\frac{l}{2}\mathbf{i}\right) - u_j\left(\mathbf{x}-\frac{l}{2}\mathbf{i}\right)\right]a_i \qquad (13.29)$$

in which subscripts i,j refer to cartesian coordinates $x_i (i = 1, 2, 3)$, u_j are the displacement components, $\mathbf{i}$ is the unit vector of axis x_i, a_i are the direction cosines of vector $\mathbf{i} = (1, 0, 0)$ or $(0, 1, 0)$ or $(0, 0, 1)$, and $\mathbf{x}$ is the coordinate vector. The mean strain may then be generalized as

$$\bar{\varepsilon}_{ij} = \tfrac{1}{2}(D_i u_j + D_j u_i) \qquad (13.30)$$

From Hypothesis I it follows that stresses σ_{ij} in a three-dimensional generalization of the imbricate continuum must depend on the mean strains $\bar{\varepsilon}_{ij}$ defined above because these strains depend only on the change of distance between points that lie a distance l apart. Consequently, the principle of virtual work for the imbricate continuum must have the form

$$\delta W = \int_{\Omega'} [(1-c)\sigma_{ij}\delta(D_i u_j) + c\tau_{ij}\delta u_{i,j} - p_i\delta u_i]dV + \delta W_b = 0 \qquad (13.31)$$

in which repeated subscripts imply summation over 1, 2, 3; a subscript preceded by a comma denotes a partial derivative; τ_{ij} are local stresses depending on displacement gradients; p_i are applied loads per unit volume; Ω' denotes the domain of the body excluding a boundary layer of thickness $l/2$; and ∂W_b is the virtual work done by the stresses within this boundary layer. Introducing a certain kind of shift operators, one can isolate the variation δu_j from the integrand in the foregoing integral and show that the continuum equations of motion (for $p_i = -\rho \ddot{u}_i$) must have the form (Bažant, 1983b)

$$(1-c)D_j\sigma_{ij} + c\tau_{ij,j} = \rho \ddot{u}_i \quad \text{(for } \mathbf{x} \in \Omega'') \tag{13.32}$$

in which

$$\sigma_{ij} = \bar{C}_{ijkm}(\bar{\varepsilon})\bar{\varepsilon}_{km} = C_{ijkm}(\bar{\varepsilon})D_m u_k \tag{13.33}$$

Here $\bar{C}_{ijkm}$ are the secant moduli which generally depend on the mean strain tensor $\bar{\varepsilon}$. Equation (13.32) holds only for points within domain Ω'' which represents the domain of the body without a boundary layer of thickness l. Within the boundary layer the continuum equations of motion are much more complicated and it is preferable to set up their discretized, finite element form directly. Similarly to σ, σ_{ij} may be called the components of the broad-range stress tensor.

Alternatively, and perhaps more rigorously, the mean strain $\bar{\varepsilon}_{ij}$ can be defined as the average of the strain gradient $u_{i,j}$ obtained by integration over a sphere of diameter l surrounding the given point. It may also be expressed as an integral of $u_i n_j$ over the surface of this sphere (where n_j are the direction cosines of an outward normal of this surface). It may also be defined by least-square fitting of a homogeneous local strain field over the surface of this sphere. All these three definitions are exactly equivalent, but they are only approximately equivalent to the definition of mean strain based on the difference operator in Equation (13.29). However, the use of Equation (13.29) is much simpler for numerical calculations, as well as for deriving the continuum equations of motion from the principle of virtual work (Bažant, 1983b).

Can the difference operators be eliminated? They can, as an approximation, if $u(x)$ is expanded in Taylor series. Then

$$\bar{\varepsilon} = D_x u \simeq u' + \frac{l^2}{24}u''' \tag{13.33a}$$

and the one-dimensional equation of motion within regions in which the tangent moduli $\bar{E}_t$ and E_t are constant becomes

$$\tilde{E}_t \frac{\partial^2 u}{\partial x^2} + (1-c)\bar{E}_t \lambda^2 \left(2\frac{\partial^4 u}{\partial x^4} + \lambda^2 \frac{\partial^6 u}{\partial x^6} \right) = \rho \frac{\partial^2 u}{\partial t^2} \tag{13.33b}$$

where

$$\tilde{E}_t = (1-c)\bar{E}_t + cE_t, \quad \lambda^2 = l^2/24 \tag{13.33c}$$

Fracture in Concrete and Reinforced Concrete 289

The foregoing equations permit obtaining solutions to one-dimensional problems analytically.

To generalize Equation (13.33a) to three dimensions, one may use the Taylor series expansion to obtain

$$D_j u_i(\mathbf{x}) \simeq u_{i,j}(\mathbf{x}) + u_{i,jkm}(\mathbf{x})\frac{1}{2V}\int_{V(\mathbf{x})}(x'_k - x_k)(x'_m - x_m)\,dV' \quad (13.33\text{d})$$

where V is a sphere constituting the representative volume, centered at $(\mathbf{x})$. The integral in the last equation may be expressed as $\delta_{km}l'^2/40$ where δ_{km} = Kronecker delta and l' = diameter of the sphere. Now we see that if we would set $l' = l$, Equation (13.33d) would not reduce to Equation (13.33a) for the case of uniaxial strain. It is preferable, therefore, to keep the coefficient $1/24$, and this may be obtained by setting $l'^2/40 = l^2/24$ as obtained in Equation (13.33a), which yields $l' = l\sqrt{5/3} = 1.29l$. Thus, the tensors of the mean displacement gradient and the mean strain may be approximated as

$$D_j u_i \simeq u_{i,j} + \frac{l^2}{24}u_{i,jkk}, \quad \bar{\varepsilon}_{ij} \simeq \varepsilon_{ij} + \frac{l^2}{24}\varepsilon_{ij,kk} \quad (13.33\text{e})$$

Note also that $\bar{\varepsilon} = \varepsilon + (l^2/24)\nabla^2\varepsilon$, etc., where ∇^2 = Laplacean. The stress difference in the equation of motion (Equation (13.32)) may be approximated as

$$D_j\sigma_{ij} \simeq \sigma_{ij,j} + \frac{l^2}{24}\sigma_{ij,jkk} \quad (13.33\text{f})$$

For the total stress, one can show that

$$S_{ij} = (1-c)(1-\lambda^2\nabla^2)C_{ijkm}(1+\lambda^2\nabla^2)\varepsilon_{km} + c\tau_{ij} \quad (13.33\text{g})$$

and the equation of motion may then be written as $S_{ij,j} = \rho\ddot{u}_i$.

The foregoing approximations by higher-order derivatives might be useful for analytical solutions.

The consequences for the blunt crack band theory can now be discussed. When the mesh size h is chosen equal to the size l of the representative volume, the finite element model of the imbricate continuum becomes equivalent to that of the classical local continuum, as used in the crack band theory. Meshes with $h > l$ (Figure 13.14(b)) make no sense for the imbricate continuum model because their characteristic length cannot be resolved. Therefore, a local continuum theory is the only meaningful basis for the finite element analysis if $h \geqslant l$. On the other hand, finite element analysis based on the local continuum theory makes no sense when $h < l$, and the imbricate continuum is needed for such a refinement of mesh (see Figure 13.14b)).

Due to the fact that for $h = l$ the finite element models for the local and the imbricate continuum coincide, the characteristic length, l, must be the same as the width of the crack band front, w_c, introduced before. This provides a method of experimental determination of l.

With the formulation of the imbricate continuum, the continuum theory appears to be put on a rigorous, continuum mechanics basis. Aside from that, the imbricate continuum may be used to calculate the detailed distribution of the smoothed macroscopic stress and strain near the fracture front, or in the vicinity of singularities in general (at concentrated loads, at corners or notches, etc.).

Before leaving this topic, we should at least briefly comment on various recent attempts to model strain-softening by means of either the classical, local continuum or the classical non-local continuum. Strain-softening has been a suspect feature in continuum mechanics ever since Hadamard pointed out that the wave propagation speeds become imaginary when the matrix of the tangent moduli ceases to be positive definite, which is the case for strain softening (Hadamard, 1903; Thomas, 1961). Hadamard's point, however, neglected the fact that different tangent moduli apply to positive and negative increments of strain. This fact was recently taken into account and it was found that the use of strain-softening with a classical, local continuum does not result in a mathematically meaningless formulation.

For example, exact stability analysis has been carried out for one-dimensional strain localization due to strain-softening (Bažant, 1976; Bažant and Panula, 1978) (as opposed to strain-localization instabilities caused by geometrical nonlinearity (Rice, 1976; Rudnicki and Rice, 1975)). It was found that the equilibrium path of the structure is unique and can be traced just as well as the equilibrium path in the buckling of the perfect column. Recently, an exact solution has been obtained for some one-dimensional wave propagation problems with strain softening (Bažant and Belytschko, 1983). Again, it was shown that the solution is unique except for some values of prescribed boundary velocities for which the solution changes discontinuously. In both the static and the dynamic problem, the strain-softening in a classical, local continuum is found to localize immediately into a layer of zero thickness in which the strains become unbounded. Furthermore, the energy consumed by strain-softening failure is found to be zero (Figure 13.16). So the problem with the use of strain-softening in the classical, local continuum theory is not that the formulation would be unsound mathematically, but that the solution is not representative of the behaviour of real strain-softening materials, in which the strain-softening regions have a finite size, and a finite energy is consumed by failures due to strain-softening.

Quite illuminating was the work of Burt and Dougill (1977). They simulated by a computer a random two-dimensional network whose joints were placed at random locations over a rectangular domain. They considered all pairs of joints with a distance less than a certain constant l, made a random selection among all such pairs, and introduced connecting pin-jointed elastic struts between each pair. The stiffnesses and strength limits of these struts were generated randomly according to prescribed normal distributions. After reaching the strength limit, the strut was assumed to fail and its internal force to drop suddenly to zero. Although, due to its neglect of shear resisting elements, this model could not have exhibited a correct elastic Poisson ratio (it was 1/3), the simulated inelastic behaviour was strikingly similar to that of concrete. Especially, strain-softening regions of finite size were obtained in these random networks. This clearly established the need to consider some sort of a non-local continuum.

In the simplest possible non-local continuum theory, and the only one which appears to have been used thus far, the stress at point x is assumed to be a function of the weighted average of strains obtained by integration over a neighbourhood of x. Other relations, particularly the continuum equations of motion in terms of stresses and the strain-displacement relations, are the same as for the classical, local continuum. It appeared, however, that this classical form of the non-local continuum theory (Kröner, 1967, 1968; Krumhansl, 1968; Kunin, 1968; Levin, 1971; Eringen and Edelen, 1972; Eringen and Ari, 1983) is unworkable for the modelling of strain-softening. Computer simulations always resulted in unstable behaviour, and did not exhibit convergence at mesh refinements (T. P. Chang), and strain-softening always localized between two adjacent points of the mesh. It was after this experience that the concept of the present imbricate continuum was conceived.

13.6 EQUIVALENT LINEAR ANALYSIS OF FRACTURE

Since uncracked concrete in tension behaves essentially elastically, the stress field surrounding the fracture process zone in concrete should be predominantly elastic. However, it corresponds to some equivalent crack length rather than the actual crack length. Thus, the idea of an equivalent crack length which has no relation to the actual crack length should allow an approximate linear analysis of concrete fracture. As it turns out, however, this is not sufficient. It is also necessary to assume that the fracture energy G_c is variable rather than constant. Now it appears that for many practical purposes the dependence of the fracture energy on structural geometry, type of loading, notch length, etc., can be neglected and only the dependence of the fracture energy on the length c of crack extension from the notch needs to be considered. Thus, the plot of G_c versus c is assumed to be unique and is called the resistance curve or R-curve. This idea, proposed by Irwin (1960) and Krafft et al. (1961), has proved to be rather useful for metals (Broek, 1974; Parker, 1981) and appears to work also quite well for concrete (Bažant and Cedolin, 1983b).

Let us briefly review the R-curve concept. We consider that the fracture energy, G_c, is a certain given function of the crack extension c from the notch or smooth surface, i.e. $G_c = G_c(c)$ in which $c = a - a_0$, a_0 = length of the notch, and a = total length of crack plus notch (Figure 13.17). The energy that must be supplied to the

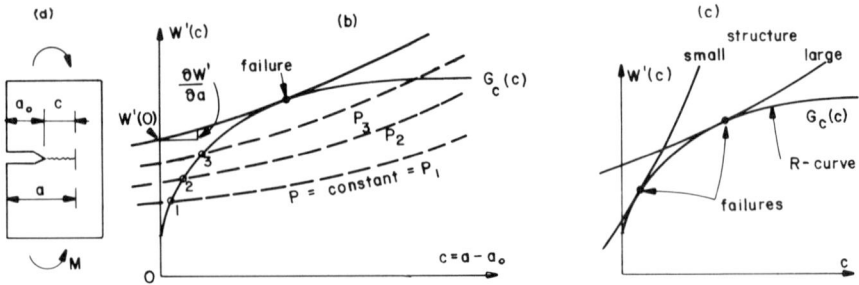

Figure 13.17 Determination failure with the help of R-curve

structure to produce the crack is $U = b \int G_c(c)\, da - W(a)$ where b is the thickness of the structure and W is the total release of strain energy from the structure or specimen. An equilibrium state of fracture occurs when no energy needs to be supplied to change a by δa and none is released, i.e. when $\delta W = 0$. Since $\delta U = (\partial U/\partial a)\delta a = 0$ in which $\partial U/\partial a = bG_c - W' = 0$ and $W' = \partial W/\partial a$, it follows that fracture equilibrium occurs when

$$W'(a) = bG_c(c) \quad \text{(equilibrium)} \quad (a = a_0 + c) \quad (13.34)$$

The equilibrium fracture state is stable if the second variation $\delta^2 U$ is positive. Since $\delta^2 U = (\partial^2 U/\partial a^2)\delta a^2$ where $\partial^2 U/\partial a^2 = b\partial G_c/\partial a - W''(a)$, and $W''(a) = \partial^2 W/\partial a^2$, the conditions for stability of fracture and the limit of stability, i.e. the failure condition, are given by

$$bG'_c(c) - W''(a) > 0 \quad \text{(stable)} \quad (13.35a)$$
$$= 0 \quad \text{(critical)} \quad (13.35b)$$

Considering structures that are geometrically similar but have dissimilar notches, and using dimensional analysis, we may write, in analogy to Equation (13.14),

$$W'(a) = \frac{P^2 g(\alpha)}{E_c bd}, \quad W''(a) = \frac{P^2 g'(\alpha)}{E_c bd^2} \quad \left(\alpha = \frac{a}{d}\right) \quad (13.36a, b)$$

in which d is the characteristic dimension of the structure, E_c is the Young's modulus, P is the applied load, and $g(\alpha)$ is a nondimensional function, common to all structures of a similar shape, which can be found for many typical specimens in handbooks and can be always determined by linear finite element analysis. Also, $g'(\alpha) = dg/d\alpha$. Substituting Equations (13.36(a,b)) into Equations (13.34)–(13.35) and eliminating P, we find the condition

$$G_c(c)g'(\alpha) - G'_c(c)g(\alpha)d > 0 \quad \text{(stable)} \quad (13.37a)$$
$$= 0 \quad \text{(critical)} \quad (13.37b)$$

Fracture in Concrete and Reinforced Concrete 293

Furthermore, noting that $dg/dc = dg/da = g'(\alpha)/d$, we find that the last form of the critical state condition is equivalent to the following maximizing condition

$$\frac{G_c(c)}{g(\alpha)} = \text{Max} \quad \left(\alpha = \frac{g_0 + c}{d}\right) \quad (13.38)$$

Thus, the value of crack extension c at failure may be simply found by a search for the maximum of the foregoing expression if functions G_c and g are known. Alternatively, the value of c at failure may be found by solving Equation (13.37b), which would normally be carried out by iterative Newton method.

Analysis of the bulk of fracture test data for concrete available in the literature (Bažant and Cedolin, 1983b) has indicated that failure loads are not very sensitive to the precise shape of the R-curve, i.e. curve $G_c(c)$. Various algebraic expressions have been used for the R-curve $G_c(c)$, including an exponential curve with a horizontal asymptote, a parabola transiting at its apex into a horizontal line, and a bilinear expression. All these expressions have been found to allow roughly equally good fits of the available scattered data. Only the initial value of G_c, the final value G_f, and the overall slope of the rising portion of the R-curve, have been found to be of importance (Bažant and Cedolin 1983b). This means that at the present level of measurement capabilities it makes no sense trying to determine the R-curve with great precision or to develop sophisticated theories for theoretical determination of the R-curve. From the convenience viewpoint, the following elliptic and parabolic curves are preferable because they facilitate the solution of failure loads;

$$G_c(c) = G_f\left[1 - \beta\left(1 - \frac{c}{c_m}\right)^2\right]^{1/2} \quad (13.39a)$$

(for $0 \leq c \leq c_m$, $G_c(c) = G_f$ for $c \geq c_m$)

$$G_c(c) = G_f\left[1 - \beta\left(1 - \frac{c}{c_m}\right)^2\right] \quad (13.39b)$$

Here G_g, c_m and β are constant material parameters. Both these functions permit finding explicit expressions for c and P at failure if the function $g(\alpha)$ is approximated linearly, $g(\alpha) = g_0 + (\alpha - \alpha_0)g_0'$ with $\alpha_0 = (a_0 + c_0)/d$. This linear approximation is always sufficient for large enough structures. Substituting the last expression and Equation (13.39a) or (13.39b) into Equation (13.37b), one can find the following equations for the crack extension c at failure;

$$c = \frac{g(\alpha)d - g'(\alpha)c_0}{g'(\alpha)(c_m - c_0) + g(\alpha)d}c_m \quad \text{(ellipse)} \quad (13.40)$$

$$(c_m - c)^2 - \left(\frac{2g(\alpha)d}{g'(\alpha)} + c_m - c_0\right)(c_m - c) + \frac{c_m^2}{\beta} = 0 \quad \text{(parabola)} \quad (13.41)$$

Here c_0 and α_0 characterize the point about which the R-curve is linearized, and if

$\alpha - \alpha_0$ is small, the approximation of $g(\alpha)$ is very good. After determining c from one of these equations, the value of P at failure may be found from Equations (13.36a) and (13.34).

Let us now check what kind of size effect is implied by the R-curve approach. For the parabolic R-curve, substitution of $P = \sigma_N bd$ and of the linear approximation for $g(\alpha)$, and elimination of c from Equations (13.36a), (13.36b), (13.34), and (13.35b), yields the equation

$$C_1 \sigma_N^4 - (C_2 + C_3 d)\sigma_N^2 + G_f = 0 \quad (P = bd\sigma_N) \quad (13.42)$$

in which

$$C_1 = \frac{b^2 c_m^2 g'^2(\alpha)}{4\beta G_f E_c^2}, \quad C_2 = \frac{b}{E_c}(c_m - c_0)g'(\alpha), \quad C_3 = \frac{b}{E_c}g(\alpha) \quad (13.43)$$

Solving Equation (13.42) for various values of d, one can construct the plot of the nominal stress at failure, σ_N, versus the characteristic dimension d; see Figure 13.18 (solid curve). We see that the resulting plot is very close to that obtained previously from dimensional analysis (Equation 13.16); see Figure 13.18 (dashed curve). This agreement lends support to the use of the simple parabolic formula for the R-curve. The function $g(\alpha)$ used in Figure 13.18 is that which corresponds to a three-point bent specimen used by Walsh (1972), for which the beam span-to-depth ratio is 4, and the notch depth to beam depth ratio is 1/3. This function is

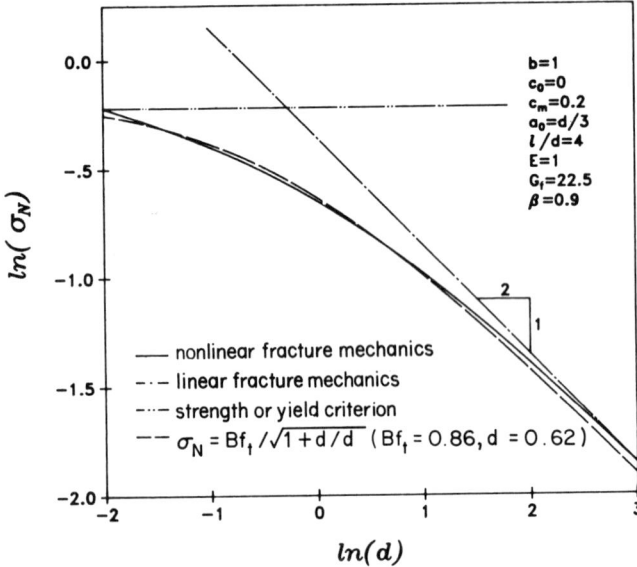

Figure 13.18 Structural size effect obtained from parabolic R-curve, and its comparison with Equation (13.16)

(Tada et al., 1973) $g(\alpha) = \pi(L/d)^2 \alpha(1.634 - 2.603\alpha + 12.30\alpha^2 - 21.27\alpha^3 + 21.86\alpha^4)^2$, where $L =$ beam span and $d =$ beam depth.

To illustrate the degree of agreement with test data which can be achieved with the parabolic R-curve, Figure 13.19 reproduces from Bažant and Cedolin (1983b) a plot of the theoretical versus measured values of P at failure, normalized with regard to the failure load P_0 calculated from the strength theory. If the theory were perfect, the data points would have to fall on a straight line through the origin, having slope 1. The deviations from such a straight line represent errors. Statistical regression analysis shows that their coefficient of variation is only 0.06, which means that the representation of test results is quite adequate. By contrast, for constant G_c the errors are much larger (Bažant and Cedolin, 1983b).

It is useful to realize the geometrical relationship of the R-curve to the family of the curves of equilibrium states (Equation (13.34)) for geometrically similar specimens of various sizes d. Introducing relative size $\lambda = d/d_a$ where d_a is the maximum size of aggregate, the curves of W' versus c at various constant values of λ, defined by Equation (13.34), form a one-parameter family of curves characterized by the equation $F(c, W', \lambda) = 0$ where $F = bG_c(c) - W'(a_0 + c, \lambda)$. Differentiating this equation with respect to c, we get $(\partial F/\partial c) + (\partial F/\partial W')W'' + (\partial F/\partial \lambda)(d\lambda/dc) = 0$, in which $\partial F/\partial c = bG'_c(c)$, $\partial F/\partial W' = -1$. Now the envelope of the family of the aforementioned curves is given by the condition $\partial F/\partial \lambda = 0$, and we see that this condition leads to the relation $bG'_c(c) - W'' = 0$, which is

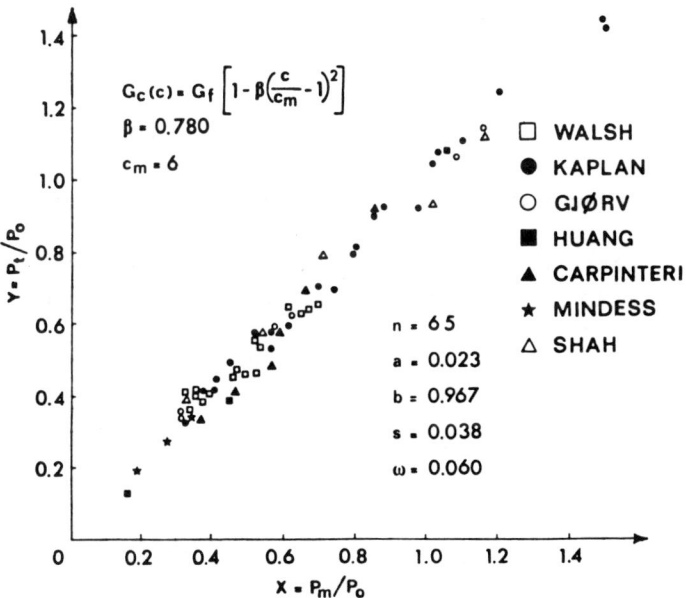

Figure 13.19 Statistical linear regression of measured failure loads versus theoretical ones (after Bažant and Cedolin, 1983b)

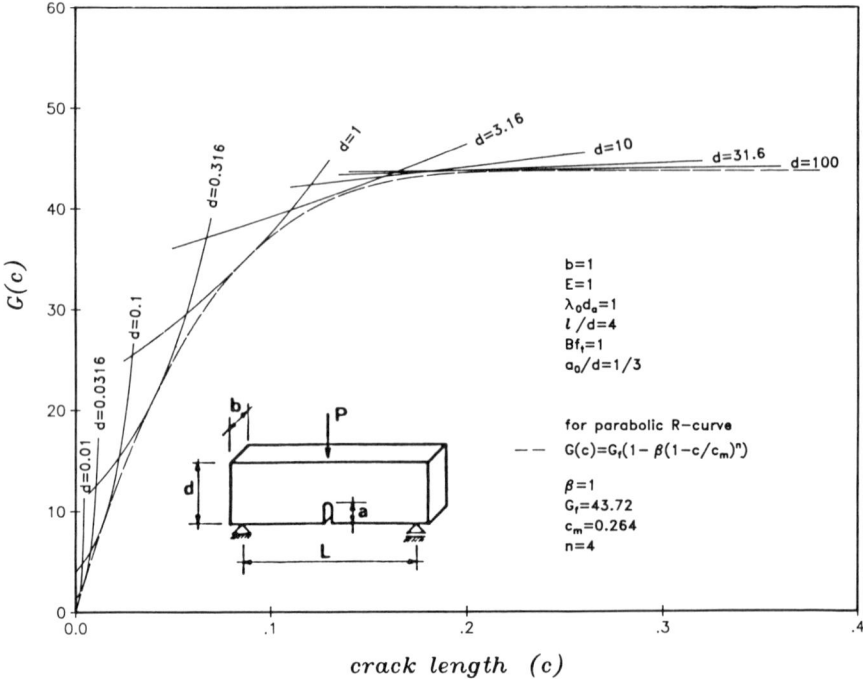

Figure 13.20 Determination of R-curve as an envelope of the curves of equilibrium states based on maximum load values for specimens of various sizes

found to be the same as Equation (13.35b) for the critical equilibrium state. Therefore, the curve of critical states may be geometrically obtained as the *envelope* of the curves of Equation (13.34) for various structure sizes; see Figure 13.20. We now show that this property permits a very simple determination of the R-curve from experimental data.

Assume that the curve $\sigma_N(\lambda)$, defined by Equation (13.16) which results from dimensional analysis, has been determined from experimental results. Thus, the failure loads (maximum loads or critical states) for various structure sizes d are $P = bd\,\sigma_N(\lambda)$ in which $\lambda = d/d_a$, and substituting this into Equation (13.34), we obtain

$$\frac{1}{b}W' = \frac{d_a}{E_c}\lambda\sigma_N^2(\lambda)g(\alpha), \qquad \alpha = \frac{a_0 + c}{\lambda d_a} \tag{13.44}$$

in which $\sigma_N^2 = B^2 f_t'^2(1 + \lambda/\lambda_0)$ from Equation (13.16). Equation (13.44) defines a family of curves of W' versus c and various constant values of the relative size λ. This family of curves is illustrated in Figure 13.20 for the three-point bent specimens used by Walsh, as mentioned before. As is seen from Figure 13.20 it is very easy to obtain the envelope of these curves graphically. This envelope represents the R-curve ($W'/b = G_c$) which exactly corresponds to the size effect according to Equation (13.16) based on dimensional analysis, for this type of specimen. Obviously, for another geometrical shape of specimen, a different R-curve is obtained in this manner. It appears, however, that the differences are not very significant in comparison with the typical scatter of experimental measurements.

To obtain the R-curve analytically, one needs to differentiate Equation (13.44) with regard to λ, which yields

$$\frac{2\lambda^2}{\sigma_N}\frac{d\sigma_N}{d\lambda} + \lambda - \frac{a_0 + c}{nd_a}\frac{g'(\alpha)}{g(\alpha)} = 0 \tag{13.45}$$

Equations (13.44) and (13.45) represent two algebraic equations for the R-curve, with λ as a parameter. For some forms of function $g(\alpha)$, parameter λ can be eliminated algebraically, yielding a single formula for the R-curve. However, this is not possible in general and a graphical construction of the envelope (Figure 13.20) is preferable.

It may be noted that the R-curve obtained graphically in Figure 13.20 can be quite closely approximated by a power curve. The approximation with an ellipse is worse.

The experimental determination of the R-curve just described is both simpler and more realistic than the existing methods. The existing methods utilize the relation $G_c = k_1 P^2 a/(E_c b^2 d^2)$, in which k_1 is a known coefficient for a given type of specimen. The values of G_c are determined either on a single specimen from loads P corresponding to various crack lengths a or on a series of specimens from critical loads P and the corresponding critical crack lengths a. In both cases, one needs to measure the crack length, which is an almost insurmountable obstacle in the case of concrete. First of all, the crack length is very difficult to define since the crack tip is blurred by a micro-cracking zone. Second, even if the crack length could be measured optically or otherwise, its usefulness is dubious because, as should be recalled, a represents an equivalent crack length rather than some actual crack length. In view of these difficulties, it has been attempted to determine a by measuring specimen compliance, either at unloading or at reloading. However, this is again of questionable relevance because at unloading or reloading the micro-cracks within the fracture process zone do not close and behave in a stiff manner, while for continued loading they have a much smaller stiffness. Therefore, the compliance measurements are likely to yield crack length values which are much too small.

All these shortcomings are avoided by the method proposed here. The scatter of measurements can be effectively removed by the use of Equation (13.16), which permits linear regression analysis in the plot of σ_N^{-2} versus λ. This plot has the vertical axis intercept $(Bf'_t)^{-2}$ and the slope $(Bf'_t)^{-2}\lambda_0^{-1}$. After determining these parameters, the R-curve is then obtained as illustrated in Figure 13.20. Measurements of crack lengths are not needed at all. Determination of material parameters entirely relies on maximum load data for geometrically similar specimens of various sizes. Such data are very easy to obtain, even in a small laboratory with the most rudimentary equipment.

13.7 CRACK SPACING AND CRACK SHEAR

For concrete structures, it is of importance to model not only isolated fractures but also systems of cracks. Of particular interest is the system of parallel equidistant cracks which forms under a uniform tensile loading of a panel reinforced by a mesh of bars, or in a beam with longitudinal reinforcement subjected to tension or bending. For the parallel crack system, it is necessary to determine the spacing of the cracks, after which the crack width can be estimated, and also the overall strain at which the cracks form. This problem has been traditionally analysed on the basis of strength criteria, coupled with the conditions of bond slip. However, the strength criteria govern only the initiation of micro-cracking, since they pertain to the peak point of the tensile stress–strain diagram. For a complete crack formation, strain-softening must reduce the stress to zero, which means that the full strain energy under the tensile stress–strain diagram must be dissipated. Therefore, complete crack formation should properly be calculated from energy fracture criteria. This approach has been taken in Bažant and Oh (1983), and simplified formulas for the crack spacing based on energy analysis were determined. Since the cracks are usually spaced rather closely compared to the maximum aggregate size, the energy balance conditions have been written for the sudden formation of the crack over its entire length. The formulas obtained from this fracture mechanics approach agreed reasonably well with the scant data on crack spacing and crack width available in the literature. Aside from energy balance, the evolution of crack spacing of a growing crack system is also governed by certain stability conditions; see Bažant, Ohtsubo and Aoh (1979).

Crack width is rather important for the shear transmission capability of cracks in concrete, which is essential for the load carrying capability of concrete structures. The shear response of cracks in concrete may be characterized by an incremental relation expressing the increments of the normal and shear stress transmitted across a crack as a function relative displacement (opening) and the tangential relative displacement (slip); see Bažant and Gambarova (1984).

ACKNOWLEDGMENT

Partial financial support under AFOSR Grant No. 83–0009 to Northwestern University is gratefully appreciated. Mary Hill deserves thanks for her excellent secretarial assistance.

REFERENCES

ASCE (1982) State-of-the-Art Report on *Finite Element Analysis of Reinforced Concrete*, prepared by a Task Committee chaired by A. Nilson, Am. Soc. of Civil Engrs., New York.

Batdorf, S. B., and Budiansky, B. (1949) 'A mathematical theory of plasticity based on the concept of slip', NACA TN 1871, April, 1949.

Barenblatt, G. I. (1959) 'The formation of equilibrium cracks during brittle fracture, general ideas and hypothesis. Axially-symmetric cracks', *Prikladnaya Matematika i Mekhanika*, **23**, 3, 434–444.

Bažant, Z. P. (1976) 'Instability, ductility and size effect in strain-softening concrete', *J. of the Engineering Mechanics Division, ASCE*, **102**, E2, 331–344.

Bažant, Z. P. (1982) 'Crack band model for fracture of geomaterials', *Proc., 4th Intern. Conf. on Numerical Methods in Geomechanics*, held in Edmonton, Alberta, Canada, June 1982 (Ed. Z. Eisenstein), Vol. 3 (invited lectures), 1137–1152.

Bažant, Z. P. (1983a) 'Size effect in brittle failure of concrete structures', Report No. 83-2/665s, Center for Concrete and Geomaterials, Northwestern University, Evanston, Illinois; also *Jour. of Engng. Mech. ASCE*, **110** (1984), 518–535.

Bažant, Z. P. (1983b) 'Imbricate continuum: variational formulation', Report No. 83-11/428i, Center for Concrete and Geomaterials, Northwestern University, Evanston, Illinois; also *J. of Eng. Mech. ASCE*, **110**, No. 10, Dec. 1984.

Bažant, Z. P. (1984) 'Mechanics of fracture and progressive cracking in concrete structures', in *Fracture Mechanics Applied to Concrete Structures* (Ed. G. C. Sih), Martinus Nijhoff Publishers, The Hague.

Bažant, Z. P., and Belytschko, T. B. (1983) 'Wave propagation in a strain-softening bar: exact solution', Report No. 83-10/401w, Center for Concrete and Geomaterials, Northwestern University, Evanston, Illinois; also *J. of Eng. Mech. ASCE*, in press.

Bažant, Z. P., and Cedolin, L. (1979) 'Blunt crack band propagation in finite element analysis', *Journal of the Engineering Mechanics Division, ASCE*, **105**, EM2, 297–315.

Bažant, Z. P., and Cedolin, L. (1980) 'Fracture mechanics of reinforced concrete', *Journal of the Engineering Mechanics Division, ASCE*, **106**, 1287–1306; with Discussion and Closure in **108**, 464–471 (1982).

Bažant, Z. P., and Cedolin, L. (1983a) 'Finite element modeling of crack band propagation', *Journal of Structual Engineering, ASCE*, **109**, ST2, 69–92.

Bažant, Z. P., and Cedolin, L. (1983b) 'Approximate linear analysis of concrete fracture by R-curves', Report No. 83-7/679a, Center for Concrete and Geomaterials, Technological Institute, Northwestern University, Evanston, Illinois; also *Jour. of Structural Engineering*, **110** (1984), 1336–1355.

Bažant, Z. P., and Gambarova, P. (1980) 'Rough cracks in reinforced concrete', *J. of the Struct. Div., Proc. ASCE*, **106**, 819–842; Discussion pp. 2579–2581.

Bažant, Z. P., and Gambarova, P. (1984) 'Crack shear in concrete: crack band microplane model', *Journal of Structural Engineering ASCE*, **110**, 2015–2035.

Bažant, Z. P., and Kim, J. K. (1984) 'Size effect in shear failure of longitudinally reinforced beams', *American Concrete Institute Journal*, **81**, 456–468.

Bažant, Z. P., and Oh, B. H. (1982) 'Rock fracture via stress–strain relations', Concrete and Geomaterials, Report No. 82-11/665r, Northwestern University, Evanston, Illinois; also *Jour. of Engng. Mechanics ASCE*, **110** (1984), 1015–1035.

Bažant, Z. P., and Oh, B. H. (1983a) 'Crack band theory for fracture of concrete', *Materials and Structures (RILEM, Paris)*, **16**, 155–177.

Bažant, Z. P., and Oh, B. H. (1983b) in 'Microplane model for fracture analysis of concrete structures', *Proc. Symp. on the 'Interaction of Nonnuclear Munitions with Structures'*, US Air Force Academy, Colorado Springs, May 1983, pp. 49–55.

Bažant, Z. P., and Oh, B. H. (1983c) 'Model of weak planes for progressive fracture of concrete and rock', *Report No. 83-2/448m*, Center for Concrete and Geomaterials, Northwestern University, Evanston, Illinois.

Bažant, Z. P., and Oh, B. H. (1983) 'Spacing of cracks in reinforced concrete', *J. of Engng. Mech. ASCE*, Vol. 109, Sept. 1983, pp. 2266–2212.

Bažant, Z. P., and Panula, L. (1978) 'Statistical stability effects in concrete failure', *J. of the Engineering Mechanics Division, ASCE*, **104**, EM5, 1195–1212.

Bažant, Z. P., and Pfeiffer, P. (in preparation) 'Finite element crack band analysis'.

Bažant, Z. P., Ohtsubo, H., and Aoh, K. (1979) 'Stability and post-critical growth of a system of cooling or shrinkage cracks', *International Journal of Fracture*, **15**, 5, 443–456.

Bažant, Z. P., Pfeiffer, P., and Marchertas, A. H. (1983a) 'Blunt crack band propagation in finite element analysis for concrete structures', Preprints 7th Int. Conf. on Structural Mechanics in Reactor Technology, Chicago, Aug. 1983.

Bažant, Z. P., Chang, T. P., and Belytschko, T. B. (1983b) 'Continuum theory for strain softening', *Report No. 83-11/428c*, Center for Concrete and Geomaterials, Northwestern University, Evanston, Illinois; also *Journal of Engineering Mechanics ASCE*, **110**, in press.

Bhal, N. S. (1968) 'Über den Einfluss der Balkenhöhe auf Schubtragfähihkeit von einfeldrigen Stahlbetonbalken mit und ohne Schubbewehrung, Dissertation, Universität Stuttgärt.

Broek, D. (1974) *Elementary Engineering Fracture Mechanics*, Noordhoff International Publishing, Leyden, Netherlands.

Brown, J. H. (1972) 'Measuring the fracture toughness of cement paste and mortar', *Magazine of Concrete Research*, **24**, 81, 185–196.

Burt, W. J., and Dougill, J. W. (1977) 'Progressive failure in a model of heterogeneous medium', *J. of the Engng. Mechanics Division, Proc. ASCE*, **103**, EM3, 365–376.

Carpinteri, A. (1980) 'Static and energetic fracture parameters for rocks and concretes', Report, Istituto di Scienza delle Costruzioni-Ingegneria, University of Bologna, Italy.

Carpinteri, A. (1981) 'Experimental determination of fracture toughness parameters K_{IC} and J_{IC} for aggregative materials', *Advances in Fracture Research*, Proc., 5th International Conference on Fracture, Cannes, France, 1981 (Ed. D. François), Vol. 4, pp. 1491–1498.

Cedolin, L., and Bažant, Z. P. (1980) 'Effect of finite element choice in blunt crack band analysis', *Computer Methods in Applied Mechanics and Engineering*, **24**, 3, 305–316.

Cedolin, L., Dei Poli, S., and Iori, L. (1983a) 'Experimental determination of the fracture process zone in concrete', *Cement and Concrete Research*, **13** (to appear).

Cedolin, L., Dei Poli, S., and Iori, L. (1983b) 'Experimental determination of the stress–strain curve and fracture zone for concrete in tension', *Proc., Int. Conf. on Constitutive Laws for Engineering Materials* (Ed. C. Desai), University of Arizona, Tucson, January.

Dugdale, D. S. (1960) 'Yielding of steel sheets containing slits', *J. Mech. Phys. Solids*, **8**, 100–108.

Entov, V. M., and Yagust, V. I. (1975) 'Experimental investigation of laws governing quasi-static development of macrocracks in concrete', *Mechanics of Solids* (translation from Russian), **10**, 4, 87–95.

Eringen, A. C., and Ari, N. (1983) 'Nonlocal stress field at Griffith crack', *Cryst. Latt. Def. and Amorph. Mat.*, **10**, 33–38.

Eringen, A. C., and Edelen, D. G. B. (1972) 'On nonlocal elasticity', *Int. J. Engng. Science*, **10**, 233–248.

Evans, R. H., and Marathe, M. S. (1968) 'Microcracking and stress–strain curves for concrete in tension', *Materials and Structures (RILEM, Paris)*, 1, Jan.–Feb., pp. 61–64.
Gjørv, O. E., Sørensen, S. I., and Arnesen, A. (1977) 'Notch sensitivity and fracture toughness of concrete', *Cement and Concrete Research*, 7, 333–344.
Hadamard, J. (1903) *Leçons sur la propagation des ondes*, Hermann, Paris, Chapter VI.
Heilmann, H. G., Hilsdorf, H. H., and Finsterwalder, K. (1969) 'Festigkeit und Verformung von Beton unter Zugspanungen', *Deutscher Ausschuss für Stahlbeton*, Heft 203, W. Ernst & Sohn, West Berlin.
Hillerborg, A., Modéer, M., and Petersson, P. E. (1976) 'Analysis of crack formation and crack growth in concrete by means of fracture mechanics and finite elements', *Cement and Concrete Research*, 6, 773–782.
Huang, C. M. J. (1981) 'Finite element and experimental studies of stress intensity factors for concrete beams', Thesis submitted in partial fulfillment of the requirements for the degree of Doctor of Philosophy, Kansas State University, Kansas.
Ingraffea, A. (1984) in *Fracture Mechanics Applied to Concrete Structures* (Ed. G. C. Sih and A. Carpinteri), Martinus Nijhoff Publishers, The Hague.
Irwin, G. R., Report of a Special Committee (1960) 'Fracture testing of high strength sheet materials', *ASTM Bulletin*, January 1960, p. 29 (also G. R. Irwin, 'Fracture testing of high strength sheet materials under conditions appropriate for stress analysis', Report No. 5486, Naval Research Laboratory, July 1960).
Kani, G. N. J. (1966) 'Basic facts concerning shear failure', Part I and Part II, *J. of ACI*, 63, 6, 675–692.
Kaplan, M. F. (1961) 'Crack propagation and the fracture of concrete', *American Concrete Institute Journal*, 58, 11.
Kesler, C. E., Naus, D. J., and Lott, J. L. (1971) 'Fracture mechanics—Its applicability to concrete', *International conference on the Mechanical Behaviour of Materials*, Kyoto, August 1971.
Kfouri, A. P., and Miller, K. J. (1974) 'Stress displacement, line integral and closure energy determinations of crack tip stress intensity factors', *Int. Journal of Pres. Ves. and Piping*, 2, 3, 179–191.
Kfouri, A. P., and Rice, J. R. (1977) 'Elastic/plastic separation energy rate for crack advance in finite growth steps', in *Fracture 1977* (Proc. of the 4th Intern. Conf. on Fracture, held in Waterloo, Ontario, June 1977) (Ed. D. M. R. Taplin), University of Waterloo Press, Vol. 1, pp. 43–59.
Knauss, W. C., 'On the steady propagation of a crack in a viscoelastic sheet; experiments and analysis', in *The Deformation in Fracture of High Polymers* (Ed. H. H. Kausch), Plenum Press, pp. 501–541.
Knott, J. F. (1973) '*Fundamentals of Fracture Mechanics*, Butterworths, London.
Krafft, J. M., Sullivan, A. M., Boyle, R. W. (1961) 'Effect of dimensions on fast fracture instability of notched sheets', *Cranfield Symposium*, 1961, Vol. I, pp. 8–28.
Kröner, E. (1967) 'Elasticity theory of materials with long-range cohesive forces', *Int. J. Solids Structures*, 3, 731–742.
Kröner, E. (1968) 'Interrelations between various branches of continuum mechanics', *Mechanics of Generalized Continua* (Ed. E. Kröner), Springer-Verlag, pp. 330–340.
Krumhansl, J. A. (1968) 'Some considerations of the relation between solid state physics and generalized continuum mechanics', *Mechanics of Generalized Continua* (Ed. E. Kröner), Springer-Verlag, pp. 298–311.
Kunin, I. A. (1968) 'The theory of elastic media with microstructure and the theory of dislocations', *Mechanics of Generalized Continua* (Ed. E. Kröner), Springer-Verlag, pp. 321–328.

Leonhardt, F., and Walther, R. (1961/1962/1963) 'Beiträge zur Behandlung der Schubprobleme im Stahlbetonbau', *Beton-u Stahlbetonbau*, **56**, 12 (1961); **57**, 2, 3, 6, 7, 8, (1962); **58**, 8, 9 (1963).
Levin, V. M. (1971) 'The relation between mathematical expectation of stress and strain tensors in elastic microheterogeneous media', *Prikladnaya Matematika i Mekhanika*, **35**, 694–701 (in Russian).
Marchertas, A. H., Kulak, R. F., and Pan. Y. C. (1982) 'Performance of blunt crack approach within a general purpose code', in *Nonlinear Numerical Analysis of Reinforced Concrete* (Ed. L. E. Schwer), Am. Soc. of Mech. Engrs., New York (presented at ASME Winter Annual Meeting, Phoenix, November 1982), pp. 107–123.
Mindess, S. (1983) 'The application of fracture mechanics to cement and concrete: a historical review', in State-of-the-Art Report of RILEM Technical Committee 50-FMC on *Fracture Mechanics of Concrete*, (Ed. F. H. Wittmann), Elsevier, Netherlands.
Mindess, S., and Diamond, S. (1980) 'A preliminary SEM study of crack propagation in mortar', *Cement and Concrete Research*, **10**, 509–519.
Mindess, S., Lawrence, F. V., and Kesler, C. E. (1977) 'The J-integral as a fracture criterion for fiber reinforced concrete', *Cement and Concrete Research*, **7**, 731–742.
Naus, D. J. (1971) 'Applicability of linear-elastic fracture mechanics to Portland cement concretes', Thesis submitted in partial fulfillment of the requirements for the degree of Doctor of Philosophy, University of Illinois at Urbana-Champaign.
Pan, Y. C., and Marchertas, A. H. (1983) Private communication, May 1983, at Argonne National Laboratory, Argonne, Illinois.
Pan, Y. C., Marchertas, A. H., and Kennedy, J. M. (1983) 'Finite element of blunt crack band propagation: a modified J-integral approach', Preprints *7th Intern. Conf. on Structural Mechanics in Reactor Technology*, Paper H, Chicago, August 1983.
Parker, A. P. (1981) *The Mechanics of Fracture and Fatigue*, E. & F. N. Spon, Ltd.—Methuen, London.
Petersson, P. E. (1980) 'Fracture energy of concrete: method of determination', *Cement and Concrete Research*, **10**, 78–89; and 'Fracture energy of concrete: practical performance and experimental results', *Cement and Concrete Research*, **10**, 91–101.
Petersson, P. C. (1981) 'Crack growth and development of fracture zones in plain concrete and similar materials', Doctoral Dissertation, Lund Institute of Technology, Lund, Sweden.
Rashid, Y. R. (1968) 'Analysis of prestressed concrete pressure vessels', *Nuclear Engng. and Design*, **7**, 4, 334–344.
Reinhardt, H. W., and Cornelissen, H. A. W. (1984), 'Post-Peak Cyclic Behavior of Concrete in Uniaxial Tensile and Alternating Tensile and Compressive Loading', *Cement and Concrete Research*, **14**, 263–270.
Rice, J. R. (1976) 'The localization of plastic deformation', in *Theoretical and Applied Mechanics*, Preprints, IUTAM Congress held in Delft, 1976 (ed. W. T. Koiter), North-Holland, Amsterdam, pp. 207–220.
Rudnicki, J. W., and Rice, J. R. (1975) 'Conditions for the localization of deformation in pressure-sensitive dilatant materials', *J. of Mech. and Physics of Solids*, **23**, 371–394.
Rüsch, M., Haugli, F. R., and Mayer, M. (1962) 'Schubversuche an Stahlbeton Rechteckbalken mit Gleichmässig Verteilter Belastung', *Deutscher Ausschuss für Stahlbeton*, Heft 145, W. Ernst und Sohn, West Berlin.
Rüsh, H., and Hilsdorf, H. (1963) 'Deformation characteristics of concrete under axial tension', *Voruntersuchungen*, Bericht Nr. 44, Munich, May.
Shah, S. P. (1984) in *Fracture Mechanics Applied to Concrete Structures* (Ed. G. C. Sih and A. Carpinteri), Martinus Nijhoff Publishers, The Hague.
Shah, S. P., and McGarry, F. J. (1971) 'Griffith fracture criterion and concrete', *Journal of the Engineering Mechanics Division, ASCE*, **97**, EM6, 1663–1676.

Sok, C., Baron, J., and François, D. (1979) 'Mécanique de la rupture appliquée au beton hydraulique', *Cement and Concrete Research*, **9**.
Swamy, R. N., and Qureshi, S. A. (1971) 'Strength, cracking and deformation similitude in reinforced T-beams under bending and shear', Part I and II, *J. of Am. Concrete Inst.*, **68**, 3, 187–195.
Swartz, S. E., Hu, K. K., Fartash, M., and Huang, C. M. J. (1981) 'Stress intensity factors for plain concrete in bending—prenotched versus precracked beams', Report, Department of Civil Engineering, Kansas State University, Kansas.
Tada, H., Paris, P. C., and Irwin, G. R. (1973) *The Stress Analysis of Cracks Handbook*, Del Research Corp., Hellertown, Pa.
Taylor, G. I. (1938) 'Plastic strain in metals', *J. Inst. Metals*, **63**, 307–324.
Taylor, H. P. J. (1972) 'The shear strength of large beams', *J. of the Structural Division, ASCE*, **98**, 2473–2490.
Thomas, T. Y. (1961) *Plastic Flow and Fracture in Solids*, Academic Press.
Walraven, J. C. (1978) 'The influence of depth on the shear strength of lightweight concrete beams without shear reinforcement', *Stevin Laboratory Report No. 5–78–4*, Delft University of Technology.
Walsh, P. F. (1972) 'Fracture of plain concrete', *The Indian Concrete Journal*, **46**, 11, 469, 470, and 476.
Wecharatana, M., and Shah, S. P. (1980) 'Resistance to crack growth in Portland Cement Composites', Report, Department of Material Engineering, University of Illinois at Chicago, Chicago, Illinois.
Witmann, F. H. (Ed.) (1983) *Fracture Mechanics of Concrete*, Elsevier, Netherlands.
Wnuk, M. P. (1974) 'Quasi-static extension of a tensile crack contained in viscoelastic plastic solid', *Journal of Applied Mechanics, ASME*, **41**, 1, 234–248.

Mechanics of Geomaterials
Edited by Z. Bažant
© 1985 John Wiley & Sons Ltd

Chapter 14

Fracture Propagation and Fracture Energy: Discusser's Report

M. P. Cleary

The discussion in this paper is divided into four main combinations of categories:

— Phenomenology and observations of fracture on the microscale (section 14.2)
— Modelling of (brittle) fracture processes on the microscale (section 14.3)
— Experimental investigation of fracture on the macroscale (section 14.4).
— Analysis of (brittle) fracture for testing and structural design (section 14.5)

A brief review is given of representative research and development in each of these areas, with particular emphasis on the relation to perceived reality. Major unresolved problems are posed for future investigation.

14.1 INTRODUCTION AND BACKGROUND

An understanding of fracture is essential to the whole spectrum of human fabrication activities, and has finally gained its rightful role in engineering research and development over the past three decades or so. The essential ingredients in fracture mechanics of materials are:

(A) A detailed knowledge of the *stress distribution* in the body of interest, at all scales of interest—often divided into micro and macro for convenience of reference, although scales obviously merge into each other asymptotically.
(b) A reliable estimate of the *material response* to the local stress level at any point, including the effects of temperature and stress history; especially, *conditions for rupture instability* (which may have to be viewed as non-local) must be well determined.

The central aspects of these ingredients are the constitutive relations for unruptured response of the materials (for any relevant loading history) and adequate criteria for the onset and propagation of rupture processes—which we comprehensively refer to as fracture. The most common parameters employed are the (perhaps incremental) material *moduli for stress analysis* and the *fracture energy required* to grow the rupture; the former have attracted considerable theoretical and experimental attention in solid mechanics for many decades (especially from the group spearheaded by William Prager, as other chapters in this volume will attest) but detailed resolution of the issues involved in fracture growth/direction criteria has been slower, despite intense scrutiny over the past decade especially (e.g. Liebowitz, 1968, Erdogan, 1976).

A major cause of this contrast is certainly the additional geometry-dependence (non-local) character of the fracture process, which tends to produce a greater variety of phenomenology and especially a strong size-effect in the response of a material sample. Although this size-effect has also been more recently realized to haunt the development of adequate constitutive relations for materials with hierarchies of microstructure (especially 'natural' materials such as rock, cement, concrete, and perhaps even ceramics), as against more homogeneous processed materials which have been solidified from a uniform (liquid) melt, it has its source of aggravation mainly in fracture-related processes: moduli dependence on scale is more readily incorporated in analyses and design.

It is thus appropriate for us to concentrate especially on the mechanics of fracture, since the major complexities of materials response must be implicitly embedded in any such study: if intrinsic materials behaviour is integrated with the presence or development of rupture processes, we should be able to deduce the overall (so-called microscale) material response—which can then be embedded in a larger-scale (so-called macroscopic) structural analysis, perhaps including the presence of a fracture which has evolved from linkage of the small-scale ruptures. This logic is the motivation for the sequence of sub-division presented next.

14.2 PHENOMENOLOGY AND OBSERVATIONS OF FRACTURE ON THE MICROSCALE

The processes involved here can be very complex but, essentially, they require some nuclei at which initiation and growth can take place. These can be stress concentration points (e.g. granular/asperity contact), inclusions such as hard or soft second phases, dislocation pile-up sites, or just weak grain-boundaries in polycrystals. Many reviews have been done on the melt-processed construction materials (e.g. Chalmers *et al.*, 1954–present) but much less has been done on the more 'natural' materials. The reasons for this may be found in the more complex and precise requirements of the expensive melt-processed materials (metals, polymers, and composites): they must withstand all conceivable states of loading, including cycles of temperature and stress (which can ratchet eventual fatigue in even the toughest materials) whereas the 'natural' materials have had lower imposed expectations—and fail even these in all too many circumstances! Cracked pavements, mine roof falls, and earthquake-prone conventional buildings illustrate the poor structural response of such solid-bonded compositions, so they are rightly the focus of much recent work, described by Bažant (Chapter 13 of this volume) for instance.

The most basic problem with materials produced by solid mixing and bonding (albeit with some chemical reactions in the cementing process) is that they typically contain intrinsic flaws (e.g. pores and cracks) and other sources of stress concentration (e.g. asperity contacts). The cementing bond can also be quite weak, but some researchers (e.g. Birchall and Kelly, 1983) claim dramatic increases in strength and toughness, even up to levels comparable with melt-processed materials, when intrinsic flaws are removed. In any case, the presence of inhomogeneity sites is clearly central and it presents two major aspects for study:

(a) How do we determine the response (and hence the strength/toughness reduction) associated with any particular/generic site?
(b) How do we characterize the distribution and interaction of such sites?

Experimental evidence to resolve these questions is not very abundant, especially on details of response around inhomogeneity sites. Some deductions can be made from macroscopically observed crack patterns (e.g. Jaeger and Cook, 1980) and more detailed electron microscopy studies have been conducted by others (e.g. Tapponier and Brace, 1976) for nominally macroscopically homogeneous stressing conditions. Actually, of course, it is well known that stresses are not uniform in testing of samples with conventional apparatus and thus the interpretations assigned can be quite misleading; in the Resource Extraction Laboratory at MIT we have been developing various techniques to achieve more uniform conditions and have especially done extensive testing (plus electron microscopy studies) on a technique called pore-pressure-induced-cracking (PPIC), in which the porous sample has internal pressure higher than confining stress-hence a tensile effective stress (which governs rupture onset). This technique, combined with detailed analysis of pore-fluid diffusion, promises to allow a detailed study of micro-crack onset and population evolution for the first time (e.g. Hess and Cleary, 1983).

The formation of a linked-up macroscopic fracture, and the associated development of crack patterns, have been observed by some researchers (e.g. Hoagland et al., 1973), and an attempt has been made to relate these processes to the observed macroscopic crack growth (Ingraffea, Chapter 12 in this volume, and overall reviews by Argon, 1983, and Bažant, Chapter 13 in this volume). However, although broad features seem consistent, there are serious shortcomings in the level of associated prediction, as we discuss next.

14.3 ANALYSIS OF FRACTURE PROCESSES ON THE MICROSCALE

Two major questions ((a) and (b) above) have been posed for experimental and theoretical analysis in the hope that they will lead to more general predictions, without the need for as much experimentation, in the future.

The first question has been chiefly regarded as a study in classical solid mechanics: the site is identified as a pore or crack or asperity in an otherwise homogeneous region and the isolated effect is worked out, using an estimated tensile strength or critical stress-intensity factor to decide on initiation/propagation of a crack from the nucleating site. Indeed, various estimates of strength reduction have been worked out, with some success (e.g. McClintock and Argon, 1968; Sikarsie, 1973), on the basis of this approach and size-effects (e.g. strength varying as sample size to a power between -0.5 and -1.0) seem to support such mechanisms as brittle cracking and effective dislocation pile-ups.

The second question has generated a greater variety of approaches. Chief among these as been a smearing out of the influence created by surrounding sites, best represented by the 'self-consistent' technique (e.g. Cleary, Chen, and Lee, 1980); the technique can be made arbitrarily precise in principle, but is limited in practice by the few available solutions for interaction between sites: special cases like two spheres are available, but an infinite 3-D array would be more useful (e.g. as used by Barr and Cleary (1982) for a 2-D array of surface cracks in the context of thermal or shrinkage cracking, see also Bažant, Chapter 13). Actually, our computer programs are now reaching the stage (e.g. Narendran and Cleary, 1982) where quite complex arrays of growing cracks can be studied and are being compared with experimental patterns observed in the laboratory (see also section 14.5 later). Thus, techniques are becoming available to correctly analyse the response of a continuum element containing a representative distribution of micro-cracks, if such a model (e.g. Kachanov, 1980) is useful, e.g. up to the onset of site/crack coalescence and macroscopic rupture formation.

A complementary approach, also needed to decide on onset of fracture at any particular site, is to assign a statistical distribution of strength to various well-defined site locations, such as grain-boundaries (e.g. McClintock and Mayson, 1976). Then, by using many available techniques such as dipole dislocation representations of each cracking event, the stress-field can be computed at each instant/load and the crack pattern can evolve in a combined deterministic/stochastic fashion. Results of this method in 3-D are still to be obtained (McClintock, private communication, 1983) but the expectation seems to have the right trend in helping to explain size-effects as being due to (statistical) sample size as well as micro-fracture mechanics, which Bažant (Chapter 13) seems to regard as exclusively rationalizing the observed data for concrete.

14.4 EXPERIMENTAL INVESTIGATION OF FRACTURE ON THE MACROSCALE

The essential features of this activity are:

(a) choice of a suitable geometry for crack growth observations;
(b) analysis of stress distributions for the chosen geometry;

Fracture Propagation and Fracture Energy 309

(c) determination/classification of conditions for fracture onset and growth;
(d) relation of observations to mechanisms operative on the microscale.

Many possible choices have evolved for specimen geometry; these are summarized by Ingraffea (Chapter 12 and Ouchterlon (1980), but we should add some others that we have found useful in the Resource Extraction Laboratory at MIT. A diametrally loaded core ('Brazilian' test, Jaeger and Cook, 1980) can be modified to include a diametral crack of various lengths—cast into artificial specimens or cut from a central hole in rock cores. This loading is simpler to apply than the internal pressure in the notched annular core (NAC), and may have more potential under confining pressure, but it does not have the desirable insensitivity to crack length in deducing toughness from the critical load at which unstable crack growth sets in.

A common attribute of these samples is the facility for stress analysis and deduction of some appropriate crack-tip-loading parameter. Extensive lists of solutions exist in the literature (e.g. Ouchterlony, 1980) for linear isotropic materials response and many anisotropic nonlinear analyses are also being conducted (e.g. Cleary and Miller (1983) for the NAC and the compact tensile specimen used extensively by us, Switchenko and Cleary (1979)). Great attention must be paid to ensure that the crack near-tip stress field is not influenced in character by the boundaries, and then its amplitude (expressed as stress-intensity factor K or energy-release-rate J) can be regarded as critical in deciding fracture onset or growth for other geometries where the same near-tip field applies—hence the name *fracture toughness testing*.

14.5 ANALYSIS OF (BRITTLE) FRACTURE GROWTH IN TYPICAL STRUCTURES

The essence of such analysis are two distinct features:

(a) determination of the stress distribution in the (cracked) structure;
(b) evaluation/imposition of an appropriate criterion for fracture onset.

Both aspects have produced a variety of approaches, some of which are discussed by Ingraffea (Chapter 12). It may be worth summarizing the major methodologies here.

(a) Two separate approaches have developed, one based on the finite element method (e.g. Rice, 1981) and another on the so-called Boundary Integral Method (e.g. Ingraffea, Chapter 12). The BIM has the advantage of requiring discretion only along boundaries, but is *practically* limited to linear/isotropic response of simple regions; whereas the FEM allows arbitrary nonlinearity and severe inhomogeneity, but requires a volume discretization and complete remeshing for any charge of boundaries. To avoid the weakness of either method, we have combined the two techniques (Annigeri and Cleary, 1982) to get a more effective scheme for analysis of fractures growing through structures (as against stationary cracks for which FEM may be the best approach (Cleary and Miller, 1983)): we have modified the conventional BIM to a surface integral equation, allowing more natural representation of growing cracks and their near-tip

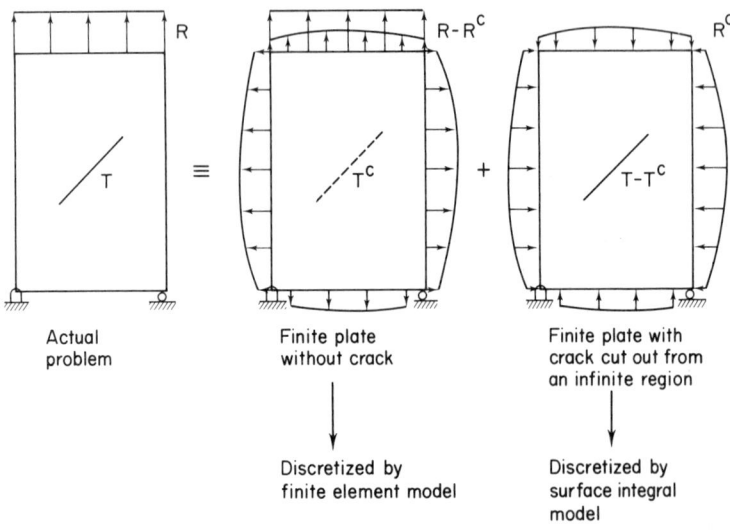

Figure 14.1

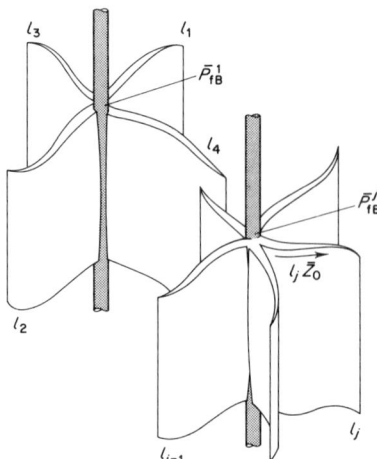

Figure 14.2 Schematic of multiple arbitrarily shaped fracture emanating from multiple wellbores

singularities, without extensive remeshing, and we have interfaced with FEM — which naturally picks up all nonlinearity and inhomogeneity (e.g. see Figure 14.1). Available general 2-D analysis codes (e.g. Annigeri, Narendran, and Cleary, 1982) are now being extended to 3-D fracture problems (e.g. Cleary, Kavvadas, and Lam, 1983).

(b) The criteria for brittle crack growth seem now to be conventionally accepted as that of a critical value for the opening stress-intensity-factor K_I at onset of propagation and that of $K_{II} = 0$ for the direction of incremental growth. We have used these criteria successfully to trace the growth of fractures observed in our laboratory apparatus (e.g. Figure 14.2, Papadopoulos, Narendran, and Cleary, 1983). Other criteria for directions have been proposed (such as maximum strain-energy density and maximum energy release rate J for a probe in the new direction of growth), but these seem to often give quite similar results (e.g. Ingraffea, Chapter 12). When the near-crack-tip stress field is no longer $1/\sqrt{r}$, then a more general J criterion must be used and even that will break down if the near-tip field has interference from boundaries or other mechanisms set in (such as slip-bands in large-scale yielding of ductile materials); such considerations are now the main concern of the extensive research effort in elastic–plastic fracture mechanics (e.g. Rice, 1981), which may require much more adaptation for application to 'natural' materials.

REFERENCES

Annigeri, B., and Cleary, M. P. (1982) 'Surface integral finite element hybrid (SIFEH) method for fracture mechanics', in *Reports of Research in Mechanics and Materials*, Dept. of Mech. Eng., MIT, REL-82-21. (See also MIT, Ph. D. Thesis 'Surface integral finite element hybrid method for localized problems in continuum mechanics', by B. Annigeri, expect Fall 1983).

Argon, A. S. (1983) 'Fracture in compression of brittle solids', report of Committee of the National Materials Advisory Board, National Academy of Sciences, *Report No. NMAB*-404, National Academy Press.

Barr, D. T., and Cleary, M. P. (1983) 'Thermoelastic fracture solutions using distributions of singular influence functions', Parts I and II, *Int. Jour. Solids & Structures*, **19**, 1, 73–91. (See also 'Thermal fracturing in impermeable geothermal reservoirs', (MS Thesis of D. T. Barr, 1980), submitted to *Journal of Geophysical Research*, 1983.)

Birchall, J. D. and Kelly, A. (1983) 'New inorganic materials', *Scientific American* **245**, 5, 104–115.

Chalmers, B., Christian, J. W., and Massalski, T. B. (1954–present) *'Progress in materials science'*, *An International Review Journal*.

Cleary, M. P. (1983) 'Applications of rock fracture mechanics', at the *Symposium on Hydraulic Fracturing and Geothermal Energy*, Tokyo and Sendai, Japan, November 1982, (Ed. S. Nemat-Nasser et al.).

Cleary, M. P. (1983) 'Modelling and development of hydraulic fracturing technology', in *Rock Fracture Mechanics* (Ed. H. P. Rossmanith), Centre Internationale des Sciences Mecaniques, Springer-Verlag, pp. 383–475.

Cleary, M. P., Chen, I. W., and Lee, S. M. (1980) 'Self-consistent techniques for heterogeneous media', *ASCE*, **106**, EM5. (Also see articles by M. M. Carroll and G. Dvorak.)

Cleary, M. P., Kavvadas, Michael, and Lam, K. Y. (1983) 'Development of a fully three-dimensional simulator for analysis and design of hydraulic fracturing', Paper No. SPE/DOE 11631, presented at *Symposium on Low Permeability Reservoirs*, Denver, March 1983.

Cleary, M. P. (1978) 'Elastic and dynamic response regions of fluid-impregnated solids with diverse microstructures', *Int. J. Solids and Structures*, **14**, 795-819.

Cleary, M. P. and Miller, B. (1983) 'Finite element analysis for fracture toughness testing', *ASCE J. Eng. Mech.*, **109**, 3, 741-755.

Costin, L., and Mecholsky, J. J. (1983) 'Time-dependent crack growth and failure in brittle rock', in *Proc. 29th US Symposium on Rock Mechanics*, Texas A. & M. University, pp. 385-394.

Erdogan, F. (Ed.) (1976) *Fracture Mechanics*, ASME, New York.

Hess, T., and Cleary, M. P. (1983) 'A study of wave velocities and attenuation in cemented materials during pore-pressure-induced-cracking', MIT Resource Extraction Laboratory, *Report No. REL-83-16*.

Hoagland, R. G., Hahn, G. T., and Rosenfield, A. R. (1973) 'Influence of microstructure on fracture propagation in rock', *Rock Mechanics*, **5**, 77-106.

Jaeger, J. C., and Cook, N. (1979) *Fundamentals of Rock Mechanics*, Halsted Press, New York.

Kachanov, M. L. (1980) 'Continuum model of medium with cracks', Special issue of the *Journal of the Engineering Mech. Div. of ASCE, Mechanics of Heterogeneous Media*, **106**, EMS.

Liebowitz, H. (Ed.) (1968) *Fracture*, Academic Press.

McClintock, F. A., and Argon, A. S. (1968) *Mechanical Behavior of Materials*, Addison-Wesley.

McClintock, F. A., and Mayson, H. (1976) 'Principal stress effects on brittle crack statistics', in *The Effects of Voids on Material Deformation* (Ed. S. C. Cowin and M. M. Carrol), ASME, New York, pp. 31-45.

Narendran, V. M., and Cleary, M. P. (1982) 'Analysis of growth and interaction of multiple hydraulic fractures', in *Reports of Research in Mechanics and Materials*, Dept. of Mech. Eng., MIT, REL-82-11. (Also paper no. SPE 12272, presented at SPE Reservoir Simulation Symposium, San Francisco, November 1983).

Ouchterlony, F. (1980) 'Review of fracture toughness testing in rock', Swedish Detonic Research Foundation, Box 32058, S-126 11, Stockholm, 1980.

Papadopoulos, J. M., Narendran, V. M., and Cleary, M. P. (1983) 'Laboratory simulations of hydraulic fracturing', Paper No. SPE/DOE 11618, presented at *Symposium on Low Permeability*, Denver, March 1983.

Rice, James R. (1981) 'Elastic-plastic crack growth', in *Mechanics of Solids: the Rodney Hill 60th Volume* (Ed. H. G. Hopkins and M. J. Sewell), Pergamon.

Sikarsie, D. (Ed.) (1973) *Rock Mechanics Symposium*, ASME, New York.

Switchenko, P. M., and Cleary, M. P., (1979) 'The thermomechanical response of oil shale', in *Reports of Research in Mech. and Materials*, Dept. of Mech. Eng., MIT. (NSF Report # ENG-77-18988-1).

Tapponier, P., and Brace, W. F. (1976) 'Development of stress-induced microcracks in Westerley granite', *Intl. Jour. Rock Mech. Min. Sci.*, **13**, 103-112.

PART VI

FLUID INFILTRATED GEOMATERIALS

Mechanics of geomaterials
Edited by Z. Bažant
© 1985 John Wiley & Sons Ltd

Chapter 15
Effect of Pore Fluid Diffusion on Deformation and Failure of Rock

J. W. Rudnicki

15.1 INTRODUCTION

The presence of pore fluid in laboratory rock samples or fissured rock masses can substantially alter the response to applied or induced loads and the conditions for failure. Pore fluid effects have been proposed as playing a role in a wide variety of geophysical and geotechnical phenomena. These include the following: migration of aftershocks (Nur and Booker, 1972; Booker, 1974); fault creep (Rice and Simons, 1976; Rice, 1979a); earthquake precursory processes (Nur, 1972; Anderson and Whitcomb, 1975; Scholz et al., 1973; Rudnicki, 1977, 1979; Rice and Rudnicki, 1979); hydraulic fracture (Rice and Cleary, 1976; Ruina, 1978; Cleary, 1979); water level changes in wells (Johnson, et al., 1973); induced seismicity (Bell and Nur, 1978; Simpson, 1976; Raleigh et al., 1976; Ohtake, 1974; Zoback and Hickman, 1982); wave speed travel time delays (Leary et al., 1979); and migration of earthquake swarms (Johnson, 1979).

The most familiar effect of the pore fluid is to reduce the effective value of the mean normal compressive stress. This effect is usually expressed in terms of so-called effective stress laws: the effect of the pore fluid on some aspect of the mechanical behaviour can be incorporated by replacing the stress by the effective stress, a linear combination of the stress and pore fluid pressure. Because the inelastic deformation of many geological materials is inhibited by an increase of hydrostatic compression, an increase in pore fluid pressure decreases the effective compressive stress and promotes further inelastic deformation. Conversely, a decrease in pore fluid pressure tends to inhibit further inelastic deformation.

Coupling of deformation with pore fluid diffusion also introduces time dependence into the response of an otherwise rate-independent solid. An elastic fluid-infiltrated solid responds more stiffly to deformations that are rapid compared to the time scale of diffusion than for deformations that are slow compared to the diffusion time. Because many geological materials change volume when sheared inelastically, pore fluid diffusion can also be coupled to inelastic deformation. A volume increase due to micro-cracking and opening of pore space, as is typical of brittle rocks deformed in compression, tends to decrease the local pore fluid pressure. If the creation of new pore space occurs slowly, the tendency for the pore fluid pressure to decrease will be alleviated by fluid mass flux. If, however, new pore space is created more rapidly than fluid mass can diffuse into it, then the pore fluid pressure decreases, increasing the effective stress, and inhibiting further inelastic deformation.

This review considers the mechanical effects of pore fluid on the deformation and failure of geological and geotechnical materials. Applications to brittle rock and, more specifically, to earth faulting are emphasized. For the purpose of this article the subject is divided into those effects that can be treated on the basis of Biot's (1941a) formulation for a linear elastic fluid-infiltrated porous solid and those that arise from inelastic volume changes. Several other recent review articles also treat this subject. Reviews by Rice (1979a, 1980, 1981) on the mechanics of earthquake rupture and precursory processes have included discussion of pore fluid effects. Rudnicki (1980), in a review of fracture mechanics applied to the earth's crust, has discussed coupled deformation diffusion effects in fault propagation, and Rudnicki (1981a) has reviewed the stabilizing effects of coupled deformation and diffusion on an inclusion model of faulting. Paterson (1978) has given a concise review of experimental work up to 1977.

15.2 LINEAR ELASTIC FLUID-INFILTRATED POROUS SOLID

The governing equations for a linear elastic fluid-infiltrated solid have been established by Biot (1941a). Although obviously an idealization of the behaviour of rocks, this theory has proved rich enough to provide insight into a wide variety of physical phenomena. At the same time, the complexity of the equations is such that the number of solutions is not great. More elaborate theories—many derived from mixture theory—have been proposed, but they offer few advantages over the Biot approach. Uncertainties about material parameters and difficulties of solution argue against pursuing overly intricate theories at this time. Until recently, very few solutions to the fully coupled Biot equations have existed. These include solutions for surface loading of half-spaces (Biot, 1941b; Biot and Clingan, 1941; McNamee and Gibson, 1960,a,b) and for the response of planar aquifers (Verruijt, 1968) and spherically symmetric pressurized cavities (Cryer, 1963). However, the possible role of coupled deformation diffusion effects in earth faulting has provided motivation to develop additional solutions.

The next subsection reviews the governing equations of the Biot theory. Rice and Cleary (1976) have given an appealing reformulation of the Biot theory and the description to be given here closely follows their point of view. After a brief subsection discussing material parameters of the Biot theory, a sampling of recent solutions and applications is given in the succeeding subsections.

15.2.1 Governing equations

Let σ_{ij} denote the total stress so that $n_i\sigma_{ij}$ is the force per unit area, including both solid and fluid phases, with unit normal n_i. Although some authors prefer to decompose the stresses into portions acting separately on the solid and fluid phases, that is unnecessary and yields no advantage. Deformations can be described by the strains of the solid matrix ε_{ij}. Two additional variables are needed to include the effects of the pore fluid. These are chosen to be the pore fluid pressure p and the fluid mass content per unit volume m. It is often convenient to express m as the product of the mass density of homogeneous pore fluid ρ and an apparent volume fraction of voids v, i.e. $m = \rho v$. The pore fluid pressure is defined as the pressure in an imaginary reservoir of homogenous pore fluid that would be needed to prevent any fluid mass flux to or from the reservoir when it is connected to a material element (Figure 15.1).

Linear relations for σ_{ij} and the deviation of m from an ambient value m_0 have the following form:

$$\sigma_{ij} = L_{ijkl}\varepsilon_{kl} - M_{ij}p \qquad (15.1)$$

$$m - m_0 = R_{ij}\varepsilon_{ij} + Qp \qquad (15.2)$$

where L_{ijkl} is a tensor of elastic moduli, M_{ij} and R_{ij} are additional constant constitutive tensors and Q is a scalar. The tensors L_{ijkl}, M_{ij} and R_{ij} have the symmetries derived from the symmetry of σ_{ij} and ε_{ij}. Because the response is elastic, the work increment can be expressed as a change in the Helmholtz

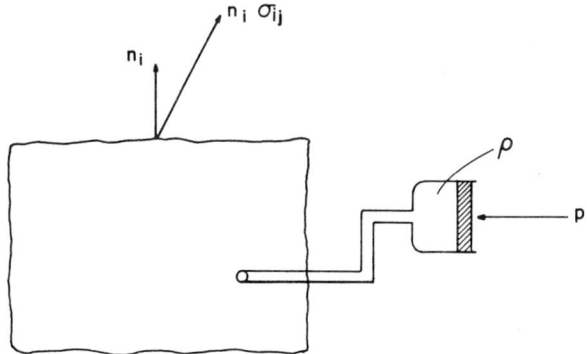

Figure 15.1 A small element of a fluid-infiltrated porous solid. The total force per unit area applied to the boundary of the element is $n_i\sigma_{ij}$ where the n_i are components of the unit normal. The pore fluid pressure is defined by imagining the element to be connected to a reservoir of pore fluid with density ρ. The pore pressure p is the pressure necessary to prevent any fluid mass flow to or from the element

function ϕ having the following differential form (Biot, 1941a, 1973) (at constant temperature):

$$d\phi = \sigma_{ij} d\varepsilon_{ij} + p d(m/\rho) \quad (15.3)$$

The Maxwell relation that follows from taking the mixed partial derivatives of the dual potential ϕ-pm/ρ with respect to ε_{ij} and p in either order is

$$\rho \frac{\partial \sigma_{ij}}{\partial p} = \frac{\partial m}{\partial \varepsilon_{ij}} \quad (15.4)$$

Substituting (15.1) and (15.2) into (15.4) yields

$$R_{ij} = \rho M_{ij} \quad (15.5)$$

When the deformation is slow enough so that alterations of the pore fluid pressure from its ambient value are eliminated by fluid mass flux, conditions are said to be *drained*. In this case, $p = 0$ and L_{ijkl} can be identified as the tensor of elastic moduli for drained deformation. The deformation is said to be *undrained* when it is too rapid to allow time for fluid mass flux to or from material elements. In this case, $m = m_0$ and (15.2) can be solved for the pore fluid pressure to give

$$p = -Q^{-1} \rho M_{ij} \varepsilon_{ij} \quad (15.6)$$

where (15.5) has been used. Substituting (15.6) into (15.1) yields

$$\sigma_{ij} = L^u_{ijkl} \varepsilon_{kl} \quad (15.7)$$

where

$$L^u_{ijkl} = L_{ijkl} + Q^{-1} \rho M_{ij} M_{kl} \quad (15.8)$$

is the tensor of elastic moduli appropriate to undrained response.

The constitutive formulation is completed by Darcy's law. In the absence of body forces, Darcy's law can be expressed as

$$q_i = -\rho \kappa_{ij} \partial p / \partial x_j \quad (15.9)$$

where q_i is the mass flow rate across a unit area with normal in the ith direction, and κ_{ij} is a (symmetric) permeability tensor.

When the response is isotropic, the tensors $M_{ij} (= \rho^{-1} R_{ij})$ and κ_{ij} are diagonal and, consequently, can be expressed as

$$M_{ij} = \zeta \delta_{ij}, \kappa_{ij} = \kappa \delta_{ij} \quad (15.10)$$

where δ_{ij} is the Kronecker delta. Furthermore, for isotropic response the tensor of elastic moduli for drained response must have the following form:

$$L_{ijkl} = G(\delta_{ik} \delta_{jl} + \delta_{il} \delta_{jk}) + (K - 2G/3) \delta_{ij} \delta_{kl} \quad (15.11)$$

where G and K are the shear and bulk moduli, respectively. The form of L^u_{ijkl} is the same as that for L_{ijkl} but the values of the shear and bulk moduli may be different

Effect of Pore Fluid Diffusion 319

for undrained and drained response. However, substitution of these forms for the modulus tensors and (15.10) into (15.8) reveals that the value of the shear modulus is the same for drained and undrained response and Q is given by

$$Q = \rho\zeta^2(K_u - K)^{-1} \tag{15.12}$$

where K_u is the bulk modulus for undrained response.

Consequently, for isotropic response, Equations (15.1), (15.2), and (15.9) can be rewritten as follows:

$$\sigma_{ij} = (K - 2G/3)\delta_{ij}\varepsilon_{kk} + 2G\varepsilon_{ij} - \zeta p \delta_{ij} \tag{15.13}$$

$$m - m_o = \zeta\rho[\varepsilon_{kk} + \zeta p/(K_u - K)] \tag{15.14}$$

$$q_i = -\rho\kappa\partial p/\partial x_i \tag{15.15}$$

In these equations the value of the density ρ is constant, as appropriate for a linear theory, and equal to the mass density of the homogeneous pore fluid as it exists in the reservoir imagined to be connected to the material element. The parameter ζ can be expressed as

$$\zeta = 1 - K/K'_s \tag{15.16}$$

where K'_s is another bulk modulus that can be identified, under some circumstances (Nur and Byerlee, 1971; Rice and Clearly, 1976), with the bulk modulus of the solid constituents. The bulk modulus for undrained response K_u satisfies $K \leqslant K_u < \infty$ where the upper limit is attained for separately incompressible solid and fluid constituents and the lower for highly compressible pore fluid. Thus, the undrained response is elastically stiffer. The scalar permeability κ is often expressed as $\kappa = k/\mu$ where μ is the fluid viscosity and k is a permeability with dimensions of length squared and usually given in units of darcies (1 darcy = 10^{-8} cm^2).

The bulk moduli for drained and undrained response can be used to derive Poisson's ratios for drained (v) and undrained (v_u) response according to the usual relation

$$v = (3K - 2G)/2(3K + G) \tag{15.17}$$

The limits of K_u require that the Poisson's ratio for undrained response v_u satisfies $v \leqslant v_u \leqslant 1/2$. If (15.13) and (15.14) are combined to obtain a relation between the pore fluid pressure and the mean normal stress for undrained response, the result is

$$p = -B\sigma_{kk}/3 \tag{15.18}$$

where

$$B = \zeta K_u/(K_u - K) \tag{15.19}$$

Thus, B and v_u can be used as alternatives to ζ and K_u. The following alternative expressions for v_u and B can be obtained from the results of Rice and Cleary

(1976):

$$v_u = \frac{3v + \zeta B(1 - 2v)}{3 - \zeta B(1 - 2v)} \tag{15.20}$$

$$B = \zeta[\zeta + v_o K(1/K_f - 1/K_s'')]^{-1} \tag{15.21}$$

where K_f is the bulk modulus of the pore fluid, v_o is the apparent porosity in the reference state and K_s'' is another bulk modulus that can, in special cases, be identified as the bulk modulus of the solid constituents.

The constitutive equations (15.13), (15.14), and (15.15), must be combined with field equations expressing compatibility, equilibrium of total stresses (in the absence of body forces) and fluid mass conservation. These are as follows:

$$\varepsilon_{ij} = \tfrac{1}{2}(\partial u_i/\partial x_j + \partial u_j/\partial x_i) \tag{15.22}$$

$$\partial \sigma_{ij}/\partial x_i = 0 \tag{15.23}$$

$$\partial q_i/\partial x_i + \partial m/\partial t = 0 \tag{15.24}$$

where u_i is the displacement of the solid matrix. Substituting (15.15) into (15.24) yields, for constant ρ and κ,

$$\rho \kappa \nabla^2 p = \partial m/\partial t \tag{15.25}$$

where $\nabla^2(\ldots) = \partial^2(\ldots)/\partial x_i \partial x_i$. Substituting (15.13) into (15.23) and using (15.22) yields

$$(K + G/3)\partial^2 u_k/\partial x_k \partial x_j + G\partial^2 u_j/\partial x_k \partial x_k - \zeta \partial p/\partial x_j = 0 \tag{15.26}$$

Differentiating this equation with respect to x_j and adding the appropriate multiple of it to (15.25) reveals that the fluid mass per unit volume m satisfies the homogeneous diffusion equation

$$\nabla^2 m = \frac{1}{c} \frac{\partial m}{\partial t} \tag{15.27}$$

where the diffusivity c is given by

$$c = \frac{\kappa(K + 4G/3)(K_u - K)}{\zeta^2(K_u + 4G/3)} \tag{15.28}$$

or, in terms of B, v and v_u as

$$c = \frac{2G\kappa B^2(1 - v)(1 + v_u)^2}{9(1 - v_u)(v_u - v)} \tag{15.29}$$

The formulation given here assumes that a description of the pore fluid pressure in terms of a single scalar variable is adequate. For very rapid deformations on the time scale of wave propagation, the pore fluid pressure will be different in differently oriented neighbouring fissures comprising a single point

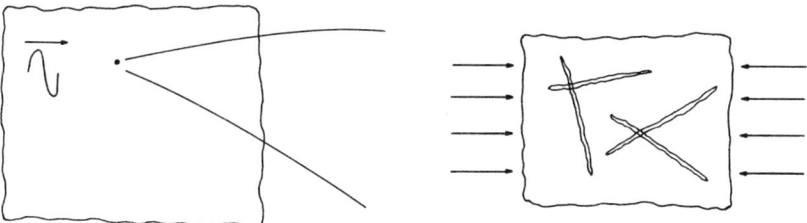

Figure 15.2 Schematic illustration indicating that a 'point' in a continuum idealization of a fissured rock is actually a representative material volume that may comprise several fissures. For very rapid deformations, for example, on the time scale of elastic wave propagation, the pressure will be different in differently oriented fissures making up the same 'point' in the continuum idealization. In these circumstances, characterization of the pore pressure in terms of a single scalar variable will be inadequate

in a continuum formulation (Figure 15.2). Consequently the term undrained conditions refers to response on time scales that, although short, are long enough to allow for local pressure equilibrium among neighbouring fissures. Estimates of relaxation times and constitutive response at shorter time scales have been given by O'Connell and Budiansky (1977) and Cleary (1978a).

15.2.2 Material parameters

Although the Biot theory introduces only two additional elastic constants, v_u and B or K_u and ζ, there have been few detailed investigations of their values. Rice and Cleary (1976) noted that if K_s' and K_s'' are assumed to be approximately equal to K_s, the bulk modulus of the solid constituents, v_u and B can be calculated from measured or inferred values of v, K_f and v_o. Using this procedure, they have tabulated results for six rocks: two granites, three sandstones, and Tennessee marble. Values of v are about 0.25 for the granites and marble, and lower, 0.12 to 0.20, for the sandstones. The undrained Poisson's ratio v_u ranges from 0.27 for Tennessee marble to 0.34 for Westerly granite. The value of B ranges from 0.51 for Tennessee marble to 0.88 for Ruhr sandstone. Roeloffs (1982) also discusses some measurements of the parameters in the Biot theory and quotes $v = 0.2$, $v_u = 0.34$ and $B = 0.51$ for Boise sandstone.

A measure of the magnitude of coupled deformation—diffusion effects is the ratio

$$\eta = (1 - v)/(1 - v_u) \qquad (15.30)$$

which figures prominently in many of the solutions to be discussed in succeeding subsections. Values for the rocks just discussed range from 1.03 for Tennessee marble and 1.04 for Charcoal granite to 1.28 for Ruhr sandstone. As noted by Rice and Rudnicki (1979), the presence of large fissures tends to increase v_u and decrease v so that the value of η *in situ* may be larger than that inferred from laboratory data, at least for low porosity rocks. They also point out that there is no direct source of the values of v_u and v *in situ*. Although measurements of seismic wave speeds can be used to obtain a value for Poisson's ratio, this value corresponds to neither v nor v_u (O'Connell and Budiansky, 1977; Cleary, 1978a): the deformations induced by the passage of waves at typical seismic frequencies are too rapid to allow for the establishment of local pore fluid equilibrium, as assumed in the Biot theory. As an indirect means of inferring *in situ* values of v and v_u, Rice (1979a) and Rice and Rudnicki (1979) have used the calculations of O'Connell and Budiansky (1974) for the elastic properties of cracked saturated solids. O'Connell and Budiansky (1974) give the elastic properties as a function of a crack density parameter. The value of this parameter can be inferred from observations of seismic wave speeds, and the results of the calculations can be used to obtain Poisson's ratio. Using this procedure and O'Connell and Budiansky's (1974) inference of the crack density parameter from measurements of wave speeds prior to the San Fernando earthquake, Rice and Rudnicki (1979) suggest that values of η in the range 1.10 to 1.25 may be representative of field conditions.

A characteristic time scale of these elastic coupled deformation diffusion effects is proportional to L^2/c where L is a length scale and c is the diffusivity. Values of c inferred by Rice and Cleary (1976) from laboratory tests on intact samples range from 10^{-4} m^2/s for the granites and marble to 1.6 m^2/s for Berea sandstone. The largest value is comparable to the value of 1.0 m^2/s suggested by Anderson and Whitcomb (1975) as being consistent with several field measurements in the vicinity of shallow earthquakes. Rice and Simons (1976) discuss this value in some detail and point out that it corresponds to plausible flow rates in fissured rock. Rice (1979a) has also inferred a value of 0.1 m^2/s from changes in water well level observed by Kovach *et al.* (1975) in response to creep events on the San Andreas fault. Values of c determined by Li (1984) from published reports of a variety of field measurements are consistent with the range from 10^{-1} to 10^2 m^2/s.

15.2.3 Plane strain shear dislocation

The solution for a suddenly introduced edge dislocation was derived by Booker (1974) for incompressible constituents and by Rice and Cleary (1976) for compressible constituents. The line of the dislocation is assumed to extend infinitely in the z-direction so that conditions of plane strain deformation apply and all field quantities are independent of z. The sudden introduction of an edge dislocation at the origin at time $t = 0$ corresponds to the following displacement discontinuity:

$$u_x(x, 0^+, t) - u_x(x, 0^-, t) = bH(-x)H(t) \qquad (15.31)$$

where u_x is the displacement in the x direction, b is the magnitude of the discontinuity and H is the unit step function (Figure 15.3). The pore fluid pressure

Effect of Pore Fluid Diffusion

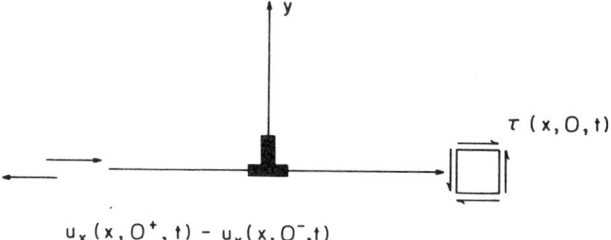

Figure 15.3 Plane strain shear (edge) dislocation corresponding to the introduction of a discontinuity in the x-direction displacement along the negative x axis. For a displacement discontinuity given by (15.31), the shear stress τ on the positive x axis is given by (15.33)

is given by

$$p(x, y, t) = \frac{GbB(1 + v_u)}{3\pi(1 - v_u)} \frac{y}{r^2} [1 - \exp(-r^2/4ct)] \quad (15.32)$$

where $r^2 = x^2 + y^2$ and the shear stress in the plane of the dislocation is given by

$$\tau(x, 0, t) = \frac{Gb}{2\pi(1 - v_u)x} \mathscr{L}(x^2/4ct) \quad (15.33)$$

where

$$\mathscr{L}(\xi) = 1 - \frac{(v_u - v)}{(1 - v)} \xi^{-1}(1 - e^{-\xi}). \quad (15.34)$$

Because $\mathscr{L}(\xi)$ has the limits $\mathscr{L}(\infty) = 1$ for short times and

$$\mathscr{L}(0) = (1 - v_u)/(1 - v) = \eta^{-1}$$

for long times, the shear stress is reduced from its initial value by a factor η^{-1}.

Nur and Booker (1972) have pointed out that the induced pore pressure predicted by (15.32) could reduce the confining stress near a recently slipped fault and provide a mechanism for the generation of aftershocks. The decrease with time of the pore pressure for $y > 0$ could cause the observed decay in numbers of aftershocks with time. Johnson (1979) has also used this solution to discuss the possibility that the migration of earthquake swarms on subsidiary faults transverse to a major transform fault could be caused by time-dependent changes of shear stress and pore fluid pressure induced by slip on the transform fault.

Booker (1974) (also, Rice (1980, 1981)) has noted that the reduction of shear stress with time predicted by (15.33) indicates that the coupling of deformation and diffusion will act to *reload* a fault subjected to a sudden stress drop. This feature is made clearer by rewriting (15.33) as follows:

$$\Delta\tau(x, 0, t) = \Delta\tau(x, 0, 0^+)\mathscr{L}(x^2/4ct) \quad (15.35)$$

where $\Delta\tau(x, 0, 0^+)$ is the stress drop at $t = 0^+$ and $\Delta\tau$ has been written for τ to emphasize the interpretation as a stress drop. Because the stress drop predicted by (15.35) decreases with time by the factor η^{-1}, the total stress on the fault will increase, possibly contributing to the generation of aftershocks. Rice (1980, 1981) has estimated the time scale of the reloading for a slipped zone of length $2a$ by superimposing dislocations of opposite sign at $x = \pm 2a/\pi$ (Figures 15.4(a), 15.4(b)) This arrangement simulates a plane strain shear crack of length $2a$ in the sense that, for drained response, the magnitude of the shear stress at $x = 0$ is related to b in the same way that the shear stress on the crack is related to the relative slip at the centre of the crack. The magnitude of the stress drop at the

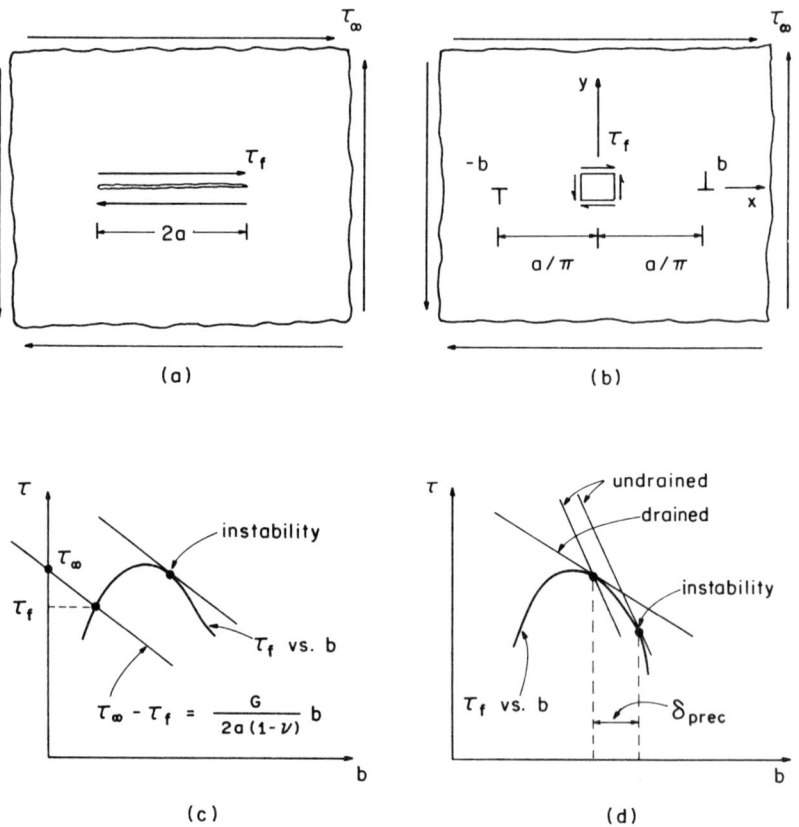

Figure 15.4 (a) Shear fault of length $2a$ sustaining a uniform friction stress τ_f. (b) Shear fault is simulated by two oppositely signed dislocations separated by $2a/\pi$. The shear stress at the centre of the two dislocations is identified with τ_f. (c) Graphical solution (after Rice, 1979a) is given by the intersection of the τ_f vs b curve with the line whose equation is (15.38). Instability occurs when this line is tangent to the τ_f vs b curve and the condition (15.39) is met. (d) Stiffer undrained response of the material surrounding the fault can delay instability and permit an additional amount of quasi-static slip δ_{prec}

Effect of Pore Fluid Diffusion

centre of the dislocations is then given by

$$\Delta\tau(t) = \Delta\tau(t = 0^+)\mathscr{L}(a^2/\pi ct)$$

Rice (1980, 1981) plots the ratio

$$\frac{\Delta\tau(t=0^+) - \Delta\tau(t)}{\Delta\tau(t=0^+) - \Delta\tau(t=\infty)} = (\pi^2 ct/a^2)[1 - \exp(-\pi^2 ct/a^2)] \quad (15.36)$$

versus a/ct and defines as characteristic the time at which the right-hand side of (15.36) equals 0.5. For $2a = 4$ km and values of the diffusivity c in the range $0.1 \text{ m}^2/\text{s}$ to $1.0 \text{ m}^2/\text{s}$, the characteristic time ranges from 3.5 to 35 days.

15.2.4 Stabilization of slip on a narrow weakening fault zone

Rudnicki (1979) has also made use of the dislocation solution to study the effects of coupling between deformation and diffusion in stabilizing rapid slip on frictional surfaces. If the magnitude of the displacement discontinuity in (15.31) is time dependent, then the stress induced in the plane of the dislocation is given by the following superposition integral:

$$\tau(x,0,t) = \frac{G}{2\pi(1-v_u)x}\int_{-\infty}^{t}\frac{\partial b}{\partial t'}(t')\mathscr{L}[x^2/4c(t-t')]\,dt'$$

If dislocations of opposite sign are again located at $x = \pm 2a/\pi$ to simulate a crack of length $2a$ and the shear stress at the centre of the dislocations is added to the uniform remotely applied stress τ_∞, the result is

$$\tau_f = \tau_\infty - \frac{G}{2a(1-v_u)}\int_{-\infty}^{t}\frac{\partial b}{\partial t'}(t')\mathscr{L}[a^2/\pi^2 c(t-t')]\,dt' \quad (15.37)$$

where τ_f is assumed to be a function of the slip b (Figures 15.4(a), 15.4(b)). If the τ_f versus b relation has a peak, then the occurrence of a rapid runaway of fault slip, interpreted as an earthquake, is possible (Figure 15.4(c)). This runaway instability occurs when the ratio of an increment of slip to an increment of farfield stress becomes unbounded. For drained response, (15.37) reduces to

$$\tau_f(b) = \tau_\infty - \frac{G}{2a(1-v)}b \quad (15.38)$$

Runaway instability occurs when the slope of the friction stress versus slip relation satisfies

$$\left(\frac{d\tau_f}{db}\right)_{\text{runaway}} = -\frac{G}{2a(1-v)} \quad (15.39)$$

Rice (1979a) has given a graphical interpretation of this instability (Figure 15.4(c)) and noted that it is analogous to the instability that occurs in a soft testing machine when the capacity of the sample to carry load decreases faster than the machine can unload. In (15.39) the magnitude of the right-hand side is the effective stiffness of the material surrounding the slip surface. If the material surrounding the slip surface is fluid infiltrated, the rapid slip that occurs as (15.39) is approached induces undrained response. For undrained response, v is replaced by v_u. Thus, the ratio of the stiffness of the surrounding material during undrained response to that for drained response is η (see (15.30)). Because $v_u > v$, this ratio exceeds unity and, consequently, the slope of the τ_f vs b curve must be more negative in order to achieve runaway instability under undrained conditions. Because the slope typically decreases with slip, the coupled deformation diffusion acts to delay instability from the point at which (15.39) is satisfied to that at which the corresponding relation for undrained response is met (Figure 15.4(d)).

For simplicity, Rudnicki (1979) assumes the following parabolic relation for τ_f versus b:

$$\tau_f(b) = \tau_p - (\tau_p - \tau_r)[(b - b_p)/(b_r - b_p)]^2 \tag{15.40}$$

The peak stress is τ_p, the slip at peak stress is b_p, and τ_r is the residual stress attained at slip b_r. For this τ_f versus slip law, the stabilizing effect of induced undrained response results in the following additional amount of slip prior to instability (Figure 15.4(d)):

$$\delta_{prec} = \frac{G(b_r - b_p)^2(v_u - v)}{4a(\tau_p - \tau_r)(1 - v)(1 - v_u)} \tag{15.41}$$

The time necessary to achieve this additional slip for a constant rate of farfield stress is called the precursor time. Rapid slipping during this precursory period may give rise to detectable effects that would make it possible to anticipate the instability.

Rudnicki (1979) has solved (15.37) with stress slip relation (15.40) numerically for slip histories in response to a prescribed constant rate of farfield stress. The calculated values of the precursor time range from less than a day to a few days for plausible ranges of the parameters. One particularly interesting feature of the results is that the precursor time can decrease with increasing length of the slip zone. Although the diffusion time $a^2/\pi^2 c$ increases with length of the slip zone and contributes to a longer precursor time, the effective stiffness of the material surrounding the slip zone is proportional to the reciprocal of the length of the zone. This decrease in effective stiffness decreases the amount of additional slip that can be achieved by pore fluid stabilization (see (15.41)) and, hence, tends to decrease the precursor time. Rudnicki (1981b) has also discussed this effect from another point of view.

15.2.5 Steadily moving shear dislocation

The expressions (15.32) and (15.33) can also be used to obtain the pore fluid pressure and shear stress generated by a dislocation moving steadily, but quasi-statically, at a speed V. If the step function time dependence in (15.31) is replaced by a Dirac delta function $\delta_{\text{DIRAC}}(t)$, the solutions are obtained from (15.32) and (15.33) by differentiation with respect to time. The result for the pore fluid pressure is

$$p(x,y,t) = \frac{B(1+v_u)}{3\pi(1-v_u)} Gb \left\{ \frac{y}{x^2+y^2}(1-e^{-r^2/4ct})\delta_{\text{DIRAC}}(t) - \frac{y}{4ct^2}e^{-r^2/4ct} \right\} \quad (15.42)$$

For introduction of the dislocation at t', t is replaced by $t - t'$ on the right-hand side of (15.42). Then, setting $x' = Vt'$ and $x = X + Vt$, and integrating from $t' = -\infty$ to $t' = t$ yields

$$p(X,y) = \frac{B(1+v_u)}{3\pi(1-v_u)} bG \left\{ \frac{y}{X^2+y^2} - \int_{-\infty}^{t} \frac{y}{4c(t-t')^2} \right.$$

$$\left. \exp\left\{ \frac{-[X+V(t-t')]^2}{4c(t-t')} \right\} dt' \right\} \quad (15.43)$$

The integral can be rearranged to yield

$$p(X,y) = \frac{B(1+v_u)}{3\pi(1-v_u)} bG \frac{y}{X^2+y^2} \{1 - (VR/2c)K_1(VR/2c)e^{-VX/2c}\} \quad (15.44)$$

where $R^2 = X^2 + y^2$ and $K_1(Z)$ is the modified Bessel function of order 1. The following result for the shear stress in the plane of the dislocation can be obtained in similar fashion:

$$\tau(X,0) = \frac{Gb}{2\pi(1-v_u)X} \left\{ 1 - \frac{(v_u-v)}{(1-v)}[(2c/VX) - (X/|X|)e^{-VX/2c}K_1(VX/2c)] \right\}$$

$$(15.45)$$

The expression (15.45) has been derived previously by Simons (1978) using integral transform techniques and Cleary (1978b) has given a plot of the term in square brackets obtained by numerical integration of (15.33). Both expressions (15.44) and (15.45) approach the undrained elastic solution in the limit $V \to \infty$.

Wesson (1981) has used the solution for a steadily moving dislocation in an ordinary elastic solid to discuss observations by Johnson et al. (1973) of water well level changes in response to creep events on the San Andreas. Wesson (1981) neglects coupled deformation diffusion effects and assumes that the pore fluid pressure is equal to the mean normal stress caused by passage of the dislocation. This is the term that remains from (15.44) in the limit $V \to \infty$ after replacing v_u by v and setting $B = 1$. An examination of (15.44) suggests, however, that coupled deformation diffusion effects can be significant (Roeloffs and Rudnicki, 1984). The pore fluid pressure from (15.44) is plotted in nondimensional form against $VX/2c$ for three values of $Vy/2c$ in Figure 15.5. For comparison, the dashed lines show

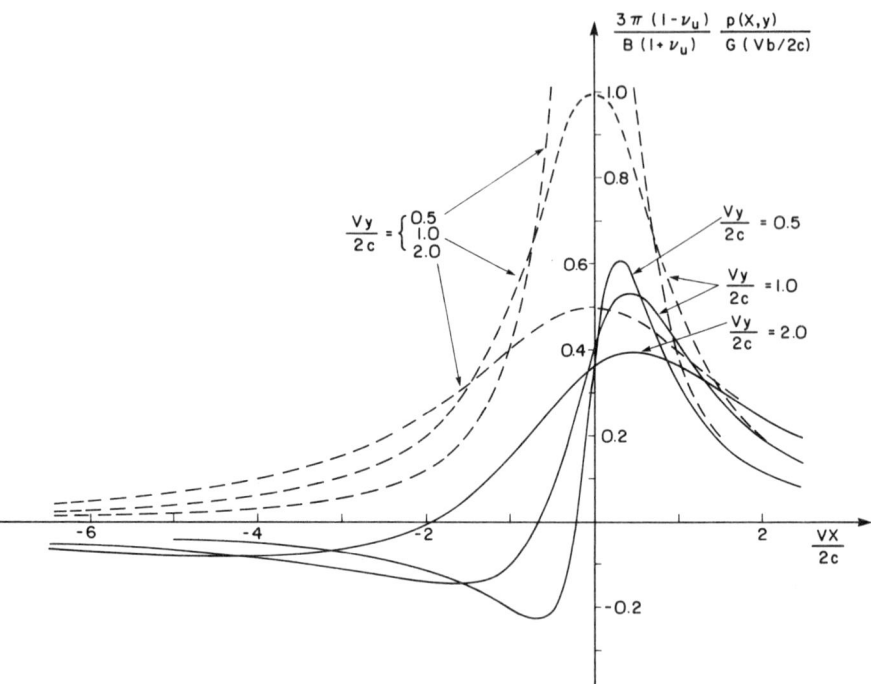

Figure 15.5 Nondimensional pore pressure charges due to a plane strain shear dislocation propagating steadily at a speed. V. Solid curves show the pressure change predicted by the fully coupled solution (15.45) at three distances from the fault. The dashed curves show the predicted response in the absence of coupled deformation diffusion effects. Because the solution is steady, reading the curves from right to left gives the time history of pressure change at a fixed distance from the fault

the result for the limiting case $V \to \infty$, which is the same as that obtained by neglecting coupled deformation–diffusion effects. Because the solution is for steady states, reading the graph from right to left gives the pore fluid pressure history experienced by a point a fixed distance from the plane of the dislocation. Coupled deformation diffusion effects not only reduce the magnitude of the pressure change but also cause the pressure change to reverse sign behind the dislocation and to approach zero through negative values rather than positive values predicted by the ordinary elastic solution. The coupled deformation diffusion effects diminish with distance from the fault, but for c in the range 0.1 to 1.0 m²/s and V in the range of 1 to 10 km/day, as observed for creep events on the San Andreas, $Vy/2c$ corresponds to 1.7 to 173 metres from the fault. For $v_u = 0.4$, $B = 0.6$ and $G = 2 \times 10^{10}$ N/m² ($= 2 \times 10^5$ bars) and V and c in the same range, the peak of the curve for $Vy/2c = 1.0$ corresponds to a pressure change of 160 to 1.6×10^4 N/m² for a slip of 1 mm.

15.2.6 Steadily propagating shear crack

Rice and Simons (1976) have examined a more realistic model of a quasi-statically propagating fault. They solve the problem of a semi-infinite, plane strain shear crack steadily moving at a speed V. The crack is loaded by shear stress τ_a applied to the crack faces over a distance L behind the crack-tip in order to simulate a finite length fault (Figure 15.6(a)). For an ordinary elastic solid, the shear stress ahead of the crack-tip is well-known (e.g. Rice, 1968) to have the form

$$\tau(X, 0) = K(2\pi X)^{-1/2}, X > 0 \tag{15.46}$$

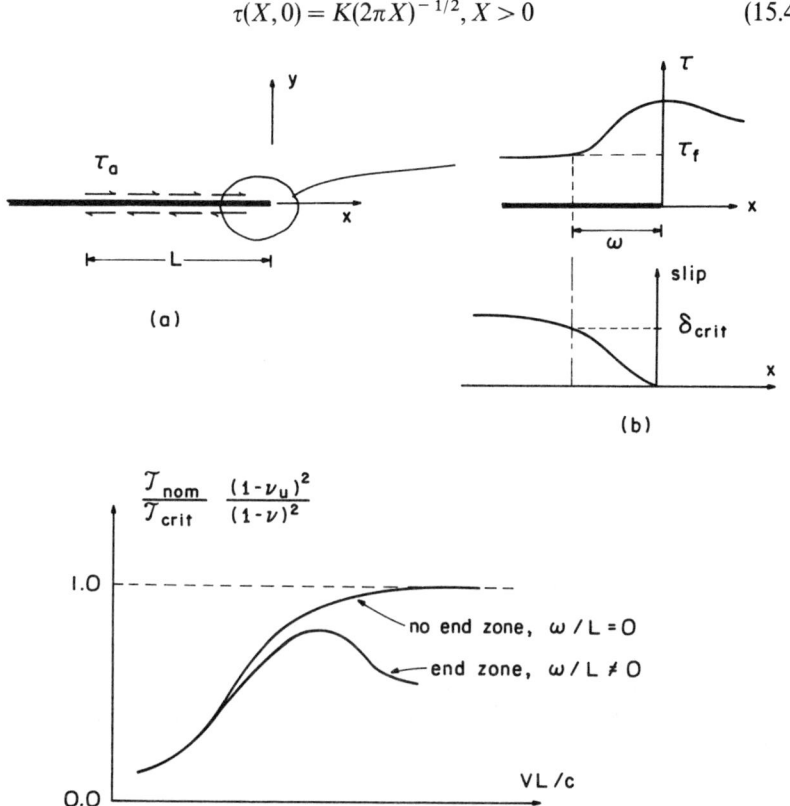

Figure 15.6 (a) Semi-infinite plane strain shear (Mode II) crack. Shear stress τ_a, which represents the difference between the applied stress and a residual friction stress τ_f, is applied over a length L behind the tip. (b) Increase of the relative slip from zero to a critical value δ_{crit} occurs in an end zone of length ω near the fault tip. Stresses resisting slip in this zone exceed the residual friction stress τ_f. (c) Schematic illustration of the results of Rice and Simons (1976) for a shear crack moving steadily at a speed V in a fluid-infiltrated elastic solid with diffusivity c, undrained Poisson's ratio ν_u and drained Poisson's ratio ν. The energy that must be supplied to the crack-tip to drive it at a speed V, $\mathcal{J}_{nom}$, exceeds that needed for crack growth in the absence of coupled deformation–diffusion effects, $\mathcal{J}_{crit}$.

where K is the stress intensity factor. For the geometry and loading considered here, K is given by

$$K = \tau_a(8L/\pi)^{1/2} \qquad (15.47)$$

Rice and Simons (1976) find that in a fluid-infiltrated elastic solid, the shear stress near the crack-tip again has the form (15.46) but the stress intensity factor is given by

$$K = K_{nom}h(VL/c) \qquad (15.48)$$

where K_{nom} is the value in (15.47) and h is a function that decreases monotonically from unity at $V=0$ to η^{-1} at $V=\infty$. Consequently, the stress at a fixed distance ahead of the crack is less in the fluid-infiltrated solid than in an elastic solid. The energy released per unit area of crack advance is related to the stress intensity factor by (Rice, 1968)

$$\mathscr{T} = (1-v)K^2/2G \qquad (15.49)$$

The drained value of Poisson's ratio enters (15.49) because the boundary condition $p=0$ on the crack faces requires that the neighbourhood of the crack-tip be drained (Rice and Simons, 1976; Simons, 1977). If it is assumed that the fracture continues to propagate as long as the energy released per unit crack advance equals some critical value, say $\mathscr{T}_{crit}$, then this criterion takes the form

$$(1-v)K^2/2G = \mathscr{T}_{crit} \qquad (15.50)$$

Substituting (15.48) and rearranging yields

$$\mathscr{T}_{nom}/\mathscr{T}_{crit} = h^{-2}(VL/c) \qquad (15.51)$$

where $\mathscr{T}_{nom} = (1-v)K^2_{nom}/2G$. The limiting value of the right-hand side of (15.51) for $V=\infty$ is η^2. Because h decreases monotonically with V, the energy that must be supplied by the applied loads, $\mathscr{T}_{nom}$, must increase in order to maintain $\mathscr{T} = \mathscr{T}_{crit}$ as crack speed increases (Figure 15.6(c)).

Rice (1966, 1979b) has emphasized that a model that idealizes the crack-tip as a singularity in the elastic stress field is appropriate only when the actual processes of material breakdown occur in a region having a size much less than any other relevant length scale. Consequently, in the fluid-infiltrated solid, the singular crack model is unsuitable at speeds for which the diffusion length c/V becomes comparable to the breakdown zone. To remedy this effect, Rice and Simons (1976) include a finite size end zone in their calculation. They then find that $\mathscr{T}_{nom}/\mathscr{T}_{crit}$ does not increase monotonically with crack speed but, instead, has a peak and then decreases (Figure 15.6(c)). For speeds greater than that corresponding to the peak in the curve, the crack increases speed without any increase in the driving stresses and hence is unstable.

Rice and Simons (1976) discuss in detail the implications of their results for pore fluid stabilization of fault creep events and of slip surface propagation in

Effect of Pore Fluid Diffusion

overconsolidated clay slopes. Specifically, they note that the propagation velocities of observed creep events on the San Andreas fault in California, one to ten kilometres per day, correspond, for realistic values of material parameters, to a range in which the curve of $\mathcal{T}_{nom}/\mathcal{T}_{crit}$ is rising stably. Furthermore, their results suggest that creep events propagating at greater velocities are not observed because they rapidly accelerate to seismic speeds.

Simons (1977) used boundary layer analysis to study further the propagation of plane strain shear cracks in fluid infiltrated solids. Ruina (1978) has studied the tensile crack problem analogous to the problem treated by Rice and Simons (1976) and has discussed its application to hydraulic fracture. Cleary (1978b) has numerically obtained solutions for steadily propagating singularities and used these to formulate integral equations for steady propagation of finite length faults or slip zones. He gives examples for several types of loadings.

15.2.7 Three-dimensional fundamental solutions

Although few three-dimensional solutions exist for the fully coupled equations, Cleary (1977) has derived the fundamental point force and fluid mass source solutions. After correcting a minor error in Cleary's expressions for a continuous fluid mass source, Rudnicki (1981b) has shown that Cleary's solution for the displacement components u_i and pore fluid pressure p due to a point force suddenly applied at the origin at $t = 0$ can be arranged in the following form:

$$Gu_i(x_1, x_2, x_3, t) = \frac{P_j}{16\pi r(1 - v_u)} \left\{ \frac{x_i x_j}{r^2} + (3 - 4v_u)\delta_{ij} \right\}$$

$$+ P_j \frac{\partial}{\partial x_j} \left[\frac{(v_u - v)}{16\pi(1 - v_u)(1 - v)} \frac{x_i}{r} u(r/\sqrt{ct}) \right] \quad (15.52)$$

$$p(x_1, x_2, x_3, t) = \frac{P_k x_k}{12\pi r^3} \frac{B(1 + v_u)}{(1 - v_u)} + P_k \frac{\partial}{\partial x_k} \left[\frac{B(1 + v_u)}{12\pi(1 - v_u)} \frac{1}{r} \mathrm{erfc}(r/\sqrt{ct}) \right] \quad (15.53)$$

where the P_j are components of the point force,

$$\mathrm{erfc}(s) = 2\pi^{-1/2} \int_s^\infty e^{-\alpha^2} d\alpha = 1 - \mathrm{erf}(s)$$

and

$$u(s) = \mathrm{erfc}(s/2) + 2s^{-2}\mathrm{erf}(s/2) - 2(s\pi^{1/2})^{-1} \exp(-s^2/4) \quad (15.54)$$

The first term in (15.52) has the form of the usual elastic solution with the undrained value of Poisson's ratio. For $t \to \infty$, $p = 0$ and (15.52) reduces to the ordinary elastic solution with the drained value of Poisson's ratio. The term in square brackets in (15.52) and (15.53) is the solution for continuous injection of fluid mass at the origin at a constant rate Q where Q is given by

$$Q = \frac{\rho c}{2G} \frac{3}{B} \frac{(v_u - v)}{(1 - v)(1 + v_u)} \tag{15.55}$$

Hence, as emphasized by Rudnicki (1981b), the second term in each equation is the response to a fluid mass dipole. Cleary (1977) has derived other singular solutions, for example the point dislocation, and has outlined their use in modelling embedded regions of inelasticity.

15.2.8 Deformation of spherical cavities and inclusions

Rice et al. (1978) have solved the problem of a spherical cavity in an infinite linear elastic fluid-infiltrated solid loaded by uniform tractions at the cavity boundary. For loading by sudden application of tractions derived from a uniform deviatoric tensor S_{ij}, the displacement at the cavity boundary $r = a$ is given by Rice et al. (1978) as

$$2Gu_i = S_{ij}x_j[\xi_u - f(ct/a^2)(\xi_u - \xi)] \tag{15.56}$$

where c is the diffusivity, $\xi = 2(4 - 5v)/(7 - 5v)$, and $\xi_u = 2(4 - 5v_u)/(7 - 5v_u)$. The function f depends on whether the condition on the pore fluid at the cavity boundary is no change in pore pressure ($p = 0$) or no fluid mass flux ($\partial p/\partial r = 0$) but, in either case, f has the limiting values $f(0) = 0$ and $f(\infty) = 1$. For no change in pore fluid pressure at the cavity boundary, Rice et al. (1978) give an explicit expression for and present a graph of f. If the cavity boundary is loaded by a uniform radial compressive stress σ_0 and a pore fluid pressure p_0, Rice et al. (1978) use the solution of Rice and Cleary (1976) to show that the displacement at the cavity boundary is independent of the pore fluid pressure and is given by

$$u_i = x_i(3/4G)\sigma_0 \tag{15.57}$$

The distribution of pore fluid pressure outside the cavity is (Rice and Cleary, 1976)

$$p = p_0(a/r)\text{erfc}[(r - a)/(4ct)^{1/2}] \tag{15.58}$$

The displacement of the cavity boundary in both (15.56) and (15.57) is compatible with a homogeneous deformation of the cavity interior. This feature made it possible for Rice et al. (1978) to use (15.56) and (15.57) to obtain relations for a spherical inclusion in a fluid-infiltrated elastic solid analogous to those obtained by Rudnicki (1977) from Eshelby's (1957) results for inclusions in ordinary elastic solids. These Eshelby relations connect the stress and strain in a homogeneous inclusion to the stress and strain applied in the farfield. For the fluid infiltrated solid, these relations are as follows:

$$G[\varepsilon_{\text{inc}}(t) - \varepsilon_\infty(t)] = (3/4)[\sigma_{\text{inc}}(t) - \sigma_\infty(t)] \tag{15.59}$$

Effect of Pore Fluid Diffusion   333

$$G[\gamma_{\text{inc}}(t) - \gamma_\infty(t)] = \gamma_{\text{inc}}(0) - \gamma_\infty(0)$$

$$= \int_0^t \{\xi_u + (\xi - \xi_u)f[c(t-t')/a^2]\}[\dot{\tau}_\infty(t') - \dot{\tau}_{\text{inc}}(t')] \, dt' \quad (15.60)$$

where σ is the mean normal stress, ε is the volume strain, τ is a shear stress, γ is a shear strain, and the subscripts '∞' and 'inc' refers to quantities in the farfield and inclusion, respectively. The form of (15.60) assumes that the response is drained at $t = 0$ and the generalization from (15.56) and (15.57) to (15.59) and (15.60) assumes that the inclusion is highly permeable so that it is a reasonable assumption to regard the pore fluid pressure there as uniform. A third relation, obtained by balancing the rate of increase of fluid mass in the inclusion with the mass flux through the cavity boundary computed from (15.58) and (15.59), is as follows:

$$\dot{m}_{\text{inc}}(t) = -(3\rho\kappa/a^2)[p_{\text{inc}}(t) - p_\infty(t)]$$

$$+ \int_{-\infty}^t \frac{a}{[\pi c(t-t')]^{1/2}} \{\dot{p}_{\text{inc}}(t') - \dot{p}_\infty(t')\} \, dt' \quad (15.61)$$

where it has been again assumed that the inclusion is relatively permeable.

Rice and Rudnicki (1979) have used (15.60) to analyse stabilizing effects of coupling between deformation and diffusion, analogous to those discussed for frictional slip, for an inclusion model of faulting (Rudnicki, 1977, 1981a) (Figure 15.7). When the inclusion shear stress is given as a function of the inclusion shear strain and the farfield response is taken to be elastic, (15.60) becomes a relation for the inclusion shear stress as a function of the applied farfield shear strain. Runaway instability corresponds to the possibility that an increment of farfield shear strain can induce an unbounded increment of inclusion shear strain. For purely drained conditions, runaway instability occurs when the slope at the τ_{inc} versus γ_{inc} curve satisfies the following condition (Figure 15.7(b)):

$$\frac{d\tau_{\text{inc}}}{d\gamma_{\text{inc}}} = -\frac{G}{\xi} \quad (15.62)$$

The magnitude of the right-hand side, G/ξ, is the effective unloading stiffness of the material surrounding the inclusion. If the response is purely undrained, the condition for runaway instability is obtained by replacing ξ by ξ_u in (15.62). Because $v < v_u$, $\xi_u < \xi$ and, hence, the unloading stiffness is greater for undrained response than for drained response. Consequently, the development of undrained response due to rapid straining as (15.62) is approached can transiently stabilize the rock mass and delay the onset of runaway instability from the time at which (15.62) is met until the time at which the corresponding condition is met in terms of the undrained response (Figure 15.7(c)). This delay time is termed the precursor time and is representative of the time period during which it may be possible to observe evidence of the impending instability.

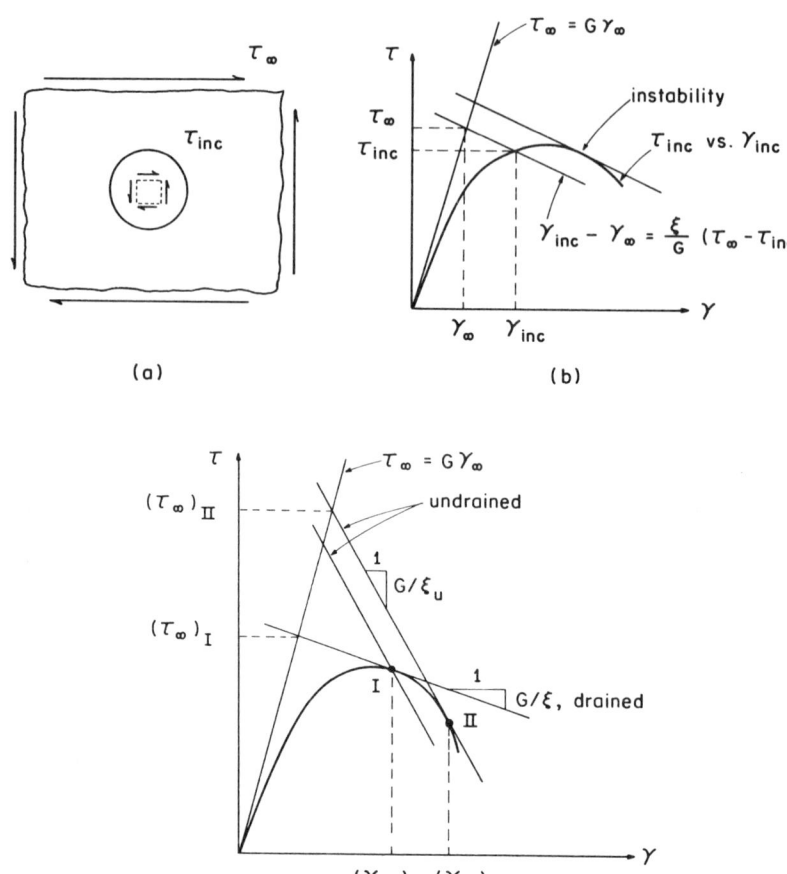

Figure 15.7 (a) Spherical inclusion in an infinite linear elastic body loaded by farfield shear stress τ_∞. (b) Graphical solution (after Rice, 1979 a) for the response of a nonlinearly deforming inclusion due to an increase of τ_∞. Results of Eshelby (1957) constrain the solution to lie on the straight line with slope $-G/\xi$ where G is the shear modulus and ξ is given the following (15.56). Runaway instability occurs when this line becomes tangent to the τ_{inc} vs γ_{inc} curve. (c) Delay of instability from I to II is caused by the stiffer undrained response of the material surrounding the inclusion. This undrained response is induced by rapid straining of the inclusion as I is approached.

Rice and Rudnicki (1979) calculate the precursor time and time history of inclusion straining for an imposed constant farfield shear strain rate and a τ_{inc} versus γ_{inc} curve that is assumed to have the following form near peak stress:

$$\tau = \tau_p - G(\gamma - \gamma_p)^2/2\lambda \qquad (15.63)$$

where γ_p is the strain at peak stress τ_p and λ describes the sharpness of the peak

Effect of Pore Fluid Diffusion   335

($\lambda^{-1} = d^2(\tau/G)/d\gamma^2$). For $\dot{\tau}_\infty = 1$ bar/year, $G = 20$ GPa and $\lambda = 0.0025$, Rice and Rudnicki (1979) present results for two values of ξ/ξ_u, 1.10 and 1.25, three values of the inclusion radius, $a = 1, 3,$ and 5 km, and two values of the diffusivity, $c = 0.1$ and 1.0 m^2/s. The results for the precursor time range from 14 to 400 days. Rice and Rudnicki (1979) note that the precursor times are a factor of two to three longer for $c = 1.0$ m^2/s than for $c = 0.1$ m^2/s and a factor of two to four longer for $a = 5$ km than for $a = 1$ km. Although the precursor times are much longer than those for the case of a narrow slip surface, precursor times estimated for a narrow, slitlike ellipsoidal inclusion with aspect ratio 18 to 1 range from 1.3 to 22 days.

15.3 COUPLING BETWEEN INELASTIC DEFORMATION AND DIFFUSION

Because inelastic deformation of geological and geotechnical materials often involves volume change, inelastic deformation can also be coupled to pore fluid diffusion. In brittle rocks, inelastic volume increase, or dilatancy, can result, even when all principal stresses are compressive, from micro-cracking in response to local tensile stresses at tips of fissures or near other inhomogeneities and from uplift in sliding over asperities on frictional surfaces (Brace et al., 1966). Inelastic volume increase in response to shear is also observed in overconsolidated clays and can result from grain rearrangement in densely packed sands or other granular materials. Although the discussion here will emphasize inelastic volume increase, shear induced compaction is observed in very porous rocks and loosely consolidated granular materials.

The next subsection outlines a framework for describing the inelastic response of brittle rock, including the effects of pore fluids, and illustrates the hardening effect of dilatancy-induced pore pressure reductions. Succeeding subsections review results on the stability of dilatantly hardened deformation and applications to earth faulting.

15.3.1 Inelastic constitutive law for brittle rock

Rice (1975) has formulated a description of inelastic deformation for the special deformation state of combined shear and uniaxial compression. Rudnicki and Rice (1975) generalized the formulation to arbitrary deformation states and magnitudes. Their framework, including the incorporation of pore fluid effects, has recently been reviewed by Rudnicki (1984b). However, it will be simplest to follow the development of Rice and Rudnicki (1979) for material subjected to shear stress τ and hydrostatic compression σ. For a stress increment $(d\tau, d\sigma)$ that involves continued inelastic deformation, the increments of shear strain $d\gamma$ and volume strain $d\varepsilon$ are as follows:

$$d\gamma = d\tau/G + d^p\gamma \tag{15.64}$$

336   *Mechanics of Geomaterials*

$$d\varepsilon = -d\sigma/K + d^p\varepsilon \tag{15.65}$$

where G is the elastic shear modulus, K is the elastic bulk modulus, and $d^p\gamma$ and $d^p\varepsilon$ are the inelastic increments of shear strain and volume strain, respectively. For elastic unloading the second terms in (15.64) and (15.65) are dropped. Because an increase in hydrostatic compression inhibits inelastic deformation in brittle rock, the inelastic increment of shear strain is taken to have the following form:

$$d^p\gamma = (d\tau - \mu d\sigma)/H$$

where μ is a friction coefficient and H is an inelastic hardening modulus. For constant hydrostatic stress H is related to the slope of the τ versus γ curve by

$$\frac{d\tau}{d\gamma} = \frac{H}{1 + H/G}, \quad d\sigma = 0 \tag{15.66}$$

Because inelastic volume strain arises from processes that accompany inelastic shear, that is, extension of micro-cracks by local tensile stresses, uplift in sliding over asperities and grain rearrangement, the inelastic volume strain increment is assumed to be given by

$$d^p\varepsilon = \beta d^p\gamma \tag{15.67}$$

where β is the dilatancy factor. (For shear-induced compaction, $\beta < 0$.) In general, μ and β, as well as G, K, and H may depend on the current state of deformation and even on the past history of deformation (although not on deformation increments), but for applications to brittle rocks it often suffices to assume that μ, β, G, and K are constant. Laboratory observations on brittle rocks suggest values of β in the range 0.1 to 0.3 and μ in the range 0.3 to 0.6 (Rice, 1975; Rudnicki and Rice, 1975; Rudnicki, 1977).

To include the effects of pore fluid pressure, the hydrostatic stress σ is replaced by the effective stress. For elastic deformation, the form of the effective stress is, as discussed earlier, $\sigma - \zeta p$ where ζ is given by (15.16). For inelastic deformation arising from extension of sharp-tipped micro-cracks and frictional sliding on fissure surfaces with small contact areas Rice (1977) has shown that the appropriate form of the effective stress is $\sigma - p$. This deduction is consistent with experimental observations on inelastic properties of brittle rock (e.g. Cornet and Fairhurst, 1974; Brace and Martin, 1968). An equation for m, the fluid mass content per unit volume, is also required. As for elastic deformation, m can be expressed as ρv, but now the increment dv is separated into an elastic portion and an inelastic portion $d^p v$. The elastic portion is calculated as in the Biot theory and, for the same conditions that $\sigma - p$ is the appropriate form of the effective stress, $d^p v = d^p\varepsilon$ (Rice, 1977). Thus, the increments of shear strain, volume strain, and fluid mass content per unit volume are as follows:

$$d\gamma = d\tau/G + [d\tau - \mu(d\sigma - dp)]/H \tag{15.68}$$

Effect of Pore Fluid Diffusion

$$d\varepsilon = -(d\sigma - \zeta dp)/K + \beta[d\tau - \mu(d\sigma - dp)]/H \quad (15.69)$$

$$\rho^{-1}dm = v\,dp(1/K_f - 1/K_s'') - \zeta(d\sigma - dp)/K + d^p\varepsilon \quad (15.70)$$

For undrained response, the change in fluid mass content per unit volume is zero. Setting $dm = 0$ in (15.70) and solving for dp yields

$$dp = \frac{-\beta K_{\text{eff}}}{H + \mu\beta K_{\text{eff}}}\,d\tau \quad (15.71)$$

where, for convenience, $d\sigma$ has been taken as zero and, using (15.16), K_{eff} can be written as follows:

$$1/K_{\text{eff}} = v(1/K_f - 1/K_s'') + (1/K - 1/K_s') \quad (15.72)$$

Hence, for dilatant inelastic deformation ($\beta > 0$) the pore fluid pressure tends to decrease. When (15.71) is substituted into (15.68), again with $d\sigma = 0$, the result is

$$d\gamma = d\tau/G + d\tau/(H + \mu\beta K_{\text{eff}}) \quad (15.73)$$

Because the hardening modulus has been augmented by the term $\mu\beta K_{\text{eff}}$, the response is stiffer and the rock mass is said to be dilatantly hardened (Figure 15.8). Dilatant hardening is a well-known phenomenon in the mechanics of granular materials and appears to have been first discussed by Reynolds (1885). Its relevance to earthquake faulting was pointed out by Frank (1965). Brace and

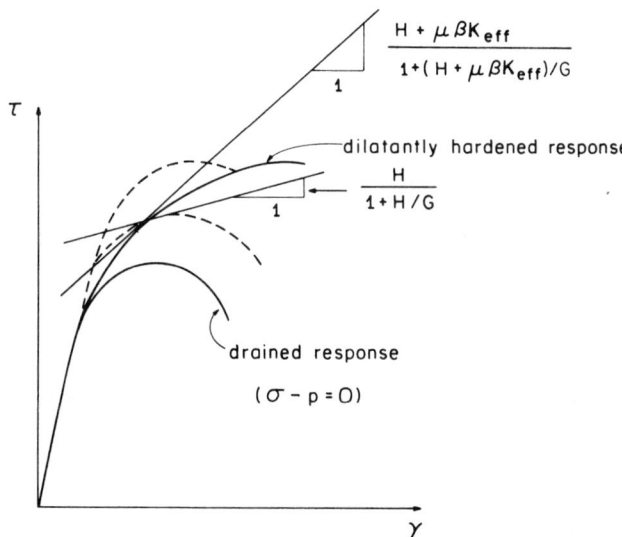

Figure 15.8 Illustration of dilatant hardening effect. The dashed curves show drained response at increasing value of the effective compressive stress, that is, for $\sigma - p > 0$.

Martin (1968) and Martin (1980) have observed dilatant hardening in axisymmetric compression tests on brittle rock.

As discussed by Rice (1975) and by Rice and Rudnicki (1979), the dilatant hardening effect can be significant for representative values of μ, β, and K_{eff}. Because the term that augments H is proportional to the product $\mu\beta$, the effect vanishes if there is no inelastic volume change or if inelastic deformation is not affected by hydrostatic stress. For highly fissured rock, K_s' and $K_s'' \gg K$ and K_f/v. Therefore, (15.72) reduces to

$$K_{\text{eff}} \sim K(1 + vK/K_f)^{-1}$$

If the pore fluid is liquid water ($K_f = 2.2\,\text{GPa}$) and the porosity is less than 10 per cent, $K_{\text{eff}} \sim K$. If K_f is substantially reduced by high temperatures, exsolution of trapped gas and pressure reduction,

$$K_{\text{eff}} \sim K_f/v$$

and dilatant hardening disappears in the limit $K_f \to 0$.

15.3.2 Stability of dilatant hardening

Rice (1975) has studied the stability of dilatant hardening for a layer subjected to a combination of uniaxial compression and pure shear. He finds that homogeneous dilatantly hardened deformation becomes unstable, in the sense that infinitesimal spatial non-uniformities will grow exponentially in time when the peak of the underlying drained stress–strain curve has been reached. That instability sets in at the peak of the drained stress–strain curve is an artifact of the special deformation state considered: in general, dilatantly hardened homogeneous deformation will become unstable when conditions for localization of deformation (Rudnicki and Rice, 1975; Rudnicki, 1983) are met in terms of the drained response and this may occur before or after the peak of the stress–strain curve.

Rice (1975) also noted, however, that for a layer with an initial finite size imperfection the development of instability would depend on the time scale of deformation. Rudnicki (1984a) has explored this possibility by considering the shear deformation of a layer containing a slightly weaker sublayer. To simplify the analysis, Rudnicki (1984a) assumes that the deformations of the sublayer and of the surrounding material are homogeneous. In addition, the rate of fluid mass flux out of the sublayer is assumed to be proportional to the difference in the pore fluid pressures of the sublayer and the surrounding material. The coefficient in this relation, which must be regarded as an empirical parameter, is expressed as c/h^2 where c is a diffusivity and h is the width of the weakened sublayer. In this case, the response becomes unstable, in the sense described by Rice (1975), when the weakened sublayer reaches the peak of its underlying drained stress strain curve. However, dilatant hardening delays the onset of final instability until the weakened sublayer reaches the peak of its undrained response curve. The time period during which the response is stabilized is again termed the precursor time.

To calculate the precursor time, Rudnicki (1984a) assumes that both the sublayer and the surrounding material have drained stress strain curves of the form (15.63), but that the peak stress in the weakened sublayer is slightly lower than that in the surrounding material. When the time scale of the applied deformation is much greater than that for fluid mass exchange between the layers, as appropriate for application to tectonic deformations, Rudnicki (1984a) obtains an asymptotic solution for the history of straining in the weakened sublayer. The resulting asymptotic prediction of the precursor time is as follows:

$$ct_{\text{prec}}/h^2 = 1.48(c\lambda/h^2\dot{\gamma}_\infty)^{1/3}\alpha^{2/3}\Delta^{-1/6}[1 + \alpha(1 - (2\Delta)^{1/2})]^{1/3} \quad (15.74)$$

where $\alpha = \mu\beta M/G$, M is an elastic modulus for one-dimensional strain, Δ is the difference in values of peak stress divided by $G\lambda$, and $\dot{\gamma}_\infty$ is the constant applied farfield strain rate. The asymptotic solution is valid for $c\lambda/h^2\dot{\gamma}_\infty \gg 1$. Note that the precursor time becomes arbitrarily large for $\dot{\gamma}_\infty \to 0$, but that the additional strain accumulating during this period goes to zero in the same limit.

For $\dot{\gamma}_\infty = 10^{-15}\,\text{s}^{-1}$, on the order of measured tectonic strain rates in southern California, $h = 1\,\text{m}$, $c = 0.1\,\text{m}^2/\text{s}$, and $\lambda = 0.0025$, $c\lambda/h^2\dot{\gamma}_\infty = 2.5 \times 10^{12}$. For $\Delta = 0.1$ and $\alpha = 0.3$, $t_{\text{prec}} \simeq 4$ hours. For $\dot{\gamma}_\infty = 10^{-6}$, on the order of laboratory strain-rates, $h = 1\,\text{cm}$, $c = 10^{-4}\,\text{m}^2/\text{s}$, and $\lambda = 0.0025$, $c\lambda/h^2\dot{\gamma}_\infty = 2.5 \times 10^3$. For $\Delta = 0.1$ and $\alpha = 0.3$, $t_{\text{prec}} \approx 10\,\text{s}$. Values of the precursor time, for both laboratory and tectonic strain-rates, are predicted to be so short that easy observation seems doubtful. Martin (1980) has, however, observed pore fluid stabilization of failure in axisymmetric compression samples of Westerly granite for periods on the order of several hundred seconds. Although there are obvious and substantial differences between the experimental configuration and that analysed by Rudnicki (1984a), the observed stabilization times are significantly longer than those predicted for comparable strain-rates and a plausible range of values for the other parameters. At least two possibilities may account for this discrepancy. The first is that analyses of non-fluid-infiltrated solids (Rudnicki and Rice, 1975) suggest that there may be significant differences in the development of localization for pure shear and axisymmetric deformation states. To investigate this possibility, Rudnicki (1983) has extended the formulation of Rudnicki (1984a) to arbitrary deformation states. Although the governing equations can be shown to have the same character as those for pure shear, no detailed results are yet available. A second, perhaps more likely, possibility is that significant stabilization due to dilatant hardening accompanies sliding after the deformation has already concentrated into a narrow zone.

15.3.3 Dilatant hardening for an inclusion model of faulting

Rice and Rudnicki (1979) have examined the effects of dilatant hardening for an inclusion model of earth faulting (Rudnicki, 1977). These effects are complementary to those due to the time dependence of the unloading stiffness discussed earlier and they are also stabilizing. Again, instability cannot occur abruptly when the condition for runaway instability in terms of the drained response is met. The rapid deformation that occurs as this condition is approached induces dilatantly hardened response of the inclusion material. Consequently, instability is delayed until (15.62) is met with the undrained slope of the τ_{inc} versus γ_{inc} curve entering the left-hand side of (15.62) (Figure 15.9). Thus, dilatant hardening delays instability by increasing the value on the left-hand side of (15.62) whereas the time-dependent unloading stiffness delays instability by increasing the magnitude of the right-hand side.

Rice and Rudnicki (1979) calculate strain histories and precursor times using the Eshelby relations (15.59) and (15.61) for a spherical inclusion. Because the time-dependent unloading stiffness effects are neglected, the drained form of (15.60), corresponding to $f = 1$, is used. The drained response of the inclusion material is assumed to be described by (15.63). For $\dot{\tau}_\infty = 0.1$ MPa/year, $G = 20$ GPa, $\lambda = 0.0025$, $\beta = 0.3$ and $K_f = 2.2$ GPa, Rice and Rudnicki (1979) calculate precursor times ranging from 55 days for an inclusion radius $a = 1$ km and diffusivity $c = 1.0 \, \text{m}^2/\text{s}$ to 1418 days for $a = 5$ km and $c = 0.1 \, \text{m}^2/\text{s}$. They also note that a factor of two decrease in the dilatancy factor β reduces the precursor time by slightly more than half, whereas a tenfold decrease in the pore fluid bulk modulus

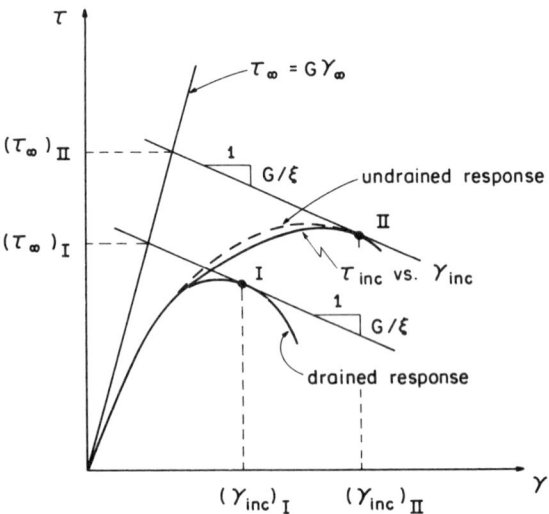

Figure 15.9 Delay of runaway instability in the inclusion model (see Figure 15.7) due to dilatant hardening of the inclusion material (after Rice, 1979a). The dashed curves show fully undrained response. Effects due to the time-dependent response of material surrounding the inclusion are neglected.

K_f reduces the precursor time by slightly less than half. Although the precursor times are generally long, Rice and Rudnicki (1979) suggest that values toward the lower end of those calculated may be more appropriate.

Rice and Rudnicki (1979) also calculate the decrease from ambient value of the pore fluid pressure in the inclusion. The predicted decrease does not appear to be large enough to cause a liquid-to-vapour phase transition in the pore fluid until very near to instability. Such a phase transition would be needed to cause the substantial changes in wave speed ratios that have been suggested as precursors to earthquakes (Nur, 1972; Anderson and Whitcomb, 1975). Consequently, the calculations do not support this possibility. Nevertheless, they do indicate that transient stabilization of incipient failure by dilatant hardening, as well as by time-dependent unloading stiffness effects, could cause a period of accelerating time-dependent strain prior to instability.

15.3.4 Dilatant hardening effects on propagating shear fractures

Rice (1973, 1979a, 1980, 1981) has analysed dilatant hardening effects in the end-zone of a steadily propagating fracture. He assumes that a uniform uplift accompanies shear in the end-zone. By assuming one-dimensional diffusion of pore fluid into the end-zone, Rice calculates the distribution of suctions induced on the fault plane. Because these suctions reduce the effective value of the stress normal to the fault plane, they inhibit propagation or, in other words, they increase the energy that must be supplied to continue advancing the fractures. Although the values of material parameters entering the analysis are uncertain, Rice demonstrates that the effect can be significant. He calculates a factor of 50 as a rough upper bound on the increase of required energy for which it is possible to maintain quasi-static propagation (Rice, 1979a). But he notes that the process may be self-limiting for two reasons: the amount of uplift decreases with increasing normal stress and the induced suctions may be large enough to cause vaporization of the pore fluid or exsolution of gases.

15.4 CONCLUDING DISCUSSION

This paper has reviewed the mechanical effects of pore fluid diffusion on the quasi-static deformation and failure of brittle rock. These effects have been illustrated by applications to earth faulting. The role of pore fluid in reducing the effective value of the confining stress in bulk samples and the normal stress across frictional surfaces has been demonstrated in laboratory experiments (e.g. Brace and Martin, 1968; Byerlee and Brace, 1972) and in field studies (e.g. Raleigh *et al.*, 1976; Zoback and Hickman, 1982). In addition, the solutions reviewed here strongly suggest that the coupling of pore fluid diffusion and deformation can introduce rate-dependence into the deformation of an otherwise rate-independent solid and make a significant contribution to the time scale of the failure process. However, few laboratory or field studies of these effects exist. Consequently, although relevant solutions of coupled-deformation and pore fluid diffusion problems have by no means been exhausted, the most pressing need is for laboratory and field data. Data are needed both to provide better estimates of material parameters and to test the accuracy of analytical predictions.

Although the time dependence introduced into the failure process is evident in the solutions reviewed here, identification of these effects *in situ* is complicated by the presence of other time-dependent processes. These include bulk viscoelastic response, coupling of crustal deformation with underlying viscous material (e.g. Anderson, 1975; Budiansky and Amazigo, 1976; Lehner *et al.*, 1981), time dependence due to slow crack growth (e.g. Das and Scholz, 1981) and time dependent frictional slip (e.g. Dieterich, 1978). The time scales of these processes are probably different, but present uncertainties in material parameters are too great to distinguish mechanisms on this basis. Observations of water level changes in calibrated wells (e.g. Kovach *et al.*, 1975) or increased use of dilatometers may be helpful in identifying coupled deformation—pore fluid diffusion. Although observations of changes in seismic wave speed ratios (Aggarwal *et al.*, 1975; Whitcomb *et al.*, 1973) have not proved to be as effective an earthquake precursor as had been hoped, such measurements may provide information about crack densities and saturation levels from which material parameters can be inferred. (However, Rice (1980) has emphasized that electro-kinetic effects, which do not depend on degree of saturation, will be more indicative of crack density.) Moreover, there is some recent evidence (Leary *et al.*, 1979) that extremely small changes in wave speed (~ 0.1 per cent) over relatively short time scales (a few days) may be caused by deformation coupled with fluid diffusion. Because of the difficulties in anticipating the time and location of spontaneous events such as earthquakes and creep, field studies in which the pore fluid pressure is actively changed by altering reservoir levels or by direct pumping (Raleigh *et al.*, 1976) will be especially useful.

Although direct extrapolation of measurements on laboratory samples to field situations in which properties can be strongly affected by large fissures is impossible, the difficulty and expense of making relevant field observations virtually ensure that laboratory measurements will continue to contribute to understanding coupled deformation–diffusion effects. Unfortunately, most laboratory experiments have been limited to investigation of the effective stress principle for various properties and have not considered the rate dependent effects of coupling of deformation with pore fluid diffusion. As has been mentioned already, relatively few determinations exist of the parameters of even the linear Biot theory. Apparently, the only laboratory investigation of coupled deformation diffusion effects in setting a time scale for the failure process is that of Martin (1980), and there is a need for additional work in this area.

Although there are limitations on the Biot approach, most notably to linear behaviour, there is general agreement that the Biot equations are a suitable approximation for many conditions of quasi-static loading. Much less agreement exists, however, for dynamic conditions. The Biot theory has been extended (Biot, 1956a, b) to describe the dynamic response of saturated porous materials and, at least in some instances, gives results in reasonable accord with observations. But it is clear that a theory that is phrased in terms of a single scalar pore pressure will be suitable for limited frequency range. Some estimates of the time scale for which the Biot theory is applicable have been given by O'Connell and Budiansky (1977) and by Cleary (1978a). Cleary (1978a) has also outlined a general formulation but, as yet, the values of parameters entering this theory are largely unknown. A better understanding of the dynamic response of saturated porous media seems essential for applications to explosive loadings, interpretation of data from geophysical explorations, and ultrasonic measurements in biological materials.

The review has been limited to discussing the mechanical effects of pore fluids, but the importance of chemical effects has been increasingly recognized. Recent experiments (e.g. see Atkinson (1982), Rice (1979a, 1980) and Rudnicki (1980) for reviews) have indicated that small amounts of pore fluid can promote slow growth of cracks in rock at speeds ranging down to 10^{-10} m/s. Typically, these results are presented in relations of the following form between the crack speed V and the tensile stress intensity factor K:

$$V = AK^n \quad \text{or} \quad V = B \exp(bK)$$

where A and n or B and b are parameters depending on the environment. Small amounts of pore fluid on frictional slip surfaces may also contribute to time-dependent behaviour by accelerating deformation at asperity contacts (Dieterich and Conrad, 1978). Moreover, much of the observed time-dependent behaviour of brittle rocks may be due to environmentally assisted growth of micro-cracks. Costin and Mecholsky (1983) have successfully used laboratory results on slow growth of a single crack to predict the time-dependent response of an intact specimen in uniaxial compression.

Environmentally assisted crack growth may be very important in establishing the long-time strength of crustal rocks, for example, on time scales comparable to the recurrence time of large earthquakes or the desired lifetime of nuclear waste repositories. Although it is not yet clear whether there is a threshold value of K below which no crack growth occurs, the existence of such a threshold is important for long-time strength predictions. Moreover, Rice (1978) has pointed out that crack healing may be an important mechanism of strength recovery on faults. Although Rice (1978) has described a theoretical framework of analysing environmentally assisted growth of single cracks, the analysis of these effects has not progressed to the point where they can be incorporated into the solution of relevant boundary value problems.

ACKNOWLEDGEMENT

Support during the preparation of this review from the US Geological Survey Earthquake Hazards Reduction Program and the NSF Geophysics Program is gratefully acknowledged.

REFERENCES

Aggarwal, Y. P., Sykes, L. R., Simpson, D. W., and Richards, P. G. (1975) 'Spatial and temporal variations in t_s/t_p and in P wave residuals at Blue Mountain Lake, New York: application to earthquake prediction', *J. Geophys. Res.*, **80**, 718–732.

Anderson, D. L., and Whitcomb, J. H. (1975) 'Time-dependent seismology', *J. Geophys. Res.*, **80**, 1497–1503.

Atkinson, B. K. (1982) 'Subcritical crack propagation in rocks: theory, experimental results and applications', *J. Struct. Geology*, **4**, 41–56.

Bell, M. L. and Nur, A. (1978) 'Strength changes due to reservoir-induced pore pressure and stresses and application to Lake Oroville', *J. Geophys. Res.*, **83**, 4469–4483.

Biot, M. A. (1941a) 'General theory of three-dimensional consolidation', *J. Appl. Phys.*, **12**, 155–164.

Biot, M. A. (1941b) 'Consolidation settlement under a rectangular load distribution', *J. of Appl. Phys.*, **12**, 426–578.

Biot, M. A. (1956a) 'Theory of propagation of elastic waves in a fluid-saturated porous solid. I. Low-frequency range', *J. Acous. Soc. Am.*, **28**, 168–178.

Biot, M. A. (1956b) 'Theory of propagation of elastic waves in a fluid-saturated porous solid. II. Higher frequency range', *J. Acous. Soc. Am.*, **28**, 178–191.

Biot, M. A. (1973) 'Nonlinear and semilinear rheology of porous solids', *J. Geophys. Res.*, **78**, 4924–4937.

Biot, M. A., and Clingan, F. M. (1941) 'Consolidation settlement of a soil with an impervious top surface', *J. Applied Phys.*, **12**, 578–581.

Booker, J. R. (1974) 'Time-dependent strain following faulting of a porous medium', *J. Geophys. Res.*, **79**, 2037–2044.

Brace, W. F., Paulding, B. W., Jr., and Scholz, C. H. (1966) 'Dilatancy in the fracture of crystalline rocks', *J. Geophys. Res.*, **71**, 3939–3953.

Brace, W. F. and Martin, R. J., III (1968) 'A test of the law of effective stress for crystalline rocks of low porosity', *Int. J. Rock Mech. Mining Sci.*, **5**, 415–436.

Budiansky, B., and Amazigo, C. (1976) 'Interaction of fault slip and lithospheric creep', *J. Geophys. Res.*, **81**, 4897–4900.

Byerlee, J. D., and Brace, W. F. (1972) 'Fault stability and pore pressure', *Bull. Seismol. Soc. Am.*, **62**, 657–660.

Cleary, M. P. (1977) 'Fundamental solutions for a fluid-saturated porous solid', *Int. J. Solids Structures*, **13**, 785–806.

Cleary, M. P. (1978a) 'Elastic and dynamic response regimes of fluid-impregnated solids with diverse microstructures', *Int. J. Solids Structures*, **14**, 795–819.

Cleary, M. P. (1978b) 'Moving singularities in elasto-diffusive solids with applications to fracture propagation', *Int. J. Solids Structures*, **14**, 81–97.

Cleary, M. P. (1979) 'Rate and structure sensitivity in hydraulic fracturing of fluid-saturated porous formations', in *Proc. U.S. Symp. Rock Mech.*, 20th Austin, Texas, (Ed. K. Gray), pp. 127–141.

Cornet, F. H., and Fairhurst, C. (1974) 'Influence of pore pressure on the deformation behaviour of saturated rocks', in *Proceedings of the Third Congress of the International Society for Rock Mechanics*, National Academy of Sciences, Washington, D. C., Vol. 1, part B, pp. 638–644.

Costin, L. S., and Mecholsky, J. J. (1983) 'Time dependent crack growth and failure in brittle rock', in *Proc. U.S. Symp. on Rock Mechanics*, 24th, College Station, Texas.

Cryer, C. W. (1963) 'A comparison of the three-dimensional consolidation theories of Biot and Terzaghi', *Quart. J. Mech. Appl. Math.*, **16**, 401–412.

Das, S., and Scholz, C. H. (1981) 'Theory of time-dependent rupture in the earth', *J. Geophys. Res.*, **86**, 6039–6051.

Dieterich, J. H. (1978) 'Time-dependent friction and the mechanics of stick–slip', *Pure Appl. Geophys.*, **116**, 790–806.

Dieterich, J. H., and Conrad, G. (1978) 'Mechanism of unstable slip in rock friction experiments', *EOS, Trans Am. Geophys. Union*, **59**, 1206.

Eshelby, J. D. (1957) 'The determination of the elastic field of an ellipsoidal inclusion and related problems', *Proc. Roy. Soc. Ser. A.*, **241**, 376–396.

Frank, F. C. (1965) 'On dilatancy in relation to seismic sources', *Rev. Geophys. Space Phys.*, **3**, 485–503.

Johnson, A. G., Kovach, R. L., and Nur, A. (1973) 'Pore pressure changes during creep events on the San Andreas fault', *J. Geophys. Res.*, **78**, 851–857.

Johnson, C. E. (1979) 'I. CEDAR—an approach to the computer automation of short-period local seismic networks. II Seismotectonics of the Imperial Valley of Southern California', Ph.D. Thesis, California Institute of Technology, 332 pages.

Kovach, R. L., Nur, A., Wesson, R. L., and Robinson, R. (1975) 'Water-level fluctuations and earthquakes on the San Andreas fault zone', *Geology*, **3**, 437–440.

Leary, P. C., Malin, P. E., Phinney, R. A., Brocher, T., and Von Colln, R. (1979) 'Systematic monitoring of millisecond travel time variations near Palmdale, CA', *J. Geophys. Res.*, **84**, 659–666.

Lehner, F. K., Li, V. C., and Rice, J. R. (1981) 'Stress diffusion along rupturing plate boundaries', *J. Geophys. Res.*, **86**, 6155–6169.

Li, V. C. (1984) 'Estimation of in-situ diffusivity' submitted to *Pure Appl. Geophysics*.

Martin, R. J., III (1980) 'Pore pressure stabilization of failure in Westerly granite', *Geophys. Res. Letters*, **7**, 404–406.

McNamee, J., and Gibson, R. E. (1960a) 'Displacement functions and linear transforms applied to diffusion through porous elastic media', *Quart. J. Mech. Appl. Math.*, **13**, 98–111.

McNamee, J., and Gibson, R. E. (1960b) 'Plane strain and axially symmetric problems of the consolidation of a semi-infinite clay stratum', *Quart. J. Mech. Appl. Math.*, **13**, 210–227.

Nur, A. (1972) 'Dilatancy, pore fluids, and premonitory variations of t_s/t_p travel times', *Bull. Seismol. Soc. Am.*, **62**, 1217–1222.

Nur, A., and Booker, J. R. (1972) 'Aftershocks caused by pore fluid flow?', *Science*, **175**, 885–887.

Nur, A., and Byerlee, J. D. (1971) 'An exact effective stress law for elastic deformation of rock with fluids', *J. Geophys. Res.*, **76**, 6414–6419.

O'Connell, R. J., and Budiansky, B. (1974) 'Seismic velocities in dry and saturated cracked solids', *J. Geophys. Res.*, **79**, 5412–5426.

O'Connell, R. J., and Budiansky, B. (1977) 'Viscoelastic properties of fluid-saturated cracked solids', *J. Geophys. Res.*, **82**, 5719–5735.

Ohtake, M. (1974) 'Seismic activity induced by water injection at Matsushiro, Japan', *J Phys. Earth*, **22**, 163–176.

Paterson, M. S. (1978) *Experimental Rock Deformation—The Brittle Field*, Springer Verlag, New York.

Raleigh, C. B., Healy, J. H., and Bredehoeft, J. D. (1976) 'An experiment in earthquake control at Rangely, Colorado', *Science*, **191**, 1230–1237.

Reynolds, O. (1885) 'On the dilatancy of media composed of rigid particles in contact, with experimental illustrations, *Phil. Mag.* (reprinted in *Papers on Mechanical and Physical Subjects* by O. Reynolds, Cambridge University Press, New York, 1901, Vol. 2, pp. 203–216.)

Rice, J. R. (1966) 'An examination of the fracture mechanics energy balance from the point-of-view of continuum mechanics', in *Proc. Int. Conf. Fract.*, 1st (Ed. T. Yokobori), Japanese Society of Strength and Fracture, Tokyo, Vol. 1, pp. 283–308.

Rice, J. R. (1968) 'Mathematical analysis in the mechanics of fracture', in *Fracture: An Advanced Treatise* (Ed. H. Liebowitz), Vol. II, Academic Press, pp. 191–311.

Rice, J. R. (1973) 'The initiation and growth of shear bands', in *Plasticity and Soil Mechanics* (Ed. A. C. Palmer), Cambridge University Engineering Department, Cambridge, England, pp. 263–274.

Rice, J. R. (1975) 'On the stability of dilatant hardening for saturated rock masses', *J. Geophys. Res.*, **80**, 1531–1536.

Rice, J. R. (1977) 'Pore pressure effects in inelastic constitutive formations for fissured rock masses', in *Advances in Civil Engineering through Engineering Mechanics*, American Society of Civil Engineers, New York, pp. 360–363.

Rice, J. R. (1978) 'Thermodynamics of the quasi-static growth of Griffith cracks', *J. Mech. Phys. Solids*, **26**, 61–78.

Rice, J. R. (1979a) 'Theory of precursory processes in the inception of earthquake rupture', *Gerlands Beitr. Geophys.*, 91–127.

Rice, J. R. (1979b) 'The mechanics of quasistatic crack growth', in *Proc. U. S. Nat. Congr. Appl. Mech. 8th, UCLA*, June 1978 (Ed. R. E. Kelley), Western Periodicals, North Hollywood, California, pp. 191–216.

Rice, J. R. (1980) 'The mechanics of earthquake rupture', in *Physics of the Earth's Interior*, Proceedings of the International School of Physics 'Enrico Fermi', Course 78, 1979 (Eds. A. M. Dziewonski and E. Boschi), Italian Physical Society, North-Holland, Amsterdam, pp. 555–649.

Rice, J. R. (1981) 'Pore fluid processes in the mechanics of earthquake rupture', in *Solid Earth Geophysics and Geotechnology* (Ed. S. Nemat-Nasser), Applied Mech. Div. Vol. 42, American Society of Mech. Eng., New York, pp. 81–89.

Rice, J. R., and Cleary, M. P. (1976) 'Some basic stress diffusion solutions for fluid-saturated elastic porous media with compressible constituents', *Rev. Geophys. Space Phys.*, **14**, 227–241.

Rice, J. R., and Rudnicki, J. W. (1979) 'Earthquake precursory effects due to pore fluid stabilization of a weakening fault zone', *J. Geophys. Res.*, **84**, 2177–2193.

Rice, J. R., Rudnicki, J. W., and Simons, D. A. (1978). 'Deformation of spherical cavities and inclusions in fluid-infiltrated elastic solids', *Int. J. Solids Struct.*, **14**, 289–303, 1978.

Rice, J. R. and Simons, D. A. (1976) 'The stabilization of spreading shear faults by coupled deformation–diffusion effect in fluid-infiltrated porous materials', *J. Geophys. Res.*, **81**, 5322–5334.

Roeloffs, E. (1982) Elasticity of saturated porous rocks: laboratory measurements and a crack problem, Ph.D. Thesis at University of Wisconsin, Madison, 157 pages.

Roeloffs, E., and Rudnicki, J. W. (1984) 'Coupled deformation–fluid diffusion effects on water level changes due to propagating creep events', to appear in *Pure Appl. Geophysics*.

Rudnicki, J. W. (1977) 'The inception of faulting in a rock mass with a weakened zone', *J. Geophys. Res.*, **82**, 844–854.

Rudnicki, J. W. (1979) 'The stabilization of slip on a narrow weakening fault zone by coupled deformation–pore fluid diffusion', *Bull. Seismol. Soc. Am.*, **69**, 1011–1026.

Rudnicki, J. W. (1980) 'Fracture mechanics applied to the earth's crust', in *Ann. reviews of Earth Planetary Sciences*, Vol. 8 (Ed. F. A. Donath), Annual Reviews Inc., Palo Alto, CA, pp. 489–525.

Rudnicki, J. W. (1981a) 'An inclusion model for processes preparatory to earthquake faulting', in *Solid Earth Geophysics and Geotechnology* (Ed. S. Nemat-Nasser), Applied Mech. Div. Vol. 42, American Society of Mech. Engn., New York, pp. 39–52.

Rudnicki, J. W. (1981b) 'On "Fundamental solutions for a fluid-saturated porous solid" by M. P. Cleary'. *Int. J. Solids Structures*, **17**, 855–857.

Rudnicki, J. W. (1983) 'A formulation for studying coupled deformation–pore fluid diffusion effects on localization', in *Proceedings of the Symposium on the Mechanics of Rocks, Soils and Ice* (Ed. S. Nemat-Nasser), *Appl. Mech. Div.* Vol. 57. American Society of Mechanical Engineers, New York, pp. 35–44.

Rudnicki, J. W. (1984a) 'Effects of dilatant hardening on the development of concentrated shear deformation in fissured rock masses', *J. Geophys. Res.*, **89**, 9259–9270.

Rudnicki, J. W. (1984b) 'A class of elastic-plastic constitutive laws for brittle rock', to appear in a special issue of the *Journal of Rheology* devoted to Geological Materials and edited by S. L. Passman.

Rudnicki, J. W., and Rice, J. R. (1975) 'Conditions for the localization of deformation in pressure-sensitive dilatant materials', *J. Mech. Phys. Solids*, **23**, 371–394.

Ruina, A. (1978) 'Influence of coupled deformation-diffusion effects on retardation of hydraulic fracture', in *Proc. US Symp. Rock Mech.*, 19th, Stateline, Nev. (Ed. Y. S. Kim), pp. 274–282.

Scholz, C. H., Sykes, L. R., and Aggarwal, Y. P. (1973) 'Earthquake prediction: a physical basis', *Science*, **181**, 803–810.

Simons, D. A. (1977) 'Boundary layer analysis of propagating mode II cracks in porous elastic solids', *J. Mech. Phys. Solids*, **25**, 99–116.

Simons, D. A. (1978) 'The analysis of propagating slip zones in porous elastic media' in *Fracture Mechanics*, Proceedings of the Symposium in Applied Mathematics of AMS and SIAM (Ed. R. Burridge), pp. 153–168.

Simpson, D. W. (1976) 'Seismicity changes associated with reservoir loading'. *Eng. Geol.*, **10**, 123–150.

Verruijt, A. (1968) 'Elastic storage of acquifers' in *Flow Through Porous Media* (Ed. R. J. M. DeWiest), Academic Press, New York.

Wesson, R. L. (1981) 'Interpretation of changes in water level accompanying fault creep and implications for earthquake prediction', *J. Geophys. Res.*, **86**, 9259–9267.

Whitcomb, J. H., Garmany, J. D., and Anderson, D. L. (1973) 'Earthquake prediction: variation of seismic velocities before the San Fernando earthquake', *Science*, **180**, 632–635.

Zoback, M. D., and Hickman, S. (1982) 'In situ study of the physical mechanisms controlling induced seismicity at Monticello reservoir, South Carolina', *J. Geophys. Res.*, **87**, 6959–6974.

Mechanics of Geomaterials
Edited by Z. Bažant
© 1985 John Wiley & Sons Ltd

Chapter 16

Effect of Pore Water and Its Diffusion in Concrete

G. Horrigmoe

16.1 INTRODUCTION

The mechanical behaviour of porous materials, including concrete, is known to depend strongly on the distribution and history of moisture content. Material parameters affected by variations in moisture content comprise practically all the major parameters employed in the analysis and design of concrete structures, e.g. strength, modulus of elasticity, thermal conductivity, moisture diffusivity, and permeability (Neville, 1977). Moreover, pore humidity is a key factor in long-time deformations of concrete, known as creep and shrinkage. Thus, reliable mathematical models that can predict the distribution of pore humidity in concrete structures is of major importance. While a large body of experimental information on moisture transfer in concrete has been assembled over many years of research, the development of prediction methods attracted relatively minor attention following the pioneering effort by Carlson (1937). It was not until the introduction of novel applications of concrete, especially in the form of pressure vessels for nuclear reactors, that research in mathematical models for predicting moisture transfer gained momentum. In these structures, a correct prediction of heat and moisture transfer is a prerequisite for safe and economical designs, and this has undoubtedly been a major contributing factor in the recent developments in mathematical models for diffusion of pore water in concrete.

Among the building materials in common use, concrete is the only material for which basic questions regarding mechanisms of deformation and constitutive relations remain unsettled. Whereas constitutive modelling for materials such as metals and polymers has been carried well into the nonlinear regime, most of the material research in concrete remains focused on the behaviour in the working stress range. In particular, constitutive relations for creep and shrinkage continue to be a question of considerable controversy in the literature (Bažant, 1975a). Since the ultimate values of these time-dependent deformations are known to be several times greater than the instantaneous elastic response, their importance as the main cause of defections, redistribution of stresses and cracking, is undisputed.

However, considerable disagreement prevails regarding the microstructural mechanisms involved in creep and shrinkage deformations. Above all, the effect of changes in pore humidity on creep and shrinkage and its constitutive modelling is far from being resolved.

The purpose of the present paper is to summarize and review the effects of diffusion of pore water in concrete, with special emphasis on mathematical models for moisture transfer. Drying of concrete under isothermal (or quasi-isothermal) conditions is a problem of widespread interest. The application of diffusion theory to model moisture transport during drying is outlined in Section 16.3. Various formulations are discussed together with the effects of advancing hydration and variable temperature. The formulation of coupled heat and moisture transfer then follows in Section 16.4. The influence of pore humidity on creep and shrinkage and the corresponding constitutive relations are briefly examined in Section 16.5.

16.2 PHYSICAL NATURE OF PORE WATER IN CONCRETE

Concrete is formed by chemical reactions between Portland cement and water. This process is generally referred to as hydration and results in a very fine gel-type structure. Typical porosities of hardened cement paste are in the range of 0.40–0.55. The major part of the voids is made up of micropores. The larger capillary pores are randomly distributed throughout the cement paste and, at least in mature and dense concretes, capillary pores can be viewed as isolated voids and not as an interconnected system. The dominant role of the micropores in cement paste produces an enormous internal surface area (of the order of $5.10^8 \, m^2$ per m^3, or approximately $200\,000 \, m^2$ per kg (Neville, 1977).

Water in concrete is held with varying degree of firmness and cannot be treated as a single quantity. First, there is *chemically bound water* which forms a part of the hydrated compounds. Next, there is *adsorbed water*, i.e. water held by surface forces in the form of films on the pore walls. Owing to the comparatively large size of the capillary pores, water in these pores is beyond the influence of surface forces and is referred to as free water or *capillary water*. The separation of water into these three categories is mainly of theoretical value since at present there is no experimental technique available by which the various forms of water can be quantitatively determined.

A more feasible way of distinguishing between different forms of water in cement paste is to divide it into two categories: evaporable and non-evaporable. The *evaporable water* is by definition the amount of water which is removed from concrete upon complete drying (usually at 105°C). Similarly, the water remaining in the dried specimen after equilibrium has been achieved is the *non-evaporable water*.

In the pores, molecules of water vapour are attracted by surface forces that retain them at the solid surface, thus forming thin adsorbed layers of water

Effect of Pore Water and Its Diffusion

molecules. The average thickness of the adsorbed layers increases with relative pore humidity as illustrated in Figure 16.1 for a micropore of variable thickness. The maximum layer thickness has been estimated to five molecules (Bažant, 1975a); hence, for a pore to accommodate complete adsorbed layers on two opposite walls, the pore thickness must at least be that of 10 molecules, or 26 Å. Provided that the thickness of the pore allows it, and that the relative humidity is sufficiently high, a capillary meniscus may develop, thereby creating a certain amount of free, capillary water within the pore.

In the pores that are less than 10 molecules thick, complete adsorbed layers of water molecules cannot develop. Such films are therefore called *hindered adsorbed layers*. Adsorbed molecules are not held permanently but retain sufficient mobility to be able to diffuse along the layers.

For any porous material in equilibrium with the surrounding air there exists a definite equilibrium moisture content that increases with increasing relative humidity. For $RH = 0$, the mass of evaporable water is zero and at $RH = 100$ per cent all pores should in principle be completely filled with water. At a given temperature the material thus possesses an equilibrium curve, or a *sorption isotherm*, defining the amount of evaporable water retained at various relative humidities. Studies of the relationship between evaporable water content in hardened cement paste and relative humidity of the ambient air can be found in the pioneering work of Powers and Brownyard (1946–47). The shape of the

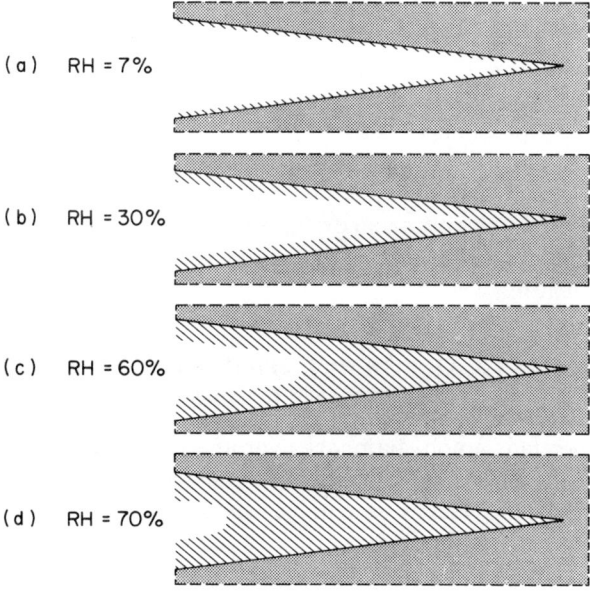

Figure 16.1 Gradual filling of a micropore due to adsorption and capillary condensation. Source Bažant (1972).

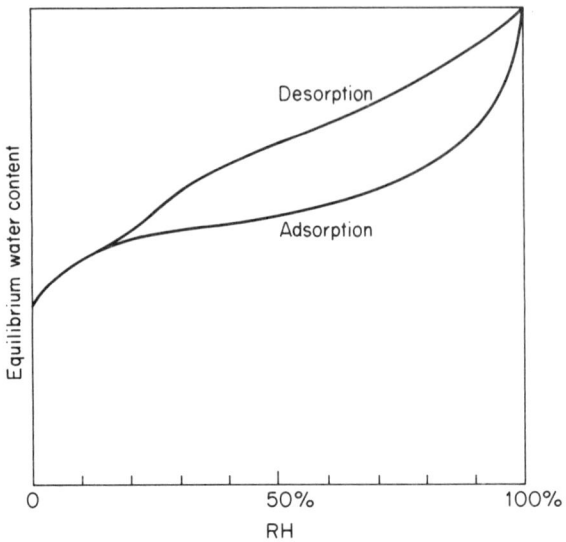

Figure 16.2　Typical sorption isochem of concrete

sorption isotherm depends on whether the material attains its equilibrium by taking up or by losing moisture, that is, by adsorption or desorption. A qualitative picture of sorption isotherms for concrete is sketched in Figure 16.2. It is seen that the adsorption isotherm deviates from that of the preceding desorption. This irreversibility is mainly caused by changes of the porosity of cement paste during drying (Young, 1982). The shape of the sorption isotherm depends on several factors, such as degree of hydration, water–cement ratio and temperature, see, for example, Nilsson (1980). It should perhaps also be noted that in actual concrete structures the material is never completely dried nor does it reach complete saturation; rather, the moisture content at a given instant is a result of successive cycles of partial drying and rewetting. Hence, the equilibrium water content becomes a function of the preceding history of moisture movement within the structure.

16.3 MATHEMATICAL FORMULATION OF DIFFUSION OF PORE WATER IN CONCRETE

Drying of concrete is a non-stationary hygrothermal process taking place in a chemically unstable material. It is generally accepted that the theory of diffusion (Welty *et al.*, 1969) provides an acceptable means of describing moisture transfer in concrete. However, it should be borne in mind that transport of moisture in a material with a complex structure like concrete is likely to take place by a number of different mechanisms. Consequently, diffusion theory can only represent an approximation to the true behaviour.

Effect of Pore Water and Its Diffusion

In general, mass transfer in porous materials can be described by a linear relationship between the flux of the diffusing substance and the potential gradient, which represents the driving force. Thus,

$$\mathbf{J} = -\tilde{c}\,\mathrm{grad}\,\mu \tag{16.1}$$

where $\mathbf{J}$ represents the mass flux per unit area and time, μ is the potential and $\tilde{c}$ a transfer coefficient characterizing the diffusion properties of the material. Although the above relation is empirical, it remains valid provided that the potential gradient is sufficiently small.

The second building block in the mathematical formulation of mass transfer is the conservation of mass, which reads,

$$\frac{\partial \mu}{\partial t} + \mathrm{div}\,\mathbf{J} + S = 0 \tag{16.2}$$

Here, t denotes time and S is a source of sink term.

Generally, moisture transfer involves simultaneous transport of energy, and in some cases transfer of momentum may also be involved. These are coupled processes that require the simultaneous treatment of each transport phenomenon involved. However, under certain simplifying assumptions, diffusion of pore water in concrete can be considered independent of other transfer processes.

16.3.1 Isothermal drying of non-saturated concrete

16.3.1.1 Linear diffusion theory

In mature concrete (negligible rate of hydration) under isothermal conditions, the driving force in the moisture diffusion may be taken as the moisture concentration gradient. Letting w denote the specific mass of evaporable water (mass per unit volume), Equation (16.1) becomes

$$\mathbf{J} = -C\,\mathrm{grad}\,w \tag{16.3}$$

where C is the diffusivity (or diffusion coefficient). This relation is known as Fick's first law. Neglecting the source term, the balance of mass equation (16.2) reduces to

$$\frac{\partial w}{\partial t} + \mathrm{div}\,\mathbf{J} = 0 \tag{16.4}$$

The differential equation governing diffusion of moisture in concrete then directly follows by combination of Equations (16.3) and (16.4),

$$\frac{\partial w}{\partial t} + \mathrm{div}(C\,\mathrm{grad}\,w) \tag{16.5}$$

For constant diffusivity, i.e. $C = C_0 =$ const., the above equation reduces to a second order, linear differential equation of the form:

$$\frac{\partial w}{\partial t} = C_0 \nabla^2 w \qquad (16.6)$$

where ∇^2 is the Laplacian operator.

Because the theory of diffusion can be applied to a number of physically different transfer phenomena the solution of the above differential equation can be obtained by well-documented procedures such as Fourier series, finite difference and finite element methods (Cranck, 1975).

Equation (16.6), generally referred to as Fick's second law of diffusion, has been preferred in most of the literature on drying of concrete. It was first used by Carlson (1937) in the 1930s. Later the linear diffusion equation (16.6) has been adopted in a number of investigations (Pickett, 1946; Hansen and Mattock, 1966; Becker and MacInnis, 1973; Hughes et al., 1966; Hancox, 1968) dealing with drying and shrinkage of concrete. The diffusion coefficients reported in these studies vary from approximately 10^{-10} to 10^{-12} m^2/s, which is a considerable difference.

An alternative formulation of the diffusion of pore water in concrete is possible. The potential μ in Equation (16.1) may be taken as the Gibbs free energy per unit mass of evaporable water. Assuming that water vapour obeys the ideal gas equation, the potential of water can be expressed as

$$\mu = \frac{R}{M} T \ln h + \mu_{\text{sat}} \qquad (16.7)$$

where R is the gas constant, M is the molecular weight of water and T is absolute temperature. The relative pore humidity h is defined by

$$h = \frac{p}{p_{\text{sat}} T} \qquad (16.8)$$

in which p is pore pressure and p_{sat} pore pressure at saturation, depending on temperature. Similarly, $\mu_{\text{sat}}(T)$ represents the value of Gibbs free energy corresponding to full saturation ($h = 1.0$). Assuming grad T to be negligible, the water flux can be expressed as

$$\mathbf{J} = -\bar{a} \operatorname{grad} h \qquad (16.9)$$

where $\bar{a}$ denotes permeability of concrete.

Diffusion of pore water in concrete is an extremely slow process and it seems reasonable to assume that thermodynamic equilibrium exists between the various phases of water within each macropore. Thus, the relation between w and h as given by the desorption isotherm (Figure 16.2) may be taken as valid for the moisture diffusion during drying of concrete. For mature concrete and constant temperature, one may write

$$\frac{\partial w}{\partial h} = \frac{1}{k(h)} \quad \text{or} \quad dh = k(h)dw \tag{16.10}$$

in which $k(h)$ is the inverse slope of the desorption isotherm $w = w(h)$, see Figure 16.2. Studies of experimental data (Powers and Brownyard, 1946–47; Nilsson, 1980) on desorption for constant temperature and water–cement ratio, reveal that the variation of the coefficient k is not too severe, except for the very high and very small values of h, which are outside the range of interest for ordinary drying. Thus, without any significant loss of accuracy, k may be taken as constant within the interval $0.90 \geqslant h \geqslant 0.25$.

Differentiating Equation (16.10) with respect to time and substituting the result into Equations (16.4) and (16.9) leads to the following differential equation:

$$\frac{\partial h}{\partial t} = \text{div}(C \text{ grad } h) \tag{16.11}$$

where diffusivity C is given by

$$C = k\bar{a} \tag{16.12}$$

Equation (16.11) represents the differential equation, in terms of h, governing drying of concrete under isothermal conditions and was developed by Bažant and Najjar (1971, 1972). If the assumption of constant diffusivity (i.e. $C = C_0$) is invoked, it reduces to the linear form equivalent to Equation (16.6). It is to be noted that the form of the diffusion equation given by Equation (16.11) was obtained by assuming that the coefficient $k = \text{const.}$, and hence this equation should not be applied in situations where this assumption is not justified. For instance, Equation (16.11) cannot be directly applied to rewetting of concrete since the slope of the adsorption isotherm deviates significantly from that during desorption.

If the boundary conditions at the surface of a drying body is that of perfect moisture transfer, this is easily implemented by requiring that

$$h = h_e \tag{16.13}$$

with h_e being the environmental humidity at the surface. For completely sealed surfaces, the normal flux of water J_n vanishes,

$$J_n = \mathbf{n} \cdot \mathbf{J} = 0 \tag{16.14}$$

where $\mathbf{n}$ is the unit outward surface normal.

The two formulations of drying defined by Equations (16.5) and (16.11) are equivalent provided that change of material parameters due to hydration is neglected. The formulation in terms of pore humidity makes it easier to incorporate the effect of aging of concrete because h is directly related to Gibbs free energy μ, cf. Equation (16.7), which is not the case with w, see Bažant and Najjar (1972). Moreover, as we have already seen, boundary conditions are most conveniently defined in terms of relative humidity. Constitutive relations for creep and shrinkage depend on pore humidity, so that the distribution of h has to be known before the time-dependent deformations can be computed. For these reasons, the differential equation for drying in terms of h, Equation (16.12), seems to have distinct advantages compared to the formulation in terms of w as defined by Equation (16.5). It should be noted that according to the definition of pore humidity (16.8) a formulation in terms of pore pressure p would be equivalent to that in terms of h.

In developing Equation (16.11) it has been tacitly assumed that concrete may be treated as an isotropic, homogeneous material. Looking at the microstructure of concrete, this certainly seems an oversimplification. Moreover, it is obvious that in concrete structures some parts will exhibit considerable anisotropy due to the casting methods employed. Hence, in a rigorous formulation diffusivity should be treated as a tensorial quantity. However, there is no experimental data available supporting such a formulation, and, in view of this, the use of a single scalar parameter to characterize diffusivity of concrete seems completely adequate.

The speed with which pore water migrates in concrete depends on the structure and distribution of pores in cement paste. Thus, diffusivity is a function of water–cement ratio and the degree of maturity, as evidenced by laboratory test.

16.3.1.2 Effect of hydration

In fresh concrete, the advancing hydration calls for certain modifications of the diffusion equation. During hydration a certain amount of evaporable water is consumed by the chemical reactions. This effect can be accounted for by adding a correction term to Equation (16.10) so that the relation between pore humidity and water content becomes

$$dh = k\,dw + dh_s \qquad (16.15)$$

where dh_s represents the drop in pore humidity caused by hydration. The function $h_s(t)$ thus represents the so-called self-desiccation taking place in a sealed specimen. The direct effect of hydration on pore humidity is relatively small since it is somewhat counterbalanced by the simultaneous reduction in porosity.

It is clear that self-desiccation, permeability and the slope of the desorption isotherm must depend on the degree of hydration. An objective measure of the maturity or degree of hydration of cement paste is the so-called equivalent curing period t_e, defined by Bažant (1975a)

$$t_e = \int dt_e = \int \beta_T(T) \cdot \beta_h(h)\,dt \qquad (16.16)$$

where β_T is assumed to obey the Arrhenius equation,

$$\beta_T = \exp\left[\frac{U_h}{R}\left(\frac{1}{T_0} - \frac{1}{T}\right)\right] \quad (16.17)$$

in which U_h is the activation energy of hydration and T_0 is a chosen reference value of the current absolute temperature T.

The effect of humidity on the equivalent curing period can be approximated by an empirical formula of the form

$$\beta_h = \{1 + [A(1-h)]^N\}^{-1} \quad (16.18)$$

which is shown graphically in Figure 16.3 for two different sets of constants, A and N.

The actual functional dependence of h_s, $\bar{a}$ and k on t_e is not dealt with here, but it is known from experiments that the effect of curing time on permeability can be severe. It should also be noted that drying inevitably causes the distribution of t_e to be non-uniform throughout the body, cf. the definition of β_h (16.18).

16.3.1.3 Effect of temperature

The dependence of pore humidity on temperature can be accounted for by augmenting Equation (16.15) by one additional term,

$$dh = k\,dw + dh_s + \kappa\,dT \quad (16.19)$$

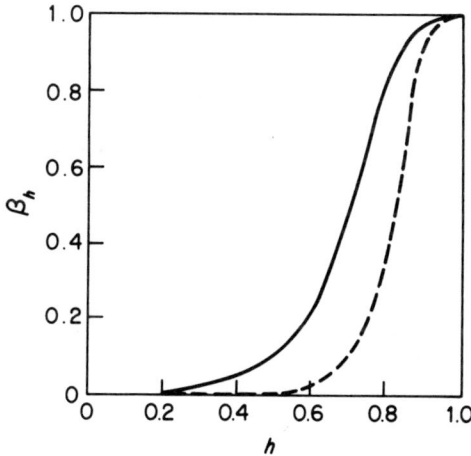

Figure 16.3 Dependence of β_h on pore humidity h according to Equation (16.18). Solid line: $A = 3.5$, $N = 4$ (ASCE, 1982); dashed line: $A = 6.0$, $N = 4$ (Bažant, 1975a)

where κ is the hygrothermal coefficient,

$$\kappa = \left(\frac{\partial h}{\partial T}\right)_{w, t_e = \text{const.}} \tag{16.20}$$

By definition, κ represents the change in h due to one degree change in temperature when w and t_e are kept constant. The variation of κ with h is shown graphically in Figure 16.4.

The differential equation governing drying of concrete, including the effects of advancing hydration and variable temperature, then becomes

$$\frac{\partial h}{\partial t} = k \, \text{div}\,(\bar{a} \, \text{grad}\, h) + \frac{\partial h_s}{\partial t} + \kappa \frac{\partial T}{\partial t} \tag{16.21}$$

where k and $\bar{a}$ should not be combined in one single term (i.e. diffusivity) because the assumption that the slope of the desorption isotherm is approximately constant may no longer be justified. Moreover, this form of the diffusion equation, which is due to Bažant and Najjar (1972), allows k and $\bar{a}$ to be directly determined from experimental data.

It is to be noted that this equation is limited to slowly varying temperatures since grad $T \approx 0$ has been assumed in the derivation of Equation (16.21). Otherwise, the temperature gradient must be included as an additional driving force in the transfer of moisture (Argyris et al., 1977a). This leads to a coupling of heat and moisture transfer and will be dealt with in Section 16.4.

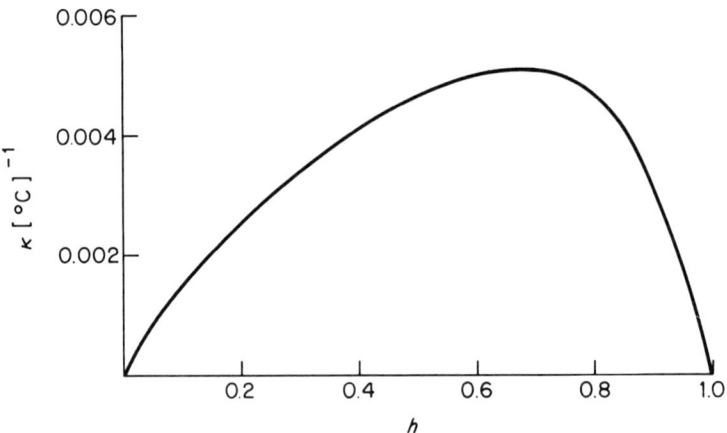

Figure 16.4 Hygrothermal coefficient κ. Source Bažant (1969)

Effect of Pore Water and Its Diffusion

16.3.1.4 Nonlinear diffusion theory

As a means of obtaining approximate solutions, the linear diffusion theory with constant diffusivity may be adequate for a great number of practical problems. Still, this simplified approach suffers from serious shortcomings. Assuming a constant value for the diffusion coefficient is contradictory to experimental evidence on the drying behaviour of concrete. Laboratory tests clearly demonstrate that in an initially fully saturated specimen subjected to drying, moisture is lost with increasing difficulty as drying progresses. Thus, diffusivity should decrease with decreasing values of pore humidity (or specific water content). This fact was noted already in the earliest theoretical investigations of drying, see, for example, Carlsson (1937). Later, the dependence of moisture conductivity on specific moisture content has been thoroughly discussed by Pihlajavaara (1963, 1965), Pihlajavaara and Kasi (1969) and Pihlajavaara and Väisänen (1965), who also carried out numerical studies based on differential equation (16.5) with diffusivity taken as a linear function of moisture content w. Recently, a more realistic functional relationship for $C(w)$ has been suggested by Sakata (1983).

The alternative nonlinear formulation in terms of pore humidity (16.12) has been explored by Bažant and Najjar (1971, 1972). This approach requires the diffusivity $C = C(h)$ to be determined by fitting of test data. However, it seems that most experimental data on drying are in the form of measured weight losses and only for a very limited number of tests has the distribution of pore humidity at various stages of drying been reported. From fits of relevant data it was concluded by Bažant and Najjar (1971, 1972) that the following relationship provided acceptable agreement with experiments:

$$C(h) = C_1 \left[\alpha_0 + \frac{1 - \alpha_0}{1 + \left(\frac{1-h}{1-h_c}\right)^n} \right] \qquad (16.22)$$

where C_1, α_0, h_c and n are constants. Bažant and Najjar (1972) found that most of the available data could be fitted by setting $\alpha_0 = 0.05$, $h_c = 0.75$ and $n = 16$. The same formula with $n = 3$ was adopted by Argyris et al. (1977a). The coefficient C_1 represents the value of diffusivity at $h = 1.0$ and is a function of temperature T and maturity t_e. A semi-empirical formula for $C_1(T, t_e)$ based on the activation energy concept has been suggested by Bažant (1975a).

The dependence of diffusivity on h is shown graphically in Figure 16.5 which also illustrates the meaning of α_0 and h_c, and the effect of varying the value of the exponent n. It is evident from Figure 16.5 that the highly nonlinear relationship between C and h may cause serious difficulties for the numerical solution of the diffusion equation (16.11). Thus, extreme care should be exercised in selecting an appropriate solution algorithm as discussed by Argyris et al. (1977a).

A plausible explanation of the severe reduction in diffusivity between $h = 0.8$

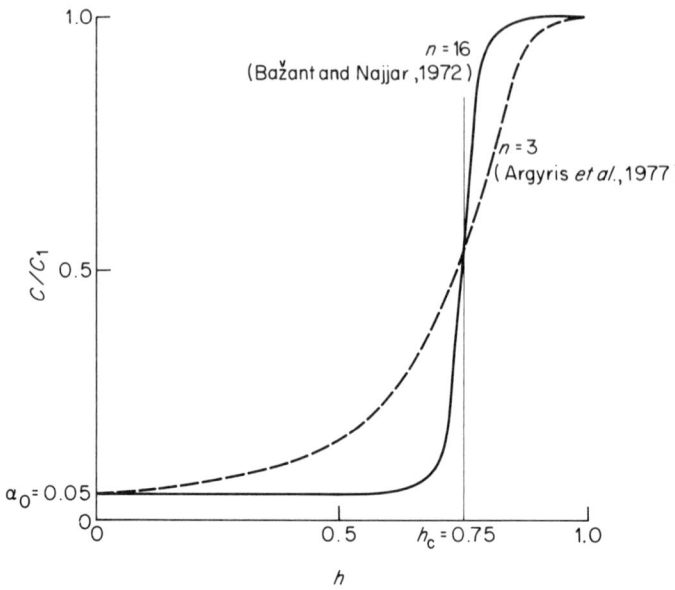

Figure 16.5   Diffusivity, C, as a function of humidity, b, according to Equation (16.22).

and $h = 0.6$ has been suggested by Bažant and Najjar (1972). At low humidities ($h < 0.6$) the migration of water molecules along the adsorbed layers is considered to be the dominant factor in the diffusion process. For humidities above $h = 0.8$ the flow of capillary water is the most important factor in the moisture transport. The interval between $h = 0.6$ and $h = 0.8$ may then be thought of as a transition zone in which flow of the less firmly held molecules in the outer layers is gradually replaced by flow of capillary water. The modelling of the transition zone is strongly dependent on the value of the exponent n, as can be observed from Figure 16.5.

In the works of Bažant and Najjar (1971, 1972) solutions obtained by the nonlinear diffusion equation (16.11) with C defined by Equation (16.22) were compared with test data. Figure 16.6 shows some results for pore humidity in the centre of a slab subjected to drying in various environmental humidities. The experimental values plotted in Figure 16.6 were reported by Abrams and Orals (1965). Also shown for comparisons are the results obtained by setting $C = C_0$ = const.

Figure 16.7 shows the distribution of pore humidity at different times for the same experiments as in Figure 16.6. The nonlinear diffusion equation is in relatively good agreement with the test data while severe discrepancies can be observed for the results obtained from the use of a constant diffusion coefficient. Although the numerical studies reported by Bažant and Najjar (1971, 1972) are limited in number, it may be concluded that prediction of drying of concrete

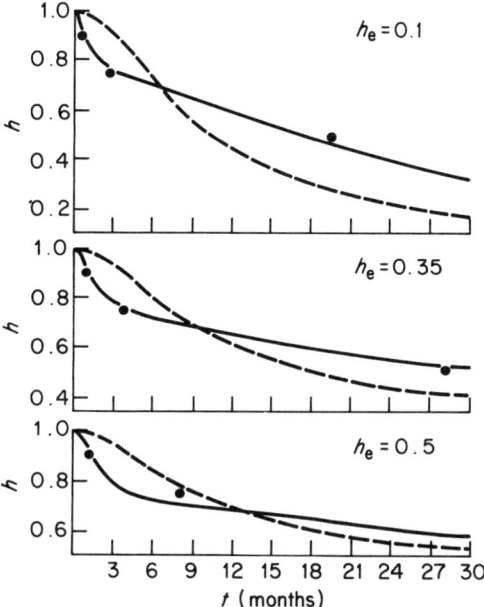

Figure 16.6 Pore humidities in centre of drying slab. Dots—experiments; solid line—nonlinear diffusion equation; dashed line—linear diffusion equation. Source Bažant and Najjar (1971)

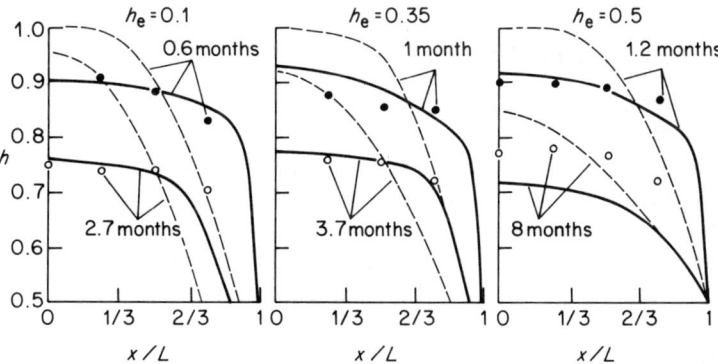

Figure 16.7 Distribution of pore humidities in drying slab. Dots—experiments; solid line—nonlinear diffusion equation; dashed line—linear diffusion equation. Source Bažant and Najjar (1971)

based on nonlinear diffusion theory gives better correlation with experimental data than does the simplified linear diffusion equation.

16.3.2 Saturated concrete

So far, only diffusion of pore water in non-saturated concrete has been considered. The differential equation governing diffusion in completely saturated concrete is similar to Equation (16.11) and can be written in the form

$$\frac{\partial p}{\partial t} = C_{sat} \nabla^2 p \qquad (16.23)$$

where p is hydraulic pressure in excess of atmospheric pressure and C_{sat} denotes diffusivity of pore water in saturated concrete.

It is possible to calculate values of diffusivity in saturated concrete from measured data on permeability. This conversion has been carried out by Murata (1965) who reported values of C_{sat} ranging from 10^{-8} to $10^{-4} m^2/s$. Compared to the normal diffusivities in non-saturated concrete these values are several orders of magnitude higher. Hence, for structures like concrete dams or concrete offshore platforms, in which both saturated and non-saturated regions may exist, a discontinuous jump between C and C_{sat} is to be expected at the interface between saturated and non-saturated concrete.

Another important aspect in massive concrete structures is the water deficiency created by hydration which leads to considerably smaller values of diffusivity than that associated with a saturated state. This can be attributed to non-saturated conditions due to air-filled voids produced by self-desiccation before the arrival of the hydraulic front. It should be noted that the use of air-entraining admixtures reinforces this effect. The diffusion equation (16.23) can be easily modified to account for the influence of self-desiccation, as discussed by Bažant (1975b).

16.4 COMBINED HEAT AND MOISTURE TRANSFER

Heating of concrete will cause increased pressures in the pores of the material, which in turn causes migration of moisture and eventual drying. At the same time, moisture transport influences heat transfer and, hence, temperature distribution throughout the body. Moreover, the existence of temperature gradients will act as a driving force of moisture transfer in addition to the gradient of moisture concentration already discussed in Section 16.3. This leads to a coupling of the fluxes of heat and mass with the gradients of temperature and moisture concentration.

Combined heat and mass transfer in concrete (and its effects of time-dependent deformations) influences structural behaviour of a variety of modern concrete structures, although this is often neglected in stress analysis and design. Of particular interest is the effect of high temperatures on the mechanical behaviour of concrete in modern prestressed nuclear reactor vessels. Another problem of great concern for these structures is the radiation shielding capacity which is strongly dependent on moisture content.

16.4.1 Material properties

Research on high-temperature effects in concrete has traditionally been focused on evaluating fire resistance of concrete structures. A major portion of this research effort has been empirical and directed towards the study of the behaviour of structural elements rather than fundamental material properties. Still, the available experimental information on high-temperature behaviour of concrete is by no means insignificant, as indicated by the recent survey (Bažant et al., 1982). A complete review of the influence of temperature on material properties is beyond the scope of this paper and the reader is referred to Bažant et al., (1982) for details and numerous references. In the following, only a few of the more fundamental aspects of high-temperature behaviour will be discussed.

It is well known that an increase of temperature accelerates drying. At room temperature it may take years for a cylindrical specimen of 150 mm diameter to dry completely while the drying time at 100°C is less than a day.

Another important effect of high temperatures is the gradual dehydration of cement paste. As temperature is raised beyond 120°C, both cement gel and calcium hydroxide are continuously decomposed and the chemically bound water is released into the pores. Dehydration increases with temperature and becomes complete at temperatures of approximately 600–850°C (Harmathy, 1970; Fischer, 1970). The chemical decomposition gradually destroys the microstructure which in turn leads to increased porosity. Harmathy (1970) reports that the porosity growth from 105°C to 850°C can be as much as 40 per cent.

It is to be noted that chemical changes take place in the aggregates as well. However, for ordinary aggregates this effect can be neglected for $T < 500°C$.

For the analysis of heat and mass transfer, information on heat capacity and thermal conductivity is needed. Owing to the presence of various constituents in concrete (cement paste, aggregates, water, air) and the phase changes that may take place in pore water, the dependence of heat capacity and thermal conductivity on temperature and moisture content is a rather complex problem with little experimental data available. The lack of quantitative test data also holds true for moisture conductivity and permeability, although these quantities can be assumed to obey the activation energy concept up to 100°C. The accelerated drying observed for temperatures above 100°C indicates a jump upwards in diffusivity as 100°C is exceeded. This may be attributed to a possible change in the mode of moisture transfer, from migration along adsorbed layers below 100°C to flow of vaporized steam at elevated temperatures (Bažant et al., 1982).

16.4.2 Analysis of heat and mass transfer

When concrete is subjected to high temperatures in a drying environment, one faces the problem of coupled heat and mass transfer in a porous body undergoing microstructural and chemical changes. The general mathematical formulation of such problems is well-established within the theory of irreversible thermodynamics (de Groot and Mazur, 1962). The influence of deformations on the fluxes of moisture and heat can be neglected, which, of course, simplifies the problem considerably. When the distribution of temperature and moisture in the body has been obtained, the resulting stresses can be calculated from the assumed constitutive relation which is a function of current values of temperature and pore humidity, and their history.

It appears that significant developments in the theory of thermal moisture transfer in porous solids have taken place in the Soviet Union. Unfortunately, very little of the results of this research has been published in Western literature, except the book by Luikov (1966). A summary of the theoretical work by Soviet scientists in this area has been prepared by the same author (Luikov, 1975). With the introduction of powerful discretization procedures such as the finite element method the numerical solution of complex problems in heat and mass transfer has been made feasible. This includes applications to transient heat conduction (Visser, 1965; Wilson and Nickell, 1966; Argyris et al., 1971; Emery and Carson, 1971; Warzee, 1974) as well as numerical modelling of drying of porous bodies (Comini and Lewis, 1968; Lewis et al., 1977; Lewis et al., 1979).

As already stated, the heat flux $\mathbf{q}$ depends not only on the temperature gradient but also on the gradient of moisture concentration w (Dufour effect). Similarly, the flux of moisture $\mathbf{J}$ is linearly related to grad w, and, in addition, $\mathbf{J}$ is a function of the temperature gradient (Soret effect). This can be written in the form

$$\mathbf{J} = -d_{11} \operatorname{grad} w - d_{12} \operatorname{grad} T$$
$$\mathbf{q} = -d_{21} \operatorname{grad} w - d_{22} \operatorname{grad} T \qquad (16.24)$$

where w is to be interpreted as the mass of all free (not chemically bound) water per unit volume of concrete. The problem is nonlinear owing to the dependence of the coefficients $d_{ij}(i, j = 1, 2)$ on moisture concentration, temperature and degree of hydration. Moreover, $d_{12} \neq d_{21}$ because grad w and grad T are not the thermodynamic driving forces associated with the fluxes $\mathbf{J}$ and $\mathbf{q}$. cf. de Groot and Mazur (1962).

Equation (16.24) leads to a system of coupled partial differential equations whose solution can only be obtained by sophisticated numerical procedures. The numerical solution is in itself a formidable task. Another difficulty stems from the identification of material parameters from test data. Given the limited experimental information available, it seems necessary to explore the possibility of simplifying Equation (16.24).

Effect of Pore Water and Its Diffusion

One such simplification can be achieved if the moisture flux is governed by a single potential $\Psi(w, T)$. Bažant (1975c) has suggested that this potential be taken as the pore water pressure p, i.e.

$$\mathbf{J} = -\frac{a}{g}\operatorname{grad} p \qquad (16.25)$$

where a is the permeability (in m/s). This is equivalent to the relation used in Section 16.3, see Equation (16.9), the gravity acceleration $g = 9.806 \text{ m/s}^2$ having been included for reasons of dimensionality only.

The Dufour effect is usually of little significance for the heat flux; hence, $d_{21} \approx 0$, and consequently,

$$\mathbf{q} = -b \operatorname{grad} T \qquad (16.26)$$

where the thermal conductivity b has been substituted for the coefficient d_{22} in Equation (16.24).

The simplifications imbedded in Equations (16.25) and (16.26) may seem drastic, and, in particular, neglecting the effect of the temperature gradient on the moisture flux is not generally justified. However, the effect of grad T is to some extent implicitly included in Equation (16.25), because we have

$$\operatorname{grad} p(w, T) = \frac{\partial p}{\partial w} \operatorname{grad} w + \frac{\partial p}{\partial T} \operatorname{grad} T$$

Formulation of the conservation laws requires careful consideration of the physical processes involved. As already discussed, dehydration of cement paste for temperatures beyond 120°C implies that chemically bound water is released into the pores. This phenomenon must be accounted for in the conservation of mass equation, which, according to Bažant and Thonguthai (1978, 1979), can be expressed as

$$\frac{\partial w}{\partial t} + \operatorname{div} \mathbf{J} - \frac{\partial w_d}{\partial t} = 0 \qquad (16.27)$$

where w_d is the mass of water released by the dehydration process.

The heat capacity of a multicomponent and multiphase system like concrete is not a single, well-defined quantity. However, a rigorous formulation in terms of contributions from the various components (solid microstructure, adsorbed water, capillary water, vapour) seems unnecessarily complex in view of the limited experimental data presently available. Thus, it may be justified to consider the average specific heat capacity c of concrete as a function of temperature only. The balance of heat equation then becomes (Bažant and Thonguthai, 1978, 1979)

$$\rho c \frac{\partial T}{\partial t} - c_w \mathbf{J} \cdot \operatorname{grad} T = -\operatorname{div} \mathbf{q} \qquad (16.28)$$

in which ρ is the mass density of concrete and c_w is the heat capacity of liquid water. The second term on the left-hand side of the above equation represents the heat supply due to moving water and may be neglected except for situations where rapid heating occurs.

To complete the formulation, a constitutive relation between pore pressure, p, water content, w, and temperature, T, in concrete is needed. An approximate, semi-empirical formula has been suggested by Bažant and Thonguthai (1978, 1979). This relation is shown diagrammatically in Figure 16.8 where the ratio of free, evaporable water content to cement content versus relative pore pressure is plotted. Below the critical point of water (374.15°C) one has to distinguish between unsaturated and saturated concrete. In Figure 16.8 the transition from saturated to nonsaturated state is modelled by a straight line.

In general, permeability is a function of temperature and humidity, i.e. $a = a(h, T)$. Below 100°C the dependence of permeability on temperature may be assumed to be governed by activation energy. Experimental observations (Chapman and England, 1977) indicate that permeability increases approximately by two orders of magnitude for temperatures above 100°C. Below the boiling point of water permeability (like diffusivity, cf. Section 16.3) decreases with decreasing relative pore humidity. Detailed expressions defining the approximate dependence of permeability on T and h can be found in the papers by Bažant and Thonguthai (1978, 1979) and only a graphical representation is included here; see Figure 16.9.

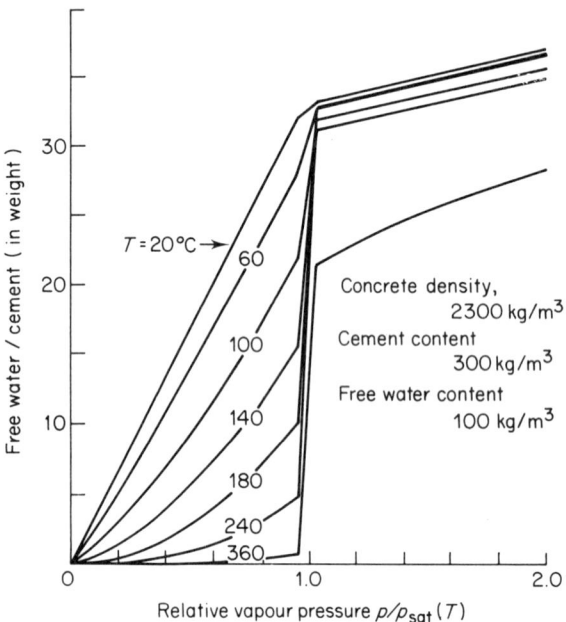

Figure 16.8 Relation between free water, w, pore pressure, p, and temperature, T. Source Bažant and Thonguthai (1978, 1979)

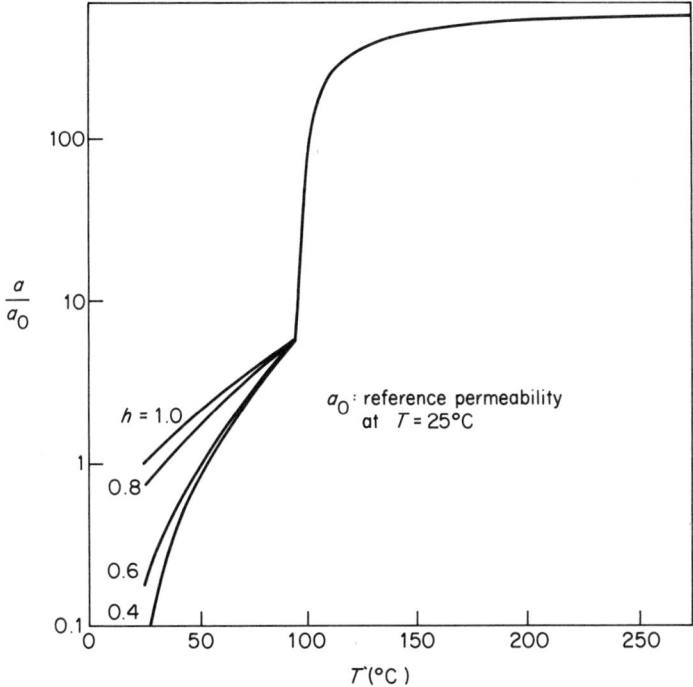

Figure 16.9 Permeability as a function of temperature and humidity. Source Bažant and Thonguthai (1978, 1979)

Spatial discretization of the field equations (16.25)–(16.28) can be performed by means of a finite element formulation of the Galerkin procedure. Within each discrete element, pore pressure and temperature are interpolated from a set of nodal values. The discretized algebraic equations of the finite element assemblage, which are computed as a sum of contributions from individual elements, can be written in the following general form;

$$\begin{bmatrix} a_{11} & a_{12} \\ a_{21} & a_{22} \end{bmatrix} \begin{bmatrix} \mathbf{p} \\ \mathbf{T} \end{bmatrix} + \begin{bmatrix} b_{11} & b_{12} \\ b_{21} & b_{22} \end{bmatrix} \begin{bmatrix} \dot{\mathbf{p}} \\ \dot{\mathbf{T}} \end{bmatrix} = \begin{bmatrix} \mathbf{h}_1 \\ \mathbf{h}_2 \end{bmatrix} \quad (16.29)$$

Here, a dot denotes differentiation with respect to time and $\mathbf{p}$ and $\mathbf{T}$ are nodal values of pore pressure and temperature, respectively. It is to be noted that the above system of equations is in general non-symmetric. This is due to the lack of symmetry in the field equations (16.25) and (16.26).

In the time domain, the differential equations (16.29) must be solved by a step-by-step procedure. In the literature, both explicit and implicit algorithms have been adopted for the numerical solution of combined heat and mass transfer in concrete. Lack of convergence and stability have been reported, especially in cases which involve transition regions between saturated and nonsaturated concrete, where the slope of the w versus p-diagram undergoes sharp changes (Figure 16.8).

Several computer programs for predicting water release and pore pressures in heated concrete already exist (Bažant and Thonguthai, 1978, 1979; Dayan, 1977; McCormack et al., 1979; Knight and Beck, 1979). These programs are essentially similar as far as the mathematical formulation of the major mechanisms controlling heat and mass transfer are concerned but differ in the choice of material parameters and their dependence on temperature and moisture content. Differences also exist between the above programs in the selection of the spatial discretization method (finite differences, finite elements) as well as numerical solution procedures. Comparisons of the computer programs with test data have been performed (Chen et al., 1980; Bažant et al., 1981a). Such evaluations are, however, somewhat inhibited by the relatively small amount of relevant experimental information available (Bažant and Thonguthai, 1978; McCormack et al., 1979; England and Ross, 1972; England and Sharp, 1971; Chapman and England, 1977). There seem to be considerable differences in the predictions of temperature, water release, and pore pressure between the above computer programs as well as in relation to test data. The closest agreement is achieved for temperature distributions and total water release. The calculated pore pressures, on the other hand, are relatively poor and leave room for considerable improvements.

16.5 TIME-DEPENDENT DEFORMATIONS

16.5.1 Creep and shrinkage

Under sustained loading, deformation of concrete increases with time. This phenomenon is known as creep and the final value of creep strain can be several times as large as the instantaneous elastic strain upon loading. In addition, concrete exhibits shrinkage (or swelling) which is a stress-independent deformation due to change in water content. In laboratory tests, creep is taken as the difference between the total deformation at a given time of a loaded specimen and the deformation of an 'identical', unloaded companion specimen stored under the same conditions during the same period of time.

Interest in creep and shrinkage has been increasing and considerable progress has been achieved in the understanding of these phenomena. However, in spite of the extensive body of literature on this subject, many basic questions regarding the constitutive modelling of creep and shrinkage remain unsettled. Undoubtedly, this is due to the complexity of concrete as an engineering material. In particular, the strong interaction with environmental conditions (temperature and humidity) which causes migration of pore water within the material, is one of the major unresolved problems in the theory of creep and shrinkage. The survey papers (Bažant, 1975a, 1982) provide an excellent insight into the complexities

Effect of Pore Water and Its Diffusion   369

involved in the development of constitutive relations for time-dependent deformations of concrete. Effects of variable temperature and humidity on creep and shrinkage is dealt with in detail by Anderson (Chapter 18 in this volume) and Wittmann (Chapter 19 in this volume) and the subsequent discussion is therefore restricted to a brief examination of the effects of pore water diffusion in existing mathematical models for predicting creep and shrinkage deformations in concrete.

The hydration of cement is accompanied by a volumetric concentration of the system cement plus water. This reduction in volume, which takes place while concrete is still in the plastic state, is commonly referred to as *plastic shrinkage* but comprises a number of different mechanism (Wittmann, 1982). *Drying shrinkage* takes place in the hardened cement paste and is caused by a removal of water from the pores of concrete due to diffusion of moisture from the interior towards the drying surface. As outlined in Section 16.3, this is an extremely slow process provided severe cracking does not occur. It can also be deduced from the discussions in Section 16.3 that drying shrinkage increases with decreasing environmental humidity and decreases when age at the start of drying is higher.

Creep of concrete under sealed conditions (no exchange of moisture with the environment) and at constant room temperature is usually referred to as *basic creep*. Owing to hydration, creep deformations are reduced with increasing age at loading. Moreover, for stresses below approximately 40 per cent of the compressive strength, creep of concrete obeys the principle of superposition, which means that the strain due to a sum of stress histories is equal to the sum of the responses due to the individual stress histories. When concrete under sustained load is allowed to dry, creep is accelerated. This is the so-called *drying creep* effect. Drying creep increases with decreasing environmental humidity and is accelerated by rapid changes in water content. Creep deformation of concrete under sustained temperature increases as the temperature level is raised (*thermal creep*). It is also worth noting that temperature cycling leads to a sharp increase in the creep deformation rate.

16.5.2 Simplified prediction methods

Most of the existing mathematical models for creep and shrinkage consist of simple algebraic formulas whose main area of application is in design of ordinary concrete structures (Bažant and Panula, 1978–79; ACI Committee 209, 1982; CEB-FIP, 1978; Rüsch et al., 1975). These formulas, which are usually semi-empirical, are capable of predicting average specimen or cross-sectional behaviour. As far as the hygrothermal effects are concerned, approximate solutions of the nonlinear diffusion equation (16.11) are to some extent incorporated in these simplified algebraic expressions. It is thus important that they are not confused with constitutive relations which are associated with local behaviour (i.e. a point) of the material. It is also clear that many of the suggested formulas for predicting creep and shrinkage deformations do not comply with the principles of invariance and objectivity on which valid constitutive relations must be based.

In a recent comparison (Müller and Hilsdorf, 1982) of some of the suggested simplified formulas with test data, it was found that these prediction methods are rather crude. Although, this may in part be attributed to the great statistical scatter of available test data, it is felt that the accuracy of the prediction methods needs to be improved.

16.5.3 Constitutive relations

16.5.3.1 Non-uniform specimen behaviour

A huge body of test data on creep and shrinkage of concrete has been published, especially during the last decade. Yet, this vast experimental information cannot be directly used as a basis for development of a quantitative theoretical model. The main reason for this is that laboratory specimens are not in a uniform state of stress during the test. The outer part of the specimen will lose moisture at a higher rate than the central core, thereby producing non-uniform distributions of pore humidity and shrinkage strains over the cross-section. As a result of the inhomogeneity of concrete, a complex three-dimensional state of stress is set up in the drying, load-free specimen. Depending on the environmental humidity and the duration of drying exposure, tensile stresses in the region closest to the surface may reach the tensile strength of concrete. The drying cracks thus formed cause redistribution of stresses within the specimen which alters the observed average deformation. In a recent investigation, Wittmann and Roelfstra (1980) studied the effect of drying-induced cracking by a simple material model combined with pore humidity distributions determined by solving the diffusion equation (16.11). The results of these computations clearly demonstrate the profound effect of cracking on the time-dependent deformation of specimens subjected to drying.

In creep tests in a drying atmosphere, the external loading superimposes a uniform compressive stress on the non-uniform stresses produced by differential shrinkage deformations. This will reduce crack formation as compared to that in the companion load-free specimen used to determine pure shrinkage deformation. Moreover, in creep tests a uniform cross-sectional deformation is imposed which also causes stress redistributions that do not occur in the shrinkage specimen (Acker, 1982). Hence, the companion 'identical' specimens used in shrinkage and creep experiments are not identical, and consequently the separation of the total deformation into shrinkage and creep components is invalid and only the total deformation can be defined. Moreover, the numerical studies carried out by Wittmann and Roelfstra (1980) indicate that no independent microstructural mechanism may be involved in drying creep, which, at least for a major part, may be explained by crack prevention caused by the applied compressive load.

To avoid formation of drying-induced cracks, environmental humidity must be gradually decreased so that large differences in pore humidity do not occur within the specimen. Calculations based on linear diffusion theory carried out by Bažant and Raftshol (1982) indicate that the required drying rate is too slow to be practically feasible. Alternatively, the gradient in pore humidity can be eliminated by reducing the thickness of the specimen. However, the critical thickness is so small (approximately 1.0 mm (Bažant and Raftshol, 1982; Chatterji *et al.*, 1981)) that specimens have to be made of cement paste rather than concrete to avoid the formation of drying cracks.

16.5.3.2 Creep laws

Creep of concrete at constant humidity and temperature may be treated within the well-established framework of aging viscoelastic materials. This leads to an integral-type creep law in which creep strain is expressed as a functional of the previous history of stress. To avoid storing the entire stress history a rate-type creep law is preferred. To this end, a generalized Maxwell chain model with age-dependent elastic moduli E_μ and viscosities η_μ may be selected, see Figure 16.10. The stress σ_μ in the μth element obeys the differential equation

$$\frac{\dot{\sigma}_\mu}{E_\mu(t)} + \frac{\sigma_\mu}{\eta_\mu(t)} = \dot{\varepsilon} - \dot{\varepsilon}_0; \quad \mu = 1, 2, \ldots, n \tag{16.30}$$

in which ε is the total strain and ε_0 is load-independent strain caused by shrinkage or thermal dilatation. A general procedure for converting the integral-type creep law into the above rate-type formulation is available in the paper by Bažant and Wu (1974a).

A simple extension of the generalized Maxwell chain model to account for variable temperature (but constant humidity, $h = 1, 0$) was suggested by Bažant and Wu (1974b). The effect of aging is included by making the material parameters E_μ and η_μ depend on equivalent hydration period t_e through Equation (16.16), rather than actual time t. Bažant and Wu (1974b) further assumed that the major creep mechanism is the diffusion of solids and water

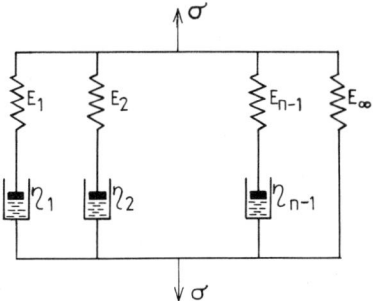

Figure 16.10 Maxwell chain model

along adsorbed layers in the microstructure. These processes can be identified by the dashpots whose viscosities may then be assumed to depend on temperature through the Arrhenius equation, i.e.

$$\frac{1}{\eta_\mu} = \frac{1}{\eta_{\mu 0}} \exp\left[\frac{U_\mu}{R}\left(\frac{1}{T_0} - \frac{1}{T}\right)\right] \qquad (16.31)$$

where U_μ is the activation energy associated with the μth dashpot and $\eta_{\mu 0}$ is a reference value of η_μ at the chosen reference temperature T_0.

In a subsequent paper (Bažant and Wu, 1974c) the same authors extended their creep model to variable humidity. This was done by separating the stress in each unit of the Maxwell chain into two parts, one representing stress σ_μ^s in solids and the other σ_μ^w representing stress in pore water, see Figure 16.11. The uniaxial constitutive relation can then be written in the form

$$\dot{\sigma}_\mu^s + \phi_{ss\mu}\sigma_\mu^s + \phi_{sw\mu}[\sigma_\mu^w - f_\mu] = E_\mu^s(\dot{\varepsilon} - \dot{\varepsilon}_0)$$
$$\dot{\sigma}_\mu^w + \phi_{ws\mu}\sigma_\mu^s + \phi_{ww\mu}[\sigma_\mu^w - f_\mu] = E_\mu^w(\dot{\varepsilon} - \dot{\varepsilon}_0) \qquad (16.32)$$

where the rate coefficients $\phi_{ss\mu}$, $\phi_{sw\mu}$, $\phi_{ws\mu}$, $\phi_{ww\mu}$ ($\mu = 1, 2, ..., n$) characterize microscopic diffusion fluxes of solid and water and are associated with the 'diffusion elements' in Figure 16.11. The instantaneous response is modelled by the springs (with moduli E_μ^s and E_μ^w) and the quantity $f_\mu(h, T)$ represents the value of σ_μ^w in adsorbed water layers in the micropores that corresponds to thermodynamic equilibrium with water in the adjacent capillary pores.

The above constitutive relation can be generalized to multiaxial stress and rewritten in incremental form so that it can be directly incorporated into nonlinear finite element programs (Bažant et al., 1981b). The chief drawback of this formulation is that the number of material parameters is rather high and available experimental information does not, at present, provide a straightforward answer to the material identification problem.

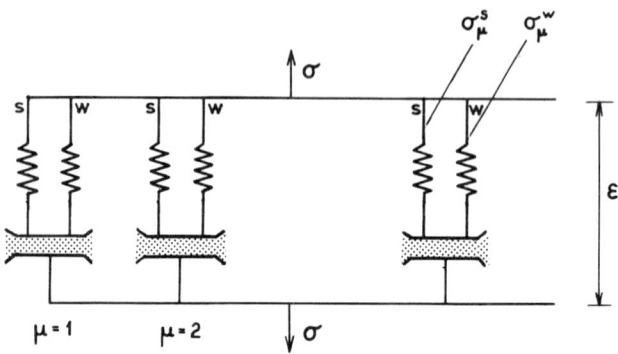

Figure 16.11 Modified Maxwell chain model for variable humidity

Time-dependent deformations of concrete, including hygrothermal effects, constitute a class of problems that can be described within the context of nonlinear thermodynamic theory. Although the physical microstructural mechanisms involved in creep and shrinkage are not yet clarified, these effects can be accounted for by the introduction of internal variables that in an average macroscopic sense represent the effects of the microstructural changes. The internal variable theory is well-established within the theory of nonlinear thermodynamics and provides a rational mathematical structure for constitutive modelling of nonlinear material problems. The adaption of this theory to creep of concrete is due to Argyris and coworkers (Argyris et al., 1976, 1977b; Pister et al., 1978). The fundamental aspects of internal variable theory of concrete creep have thus been founded but further refinement in the form of constitutive assumptions and optimal selection of suitable internal variables as well as their identification from test data, is needed.

16.6 CONCLUDING REMARKS

In concrete, pore water and its diffusion plays an important role in the understanding of time-dependent deformations which are strongly influenced by hygrothermal conditions. Constitutive relations for shrinkage and creep require the spatial distribution of pore humidity to be known *a priori* before the deformation problem can be solved. Thus, the need for accurate and reliable mathematical models for predicting pore humidity is evident. There is ample experimental evidence supporting the use of nonlinear diffusion theory in which diffusivity depends on relative pore humidity, but further test data is required to determine the functional dependence of moisture diffusivity on pore humidity. In particular, the effects of water–cement ratio and maturity need to be studied in more detail.

Heat and moisture transfer in concrete at elevated temperature can be described by standard coupled differential equations in terms of temperature and water content. The introduction of pore pressure as a variable leads to a relatively simple and computationally efficient formulation. Constitutive assumptions regarding the relationship between pore pressure, water content and temperature will require more extensive test data to be completely verified. This is also the case with permeability and its dependence on temperature and pore humidity.

Over the years, various mechanisms have been suggested to explain shrinkage and creep deformation of concrete, but no single mechanism has gained exclusive acceptance. Recently, attention has been focused on the effect of moisture diffusion and associated cracking on the observed time-dependent deformations of concrete specimens. The results of this research have exposed the need for a more careful evaluation of existing test data since drying specimens are in a spatial non-uniform state of pore humidity, strain and stress. This also emphasizes the importance of continued theoretical research since a constitutive theory is necessary for a proper understanding and evaluation of experiments. It is felt that significant progress in the constitutive modelling of creep and shrinkage of concrete can best be achieved by a coordinated effort that unifies experimental research, continuum mechanics and numerical analysis.

REFERENCES

Abrams, M. S., and Orals, D. L. (1965) 'Concrete drying methods and their effect on fire resistance', *Moisture of Materials in Relation to Fire Tests*, American Society for Testing Materials, STP No. 385, 52–73.

ACI Committee 209 (1982) 'Prediction of creep shrinkage, and temperature effects in concrete structures', *Design for Creep and Shrinkage in Concrete Structures*, ACI, SP-76, 193–300.

Acker, P. (1982) 'Drying of concrete: consequences for the evaluation of creep test', *Fundamental Research on Creep and Shrinkage of Concrete* (Ed. F. H. Wittmann), Martinus Nijhoff Publishers, The Hague, pp. 149–169.

Argyris, J. H., Balmer, H., Doltsinis, J. St., and Willam, K. (1971) 'Finite element analysis of thermomechanical problems', *Proceedings of the Third Conference on Matrix Methods in Structural Mechanics*, Wright-Patterson Air Force Base, Dayton, Ohio, AFFDL-TR-71-160, 729–772.

Argyris, J. H., Pister, K. S., and Willam, K. J. (1976) 'Thermomechanical creep of aging concrete—a unified approach', IABSE, *Publ.* 36-1.

Argyris, J. H., Warnke, E. P., and Willam, K. J. (1977a) 'Berechnungen von Temperatur—und Feuchtefeldern in Massivbauten nach der Methode der Finiten Elemente', *Deutscher Ausschuss für Stahlbeton*, Heft 278.

Argyris, J. H., Pister, K. S., Szimmat, J., and Willam, K. J. (1977b) 'Unified concepts of constitutive modelling and numerical solution methods for concrete creep problems', *Computer Methods in Applied Mechanics and Engineering*, **10**, 199–246.

ASCE (1982) 'Finite element analysis of reinforced concrete', State-of-the art report prepared by the Task Committee on Finite Element Analysis of Reinforced Concrete Structures of the Structural Division Committee on Concrete and Masonry Structures, American Society of Civil Engineers.

Bažant, Z. P. (1969) 'Thermodynamic theory of concrete deformation at variable temperature and humidity', Department of Civil Engineering, University of California, Berkeley, *Report no.* 69–11.

Bažant, Z. P. (1972) 'Thermodynamics of interacting continua with surfaces and creep analysis of concrete structures', *Nuclear Engineering and Design*, **20**, 477–505.

Bažant, Z. P. (1975a) 'Theory of creep and shrinkage in concrete structures: a precis of recent developments', *Mechanics Today* (Ed. S. Nemat-Nasser), Pergamon Press, Oxford, Vol. 2, pp. 1–93.

Bažant, Z. P. (1975b) 'Pore pressure, uplift and failure analysis of concrete dams', *Proceedings of the International Symposium on Criteria and Assumptions for Numerical Analysis of Dams*, Swansea, September 8–11, 1975, pp. 782–808.

Bažant, Z. P. (1975c) 'Some questions of material inelasticity and failure in the design of concrete structures for nuclear reactors', Preprints of the *Third International Conference on Structural Mechanics in Reactor Technology*, London, September 1975, Paper No. H1/1.

Bažant, Z. P. (1982) 'Mathematical models for creep and shrinkage of concrete', *Creep and Shrinkage in Concrete Structures* (Ed. Z. P. Bažant and F. H. Wittmann), John Wiley & Sons Ltd., pp. 163–225.

Bažant, Z. P., and Najjar, L. J. (1971) 'Drying of concrete as a nonlinear diffusion problem', *Cement and Concrete Research*, **1**, 461–473.
Bažant, Z. P., and Najjar, L. J. (1972) 'Nonlinear water diffusion in nonsaturated concrete', *Materials and Structures (RILEM, Paris)*, **5**, 25, 3–20.
Bažant, Z. P., and Panula, L. (1978–79) 'Practical prediction of time-dependent deformations of concrete', *Materials and Structures (RILEM, Paris)*, Part I: 'Shrinkage', **11**, 65, 307–316; Part II: 'Basic creep', **11**, 65, 317–328; Part III: 'Drying creep', **11**, 66, 415–424; Part IV: 'Temperature effect on basic creep', **11**, 66, 424–434; Part V: 'Temperature effect on drying creep', **12**, 69, 169–183.
Bažant, Z. P., and Raftshol, W. J. (1982) 'Effect of cracking in drying and shrinkage specimens', *Cement and Concrete Research*, **12**, 209–226.
Bažant, Z. P., and Thonguthai, W. (1978) 'Pore pressure and drying of concrete at high temperature', *Proceedings of the ASCE, Journal of the Engineering Mechanics Division*, **104**, EM5, 1059–1079.
Bažant, Z. P., and Thonguthai, W. (1979) 'Pore pressure in heated concrete walls: a theoretical prediction', *Magazine of Concrete Research*, **31**, 107, 67–76.
Bažant, Z. P., and Wu, S. T. (1974a) 'Rate-type creep law of aging concrete based on Maxwell chain', *Materials and Structures (RILEM, Paris)*, **7**, 45–60.
Bažant, Z. P., and Wu, S. T. (1974b) 'Thermoviscoelasticity of aging concrete', *Proceedings of the ASCE, Journal of the Engineering Mechanics Division*, **100**, EM3, 575–597.
Bažant, Z. P., and Wu, S. T. (1974c) 'Creep and shrinkage law for concrete at variable humidity', *Proceedings of the ASCE, Journal of the Engineering Mechanics Division*, **100**, EM6, 1183–1209.
Bažant, Z. P., Chern, J.-C., and Thonguthai, W. (1981a) 'Finite element program for moisture and heat transfer in heated concrete', *Nuclear Engineering and Design*, **68**, 61–70.
Bažant, Z. P., Rossow, E. C., and Horrigmoe, G. (1981b) 'Finite element program for creep analysis of concrete structures', Preprints of the *Sixth International Conference on Structural Mechanics in Reactor Technology*, Paris, August 17–21, 1981, Paper No. H2/1.
Bažant, Z. P., Chern, J. C., Abrams, M. S., and Gillen, M. P. (1982) *Normal and refractory concretes for LMFBR applications*, Vol. 1: 'Review of literature on high-temperature behavior of Portland cement and refractory concretes', Electric Power Research Institute, EPRI NP-2437.
Becker, N. K., and MacInnis, C. (1973) 'A theoretical method for predicting the shrinkage of concrete', *Journal of the American Concrete Institute*, **70**, 652–657.
Carlson, R. W. (1937) 'Drying shrinkage of large concrete members', *Journal of the American Concrete Institute*, **33**, 327–336.
CEB-FIP (1978) *Model Code for Concrete Structures*, Comité Euro-International du Beton, Paris.
Chapman, D. A., and England, G. L. (1977) 'Effects of moisture migration on shrinkage, pore pressure and other concrete properties', Preprints of the *Fourth International Conference on Structural Mechanics in Reactor Technology*, San Francisco, August 1977, Paper No. H5/3.
Chatterji, S., Thaulow, N., and Christensen, P. (1981) 'Formation of shrinkage cracks in thin specimens of cement paste', *Cement and Concrete Research*, **11**, 155–157.
Chen, K. H., Gluekler, E. L., Lam, S. T., and Shippey, V. S. (1980) 'Comparison of mechanistic codes for predicting water release from heated concrete', General Electric Company, *Report No. GEFR-00521*.
Comini, G., and Lewis, R. W. (1976) 'A numerical solution of two-dimensional problems involving heat and mass transfer', *International Journal of Heat and Mass Transfer*, **19**, 1387–1392.

Cranck, J. (1975) *Mathematics of Diffusion*, Oxford University Press, London.
Dayan, A. (1977) *COWAR-2 User's Manual*, General Electric Company, GEFR-00090 (L).
Emery, A. F., and Carson, W. W. (1971) 'An evaluation of the use of the finite-element method in the computation of temperature', *Transactions of the ASME, Journal of Heat Transfer*, 136–145.
England, G. L., and Ross, A. D. (1972) 'Shrinkage, moisture, and pore pressure in heated concrete', *Concrete for Nuclear Reactors*, ACI Special Publication, SP-34, pp. 883–906.
England, G. L., and Sharp, T. J. (1971) 'Migration of moisture and pore pressure in heated concrete', Preprints of the *First International Conference on Structural Mechanics in Reactor Technology*, Berlin, 1971, Paper No. H2/4.
Fischer, R. (1970) 'Über das Verhalten von Zementmörtel und Beton bei Hoheren Temperaturen', *Deutscher Ausschuss für Stahlbeton*, Heft 214.
de Groot, S. R., and Mazur, P. (1962) *Non-Equilibrium Thermodynamics*, North-Holland Publishing Company, Amsterdam.
Hansen, T. C., and Mattock, A. H. (1966) 'Influence of size and shape of member on the shrinkage and creep of concrete', *Journal of the American Concrete Institute*, **63**, 267–289.
Hancox, N. L. (1968) 'The role of moisture diffusion in the drying of cement paste under the influence of temperature gradients', *British Journal of Applied Physics*, Series 2, **1**, 1769–1777.
Harmathy, T. Z. (1970) 'Thermal properties of concrete at elevated temperatures', *ASTM Journal of Materials*, **5**, 47–74.
Hughes, B. P., Lowe, I. R., and Walker, J. (1966) 'The diffusion of water in concrete at temperatures between 50 and 95°C', *British Journal of Applied Physics*, **17**, 1545–1552.
Knight, R. L., and Beck, J. V. (1979) 'Model and computer code for energy and mass transport in decomposing concrete and related materials', *Proceedings of the International Meeting on Fast Reactor Safety Technology*, Seattle, Washington, August 19–23, 1979.
Lewis, R. W., Strada, M., and Comini, G. (1977) 'Drying-induced stresses in porous bodies', *International Journal for Numerical Methods in Engineering*, **11**, 1175–1184.
Lewis, R. W., Morgan, K., and Thomas, H. R. (1979) 'Drying-induced stresses in porous bodies–an elastoviscoplastic model', *Computer Methods in Applied Mechanics and Engineering*, **20**, 291–301.
Luikov, A. V. (1966) *Heat and Mass Transfer in Capillary–Porous Bodies* (translated from Russian), Pergamon Press, Oxford.
Luikov, A. V. (1975) 'Systems of differential equations of heat and mass transfer in capillary–porous bodies (review)', *International Journal of Heat and Mass Transfer*, **18**, 1–14.
McCormack, J. D., Postma, A. K., and Schur, J. A. (1979) 'Water evolution from heated concrete', Hanford Engineering Development Laboratory, *Report No. HDL-TIME 78-87*.
Müller, H. S., and Hilsdorf, H. K. (1982) 'Comparisons of prediction methods for creep coefficients of structural concrete with experimental data', *Fundamental Research on Creep and Shrinkage of Concrete* (Ed. F. H. Wittmann) Martinus Nijhoff Publishers, The Hague, pp. 269–289.
Murata, J. (1965) 'Studies on the permeability of concrete', *Materials and Structures (RILEM, Paris)*, **29**, 47–54.
Neville, A. M. (1977) *Properties of Concrete*, 2nd edn, Pitman Publishing Ltd.
Nilsson, L.-O. (1980) 'Hygroscopic moisture in concrete—drying, measurements and related material properties', Lund Institute of Technology, Division of Building Materials, Lund, Sweden, *Report No. TVBM-1003*.

Pickett, G. (1946) 'Shrinkage stresses in concrete', *Journal of the American Concrete Institute*, **42**, 165–195 and 361–392.

Pihlajavaara, S. E. (1963) 'Notes on the drying of concrete', State Institute for Technical Research, Helsinki, *Publication No. 74*.

Pihlajavaara, S. E. (1965) 'On the main features and methods of investigation of drying and related phenomena in concrete', Ph. D. Thesis, State Institute for Technical Research, Helsinki, *Publication No. 100*.

Pihlajavaara, S. E., and Kasi, S. S. H. (1969). 'An approximate solution of a quasi-linear diffusion problem', State Institute for Technical Research, Helsinki, *Publication No. 153*.

Pihlajavaara, S. E., and Väisänen, M. (1965) 'Numerical solution of diffusion equation with diffusivity concentration dependent', State Institute for Technical Research, Helsinki, *Publication No. 87*.

Pister, K. S., Argyris, J. H., and Willam, K. J. (1978) 'Creep and shrinkage of aging concrete', *Douglas McHenry International Symposium on Concrete Structures*, ACI SP-55, pp. 1–30.

Powers, T. C., and Brownyard, T. L. (1946–47) 'Studies of the physical properties of hardened cement paste', *Journal of the American Concrete Institute*, **18**, 101–132, 249–336, 469–504, 549–602, 669–712, 845–880, 933–992.

Rüsch, H., Jungwirth, D., and Hilsdorf, H. (1973) 'Kritische Sichtung der Verfahren zur Berücksichtigung der Einflüsse von Kriechen und Schwinden des Betons auf das Verhalten der Tragwerke', *Beton- und Stahlbetonbau*, **68**, 49–60, 76–86, 152–158.

Sakata, K. (1983) 'A study on moisture diffusion in drying and drying shrinkage of concrete', *Cement and Concrete Research*, **13**, 216–224.

Visser, W. (1965) 'A finite element-method for the determination of non-stationary temperature distribution and thermal deformations', *Proceedings of the First Conference on Matrix Methods in Structural Mechanics*, Wright-Patterson Air Force Base, Dayton, Ohio, AFFDL-TR-66-80, 925–943.

Warzee, G. (1974) 'Finite element analysis of transient heat conduction, application of the weighted residual process', *Computer Methods in Applied Mechanics and Engineering*, **3**, 255–268.

Welty, J. R., Wicks, C. E., and Wilson, R. E. (1969) *Fundamentals of Heat and Mass Transfer*, John Wiley & Sons.

Wilson, E. L., and Nickell, R. E. (1966) 'Application of the finite element method to heat conduction analysis', *Nuclear Engineering and Design*, **4**, 276–286.

Wittmann, F. H. (1982) 'Creep and shrinkage mechanisms', in *Creep and Shrinkage in Concrete Structures* (Ed. Z. P. Bažant and F. H. Wittmann), John Wiley & Sons Ltd., pp. 129–161.

Wittmann, F. H., and Roelfstra, P. E. (1980) 'Total deformation of loaded drying concrete', *Cement and Concrete Research*, **10**, 601–610.

Young, J. F. (1982) 'The microstructure of hardened Portland cement paste', *Creep and Shrinkage in Concrete Structures* (Ed. Z. P. Bažant and F. H. Wittmann), John Wiley & Sons, pp. 3–49.

Mechanics of Geomaterials
Edited by Z. Bažant
© 1985 John Wiley & Sons Ltd

Chapter 17

Mechanics of Fluid-Saturated Geomaterials: Discusser's Report

V. N. Nikolaevsky

17.1 INTRODUCTION

Mechanics of fluid-saturated geomaterials is now a developed chapter of continuum mechanics which is based on a wide scope of experimental studies and practical applications. However, widening applications are permanently revealing new problems and attracting scientists who introduce concepts different from the classical ideas formulated by H. Darcy, N. E. Zhukovsky, K. Terzaghi, J. I. Frenkel, and M. A. Biot. In this report a thermodynamically unified discussion of these concepts is given. The reader may note the pronounced differences of the theories of poroelasticity and of dilatancy formulated here from the description presented by J. W. Rudnicki and G. Horrigmoe. An alternative theory is given for the phenomenon which is known as the corrosion creep. The usage of the dilatancy theory concepts for the estimations of rock porosity and permeability generated by explosions or by tectonic waves represents a new field of studies.

17.2 DYNAMICS OF SATURATED GEOMATERIALS

Saturated geomaterials can be considered as a multiphase mixture of a solid porous matrix and of fluids contained in the pore space. Therefore, the elementary volume ΔV is practically a sum $(\Delta V_1 + \Delta V_2)$ of the solid material volume, ΔV_1, and the pore volume, ΔV_2. According to a continuum description, each phase exists in a space macropoint (ΔV) and its content is characterized by the porosity $m = m_2 = \Delta V_2/\Delta V$, and $m_1 = 1 - m = \Delta V_1/\Delta V$. So, the porosity is simply a void concentration.

Therefore, the porosity as a phase concentration is governed by mass balance equations:

$$\frac{\partial}{\partial t}m_\alpha \rho_\alpha + \frac{\partial}{\partial x_i}m_\alpha \rho_\alpha v_i^{(\alpha)} = 0 \tag{17.1}$$

Here $v_i^{(\alpha)}$ = velocity, ρ_α = density of the phase $\alpha(\alpha = 1, 2)$. The momentum equations for the medium as a whole and for the fluid phase can be formulated as

$$\rho_1(1-m)\frac{d_1 v_i^{(1)}}{dt} + \rho_2 m \frac{d_2 v_i^{(2)}}{dt} = \frac{\partial \Gamma_{ij}}{\partial x_j} + \rho_1(1-m)g_i + \rho_2 m g_i$$

$$\rho_2 m \frac{d_2 v_i^{(2)}}{dt} = -\frac{\partial mp}{\partial x_i} + \Psi_i \qquad (17.2)$$

in which $\gamma_i^{(\alpha)} = \rho_\alpha g_i$, g_i = free fall acceleration;

$$\Gamma_{ij} = (1-m)\sigma_{ij} - mp\delta_{ij} = \sigma_{ij}^f - p\delta_{ij} \qquad (17.3)$$

which represents the total stress; and according to Zhukovsky (1889) the force of phase interaction

$$\Psi_i = p\frac{\partial m}{\partial x_i} + R_i \qquad (17.4)$$

is introduced. Here R_i is a viscous resistance to the fluid flow through the geomaterial, σ_{ij} = true solid stress, distinct from Terzaghi's (1925) effective stress σ_{ij}^f. The latter can be easily measured as a difference of the load and the pore pressure, p.

For the formulation of rheologic laws one has to use at first thermodynamic analysis. The balances of total energy have the following form

$$\frac{\partial}{\partial t}\rho_1(1-m)\left(\varepsilon_1 + \frac{v_i^{(1)} v_i^{(1)}}{2}\right)$$

$$+ \frac{\partial}{\partial x_i}\left\{\rho_1(1-m)\left(\varepsilon_1 + \frac{v_i^{(1)} v_i^{(1)}}{2}\right)v_i^{(1)} - (1-m)\sigma_{ij}v_j^{(1)} + j_i^{(1)}\right\} - \frac{\delta W}{dt} + Q = 0$$

$$\frac{\partial}{\partial t}\rho_2 m\left(\varepsilon_2 + \frac{v_i^{(2)} v_i^{(2)}}{2}\right)$$

$$+ \frac{\partial}{\partial x_i}\left\{\rho_2 m\left(\varepsilon_2 + \frac{v_i^{(2)} v_i^{(2)}}{2}\right)v_i^{(2)} + mpv_i^{(2)} + j_i^{(2)}\right\} + \frac{\delta W}{dt} - Q = 0 \qquad (17.5)$$

where ε_α = internal energy, $j_i^{(\alpha)}$ = heat flux in the phase α, Q = interphase heat flux. The interphase work $\delta W/dt$ has to be postulated (Nikolaevsky et al., 1970) in the form characteristic for a heterogeneous mixture

$$\frac{\delta W}{dt} = p\frac{\partial m}{\partial t} + R_i v_i^{(1)} \qquad (17.6)$$

Note that for a multicomponent mixture one has to use the relations $\Delta V_1 = \Delta V_2 = \Delta V$, $\Psi_i \equiv R_1$, $\delta W/\partial t = R_i v_i^{(1)}$, and then partial pressures $p_1(1-m)p$, $p_2 = mp$ correspond to Dalton's law.

The usage of kinetic phase balances, which correspond to the momentum balances (17.2)–(17.4) together with Equation (17.5), gives the Gibbs thermodynamic relations

$$\rho_1(1-m)d_1\varepsilon_1 - \sigma_{ij}^f d_1 e_{ij} + \rho_1(1-m)p d(1/\rho_1) = \rho_1(1-m)T_1 d_1 s_1$$

$$\rho_2 m d_2\varepsilon_2 + \rho_2 m p d_2(1/\rho_2) = \rho_2 m T_2 d_2 s_2 \qquad (17.7)$$

Here T_α, s_α = phase temperature and entropy, e_{ij} = matrix strain, and the differentials are determined according to the phase substantial (material) derivatives

$$\frac{d_\alpha}{dt} = \frac{\partial}{\partial t} + v_i^{(\alpha)}\frac{\partial}{\partial x_i} \qquad (17.8)$$

The heat flux balance has the following form

$$(1-m)\rho_1 T_1 d_1 s_1 = -(\partial j_i^{(1)}/\partial x_i) + Qdt$$

$$m\rho_2 T_2 d_2 s_2 = -(\partial j_i^{(2)}/\partial x_i) - Qdt + R_i(v_i^{(2)} - v_i^{(1)})dt \qquad (17.9)$$

which means that dissipation takes place only in the fluid phase because of its resistance to relative motion. So, the phase deformation is fully recoverable. If one uses the thermodynamic formalism for the equation of production of the entropy of the whole medium, $S = (1-m)\rho_1 s_1 + m\rho_2 s_2$, the following simple relations ensue

$$R_i = r(v_i^{(2)} - v_i^{(1)}), r = \mu\frac{m}{k}f(Re), Q = \mathcal{H}\left(\frac{1}{T_1} - \frac{1}{T_2}\right) \qquad (17.10)$$

Here the phase heat fluxes are supposed to be equal to zero, r = some resistance coefficient, μ = fluid viscosity, k = matrix permeability, $Re = \sqrt{k/m}\,|v_i^{(1)} - v_i^{(2)}|/\mu$ = Reynolds number. The typical functions $f(Re)$ were published and the factor $\mathcal{H}$ was estimated (Nikolaevsky et al., 1970). The case $f \equiv 1$ corresponds to Darcy's law.

According to the Gibbs relations (17.7), one can write the following Hooke's law

$$\sigma_{ij}^f = \left(K - \frac{2}{3}G\right)e\delta_{ij} + 2Ge_{ij} + \beta_1 Kp\delta_{ij} + \alpha_1 K(T_1 - T_0)\delta_{ij} \qquad (17.11)$$

which is analogous to well-known thermoelastic relation. Here K = matrix bulk modulus, G = matrix shear modulus, β_α = material compressibilities, α_α = coefficients of thermal expansion; and so

$$\frac{\rho_1}{\rho_{10}} = 1 - \frac{1}{3}\beta_1\sigma_{ij}\delta_{ij} - \alpha_1 T_1; \quad \frac{\rho_2}{\rho_{20}} = 1 + \beta_2 p - \alpha_2 T_2 \qquad (17.12)$$

So, the true stress σ_{ij} governs the density changes and the effective stress σ_{ij}^f governs the repacking of solid grains. As a result, the linear variant of the theory has the following form of mass and momentum balance equations

$$\frac{\partial m}{\partial t} - \beta_1(1-m_0)\frac{\partial p}{\partial t} - (1-m_0)\frac{\partial v_i^{(1)}}{\partial x_i} - \beta_1\frac{\partial \sigma^f}{\partial t} + \alpha_1(1-m_0)\frac{\partial T_1}{\partial t} = 0$$

$$\frac{\partial m}{\partial t} + \beta_2 m_0 \frac{\partial p}{\partial t} + \alpha_2 m_0 \frac{\partial T_2}{\partial t} + m_0 \frac{\partial v_i^{(2)}}{\partial x_i} = 0 \quad (17.13)$$

$$(1-m)\left(\rho_1 \frac{\partial v_i^{(1)}}{\partial t} - \rho_2 \frac{\partial v_i^{(2)}}{\partial t}\right) - \frac{\partial \sigma_{ij}^f}{\partial x_j} - \frac{\mu}{k}(v_i^{(2)} - v_i^{(1)}) = 0$$

$$\rho_2 \frac{\partial v_i^{(2)}}{\partial t} + \frac{\partial p}{\partial x_i} + \frac{\mu}{k}(v_i^{(2)} - v_i^{(1)}) = 0 \quad (17.14)$$

which has to be augmented by the equations of heat flux balances, also in their linear variant (Nikolaevsky et al., 1970). One can see that there is no need in the constitutive equation for a porosity. The porosity is fully determined by the two mass balances.

The simple thermodynamic approach leads to the basic mathematical model of poroelasticity, which closely depends on experimental measurements of the real geomaterial parameters. The classical Darcy's law is a combination of a non-inertial momentum balance for the fluid and of the Onsager relation for the phase interaction force. The system under consideration reduces to the simple diffusion equation only in some particular quasi-static cases (Geertsma, 1957; Nikolaevsky et al., 1970). Moreover, the direct problem formulation for the stress functions is necessarily accompanied by an initial condition change because of the existence of two compression waves.

The poroelastic model given here is more convenient than Biot's equations (1956), firstly because of the usage of physically determined coefficients with the ratio $\beta_1 K$ as a small parameter among them. Secondly, our approach shows the way for a proper nonlinear generalization which is necessary, for instance, in simulating explosion action.

A review of problems solved for some typical applications (in seismic action and in oil and gas recovery) was given earlier (Nikolaevsky et al., 1970). Because of thermally active properties of gases and oils with gas dissolved, the thermoelastic effects are sometimes essential.

Let us mention some additional results. It is worthy to note the problem of the stress–strain state in the vicinity of a well used to pump a fluid from the layer (Nikolaevsky and Ramazanov, 1982). Barzam (1980) used the Thompson–Haskell matrix methods for the problems of seismic wave reflection from the saturated layers. He found that Biot's apparent additional mass in the momentum equations led to some unreasonable resonance. Migunov (1981) extended Frenkil's study (1944) of electric fields generated in saturated soils under seismic action. Foda and May (1983) considered the Rayleigh wave in a saturated halfspace.

17.3 GEOMATERIALS WITH PORES AT PARTIAL SATURATION

In this case the surface Gibbs phase has to be taken into account. If one supposes that it moves with the solid matrix, then the phase momentum equations have the following form (Vedernikov and Nikolaevsky, 1978)

$$\rho_1(1-m)\left(\frac{\partial v_i^{(1)}}{\partial t} + \frac{\partial v_i^{(1)} v_j^{(1)}}{\partial x_j}\right)$$

$$= \frac{\partial \sigma_{ij}^f}{\partial x_j} - (1-m)\frac{\partial P}{\partial x_i} - m(p_2 - p_3)\frac{\partial \theta}{\partial x_i} + R_i^{(1)} + (1-m)\rho_1 g_i$$

$$\rho_2 m\theta\left(\frac{\partial v_i^{(2)}}{\partial t} + \frac{\partial v_i^{(2)} v_j^{(2)}}{\partial x_i}\right) = -m\theta\frac{\partial p_2}{\partial x_i} + R_i^{(2)} + m\theta\rho_2 g_i$$

$$\rho_3 m(1-\theta)\left(\frac{\partial v_i^{(3)}}{\partial t} + \frac{\partial v_i^{(3)} v_j^{(3)}}{\partial x_j}\right) = -m(1-\theta)\frac{\partial p_3}{\partial x_i} + R_i^{(3)} + m(1-\theta)\rho_3 g_i \quad (17.15)$$

where θ = saturation of the pores by phase $\alpha = 2$, $R_i^{(\alpha)} = \sum_{\beta \neq \alpha} R_i^{(\alpha\beta)} R_i^{(\alpha\beta)}$ — interaction force between phases α and β; $p_\alpha(\alpha = 2, 3)$ — phase pressures. The effective stress

$$\sigma_{ij}^f = \Gamma_{ij} + P\delta_{ij} = (1-m)(\sigma_{ij} + P\delta_{ij}), \quad P = \theta p_2 + (1-\theta)p_3 \quad (17.16)$$

governs repacking of the solid grains. Other variants of such a stress

$$\sigma_{ij}^f(2) = \Gamma_{ij} + p_2\delta_{ij}, \quad \sigma_{ij}^f(3) = \Gamma_{ij} + p_3\delta_{ij} \quad (17.17)$$

correspond to repacking of the solid grains together with particles of the third or the second phases. Experimentally this question was studied by Fredlund and Morgenstern (1977).

The energy balance for fluid phases ($\alpha = 2, 3$) can be written in the following manner

$$m_\alpha\left\{\rho_\alpha\frac{d_\alpha \varepsilon_\alpha}{dt} + p_\alpha\frac{\partial v_i^{(\alpha)}}{\partial x_i}\right\} + p_\alpha\frac{d_\alpha m_\alpha}{dt} = \frac{\partial m_\alpha j_i^{(\alpha)}}{\partial x_i} - \sum_\beta R_i^{(\alpha\beta)}(v_i^{(\alpha)} - v_i^{(\beta)}) - Q_\alpha \quad (17.18)$$

where Q_α = volume heat source, $m_2 = m\theta$, $m_3 = m(1-\theta)$. Because of the phase mass balances

$$\frac{\partial m_\alpha \rho_\alpha}{\partial t} + \frac{\partial m_\alpha \rho_\alpha v_i^{(\alpha)}}{\partial x_i} = 0 \quad (17.19)$$

the equation of internal energy of the solid matrix has the form

$$(1-m)\left\{\rho_1\frac{d_1\varepsilon_1}{dt} - \sigma_{ij}^{(1)}\frac{\partial v_i^{(1)}}{\partial x_j}\right\} + P\frac{d_1(1-m)}{dt} = \frac{\partial(1-m)j_i^{(1)}}{\partial x_i} - Q_1 \quad (17.20)$$

Some further transformations give the Gibbs relation if one associates the energy, ε_c, and the entropy s_c of the capillary layers within the solid phase

$$(1-m)\rho_1 \frac{d_1\varepsilon_1}{dt} + m\frac{d_1\varepsilon_c}{dt} + \frac{\sigma_{ij}^f}{2}\left(\frac{\partial v_i^{(1)}}{\partial x_j} + \frac{\partial v_j^{(1)}}{\partial x_i}\right) + P\frac{d_1}{dt}\frac{1}{\rho_1} + p_c\frac{d_1\theta}{dt}$$

$$= (1-m)\rho_1 T_1 \frac{d_1 s_1}{dt} + mT_c \frac{d_1 s_c}{dt} \qquad (17.21)$$

$$m_\alpha \rho_\alpha \frac{d_\alpha \varepsilon_\alpha}{dt} + p_\alpha \frac{d_\alpha}{dt}\frac{1}{\rho_\alpha} = T_\alpha \rho_\alpha \frac{d_\alpha s_\alpha}{dt}$$

Here T_c = capillary layer temperature. If $T = T_\alpha = T_c$, then the elastic strains will be generated by σ_{ij}^f, mean fluid pressure P and capillary pressure $p_c(\theta)$. The latter was measured experimentally by Leverett (1941). Correspondingly, the matrix deformation e_{ij} consists of three additive components e_{ij}^f, e_{ij}^p and e_{ij}^θ. Then Hooke's law has the generalized form

$$\sigma_{ij}^f = \left(K - \frac{2}{3}G\right)e_{ij}\delta_{ij} + 2Ge_{ij} + \beta_1 KP\delta_{ij} + \alpha K(T_1 - T)\delta_{ij}$$

$$+ \beta_\theta K[p_c(\theta) - p_c(\theta_0)]\delta_{ij} \qquad (17.22)$$

where β_θ = factor of geomaterial swelling ($\beta_\theta > 0$) or shrinkage ($\beta_\theta < 0$).

Again, according to the thermodynamic formalism one can get the necessary rheologic relations. Among them, the following ones may be discussed;

$$\frac{d_1\theta}{dt} = \frac{p_2 - p_3 - p_c(\theta)}{\tau_c} \qquad (17.23)$$

$$R_i^{(\alpha)} = -\sum_{\beta \neq \alpha} a_{\alpha\beta}(v_i^{(\alpha)} - v_i^{(\beta)}) \qquad (17.24)$$

The first one shows the relations among the fluid phase pressures at times which do not suffice for local capillary equilibrium. The coefficients $a_{\alpha\beta}$ are in practice determined for multiphase fluid flows in a motionless geomaterial. Then the inertial terms are negligible and the momentum balance equations (17.15) gives the following generalization of the Darcy's law

$$-m\theta\frac{\partial p_2}{\partial x_j} = a_{21}v_j^{(2)} + a_{23}(v_j^{(2)} - v_j^{(3)})$$

$$-m(1-\theta)\frac{\partial p_3}{\partial x_j} = a_{31}v_j^{(3)} + a_{32}(v_j^{(3)} - v_j^{(2)}) \qquad (17.25)$$

Here the coefficient $a_{23} = a_{32}$ describes the effect of motion of one fluid phase due to pressure gradient in another fluid phase. If $a_{23} = a_{32} = 0$, then (17.25) corresponds to the well-known concepts (Muskat and Meres, 1936) of a relative phase permeability;

$$w_j^{(2)} = m\theta v_j^{(2)} = -\frac{k}{\mu_2}f_2(\theta)\frac{\partial p_2}{\partial x_i}, \quad f_2(\theta) = \frac{m^2\theta^2}{a_{21}}\frac{\mu_2}{k}$$

$$w_j^{(3)} = m(1-\theta)v_j^{(3)} = -\frac{k}{\mu_3}f_3(\theta)\frac{\partial p_3}{\partial x_j}, \quad f_3(\theta) = \frac{m^2(1-\theta)^2}{a_{31}}\frac{\mu_3}{k}$$

(17.26)

Measurements show that the coefficients $a_{\alpha 1}/\mu_\alpha$ are approximately constant for a given geomaterial texture.

The presently formulated variant of the dynamic model was used for studying seismic waves in saturated rocks with small gas bubles (Nikolaevsky et al., 1970). Although the local non-equilibrium effects due to bubble oscillations are limited by pore dimensions, they can be taken into account if one avoids the Onsager quasi-equilibrium relations.

From the generalized Hooke's law one can explain the appearance of fractures at a soil surface after a drought. Indeed, if saturation θ decreases then a capillary pressure grows and under the given total strains e_{ij} it means the appearance of tensile effective stresses σ_{ij}^f. If these are large, cracks can be produced (Vedernikov and Nikolaevsky, 1978).

The observations of sound emissions of sands of various saturations due to waves of ultra-high frequencies are striking (Vilchinska, 1982). It was observed that the tail part of a propagating wave has a low-frequency spectrum, and that this spectrum corresponds to acoustical emissions of the same sand under quasi-static loading. If $\theta = 0$, then only this part of the wave is seen and its velocity is equal to the second compression wave velocity. At $\theta = 1$, only the frontal high-frequency, part of the wave can be visible. Its velocity corresponds to the first compression wave (according to the Frenkil–Biot classification) At other values θ, both waves are observable at distances up to 1 metre.

17.4 STRENGTH AND CORROSION DILATANCY OF GEOMATERIALS

The strength conditions for soft soils saturated with fluids are always formulated in terms of σ_{ij}^f according to Terzaghi (1925). The analogous role of the effective stresses for saturated rocks was shown by Handin et al. (1963). Beyond the elastic limit, non-recoverable strains appear and they exhibit a dilatancy effect (Reynolds, 1885). In the well-known paper by Drucker and Prager (1952), it was suggested that the plasticity theory be used for describing dilatancy. On the basis of their model, Shield (1955) obtained the essential result that in dilating media a tangential discontinuity had to be accompanied by a normal discontinuity. This means that dilatancy leads to a finite width of the slip band (or the band of localization of deformations). However, the use of the flow rule associated with the solid friction on the limit surface led to an overestimation of dilatancy effects. During the last fifteen years the demand for an associated flow rule was an obstacle for the development of an adequate mathematical model.

To formulate such a model, it was necessary to use independently the limit condition of solid friction for stresses and the dilatant condition for plastic strain increments (Nikolaevsky, 1967, 1971). An inequality of the internal solid friction coefficient, α, and the experimentally measured dilatancy rate, Λ, means a non-associated flow rule. The physical reasoning for such a formation was discussed elsewhere (Nikolaevsky and Rice, 1979).

If the geomaterial is wet, the inelastic and dilatant strains will develop slowly in time. This effect is termed the corrosion creep (Das and Scholtz, 1981) and for its interpretation the traditional ideas of rupture thermo-activation at microlevel are used (Cruden, 1970). However, the bands of strain localization known from the theory of dilatant plasticity (Rudnicki and Rice, 1975) were observed by Kurita et al. (1983) at the stage of tertiary (nonsteady) creep. Moreover, the time needed for the creep rupture of a specimen is in a good correlation with the time of diffusion of moisture in it (Nikolaevsky, 1982). From the geological description, we know that it is possible to distinguish the cases of wet and dry formations of slip bands. Therefore, let us try to develop an adequate theory as an analogy of the thermoplasticity theory.

The decrease of the fracture toughness and of the solid friction due to saturation growth can be taken into account by a proper dependence of cohesion Y and of α on θ. The last variable is interpreted here as the moisture content. The increment of strain de_{ij} is equal to the sum of elastic de_{ij}^e and plastic de_{ij}^p parts. We have

$$de_{ij} = de_{ij}^e + de_{ij}^p, \quad de_{ij}^e = Q_{ijkl}^e d\sigma_{kl}$$

$$de_{ij}^p = \left(\sigma_{ij} + \frac{2}{3}\Lambda Y \delta_{ij} + \sigma\delta_{ij} + \frac{2}{3}\Lambda\alpha\sigma\delta_{ij}\right)d\lambda, \quad (17.27)$$

$$\sigma = -\frac{1}{3}\sigma_{ij}\delta_{ij}$$

in which Q_{ijkl}^e = elastic modulus tensor, and $d\lambda \geq 0$ in the case of active loading, that is, if

$$\Phi_\sigma \equiv \frac{2}{\sqrt{3}}\tau\mathcal{H} - \alpha(\chi,\theta)\sigma - Y(\theta,\chi) = 0, \quad \mathcal{H} = \text{sgn } d\gamma^p$$

$$d_*\Phi_\sigma = (\partial\Phi_\sigma/\partial\sigma)d\sigma + (\partial\Phi_\sigma/\partial\tau)d\tau + (\partial\Phi_\sigma/\partial\theta)\,d\theta > 0$$

(17.28)

Here χ = hardening (softening) parameter, $d\gamma^p$ = plastic shear increment. If $\Phi_\sigma <$ 0, unloading takes place. If $\Phi_\sigma = 0$, but $d_*\Phi_\sigma = 0$, it is the neutral loading. In both cases $d\lambda = 0$. The situation $\Phi_\sigma = 0$, $d_*\Phi_\sigma < 0$ corresponds to a plastic softening process, and $d\lambda > 0$ (although the stress level is decreasing).

Practical measurements lead to functionals $\tau = \tau(\gamma,\theta)$, $e = e(\gamma)$ and also to the yield function $\tau(\sigma, \theta)$ at θ = const. They can be written in differential forms which are an extension of Equations (17.28)

$$de = -\frac{d\sigma}{K} + \frac{\Lambda}{H}\left(\frac{2}{\sqrt{3}}\mathcal{H}d\tau - \alpha d\sigma - M\,d\theta\right)$$

$$d\gamma = \frac{d\tau}{G} + \frac{1}{H}\left(\frac{2}{\sqrt{3}}\mathcal{H}d\tau - \alpha d\sigma - M\,d\theta\right) \quad (17.29)$$

One can see that if θ = const., the flow rule (17.27)–(17.28) is identical to the one suggested in the paper by Nikolaevsky (1971), and the stress–strain relation (17.29) to the formulation suggested by Rudnicki and Rice (1975) for studying the band localization criterion. Their work was substantially augmented by Garagash (1982) who showed that uniform dilatant flows inside a layer can be also unstable and tend to form periodical hexagons. In geology they are known as boudine structures.

We may add diffusion transfer to the moisture convection

$$\frac{\partial \theta}{\partial t} + \frac{w}{m}\frac{\partial \theta}{\partial x} = \frac{\partial}{\partial x}D\frac{\partial \theta}{\partial x} \quad (17.30)$$

where D = diffusion coefficient. The mass balances (17.13) will give the necessary relations

$$\frac{\partial w}{\partial x} = -\frac{\partial m}{\partial t} = -(1-m)\frac{\partial e}{\partial t} \quad (17.31)$$

Garagash and the author considered the simple case of stability of corrosion dilatancy of rock in presence of uniformly distributed stresses $\tau(t), \sigma(t)$. Such a situation closely approximates the case of fully saturated rock masses under the action of p and σ^f, a study of whose stability was made by Rice (1975). So, suppose that at the instant t_0 the moisture content θ_0 suffers a small disturbance $\Delta\theta$, i.e. $\theta = \theta_0(t) + \Delta\theta$, in which $\theta_0, m_0, \ldots$ correspond to the reference state and

$$\Delta\tau = \Delta\sigma = 0, \qquad e = e_0 + \Delta e, \qquad \gamma = \gamma_0 + \Delta\gamma$$
$$m = m_0 + \Delta m, \qquad w = w_0 + \Delta w, \qquad D = D_0 + \Delta D \quad (17.32)$$

For a one-dimensional situation, one can get

$$\frac{\partial \Delta\theta}{\partial t} = D_0\frac{\partial^2 \Delta\theta}{\partial x^2} - \frac{w_0}{m_0}\frac{\partial \Delta\theta}{\partial x} - \frac{\dot\theta_0}{m_0}\Delta m \quad (17.33)$$

where $\dot\theta_0$ corresponds to the porosity change $m_0(t)$ of the reference state and

$$\Delta m = (1-m_0)\Delta e, \quad \Delta e = \Lambda\Delta\gamma = -\Lambda(M/H)\Delta\theta$$

If the reference state is changing much slower than the disturbances, then

$m_0, w_0, \ldots$ can be treated approximately as constants. The solution

$$\Delta\theta = \frac{1}{\sqrt{2\pi t}} \exp\left\{-\frac{(x - w_0 t/m_0)^2}{4D_0 t} + \frac{1 - m_0}{m_0} \Lambda \frac{M}{H} \dot\theta_0 t\right\} \quad (17.34)$$

corresponds to the point initial disturbance $\Delta\theta(0, x) = \delta(x)\sqrt{2\pi}$ where $\delta(x)$ is the Dirac delta function.

The criterion of unlimited growth of the shear disturbance $\Delta\gamma$, corresponding to (17.34), is defined at point $x = 0$ by the condition

$$\frac{\partial \Delta\gamma}{\partial t} = -\frac{M}{H}\frac{\partial \Delta\theta}{\partial t} \geqslant 0 \quad (17.35)$$

Such unlimited growth will take place at the instant $t = t^*$, where

$$t^* = \frac{4D}{4D\dfrac{1 - m_0}{m_0}\Lambda\dfrac{M}{H}\dot\theta_0 - \dfrac{w_0^2}{m_0}} \quad (17.36)$$

and therefore

$$\frac{M}{H}\left(\frac{4\Lambda\dot\theta_0 D}{w_0^2}\frac{m_0}{1 - m_0}\right) > 1 \quad (17.37)$$

Because $M < 0$, unlimited shear growth can occur within the ranges $H < 0$, $\Lambda\dot\theta_0 > 0$ or $H > 0$, $\Lambda\dot\theta_0 < 0$. In the first case, an intensive loosening is more necessary for instability than a decreasing of specific moisture content, and, in the second case, intensive consolidation more than moisture increase.

17.5 ROCK POROSITY AND PERMEABILITY INDUCED BY EXPLOSIONS AND EARTHQUAKES

The dilatant plastic flow rule (17.27) gives the following integral for the case of a centrally symmetric motion (Nikolaevsky, 1967)

$$v(r, t) = C(t) r^{-n}, \quad n = \frac{2\sqrt{3} + \Lambda\mathcal{H}}{\sqrt{3} - \Lambda\mathcal{H}} \quad (17.38)$$

if $\Lambda = \text{const}$. Such a flow is realized at contained underground explosions; r is the distance from the explosion point. The values $n < 2$ correspond to pore space increase and $n > 2$ to its decrease. Indeed, the mass balance (17.1) and the integral (17.38) gives for porosity the integral (Dunin and Sirotkin 1977)

$$m/m_0 = 1 - r^{n-2} f(x), \quad x = r^{n+1} - (n + 1)\int C(t) \, dt \quad (17.39)$$

if $\rho_1 = \text{const}$. Here $f(x)$ and $C(t)$ are arbitrary functions. The corresponding analytical solutions of a contained explosion, which generalize the well-known earlier solution (Chadwick, Cox and Hopkins, 1964), were published by many authors in *Soviet Applied Mechanics and Technical Physics* (PMTF) during 1977–83 and by the writer's postgraduate student E. L. Samarov for the case of a sign change in the dilatancy rate Λ (1983) and for a geomaterial with fully saturated pore space (1982). In this solution it was supposed that the 'undrained' conditions are realized for the plastic motion during the explosion, that is, $v_i^{(1)} = v_i^{(2)}$. Then the case $\Lambda = 0$ corresponds to a decreasing pore pressure p and an increasing effective pressure σ^f. At $\Lambda < 0$, the opposite takes place.

However, only numerical computer simulations can give a quantitative correspondence of the post-explosion parameters with the main experimental data. It is found (Bovt and Nikolaevsky, 1981) that a characteristic breakage of the curves of the radial distribution of the peak particle velocity and pressure with the distance coincides with the outward boundary r_τ of the plastic zone. Dilatant plasticity of low-porosity rocks corresponds to crack or void generation ($\Lambda > 0$), and the appearance of an essential permeability plays a role of its indicator. Indeed, the breakage of the curve max v_r and the outer boundary of the permeable zone coincide ($\bar{r}_k = \bar{r}_\tau = 1$ m/kg$^{1/3}$, where $\bar{r} = r/W^{1/3}$, W = explosion yield). The permeability becomes up to million times higher than the initial one (10^{-5} darcy) and it is distributed according to the law $(r/a)^{-5}$ where a = explosion cavity radius (McKee *et al.*, 1982). The porosity is also much higher than the initial one. The wave profiles have two maxima. The first one corresponds to the propagating wave front, and the second one to the maximum residual stress in the vicinity of the cavity. The second one later plays the role of a spherical arch.

Porous dry geomaterials are consolidated by explosions to some critical level and then the loosening process begins, i.e. a change of the sign of Λ actually takes place. The corresponding spherical zone of maximum porosity decrease is observed in experiments (Bovt and Nikolaevsky, 1981). In Figure 17.1, the post-explosion porosity ($\Delta m = -\Delta \rho$) distribution and the P-wave in the radial ($\parallel$) and hoop ($\perp$) directions are given for a cemented sand with $m_0 = 25$ per cent. The first two curves (1a, 1b) are correlated with each other but the plot (1c) is correlated with the permeability decrease data (1d). The explosion wave obviously generates an anisotropy of the velocity, and the permeability defined by the air fluxes through the well appears to be everywhere smaller than its initial value ($k_0 = 0.15$ darcy). It means that dilatant loosening in the zone $0.16 \leqslant \bar{r} \leqslant 0.32$ is suppressed by additional milonitization of matrix grains. The fragmentation due to the explosion, which generates high permeability of intact rock in a porous geomaterial, is not essential. The value $m_0 = 15$ per cent appears to play a marginal role. This analysis is supported by independent studies of the explosion rupture zone in sandstone, performed with a scanning microscope (Durham, 1981). Also, non-monotonous effects were observed by acoustic soundings at some explosions in sedimentary rocks *in situ*.

The post-explosion plastic zone in a porous geomaterial is limited by the

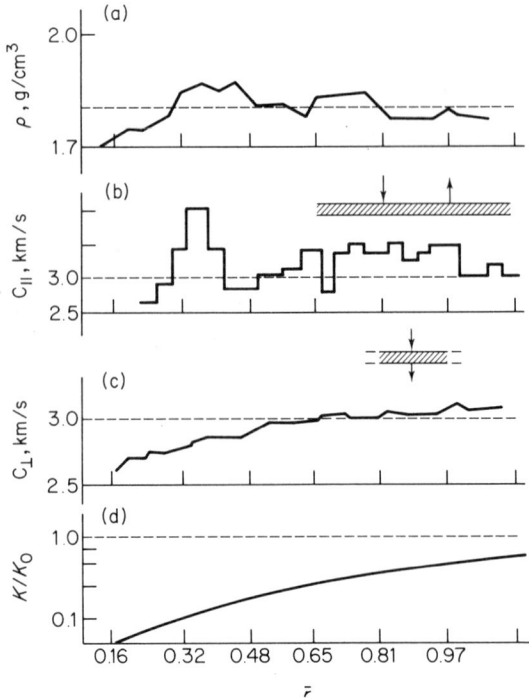

Figure 17.1 Explosively induced changes of porous media

coordinate $\bar{r} = 0.48 \, \text{m/kg}^{1/3}$ and corresponds to the outward radius of dilatant consolidation. For explosions in sands of the same porosity ($m_0 = 25$ per cent), the curve max v_r has a discontinuity at $\bar{r} = 3 \, \text{m/kg}^{1/3}$. However, these data are in agreement if one introduces a true dimensionless coordinate, $R = r(\rho c^2/W^{1/3})$, which takes into account the difference of acoustic rigidity of the geomaterial. It appears that $R_\tau = 12.4 \, (\text{m MPa/kg}^{1/3})$. The outward boundary of the permeability decrease zone takes place further at $\bar{r} = 2$ (or $R = 50$). On other hand, in soils with $\rho c^2 = 5.94 \, \text{MPa}$, explosions generate so-called sublimit plastic deformations (which are of the order of elastic strains) up to $\bar{r} = 6$, that is, up to $R = 50$. One may conclude that the permeability is sensitive to very small residual strains of the pore space.

In saturated porous geomaterials, contained explosions generate a non-monotonic permeability change (Bovt and Nikolaevsky, 1981); a small decrease in a consolidation zone ($0.42 \leqslant \bar{r} \leqslant 1.0$) is followed by a substantial decrease in the zone of loosening ($0.2 \leqslant \bar{r} \leqslant 0.42$). The data for an explosion of an 8 kiloton yield charge in an oil reservoir (Bakirov et al., 1981) show that there is one more characteristic zone ($1.0 \leqslant \bar{r} \leqslant 4.0$) of minor enhancement of the underground flow parameters (up to 50 per cent). The bottom-pressure build-up curves at wells in this zone are of the kind typical for media with double porosity (Warren and Root, 1963). The release of rare gases, just before earthquake events may be an evidence of freshly appearing cracks.

One may expect that the mechanical effect of the explosion will be different for oil and water influxes to the wells. Recently Kissin (1983) has found that prior to the occurrence of earthquakes the water percentage in active oil wells grows up to 100 per cent. On the other hand, Wilson (1956) observed earlier that the relative permeability of the water phase decreases under confining pressures. Therefore, the earthquakes were preceded by an unloading of the reservoir in the aforementioned cases. The effects of changing well outfluxes and bottom pressures observed before and after the earthquakes are well known and they will not be discussed here.

17.6 WATER AND LITHOSPHERE STRUCTURE

Water is the main element not only for the life on the Earth but it also determines the nature of the Earth materials. Water plays the role of a transport agent for the creation of natural resources. It is well established that water percolation influences the intrinsic relaxation time of earthquake sources. However, the question appears whether there is a limit depth for water percolation into the lithosphere. And where can free water disappear or appear again due to the creation or decay of water-contained minerals? The answer depends on the estimation of the limit depth of open, that is, dilatant cracks.

Consider the idealized diagram in Figure 17.2 which involves the well-known experimental data by W. Brace, J. Tullis, R. Schock *et al.*; see for instance the book by Paterson (1978). One can see that the transition from a cataclastic or pseudoplastic state to true plasticity, in other words from strains due to creation and motion of crack nets to an inelastic dislocation mechanism, corresponds exactly to the PT-conditions at the lower boundary of the normal continental crust (35 km depth, or 1 GPa and 600 °C), provided that vertical pressure σ_z is supposed to coincide with confining pressure P_c in triaxial rock tests (Nikolaevsky, 1979). The true plasticity implies crack localization inside individual mineral grains, that is a virtual impermeability for water or water vapour. It is a condition of 'dryness' of the upper mantle. Therefore, in accordance with the Green–Ringwood phase diagram (curve 2 in Figure 17.3), eclogite rocks can exist below the Mohorovicic boundary. In Figure 17.3, curve 1 corresponds to the depth of water-permeable states, the black point to the Moho thermodynamic state, curve 3 to a supposed geotherm and the state 4 corresponds to garnet granulite intermittent phase.

In the crust above the Moho, the water vapour can circulate and gabbro-amphibolites and granites are stable. The idea that the Moho is determined by the permeability for water (Nikolaevsky, 1979) explains the relatively small thickness of a normal ocean crust, in view of the loss of its permeability due to appearance

Mechanics of Geomaterials

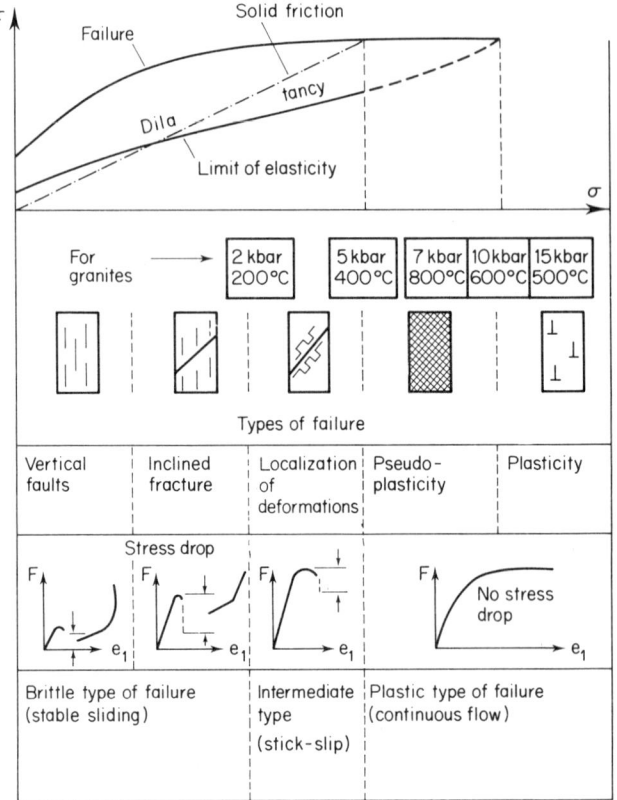

Figure 17.2 Dependence of rock rupture on PT conditions

of dislocation plastic state in the 3d serpentinite layer. It can be seen that the oceanic crust bottom is in extremely good correspondence with the isobar of 0.2–0.25 GPa which was predicted by experiments (Raleigh and Paterson, 1968). The upper mantle contains periodotites because ocean water supply is prevented and the Hess reaction (olivine + water = serpentinite) appears to be impossible. Some geophysical implications such as bands of secondary paleomagnetism, double seismic focal zones in subducting plates, and the structure of ophiolite masses, were discussed by Karakin et al. (1982).

In the regions of young mountains, the boundary of a truly plastic rock state and of a decrease of permeability (curve 1 in Figure 17.4) is above the Moho, and the Moho states correspond to the crossing of the Green–Ringwood diagram by the geotherm curve 3. The Moho depth is equal to 50–70 km (the variants 4 or 5). According to Figure 17.4, plastic states above the Moho guarantee the absence of water in the lower part of the crust. The additional seismic boundary 7 can be explained by the appearance of garnet granulite phase 6. If a steep geothermal

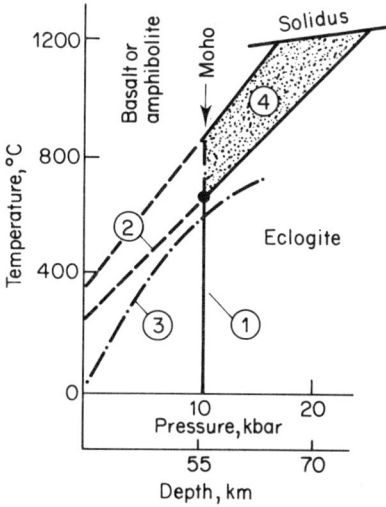

Figure 17.3 Thermodynamic conditions for normal continental crust

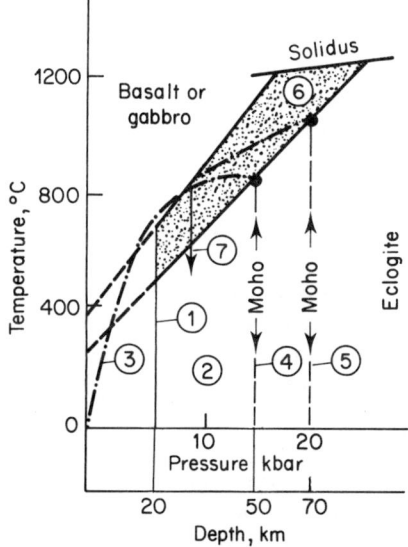

Figure 17.4 Thermodynamic conditions for regions of young mountains

curve is combined with water presence in the Moho then the Earth crust thickness is of the order of 25–27 km (for example, in the Pannonia Basin).

Comparison of the seismic boundaries inside normal continental earth crust (curve 1 in Figure 17.5) with the data on the transition of rupture types (Figure 17.2) also gives a striking agreement. The Conrad boundary (symbol C)

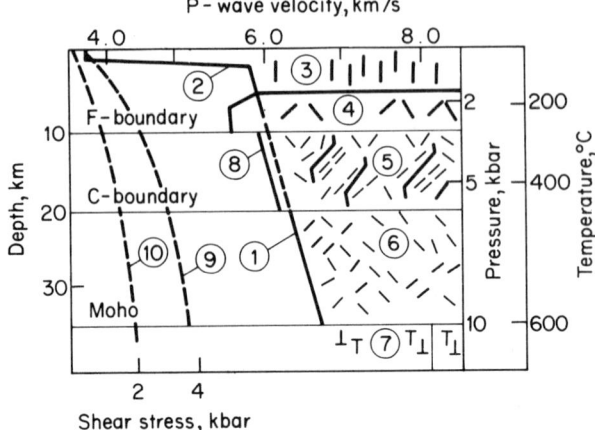

Figure 17.5 Seismic boundaries and dilatant fractures of the Earth's crust

corresponds to the transition between cataclastic flow states 6 and localization of deformation 5 zones. There are hypocentres of large crust earthquakes at this boundary (Sibson, 1982). In the cataclastic zone 6, granite strength is less than the solid friction resistance at two surfaces in contact, and there is no stick–slip phenomenon here. The stick–slip zone coincides with the range of localization of deformations zone 5. Therefore, the nature of the stick–slip is a competition between a solid friction and an intact geomaterial rupture.

Near the ground level (zone 3), crust faults are close to a vertical position but deeper they are inclined (zone 4). The seismic area exploration of the Ukrainian shield (Sharov and Grechishnikov, 1982) shows that even individual faults of the crystalline basement reflect common features: a nearly vertical position changes to an inclined position with increasing depth. Near the Conrad boundary, the faults become close to a horizontal position, which corresponds to the condition $\sigma_x \sim \sigma_y > \sigma_z$. Below the C-boundary, the faults become seismically invisible because of a general cataclastic crack net. The well-known waveguides are lying above the Conrad boundary (in Figure 17.5—above the Fortsch boundary), which appears to correspond to the maximum of dilatant void space. Curve 8 corresponds to seismic velocities of intact granites. The pores influence the seismic velocities of sedimentary rocks (curve 2). Curve 9 shows shear stresses necessary for maintenance of dilatant voids without consideration of temperature effects. Curve 10 shows that this requires a shear stress of 0.2 GPa for opening of dilatant cracks or for cataclastic flows in the Earth's crust.

Geomaterials of the upper mantle, that is of the lower part of the lithosphere, can also have cracks and voids because they are more rigid than the crust geomaterials. But the upper mantle is isolated from the meteoritic waters by the plastic screen at the Moho. So, the Earth crust has to be a reservoir somewhere connected with the underground water basin in the sedimentary layers. The proof of such a conclusion can be found in the deep drilling data, in anomalously low pore pressures of some oil and gas reservoirs at the depths of crystalline basement surface and even by a change of the oil reservoir regime in phase with some geotectonic periodicity (the Sakhalin island case). The very striking fact of a pulsating lake Chany (Smirnova and Shnitikov, 1982) has to be mentioned here. Its water level changes drastically with period of 10–12 years. For years, scientific observations of the effects of vaporization and external water inflows were conducted, and there was a substantial imbalance d for which the same periodicity was evident (Figure 17.6).

An analogous periodicity has been found recently by A. S. Malamud and the author for the Hindukush mantle earthquakes. It appears that the hodograph of hypocentres (at depth H, km, and at time t, years) of the class $K = 12$ exhibits the aseismic band 3 (Figure 17.6), the depth of which is changing with the same period, i.e. 10–12 years. In the writer's opinion, the aseismic band may be explained by a transition of the geomaterials of the subducting plate (or of two continental plates of equal strength) into a plastic state or into another physical phase. Then the 30 km amplitude of the band depth is equal to 1 GPa change of vertical pressure level. Even if one supposes that such a pressure increase leads to a periodic dilatant change of seismic velocities up to 10 per cent, then the depth amplitude would be 15 km, and the pressure amplitude 0.5 GPa. Taking into account the inclination of the phase transition curve may reduce this estimate to 0.1 GPa.

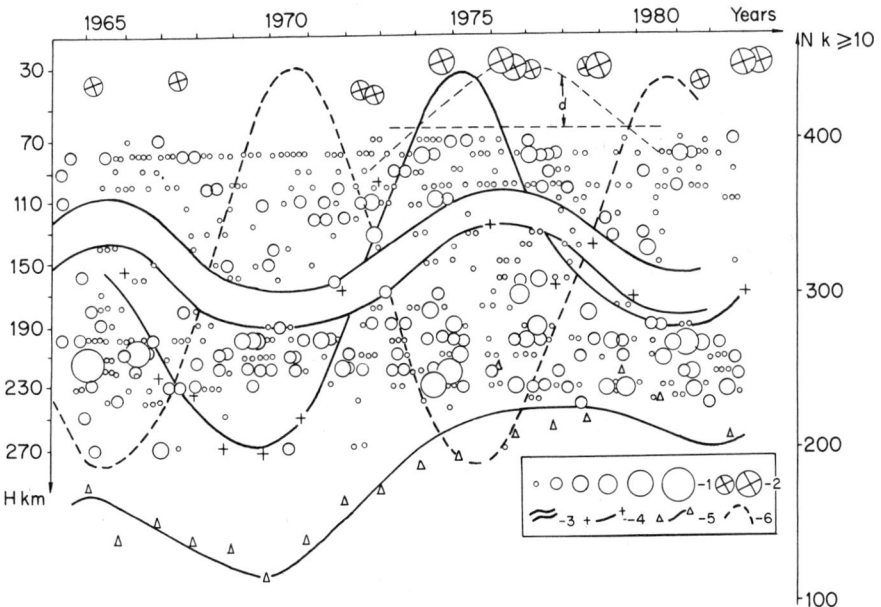

Figure 17.6 Hindukush aseismic band and periodicity phenomenon

These values correspond to continental drift but they are related to a periodic force mechanism, that is, the shear tractions applied to the continental plates from the astenosphere flows over zones of 5–10 thousand kilometres also must be periodical. Moreover, this idea is supported by the fact that, during the years of an increasing subducting force, the largest earthquakes (energetic class $K = 16$–17)—the largest symbols 1 in Figure 17.6—take place when the subducting plate end stops. Also, during these years of large crust earthquakes (2), high annual frequencies of the mantle (4) and the crust (5) earthquakes ($K \geqslant 10$) are observed. The sun activity curve 6 is given here for comparison. The years of increasing subducting tractions correspond to assumed water flows into Chany Lake ($d > 0$) from the fault net.

Temporal changes of seismic-wave velocities and also of their attenuation were revealed (Gamburzeva et al., 1982) by the data of seismic signals generated by large explosions and recorded at the USSR seismic stations during a period of 14 years. These temporal variations appear to be periodic (with 2–3 and 6 year periods) and they take place in seismically active regions as well as in vast aseismic regions of Russian and Kazakh platforms. Periodicity, and even antiphase changes, of the same type were found (Malamud and Nikolaevsky, 1983) for the distribution of the frequency of small earthquakes ($K \geqslant 10$) above and below the Hindukush aseismic band. This fact is an evidence of tectonic waves with a 90 km length. Because the same antiphase changes were noted for different layers of the crust in the Dushanbe–Vahsh district, it may be concluded that these waves are of the bending type. According to the available data, the same situation appears to exist in Pamir and Vranch regions.

17.7 THEORY OF TECTONIC WAVES BENDING THE LITHOSPHERE

First of all, let us assume that the viscous astenosphere is incompressible.

$$\frac{\partial v}{\partial x} + \frac{\partial w}{\partial z} = 0 \quad (17.40)$$

Then the dependence of horizontal velocity v on x and t means that the vertical velocity w must be inequal to zero. So, the vertical displacement component η at the contact ($z = h$) between the astenosphere and the lithosphere will be essential. We shall seek a velocity field of the following type

$$v = \left(v_0 + \frac{\partial u}{\partial t}\right)\frac{z}{h} + \left(B_0 + \frac{\partial \phi}{\partial t}\right)\left(\frac{z}{h}\right)^2 - \frac{\partial \phi}{\partial t}\left(\frac{z}{h}\right) \quad (17.41)$$

and $w(x, h, t) = \partial \eta / \partial t$ will be found from Equation (17.40). This equation must be integrated over the cross-section ($h \leqslant z \leqslant 0$) of the astenosphere flow, the bottom ($z = 0$) of which is supposed to be impermeable for the astenosphere fluid and rigid:

$$\frac{\partial \eta}{\partial t} + \frac{h}{2}\frac{\partial^2 u}{\partial x \partial t} - \frac{h}{6}\frac{\partial^2 \phi}{\partial x \partial t} = 0 \qquad (17.42)$$

The momentum equation of such a flow

$$\frac{\partial p}{\partial x} = \frac{\partial \tau}{\partial z}, \quad \tau = \mu\left(\frac{\partial v}{\partial z} + \frac{\partial w}{\partial x}\right) \qquad (17.43)$$

gives the possibility to relate the mean pressure $\langle p \rangle$ at the cross-section with shear traction τ at the flow boundaries

$$\frac{\partial}{\partial x}h\langle p \rangle = \tau(h) - \tau(0) \equiv \frac{2\mu}{h}\left(B_0 + \frac{\partial \phi}{\partial t}\right) + \mu\frac{\partial^2 \eta}{\partial x \partial t} \qquad (17.44)$$

The lithosphere can be treated as a thin plate under the actions of compression force $N = HE(\partial u/\partial x)/(1-v^2)$, bending moment $M = H^3 E K_0 \partial^2 \eta/\partial x^2$, $K_0^{-1} = 12(1-v^2)$, its own weight $q = \gamma(H + \eta)$, shear $\tau(h)$ and normal $p(h) \approx \langle p \rangle$ force from the astenosphere. Here H is the thickness of lithosphere and v is Poisson's ratio. The balance of bending moments means that

$$\langle p \rangle = q + \{EH^3 K_0(\partial^4 \eta/\partial x^4) + \partial(N\partial \eta/\partial x)/\partial x\} \qquad (17.45)$$

Equations (17.44) and (17.45) give the equation for the vertical displacements of the lithosphere

$$L\frac{\partial \eta}{\partial x} + H^3 K_0 \frac{\partial^5 \eta}{\partial x^5} + \frac{\partial^2}{\partial x^2}\left(\frac{N}{E}\frac{\partial \eta}{\partial x}\right) = \frac{2\theta}{h^2}\frac{\partial \phi}{\partial t} + \frac{\theta}{h}\frac{\partial^2 \eta}{\partial x \partial t} + \frac{2\theta}{h^2}B_0 \qquad (17.46)$$

where $\theta = \mu/E$ is a characteristic relaxation time of the system 'lithosphere + astenosphere', $L = \gamma/E$. The equation for horizontal displacements of the lithosphere

$$\frac{1}{1-v^2}\frac{\partial^2 u}{\partial x^2} = \frac{\theta}{hH}\left(\frac{\partial u}{\partial t} + \frac{\partial \phi}{\partial t}\right) + \frac{\theta}{h}\frac{\partial^2 \eta}{\partial x \partial t} - \frac{\theta}{hH}(v_0 + B_0) \qquad (17.47)$$

is a generalization of the Elsasser equation (1969) due to the vertical displacement effect ($\eta \neq 0$) and to the associated astenosphere flow ($\phi \neq 0$). The influence of stationary flows (v_0, B_0) is also essential.

The Young quasi-static modulus E for granites changes from the value of $5 \cdot 10^{10}$ Pa to $5 \cdot 10^9$ Pa depending on the growth of dilatant cracks. The modulus estimated by the P-wave velocity is 10 times higher. The range for the astenosphere viscosity is $10^{11} \leqslant \mu \leqslant 10^{12}$ (Pa·year). Also, $2 \leqslant \theta \leqslant 200$ (years). Let us discuss the motions for which $u \sim \phi \sim 0.1$ (m), $v_0 \sim B_0 \sim 0.1$ (m/year), $h \sim H \sim 10^5$ (m). According to Smoluchowsky (1909), we may neglect the buckling term (the 3d term on the left-hand side of Equation (17.45)).

The system of equations discussed here may be converted to the following form

$(v^2 \ll 1)$:

$$\frac{Hh}{\theta}\frac{\partial^3 u}{\partial x^3} = 4\frac{\partial^2 u}{\partial x \partial t} + \frac{6}{h}\frac{\partial \eta}{\partial t} + h\frac{\partial^3 \eta}{\partial x^2 \partial t}$$

$$\frac{12\theta}{h}\frac{\partial \eta}{\partial t} + 6\theta\frac{\partial^2 u}{\partial x \partial t} + \theta h\frac{\partial^3 \eta}{\partial x^2 \partial t} = H^3 h K_0 \frac{\partial^6 \eta}{\partial x^6} \qquad (17.48)$$

The solution of the type $u = u_0 \exp i(\kappa x - \omega t)$ corresponds to the dispersion equation

$$2(6 + 3n - 2nH_0)\zeta^2 + i\zeta(12H_0 n + 4K_0 H_0^3 n^3 - H_0^2 n^2) - H_0^4 K_0 n^4 = 0$$

$$n = K^2 H^2 = H^2/\lambda^2, \quad \zeta = \omega\theta, \quad H_0 = h/H \qquad (17.49)$$

which has pure imaginary roots,

$$\zeta = -i\frac{4K_0 n^3 + 12n - n^2}{4(6 + n)} \pm i n\sqrt{D}$$

$$D = \left(\frac{4K_0 n^2 + 12 - n}{4(6 + n)}\right)^2 - \frac{K_0 n^2}{2(6 + n)}, \quad \omega = -i\frac{B(n)}{\theta} \qquad (17.50)$$

if $K_0 \sim 0.1$, $n \leqslant 1$, $H_0 = 1$; $B(n)$—real number. This means that the centres of disturbances (with $\lambda \geqslant 90$ km) are not propagating, that is, such waves are standing waves. It must be emphasized that the system of equations needs to be changed for waves with smaller lengths.

Tectonic waves can be moving because of the energy supply from the stationary astenosphere flows. The supporting mechanism is mathematically close to the case of surface capillary waves on flowing films (Kapitza, 1948). The running impulse solution depends on the coordinate $\xi = x - Vt$. Then the system of Equations (17.48) becomes as follows

$$\frac{dw}{d\xi} + \frac{\beta w}{H} = \frac{\beta(v_0 + B_0)}{4VH} - \frac{\beta h}{4H}\frac{d^2\eta}{d\xi^2} - \frac{6(\eta - \eta_0)}{4Hh} \equiv F(\xi)$$

$$K_0 H^3 \frac{d^5\eta}{d\xi^5} + \frac{\beta}{4}\frac{d^2\eta}{d\xi^2} + \frac{\beta}{2}\left(\frac{3w}{h} + \frac{6(\eta - \eta_0)}{h^2}\right) = \frac{\beta}{2h}\frac{B_0}{V} \qquad (17.51)$$

where $w = du/d\xi$, $\beta = 4\theta V/h$. The expression

$$w(\xi) = \exp\left(-\frac{\beta\xi}{H}\right)\int_0^\xi F(\xi)\exp\left(\frac{\beta\xi}{H}\right)d\xi + w_0 \exp\left(-\frac{\beta\xi}{H}\right) \qquad (17.52)$$

is a solution of the first of Equations (17.51). The use of (15.52) and the choice of the tectonic wave velocity as

$$V = \frac{h}{\eta_0}(3v_0 - B_0) \qquad (17.53)$$

allow us to transform the second equation (17.51) to the following one ($\xi = H\xi_*$)

$$\left(\beta + \frac{d}{d\xi_*}\right)\left(K_0 \frac{d^5\eta}{d\xi_*^5} + \frac{\beta}{4}\frac{d^2\eta}{d\xi_*^2} + 3\beta H_0^2 \eta\right) - \frac{3}{2}\beta H_0\left(\frac{\beta}{4H_0}\frac{d^2\eta}{d\xi_*^2} + \frac{3}{2}\beta H_0 \eta\right) = 0$$

(17.54)

Here a reference displacement η_0 and deformation w_0 at the front of the tectonic wave can be assumed as 0.1 m and 10^{-5}–10^{-6}, correspondingly. Then $V = 10$–100 km/year. For the solutions of an exponential type, Equation (17.54) can be approximated by the following one

$$\frac{d\eta}{d\xi} = -\frac{\beta}{4}\eta, \quad \eta = \eta_0 \exp\left(-\frac{\xi}{\lambda}\right), \quad \lambda = \frac{Hh}{\theta V}$$

(17.55)

The solution is decaying at $\xi \to \infty$.

One can expect essential changes of the behaviour of the tectonic waves due to the finite thickness of the lithosphere, to its rheological stratification and to waveguides, which can cause attenuation of bending oscillations in the upper part of the Earth's crust.

REFERENCES

Bakirov, A. A., Bakirov, E. A., Vinogradov, V. N. et al. (1981) *Applications of Nuclear Explosions in Oil Industrial Recovery*, Moscow, Nedra.

Barzam, V. A. (1980) 'The apparent additional mass effect and waves in poroelastic layered media', *Lett. Appl. Engng. Sci.*, **18**, 641–649.

Biot, M. A. (1956) 'Theory of propagation of elastic waves in a fluid-saturated porous solids', *J. Acoust. Soc. Amer.*, **28**, 168–186.

Bovt, A. N., and Nikolaevsky, V. N. (1981) 'Dilatancy and mechanics of underground explosion', *Advances of Sci. Techn. (Itogi nauki i techniki), ser. Mech. Def. Solids*, **14**, 129–169.

Chadwick, P., Cox, A. D., and Hopkins, H. G. (1964) 'Mechanics of deep underground explosions', *Phil. Trans. Roy. Soc., London, A*, **256**, 1070.

Cruden, D. M. (1970) 'A theory of brittle creep in rock under uniaxial compression', *J. Geophys. Res.*, **75**, 17, 3431–3442.

Das, Sh., and Scholtz, C. H. (1981) 'Theory of time dependent rupture in the Earth', *J. Geophys. Res.*, **56**, B7, 6039–6056.

Drucker, D. C., and Prager, W. (1952) 'Soil mechanics and plastic analysis or limit design', *Quart. Appl. Math.*, **10**, 2, 157–165.

Dunin, S. Z., and Sirotkin, V. K. (1977) 'Expansion of gas cavity in a brittle rock with account of its dilatancy', *Sov. Appl. Mech. Techn. Phys. (PMTF)*, 4, 106–109.

Durham, W. B. (1981) 'Direct observation of explosively induced damage in sandstone with application to reservoir simulation', in *Scanning electron microscopy*, **1**, 585–594.

Elsasser, W. H. (1969) 'Convection and stress propagation in the upper mantle', in *Applications of Modern Physics to the Earth Planeta Interior*, John Wiley, New York, pp. 223–246.

Foda, M. A., and Mei, Ch. C. (1983) 'A boundary layer theory for Rayleigh waves in a porous fluid-filled half space', *Int. J. Soil Dynamics Earthquake Engng.*, **2**, 62–65.

Fredlund, D. G., and Morgenstern, N. R. (1977) 'Stress state variables for unsaturated soils', *J. Geotechn. Engng. Div.*, **103**, 5.
Frenkil, J. I. (1944) 'On the theory of seismic and seismically electric phenomena in wetted soils', *Trans. (Izvestia) Acad. Sci. USSR, ser. geophys. geograph.*, 8, 134–149.
Gamburzeva, N. G., Luket, E. I., Nikolaevsky, V. N., Oreshin, S. I., Pasechnik, I. P., Peregonzeva, V. E., and Rubinstein, Ch. D. (1982) 'Periodical variations of seismic waves at sounding of lithosphere by large explosions', *Proc. (Doklady) USSR Acad. Sci.*, **266**, 6, 1349–1359.
Garagash, I. A. (1982) 'Generation of cell texture in elasto-plastic media with solid internal friction and dilatancy', *Proc. (Doklady) USSR Acad. Sci.*, **266**, 1, 59–62.
Geertsma, J. (1957) 'A remark on the analogy between thermoelasticity and the elasticity of saturated porous media', *J. Mech. Phys. Solids*, **6**, 13–16.
Handin, J., Hager, R. V., Friedman, M., and Feather, J. N. (1963) 'Experimental deformation of sedimentary rocks under confining pressure: pore pressure tests', *Amer. Assoc. Petr. Geolog. Bull.*, **47**, 715–755.
Kapitza, P. L. (1948) 'Wave motions of thin layers of viscous fluids', *J. Exptl. Theoret. Phys., (JETF)*, **18**, 1, 3–28.
Karakin, A. V., Lobkovsky, L. I., and Nikolaevsky, V. N. (1982) 'Generation of serpentinite layer of oceanic crust and some geological–geophysical implications', *Proc. (Doklady) USSR Acad. Sci.*, **265**, 3, 572–576.
Kissin I. G. (1983) 'On anomalous variations of water percentage in oil well production before earthquakes', *Proc. (Doklady) USSR Acad. Sci.*, **249**, 4, 817–821.
Kurita, K., Swanson, P. L., Getting, I. C., and Spetzler, H. (1983) 'Surface deformation of Westerly granite during creep', *Geophys. Res. Lett.*, **110**, 1, 75–78.
Leverett, M. C. (1941) 'Capillary behavior in porous solids', *Trans. AIME*, **142**, 152.
McKee, C. R., Jacobson, R. H., and Way, S. C. (1982) 'Design criteria for *in situ* mining of hard rock ore deposits'. *In situ*, **6**, 3, 179–230.
Malamud, A. S., and Nikolaevsky, V. N. (1983) 'Periodicity of Pamir–Hindukush earthquakes and tectonic waves in subducting lithosphere plates'. *Proc. (Doklady) USSR Acad. Sci.*, **269**, 5, 1075–1078.
Migunov, N. I. (1981) 'On propagation of elastic P-waves in soils with electrokinetical properties', *Trans. (Izvestia) Acad. Sci. USSR, Phys. Earth*, 3, 47–54.
Muskat, M., and Meres, M. W. (1936) 'Flow of heterogeneous fluids through porous media', *Physics*, **7**, 346.
Nikolaevsky, V. N. (1967) 'On the relation of volume and plastic deformations and on shock waves in soft soils', *Proc. (Doklady) USSR Acad. Sci.*, **177**, 3, 542–545.
Nikolaevsky, V. N. (1971) 'Governing equation of plastic deformation of granulated media', *Sov. Appl. Math. Mech. (PMM)*, **35**, 6, 1070–1082.
Nikolaevsky, V. N. (1979) 'The Mohorovicic boundary as limit depth of brittle–dilatant state of rocks', *Proc. (Doklady) USSR Acad. Sci.*, **249**, 4, 817–821.
Nikolaevsky, V. N. (1982) 'Deformations of geomaterials and porous media', *Trans. (Izvestia) Acad. Sci. USSR. Mech. Solids*, 2, 96–109.
Nikolaevsky, V. N., and Ramazanov T. K. (1982) 'Stress–strain state of the layer with account of water flow', *Phys. Techn. Probl. Mineral Res. Recovery*, 5, 37–45 (in Russian).
Nikolaevsky, V. N., and Rice, J. R. (1979) 'Current topics in non-elastic deformation of geological materials', in *High-pressure Sci. & Techn.*, **2**, 455–464.
Nikolaevsky, V. N., Basniev, K. S., Gorbunov, A. T., and Zotov, G. A. (1970) *Mechanics of Saturated Porous Media*. Moscow, Nedra (in Russian).
Paterson, M. S. (1978) *Experimental rock deformation. The brittle field*. Berlin, Heidelberg, New York, Springer Verlag.
Raleigh, C. B., and Paterson, M. S. (1969) 'Experimental deformation of serpentinite and its tectonic implications', *J. Geophys. Res.*, **70**, 3965–3985.

Reynolds, O. (1885) 'On the dilatancy of media composed of rigid particles in contact', *Phil. Mag.*, ser. 5, **20**, 127, 469–481.
Rice, J. R. (1975) 'On the stability of dilatant hardening for saturated rock masses', *J. Geophys. Res.*, **80**, 11, 1531–1536.
Rudnicki, J. W., and Rice, J. R. (1975) 'Conditions for the localization of deformation in pressure-sensitive dilatant materials', *J. Mech. Phys. Solids*, **23**, 6, 371–394.
Samarov, E. L. (1982) 'Spherical shock wave in dilating water-saturated soils', in: *Numerical Methods of theory of elasticity and plasticity*, Novosibirsk, USSR Academy of Sciences, 97–108.
Samarov, E. L. (1983) 'Expansion of explosion cavity in dilatant plastic soils', *Trans. (Izvestia) Acad. Sci. USSR, Phys. Earth*, 2, 68–76.
Sharov, V. I., and Grechishnikov, G. A. (1982) 'On behavior of tectonic faults at different depth levels of earth crust according to method of reflected waves', *Proc. (Doklady) USSR Acad. Sci.*, **263**, 2, 412–416.
Shield, R. T. (1953) 'Mixed boundary value problems in soil mechanics', *Quart. Appl. Math.*, **11**, 1, 61–75.
Sibson, R. H. (1982) 'Fault zone models, heat flow and the depth distribution of earthquakes in the continental crust of the United States', *Bull. Seism. Soc. Amer.*, 1, 159–163.
Smoluchowski, M. (1909) 'Einige Bemerkungen über die physikalischen Grundlagen der theorien der Gebirgsbildung', *Kosmos, Lwow*, **34**.
Terzaghi, K. (1925) *Erdbaumechanik*, Wien, F. Deuticke.
Vedernikov, V. V., and Nikolaevsky, V. N. (1978) 'The equations of mechanics of porous media saturated by two-phase fluid', *Trans. (Izvestia) Acad. Sci. USSR, Mech. Fluid & Gas*, 5, 165–168.
Vilchinska, N. A. (1982) 'Repacking' wave in sands and acoustical emission', *Proc. (Doklady) USSR Acad. Sci.*, **262**, 3, 568–572.
Warren, J. E., and Root, P. J. (1963) 'The behavior of naturally fractured reservoirs', *J. Soc. Petrol. Eng.*, **3**, 3, September.
Wilson, I. W. (1956) 'Determination of relative permeability under simulated reservoir conditions', *AICHE J.*, March.
Zhukovsky, N. E. (1889) 'Theoretical study of underground water motions', in *Selected Papers*, Vol. 7, Moscow, Gosizdat (1937).

PART VII

CREEP, SHRINKAGE, AND AGEING

Mechanics of Geomaterials
Edited by Z. Bažant
© 1985 John Wiley & Sons Ltd

Chapter 18
Creep and Thermal Effects in Ageing Solids

C. A. Anderson

18.1 INTRODUCTION

Professor Prager devoted much of his life to the development of simple material models to represent the behaviour of solids beyond the elastic range. Thus, his name is associated closely with the development of modern theories of plasticity (for example, the perfectly plastic solid). In addition he later became interested in numerical methods and computing and how numerical techniques could be used to provide solutions to problems in the field of solid mechanics. In the last 20 years we have seen a tremendous growth in the field of computational solutions in mechanics—promoted primarily by advocates of the finite element method—and today numerical methods allow us to solve multidimensional problems of solid mechanics where the materials obey complex constitutive laws. Thus, with the advent of numerical methods and large capacity computing machines to solve complex analytical problems, there is a re-awakening to the need to develop physically based constitutive laws to represent real material behaviour.

The purpose of this chapter is to describe constitutive models that have been developed for rocks and concrete to represent creep, thermal, stress, and ageing effects. Also described in the paper are how these constitutive models are being used to numerically predict the behaviour of solids in such diverse situations as ageing creep of concrete structures and creeping mantle convection. In all cases low strain rate, compressive material behaviour will be the main interest.

For the time-dependent behaviour of concrete under long-term loads, viscoelastic models have been developed and applied to the numerical analysis of the behaviour of concrete structures. A commonly applied theory for concrete structures is the linear, ageing viscoelastic theory for which the superposition principle applies. Section 18.2 discusses the theory for ageing concrete creep including expansion of the creep functional in a finite Dirichlet series that enables the numerical solution of practical concrete creep problems. A numerical example of calculating the creep of a concrete structure is illustrated. How the theory is altered to account for temperature is also discussed.

Section 18.3 examines temperature and stress effects in creep of rocks with the ultimate goal of developing a capability to realistically model material flow, temperature and deformation in the Earth's crust and mantle. The major difficulty in modelling geodynamic phenomena today is the complex rheological behaviour of rocks. Depending on conditions of temperature, pressure and stress, rocks can act as elastic or viscoplastic solids, brittle solids subjected to fracture, or even viscous fluids. A successful predictive capability must be able to cope with these contrasting rheological properties within a single computation because the interaction between materials which behave differently (for example, the coupling of an elastic lithosphere to an underlying viscous mantle) is an essential aspect of most unsolved geological problems.

However, the fluid-like mantle itself presents difficulties for numerical modelling. The rheology is that of a power-low fluid with a temperature, stress and pressure dependent effective viscosity. The rheological contrasts between lithosphere and underlying mantle can partly be accounted for by the relatively large temperature change across the surface thermal boundary layer of a power law fluid with highly temperature dependent effective viscosity. This will produce huge effective viscosity variations across the surface boundary layer and a nearly rigid cold upper lithosphere. Large variations in effective viscosity will also occur across any other boundary layers that may exist in the mantle, for example, at the core–mantle interface or at interfaces between convection cells in a vertically layered mantle. It is possible that abrupt changes will take place in the creep behaviour of mantle rocks when they move through the major solid–solid phase transitions at depths of about 400 and 700 km. The large effective viscosity variations and changes in rheological behaviour that are expected to occur across the narrow boundary layers and transition zones of a vigorously convecting mantle exacerbate the demands on resolution required of numerical simulations.

Among the presently available numerical approaches, the method of finite elements offers the most promise of eventually coping with these enormous difficulties. Not only can arbitrary geometries be described with finite element meshes, but they can also be readily designed with relatively coarse resolution almost everywhere and high resolution only where required in thin boundary layers. This can represent a considerable savings in computer time over traditional finite difference schemes with uniform grids. The rheology of each element in a mesh can be individually specified thereby facilitating the simultaneous incorporation of contrasting rheologies into a single flow or creep calculation. Finite element schemes have other advantages for mantle convection modelling. The ability of elements to deform will permit realistic simulations of surface deformation above a convecting medium with flexural rigidity near its surface. The adaptability of finite elements to a Lagrangian time stepping method makes it possible to study convection in an inhomogeneous mantle and the convective mixing and dispersal of chemically identifiable species.

The basic equations for creeping flow of rocks, including energy transport, are described in Section 18.3 together with the currently accepted creep models for rocks subjected to high temperature ($T > 0.5\,T_m$, where T_m is the liquidus temperature) and stress. The basic equations have been implemented in a two-dimensional finite element code, which is currently being used to study Rayleigh–Benard convection in a model of the Earth's upper mantle. The finite element model is characterized by meshes that provide high resolution in very thin boundary layers. Calculations are illustrated to show the effect of temperature on mantle creep for the cases of base heating and combined internal and base heating, which is the case most relevant to the Earth's mantle. Extension to stress and temperature-dependent creep behaviour of the material making up the Earth's mantle is discussed as well as how ageing effects are important and can be factored into such models.

18.2 CREEP, THERMAL, AND AGEING EFFECTS IN CONCRETE

The physical mechanisms underlying the creep of concrete, just now being fully understood, are very complex (Bažant, 1975; Wittmann, 1982) and their complete incorporation into numerical codes for the prediction of the creep of complex concrete structures is still impossible. Useful numerical results can often be obtained though, by using simplified creep constitutive models that do not yet incorporate the complicating effects of moisture change, temperature change, and an extended working range of stress and strain. In this section we will first present the simplest such constitutive model, the so-called linear ageing viscoelastic model, that has been proposed and often used for the numerical creep analysis of massive concrete structures. Later in the section we will discuss the effect of temperature and how this effect is accommodated in the constitutive law.

18.2.1 The linear ageing model

The characteristic of concrete that distinguishes it from the traditional viscoelastic material is the ageing effect. Thus, as a function of time the constitutive law is changing through the chemical action of hydration, and it is important to its creep response as to when in its lifetime an ageing viscoelastic structure is loaded. Fortunately, experiments indicate that the response due to an increment in load is independent of all other past load increments, and the superposition principle applies. Thus, for small increments in stress $d\sigma = \dot{\sigma}(t')dt'$ occurring at t' measured from the time of casting, we have

$$\varepsilon(t) = \int_0^t J(t, t')\dot{\sigma}(t')dt' \qquad (18.1)$$

where $J(t, t')$ is called the creep function.

Numerical creep analysis based on the stress–strain law of Equation (18.1) may be performed by subdividing the total time interval of interest into time steps Δt

and discrete times $t_r(r = 1, 2, ...)$. The integral in Equation (18.1) can then be approximated by finite sums involving incremental stress changes over the time steps. Details of this method, which is generally applicable to any form of the creep function $J(t, t')$, are given in Bažant (1975). Because the numerical method results in extensive storage and computational requirements, it has been superseded by methods that involve approximating the creep function $J(t, t')$ by a Dirichlet series and thus tying the constitutive model physically to Kelvin chain models (or Maxwell chain models if the relaxation function is approximated), which then yield the structural equations.

The creep function $J(t, t')$ is approximated by a series of real exponentials of the form

$$J(t, t') = \frac{1}{E(t')} + \sum_{i=1}^{n} \frac{1}{E_i(t')} \{1 - \exp[-(t - t')/\tau_i]\} \qquad (18.2)$$

in which τ_i are constants called retardation times and E_i are ageing coefficients. When this function is introduced into the superposition integral (Equation (18.1)) the integrand degenerates into the product of a function of t' and a function of t. The latter function does not involve the variable of integration and can be extracted from the integral, leaving only an integration of functions that are independent of t. Thus, at each new time step, it is only necessary to compute the change in value of the integral from the last time step rather than from the time of initial loading, as is required in a general case. Using this method Bažant and Wu (1973) describes a completely stable numerical method for obtaining computational solutions to concrete structural creep problems.

Restricting ourselves to situations of one-dimensional stress, it can be shown that the Dirichlet series is the solution to the system of ordinary differential equations

$$\dot{\varepsilon} = \dot{\sigma}/E + \sum_{i=1}^{n} \dot{\varepsilon}_i$$

$$\dot{\varepsilon}_i = \sigma_i/\eta_i \qquad i = 1, 2, ..., n \qquad (18.3)$$

$$\dot{\sigma} = \dot{\sigma} + E_i \dot{\varepsilon}_i \qquad i = 1, 2, ..., n$$

when a unit step stress $\sigma(t)$ is applied at time t'. This system of equations, which corresponds to the physical system shown at the top of Figure 18.1 and with variables E, E_i, η_i as shown, is called the Kelvin chain model. Since the formulation defined by Equation (18.3) states the relations between the rates of stress and strain, it is referred to as a rate-type formulation.

Another formulation of the viscoelasticity problem is through the use of the relaxation function $G(t, t')$ rather than the creep compliance function $J(t, t')$,

$$\sigma(t) = \int_0^t G(t, t') \dot{\varepsilon}(t') dt' \qquad (18.4)$$

If the relaxation function is expanded in a Dirichlet series and substituted into

Creep and Thermal Effects in Ageing Solids

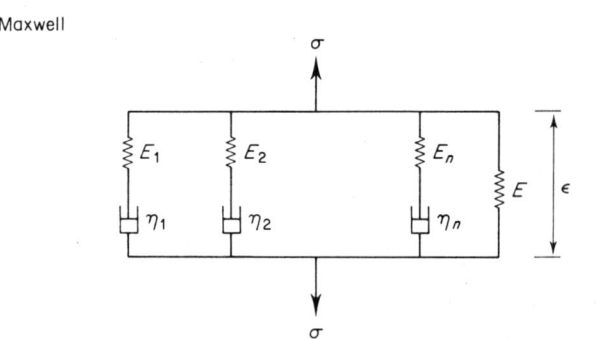

Figure 18.1 Kelvin and Maxwell chains for ageing viscoelastic solids

Equation (18.4), it can be shown that the differential equations (for one-dimensional stress)

$$\sigma = \sum_{i=1}^{n} \sigma_i$$

$$\dot{\varepsilon} = \frac{\dot{\sigma}_i + \sigma_i/\tau_i}{E_i} \qquad i = 1, 2, \ldots, n \qquad (18.5)$$

are obtained (Bažant and Wu, 1974). This system of rate equations corresponds to the physical system shown at the bottom of Figure 18.1 and is called the Maxwell chain model. The quantity, $\tau_i = \eta_i/E_i$, is the relaxation time of the ith unit of the Maxwell chain. As with the Kelvin chain model described previously, incremental stress–strain laws for the Maxwell chain model can be formulated. As with the Kelvin model the stress history in the incremental law is defined by a recurrence relation, and the need to sum the complete stress history at each time step is eliminated.

Kelvin and Maxwell chain models can be used interchangeably to solve creep problems provided the relaxation function G can be determined from J, or vice versa. For example if G is smooth and there is no ageing effect then one can solve

$$G(0)J(t) + \int_0^t G(s)J(t-s) = 1 \qquad (18.6)$$

to obtain the creep function J while for ageing material the general relation Equation (18.1) must be solved. In general, advantages accrue to the Maxwell model when the effects of temperature and humidity change are included since these involve summing stored stress variables.

18.2.2 Modelling temperature effects in ageing viscoelastic solids

Accommodating the Kelvin chain model to pore moisture and temperature changes is not possible based on the underlying physics of the creep mechanism for concrete. On the other hand, as shown in Bažant (1975), incorporating the temperature effect into the creep law corresponding to the Maxwell model is rather simple. Equation (18.5) is rewritten as

$$\sigma = \sum \sigma_i \qquad \dot{\varepsilon} = \frac{\dot{\sigma}_i}{E_i(t_e)} + \frac{\sigma_i}{\eta_i(t_e)} \qquad (18.7)$$

where η_i is the age-dependent viscosity associated with the ith Maxwell unit, which equals $\tau_i E_i$ at constant temperature T. Since creep is a thermally activated process, it is known from physics that η_i should depend on temperature according to the Arrhenius equation,

$$\frac{1}{\eta_i} = \frac{1}{\eta_0} \exp\left[\frac{U_i}{R}\left(\frac{1}{T_0} - \frac{1}{T}\right)\right] \qquad (18.8)$$

where T_0 is the reference temperature, η_0 is the value of η at T_0, R is the universal gas constant, and U_i is the activation energy of the ith Maxwell unit. The effect of temperature on ageing is represented by making E_i and η_i dependent on t_e rather than on t, where Δt_e is an equivalent hydration period that yields the same degree of hydration at temperature T as occurs during a period Δt at temperature T_0. This equivalent time is given by

$$\Delta t_e = \beta_T \Delta T \qquad (18.9)$$

with

$$\beta_T = \exp\left[\frac{U_h}{R}\left(\frac{1}{T_0} - \frac{1}{T}\right)\right] \qquad (18.10)$$

in which U_h equals the activation energy for hydration.

18.2.3 Application

The constitutive equations for the Kelvin or Maxwell models of ageing viscoelastic materials have been incorporated into several US and European concrete creep analysis computer codes. In particular Anderson (1982) describes how the constitutive laws are incorporated in a large finite element code for the creep analysis of complex three-dimensional concrete structures, and many

Creep and Thermal Effects in Ageing Solids 411

examples are given. Below is illustrated one such calculation that indicates the current state of the art in concrete creep analysis. Temperature effects were not considered in this problem.

Figure 18.2 illustrates a concrete ring that is post-tensioned by two cables on the exterior surface as shown. In this problem the steel cables are elastic whereas

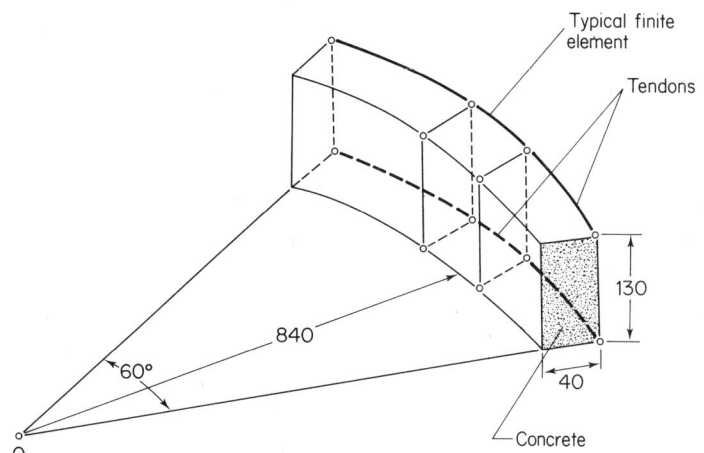

Figure 18.2 Post-tensioned concrete ring finite element model

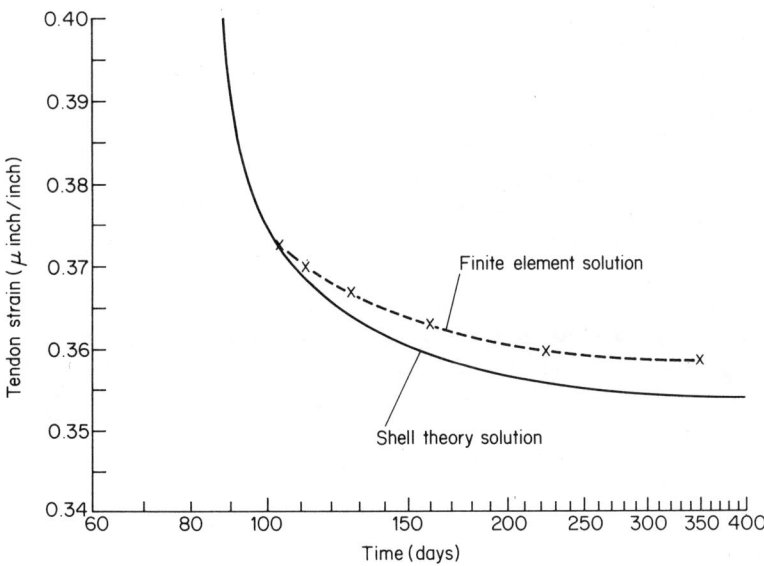

Figure 18.3 Variation in post-tensioning strain with time

the concrete creeps according to the viscoelastic ageing relation of the Dirichlet series form given in Equation (18.2). A two-term approximation was used. For this example the coefficients $E_i(t')$, $i = 1, 2$ are given by

$$\frac{1}{E_i(t')} = a_i + b_i(t')^{n_i} \tag{18.11}$$

where $a_1 = a_2 = 7.5 \times 10^{-9}$, $b_1 = b_2 = 0.233 \times 10^{-6}$ and $n_1 = n_2 = -0.333$. The relaxation times were taken to be $\tau_1 = 5.6d$ and $\tau_2 = 56d$, respectively.

Figure 18.3 illustrates the variation of the post-tensioning strain in the cable as a function of time out to $400d$ starting from an initial strain in the cable of 0.0005 applied at $90d$ from concrete casting. Also shown for comparison is a shell solution to this problem for a complete composite concrete steel shell with the same geometry, material quantity, and material properties as in the finite element model. At $350d$ there is a 1.3 per cent difference in the finite element cable strain and the cable strain predicted by the shell theory solution. Further details are given in Anderson (1982).

18.3 CREEP, THERMAL, AND AGEING EFFECTS IN ROCKS

There are many phenomena in the Earth that testify to the importance of long-term creep processes. These include geoidal flattening of the Earth as a whole, plate tectonic motions in the outer hundred kilometres of the Earth driven by thermal creep of the underlying mantle, and episodal creep events along major faults down to grain-scale ductile flow. Fully understanding these phenomena will eventually require the development of constitutive models that take into account temperature, stress, and ageing effects on the creep rates of rocks and implementation of these constitutive models in numerical procedures that can predict the large-scale phenomena that are being observed. These models can then be calibrated using the observed data such as surface velocity, heat transfer, and surface deformation. Finally, with a well-calibrated model basic questions of time history of evolution of the Earth's structure and questions of ageing can be more confidently addressed. This section briefly describes constitutive laws that have been proposed for modelling high temperature creep of rocks causing mantle convection as well as detailed calculations with a finite element code that describe high Rayleigh number convection of the mantle from combined internal and base heating.

The elastic and inelastic behaviour of rocks at low (near surface) stresses and temperatures is dominated by fractures on all scales. Here compaction and dilation are important processes with the resulting inelastic strains mainly due to primary and tertiary creep. However, at absolute temperatures greater than one-half the melting temperature, most large-scale ductile deformation of rocks occurs as secondary or steady-state creep. Here the flow involves plastic deformation resulting from dislocation motion (Heard, 1982).

Useful laboratory steady-state creep results for uniaxial stress are available for a number of rocks over wide ranges of strain-rate, temperature and driving stress. Among these are ice, halite (rock salt), marble, limestone, dolomite, anhydrite, quartzite, granite, dunite, pyroxenite, and peridotite. All of these steady-state creep results are well-fit by a relationship of the form (Heard, 1982; Carter, 1976)

$$\dot{\varepsilon} = \frac{A}{T}\mu \exp[-(Q+pV^*)/RT]\left(\frac{\sigma}{\mu}\right)^N \qquad (18.12)$$

where Q and V^* are activation energies and volumes, μ is the shear modulus and A and N are material constants. In Equation (18.12) R is the universal gas constant and T, p and σ are absolute temperature, pressure, and driving stress respectively. In many cases Equation (18.12) can be simplified to the Weertman form

$$\dot{\varepsilon} = B \exp[-(Q+pV^*)/RT]\sigma^N \qquad (18.13)$$

where B is a material constant. For the previously mentioned solids values of Q vary from 120 to 500 kJ/mol and N from 3 to 9. Data on V^* are quite limited.

Numerical calculation using the constitutive model Equation (18.13) have been presented for the creep behaviour of underground cavities (Anderson, 1976) and geotechnical structural models that represent rifts and rock folds (Anderson and Bridwell, 1980; Bridwell and Anderson, 1981) usually driven by direct stresses caused by applied loads or material density discontinuities. Also, in Anderson (1981) analytical solutions are given for an underground spherical cavity for the case where the surrounding medium is a constant viscosity solid. Currently accepted models for plate tectonic motion, on the other hand, are driven not by direct stresses but by drag forces developed during slow creep or convection of the underlying mantle. The equations describing this process of mantle convection are inherently nonlinear even for constant viscosity situations, and have in the past been solved by boundary layer or finite difference numerical techniques (Schubert et al., 1976; Jarvis and Peltier, 1982). However, these methods often do not handle the physically interesting cases of non-constant (for example, stress, temperature, or pressure dependent creep given by Equations (18.12) or (18.13)) viscosity or the situation of internal heating of the material of the mantle caused by radioactive decay. The method of analysis described below does not possess these restrictions.

We now reconsider the problem of Rayleigh–Benard convection in a square smooth box. The basic equations that describe the steady creep of the material of the box under thermally induced buoyancy forces are the equilibrium equations, the constitutive law for the relation between deviatoric stresses (s_x, s_y, τ_{xy}) and velocity (u, v) gradients, the incompressibility equation, and the energy transport equation. Written with respect to an Eulerian reference frame and using the usual

Boussinesq approximation, these equations are

$$\frac{\partial s_x}{\partial x} + \frac{\partial \tau_{xy}}{\partial y} - \frac{\partial p}{\partial x} = 0$$

$$\frac{\partial \tau_{xy}}{\partial x} + \frac{\partial s_y}{\partial y} - \frac{\partial p}{\partial y} = -\rho g \alpha (T - T_0) \tag{18.14}$$

$$\frac{\partial u}{\partial x} = \frac{1}{2\mu} s_x, \frac{\partial v}{\partial y} = \frac{1}{2\mu} s_y, \frac{\partial u}{\partial y} + \frac{\partial v}{\partial x} = \frac{1}{\mu} \tau_{xy} \tag{18.15}$$

where α is the bulk coefficient of expansion and where $\mu = \mu(T, \sigma_{ef}, p)$ is the viscosity as a function of temperature and octahedral shearing stress, and pressure. The viscosity in Equation (18.15) can be derived from Equations (18.12) or (18.13). For details see Anderson and Bridwell, (1980). Also,

$$\frac{\partial u}{\partial x} + \frac{\partial v}{\partial y} = 0 \tag{18.16}$$

is the incompressibility requirement and

$$\rho C_p \left(u \frac{\partial T}{\partial x} + v \frac{\partial T}{\partial y} \right) = k \nabla^2 T + Q \tag{18.17}$$

is the energy transport equation where ρ is the density, C_p is the heat capacity, k is the thermal conductivity and Q is the internal volumetric heating. In the above equations terms of the order of products of velocity have been neglected since velocities can be shown to be small ($\sim 10^{-10}$ m/s) in mantle convection.

The equations embodied in (18.14)–(18.17) have been discretized by the finite element method using a six-node triangle with quadratically interpolated velocities and linearly interpolated pressures and implemented in a computer program (Sato and Thompson, 1976; Thompson, 1979). This computer program, called MANTLE, has been compiled on the Los Alamos computer system, modified for random access memory, energy checks and balances, graphics, and other features and used in the numerical studies of creep and thermal effects in the Earth's mantle that are described below. The nonlinear terms in Equation (18.17) were handled by iterating between the flow equations, which determine u, v and p, and the energy transport equation, which determines T.

The problem that we have investigated is represented by a two-dimensional square box containing a creeping solid (or a very viscous fluid) that is heated on its under side and internally, cooled on the top, and insulated on the vertical sides. All boundaries are mechanically rigid and smooth. When the temperature difference between the bottom and top is small enough, energy transport occurs only by conduction in the solid. However, as the bottom temperature (or flux) is raised, a critical temperature is observed (both experimentally and analytically) at which the material in the box begins to convect heat as well as conduct it. In the

Creep and Thermal Effects in Ageing Solids

absence of internal heating the critical temperature difference for onset of flow of a constant viscosity solid in a square smooth box of the length L can be determined analytically by the critical Rayleigh number R_c,

$$R_c = \frac{\alpha g L^3}{\kappa \nu}(T_B - T_0) = 779.273 \qquad (18.18)$$

where κ is the thermal diffusivity, ν is the kinematic viscosity and $T_B - T_0$ is the temperature difference. When internal heating is accounted for, the critical temperature difference must be determined numerically.

Figure 18.4 illustrates the problem geometry, material properties, and boundary conditions for the bottom heated convection case. Figure 18.5 illustrates four of the finite element meshes that were used in this study. Material properties and the dimensions of the box were selected so that a base temperature of one degree corresponds to the critical Rayleigh number. For all bottom heated cases the velocity field consisted of a unicellular vortex flow. Figure 18.6 illustrates the horizontally averaged temperatures in a constant viscosity convecting solid using the variable meshes of Figure 18.5 and where the Rayleigh numbers are 1000, 10 000, and 100 000 times the critical value (779.293). For these Rayleigh numbers it can be seen that there are increasingly narrow boundary layers along the top and bottom edges of the box (this is true of the vertical edges,

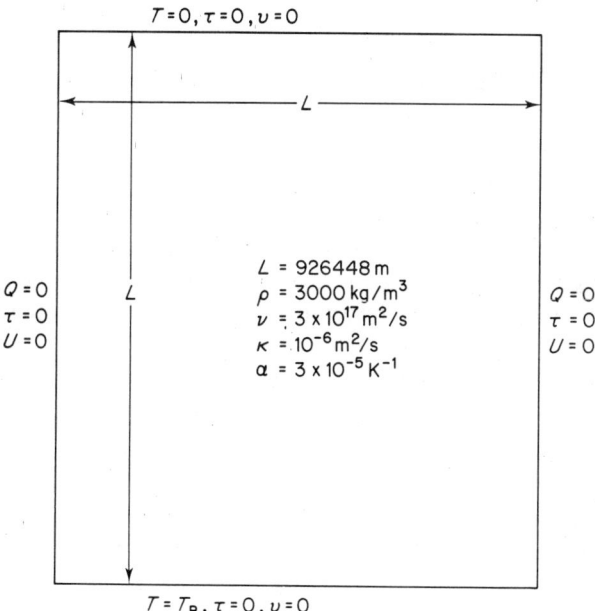

Figure 18.4 Calculational model for bottom heated Rayleigh–Benard convection

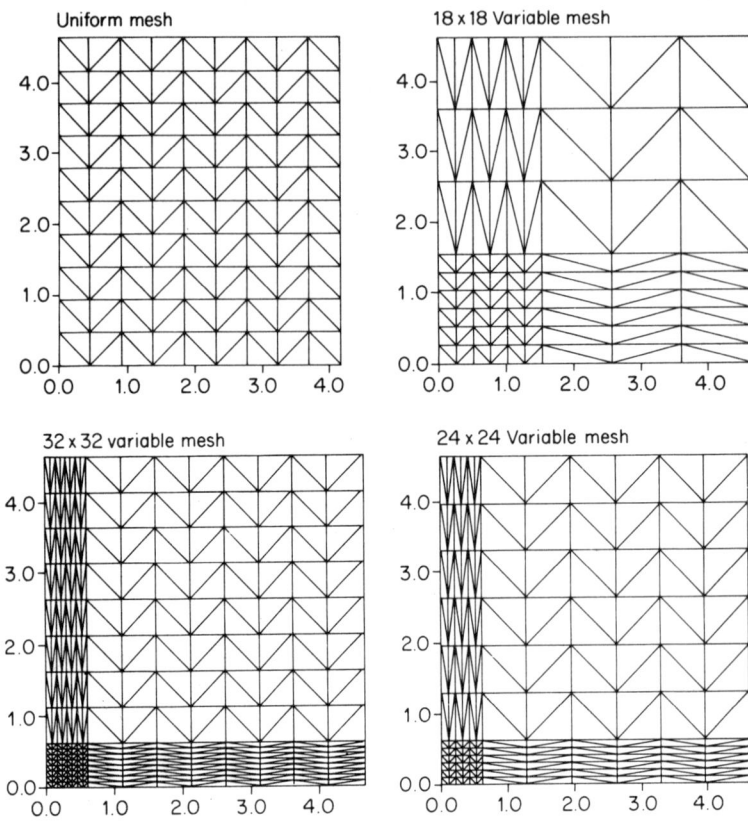

Figure 18.5 Uniform and variable finite element meshes used to study Rayleigh–Benard convection. Only one-quarter of the mesh is shown

also) where the temperature gradients are high, and a central core of material that is nearly isothermal and at the average temperature of the solid. Figure 18.7 shows the contour plots of the temperature field along the left boundary for these three cases and Table 18.1 gives some selected results we have obtained so far for high Rayleigh number constant viscosity convection. Also shown are the results

Table 18.1 Summary of numerical results for high Rayleigh number bottom heated thermal convection

	Finite element		Finite difference	
R_a/R_c	Mesh spacing	Nusselt number	Mesh spacing	Nusselt number
100	18 × 18	9.730	48 × 48	9.625
1000	18 × 18	20.289	96 × 96	20.13
10 000	24 × 24	42.17	96 × 96	40.7
100 000	32 × 32	87.75	—	—

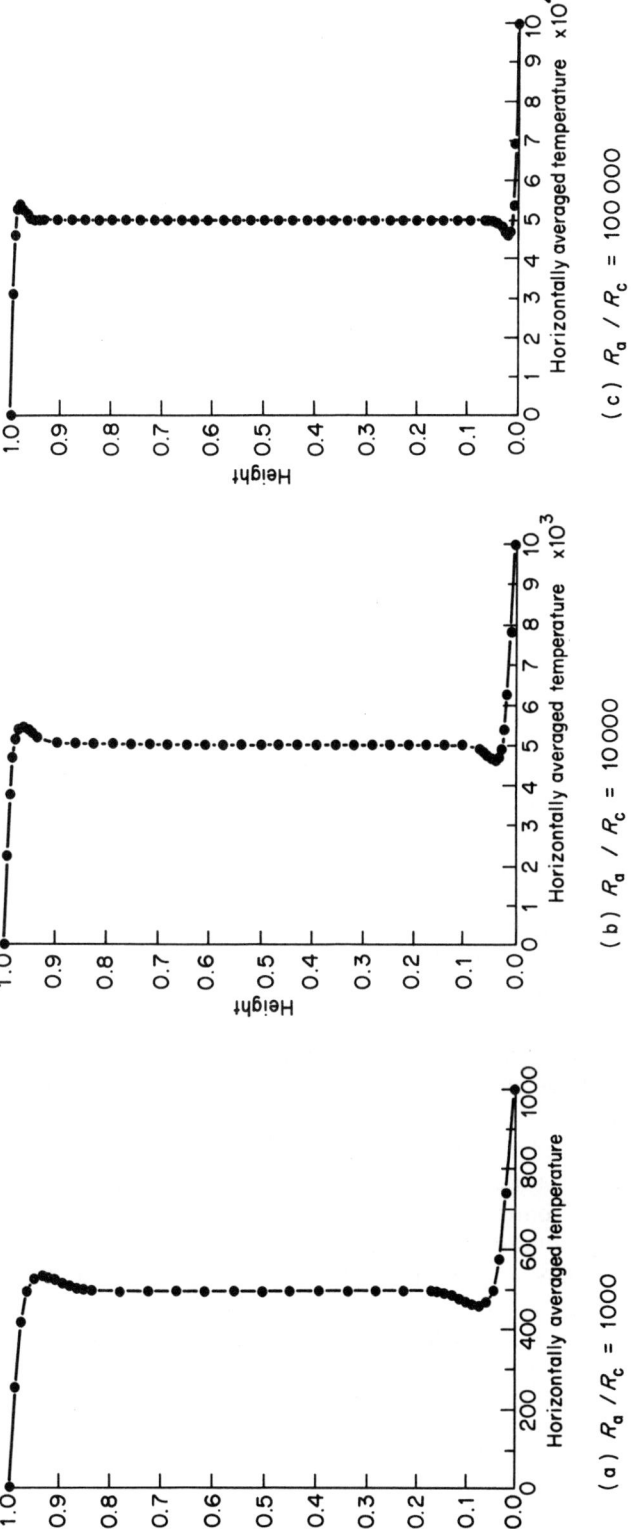

Figure 18.6 Horizontally averaged temperature profiles for constant viscosity solid heated from below

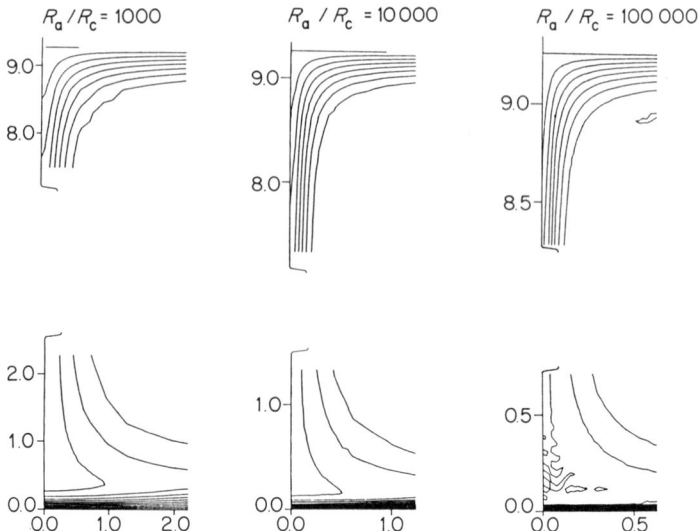

Figure 18.7 Temperature contour plots showing development of narrow boundary layers for $R_a/R_c = 1000$, $10\,000$ and $100\,000$ respectively

of Jarvis and Peltier (1982) for uniform finite difference meshes. The comparison is based on the Nusselt number, which is the ratio of the total heat transmitted out of the top of the box to the heat that would be transmitted if only conduction were occurring. The quantity R_a/R_c is the ratio of the actual to critical Rayleigh number. Details of this study of high Rayleigh number bottom heated mantle convection using variable finite element meshes for resolving thermal boundary layers are given in Schubert and Anderson (1985).

Thermal effects on the material creep behaviour were also investigated. Here the aim is a systematic study of how convection is influenced by increasingly large viscosity variations within the creeping solid. Since, as shown above, temperature variations are confined to narrow boundary layers at high Rayleigh numbers, there will be large effective viscosity variations within these boundary layers. Again the variable mesh capability of a finite element approach is a crucial factor in resolving these large viscosity variations.

Variation of the effective viscosity of mantle rocks as a function of depth is a subject of some controversy. For our purposes we have taken a viscosity variation with temperature of the Arrhenius form (that is, of the form of Equation (18.13) with $V^* = 0$ and $N = 1$). We have selected the parameters B and Q/R so that the effective viscosity at the top of the model, where $T = 0°C$ in all cases, is 1.2×10^{21} Pa-s while at $T = 1000°C$ we assumed the viscosity to be three orders of magnitude less, which is in agreement with certain geophysical data on variation of rock viscosity with depth in the mantle. Calculations with MANTLE were then carried out for the bottom heated thermal convection case with base tempera-

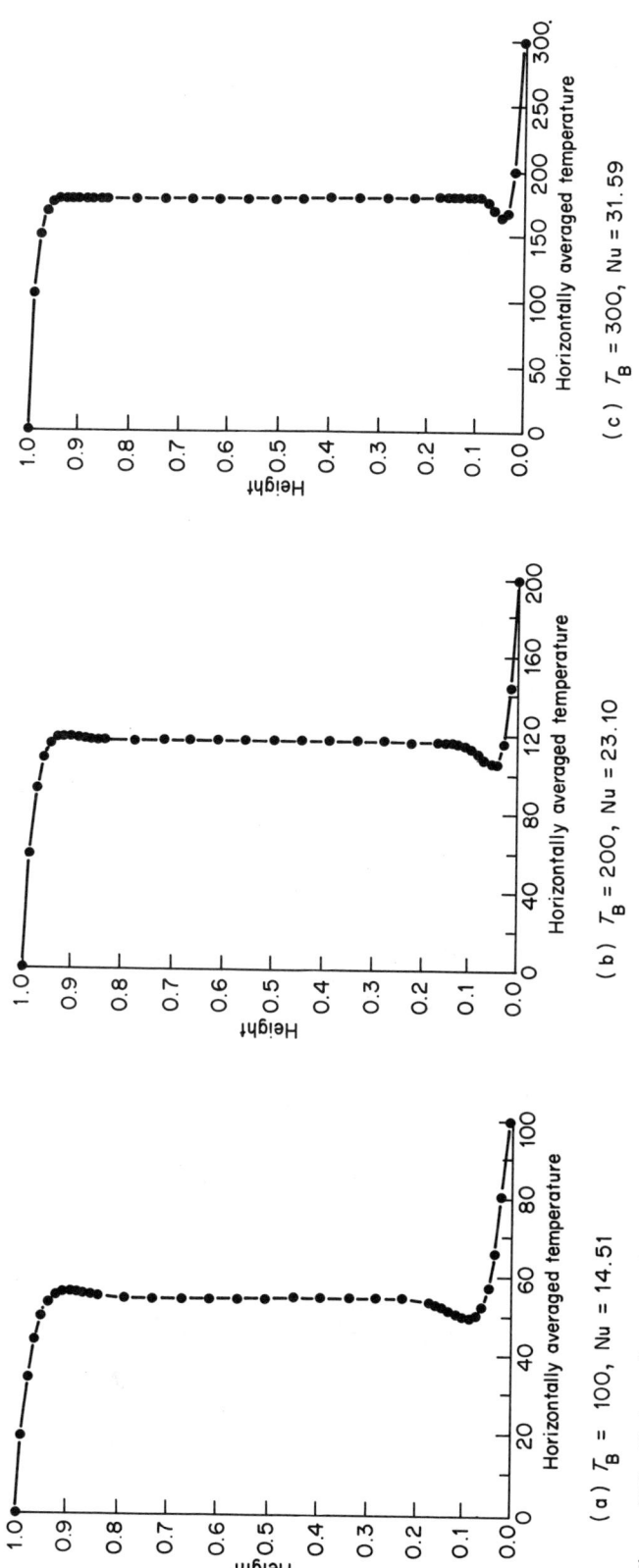

Figure 18.8 Horizontally averaged temperature profiles for temperature-dependent viscosity solid heated from below with base temperatures $T_B = 100$, 200, and 300 respectively

tures equal to 100, 200, and 300 and using the 18 × 18 variable mesh of Figure 18.5. This corresponds to factors of 10, 40, and 100 decrease in viscosity from top to bottom, respectively. The results are summarized in Table 18.2. Here is given the average temperature of the isothermal central core, the viscosity corresponding to that temperature, and the value of R_a/R_c corresponding to the bulk solid viscosity and the base temperature as calculated by Equation (18.18).

The horizontally averaged temperature profiles are shown in Figure 18.8, and Figure 18.9 shows the temperature profiles for the case of bottom temperature equal to 200. Comparing with the results for the constant viscosity cases shown in

Table 18.2 Summary of numerical results for bottom heated thermal convection with temperature dependent viscosity solid

Bottom temperature	Core temperature	Average viscosity	Nusselt number	R_a/R_c
100	55.0	0.28×10^{21} Pa-s	14.506	430
200	116.3	0.089×10^{21} Pa-s	23.103	2700
300	180.5	0.037×10^{21} Pa-s	31.589	10000

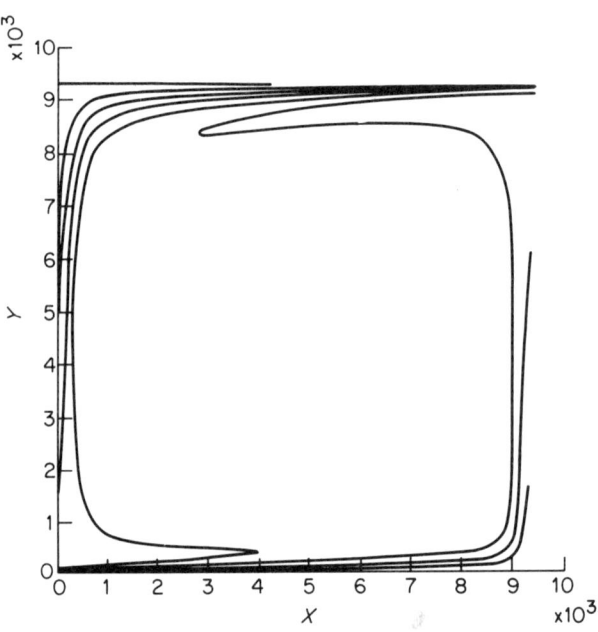

Figure 18.9 Temperature contours for temperature-dependent viscosity solid heated from below with $T_B = 200$

Figure 18.6 and Table 18.1, we see a significant difference in the form of convection caused by the temperature-dependent creep behaviour of the mantle material; the boundary layers become more concentrated at the warmer base of the model as expected, the isothermal region enlarges, and the bulk temperature of the isothermal region is higher than the mean of the base and top temperatures.

We have also carried out studies of convection of the mantle with combined heating from within and below. Here the critical Rayleigh number for convection is not known *a priori* and must be determined numerically. Several cases of varying ratios λ of internal heating to total heating have been investigated. Results will be illustrated for the case $\lambda = 0.6$ where the critical internal heating rate was determined to be 3.1×10^{-12} W/m^3. Table 18.3 illustrates some results for higher internal heating rates using the uniform mesh shown in Figure 18.5.

Whereas the bottom heated case showed only unicellular convection up to very high Rayleigh numbers, the combined heating case often produced solutions composed of one or two convection cells and with two or three boundary layers respectively. Figure 18.10 illustrates the one and two cell temperature profiles for $R_a/R_c = 645$. Figure 18.11 again illustrates the horizontally averaged temperatures for the two solutions obtained at $R_a/R_c = 645$. Calculations were also carried out with the temperature-dependent creep law described previously to ascertain the effect of coupling the thermal profile to the effective viscosity of the mantle material. These results for the heating conditions corresponding to $R_a/R_c = 645$ of Table 18.3 are shown in Figure 18.12. Only the converged two-cell solution was obtained at $R_a/R_c = 645$. Again a significant change in the Nusselt numbers (9.311 vs 7.812) and the temperature profiles is observed.

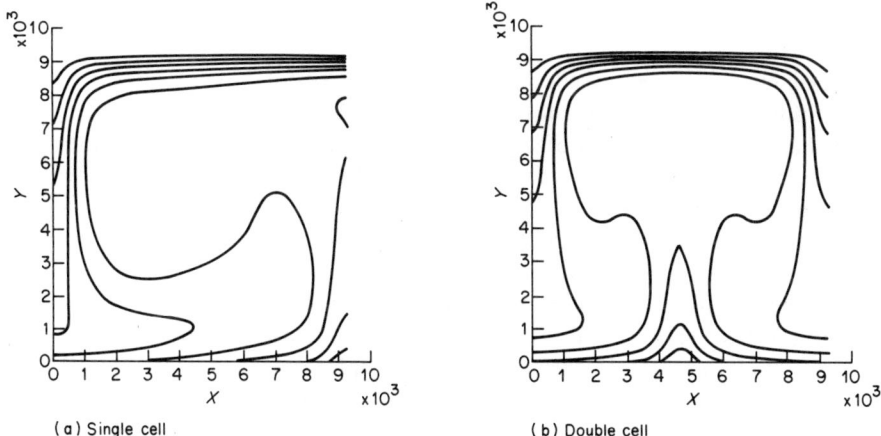

Figure 18.10 One- and two-cell temperature profiles for combined internal and base heating and $R_a/R_c = 645$. Constant viscosity solid

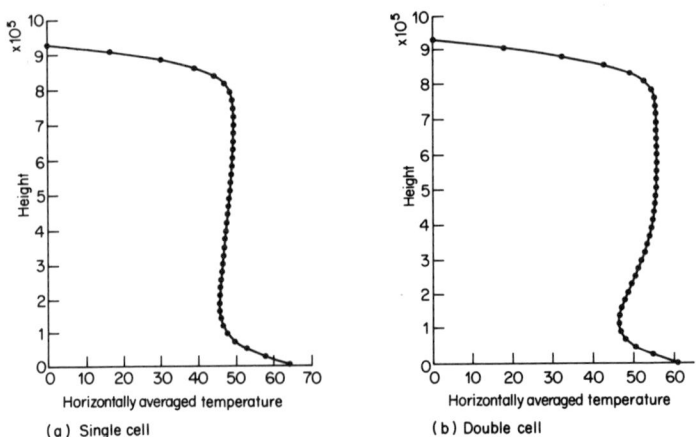

Figure 18.11 Horizontally averaged temperature profiles for combined internal and base heating case with $R_a/R_c = 645$. Constant viscosity solid

Table 18.3 Summary of numerical results for combined internal and base heating $\lambda = 0.6$

R_a/R_c	Mesh spacing	Nusselt number	
		One-cell	Two-cell
161	20 × 20	6.208	5.335
322	20 × 20	7.198	6.487
645	20 × 20	8.204	7.812
968	20 × 20	8.781	8.690
1290	20 × 20	—	9.361
1613	20 × 20	—	9.91

The inclusion of thermal effects on the creep behaviour of rocks will allow us to now address the lithospheric thinning problem. The erosion of the Earth's crust by a thermal plume at its base is undoubtedly an important aspects of continental rifting and the creation of oceanic swells. In this calculation a hot plume will impinge on a thick continental lithosphere and will erode the lithosphere by warming it and carrying away lithospheric material made mobile by virtue of its higher temperature as shown schematically in Figure 18.13. The finite element approach will allow a concentration of elements in the transition region between the lithosphere and mantle to resolve the large viscosity variation across this region. Thinning rates, surface heat flow and uplift rates will be calculated as a function of plume strength; these quantities are geological observables and can accordingly be used to constrain model parameters.

Incorporation of ageing effects in mantle convection and its interaction with the Earth's crust would also be an important consideration. For example, the

Creep and Thermal Effects in Ageing Solids 423

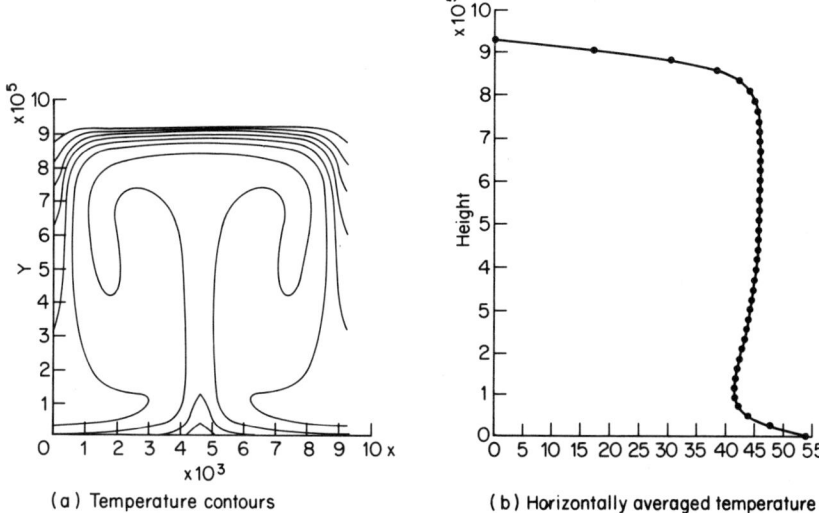

(a) Temperature contours

(b) Horizontally averaged temperature

Figure 18.12 Contour temperature plot showing plume development and horizontally averaged temperature profile for combined heating case with $R_a/R_c = 645$. Temperature-dependent solid

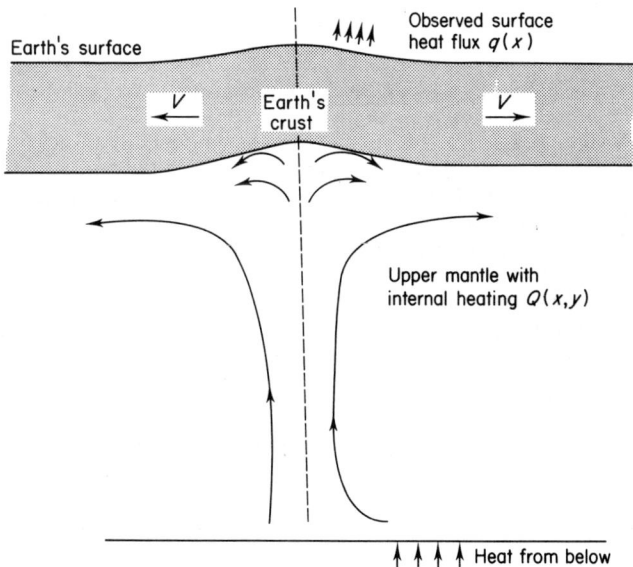

Figure 18.13 Model for studying lithospheric thinning

carbonate rocks thought to make up the mantle could change from the carbonate phase to a chemically reduced material with the attendant release of CO_2 which would significantly affect material (including creep) properties. Partial melt of mantle materials or precipitation of material from a molten state would also affect material properties significantly. Thus, ageing effects, while difficult to incorporate into the current mantle creep models, are important considerations for the prediction of the structural behaviour of the Earth's mantle and crust.

REFERENCES

Anderson, C. A. (1976) 'An investigation of the steady creep of a spherical cavity in a half-space', *J. Appl. Mech.*, **98**.

Anderson, C. A. (1981) 'Boussinesq–Papkovich functions for creep around a spherical cavity or rigid inclusion in a gravity-loaded half-space', *J. Appl. Mech.*, **28**.

Anderson, C. A. (1982) 'Numerical creep analysis of structures', in *Creep and Shrinkage in Concrete Structures*, John Wiley & Sons, pp. 269–303.

Anderson, C. A. and Bridwell, R.J. (1980) 'A finite element method for studying the transient nonlinear creep of geological structures', *Int. J. for Num. and Anal. Methods in Geomechanics*, **4**, 255–276.

Bažant, Z. P. (1975) 'Theory of creep and shrinkage in concrete structures: a precis of recent developments', *Mech. Today*, **2**, 1–93.

Bažant, Z. P., and Wu, S. T. (1973) 'Dirichlet series creep function for aging concrete', *J. Eng. Mech. Div. ASCE*, **99**, EM2, 367–387.

Bažant, Z. P., and Wu, S. T. (1974) 'Rate-type creep law of aging concrete based on Maxwell chain', *Mater. Constr. (Paris)/Mater. Struct.*, **7**, 45–60.

Bridwell, R. J., and Anderson, C. A. (1981) 'Thermomechanical models of the Rio Grande rift', in *Mechanisms of Continental Drift and Plate Tectonics*, Academic Press, pp. 41–59.

Carter, N. L. (1976) 'Steady flow of rocks', *Rev. Geophys and Space Phys.*, **14**, 301–360.

Heard, H. (1982) 'Flow of rocks at depth', Viscoplasticity Workshop Abstracts, Los Alamos National Laboratory, September 1–3.

Jarvis, G. T., and Peltier, W. R. (1982) 'Mantle convection as a boundary layer phenomenon', *Geophys. J. R. Ast. Soc.*, **68**, 389–427.

Sato, A., and Thompson, E.G. (1976) 'Finite element models for creeping convection', *J. Comp. Phys.*, **22**, 229–244.

Schubert, G., and Anderson, C. A. (1985) 'Finite element calculations of very high Rayleigh number thermal convection', *Geophys. J. R. astr. Soc.*, **80**.

Schubert, G., Froidevaux, C., and Yuen, D. A. (1976) 'Oceanic lithosphere and asthenosphere: thermal and mechanical structures', *J. Geophys. Res.*, **81**, 3625–3640.

Thompson, E. G. (1979) 'MANTLE: a finite element program for thermal–mechanical analysis of mantle convection', final report on NASA Grant NGR 06-002-191.

Wittmann, F. H. (1982) 'Creep and shrinkage mechanisms', in *Creep and Shrinkage in Concrete Structures*, John Wiley & Sons, pp. 129–161.

Mechanics of Geomaterials
Edited by Z. Bažant
© 1985 John Wiley & Sons Ltd

Chapter 19
Deformation of Concrete at Variable Moisture Content

F. H. Wittmann

19.1 INTRODUCTION

In modern computerized structural analysis realistic material laws may be taken into consideration. This fact is a real challenge for materials scientists and engineers. Detailed knowledge of structural materials has become an urgent necessity.

In most cases concrete is subjected to mechanical loading under drying conditions. As the drying process is extremely slow, concrete members with usual dimensions do not reach equilibrium for many years. To predict the time-dependent deformation of concrete under load with simultaneous drying, laboratory tests are usually carried out on loaded and unloaded companion specimens. In this way the so-called basic creep and the drying shrinkage can be determined respectively as deformations of a sealed loaded specimen and of a drying but unloaded specimen. It has been found at an early stage, that the total deformation of a drying and loaded specimen is bigger than the sum of the two deformations measured separately.

In Figure 19.1 this situation is schematically shown. The increased deformation is usually explained by special mechanisms such as drying creep or load induced shrinkage (Wittmann, 1982a).

An analogous situation can be provided if the temperature of concrete is changed under load. Consequently both hygral and thermal changes are supposed to cause transient creep (Schneider, 1982).

In this contribution we shall try to explain the material behaviour on the basis of real mechanisms. It will become evident that micromechanics of the composite structure is a necessary key for a better understanding of the macroscopically observed deformation. Finally the significance of empirical terms such as drying creep will be questioned.

This rational approach makes it necessary to subdivide the composite structure of concrete into three levels. Therefore we introduce the term 'Three-level-approach', i.e. the abbreviation TL-approach. We are going to show that

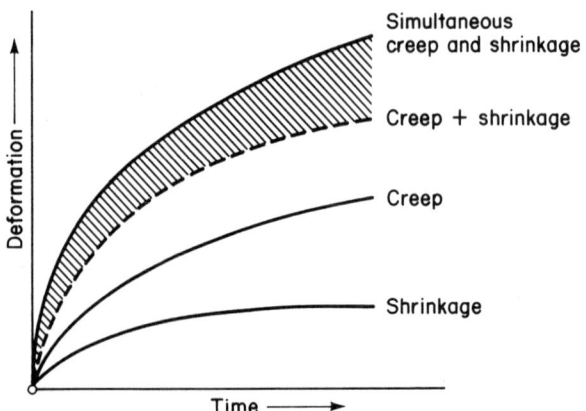

Figure 19.1 Schematic representation of creep, shrinkage, and simultaneous creep and shrinkage. The increased deformation which is observed when creep and shrinkage take place simultaneously is indicated by the shaded area

advanced experimental and numerical methods can be applied simultaneously and that the resulting combination provides us with new and promising results.

19.2 TL-APPROACH

19.2.1 Hierarchic system of three structural levels

Concrete is a heterogeneous material. To characterize its structure it has proved to be advantageous to introduce different levels (Wittmann, 1982b). The hydration products of cement form a matrix in which the aggregates are embedded. We do not yet understand the structure of the hydration products in all its details. Therefore most of the knowledge available concerning the microstructure is summarized in several models. These models represent the microlevel and are called materials science models.

The behaviour of the composite material cannot be linked directly to mechanisms of the microstructure. The effect of pores, inclusions and geometry has to be taken into consideration as well. These particularities are introduced on the mesolevel and the resulting models are called materials engineering models.

Finally, by using the results of the models of the micro- and mesolevel, material laws for the behaviour of concrete can be derived. This is the aim of the work done on the macrolevel and the corresponding engineering models. These models should preferably be presented in such a form that they can be used immediately in advanced numerical analysis.

Characteristic features, mechanisms and models associated with the three

Table 19.1 Three structural levels of concrete and corresponding characteristic features, mechanisms and models

Structural levels	Characteristic features	Mechanisms	Type of models
Microlevel	Structure of hardened cement paste Xerogel	Particle displacement Capillary tension Disjoining pressure Surface free energy	Materials science models
Mesolevel	Pores Cracks, Inclusions	Crack formation and extension Differential stresses	Materials engineering models
Macrolevel	Quasi-homogeneous structural elements	Apparent mechanisms	Structural engineering models Material laws

structural levels are compiled in Table 19.1. Details will be briefly described in the following sections.

19.2.2 Microlevel: hardened cement paste

It is very difficult to investigate the microstructure of hardened cement paste. Most of the classical methods, such as X-ray analysis, can hardly be applied. Early sorption measurements and later electron microscopy proved that the hydration products are in a highly dispersed state with an internal surface of some hundred square metres per gram. Such a coherent solid built up essentially by colloidal particles and having a limited water content can suitably be called a xerogel.

All properties of a xerogel are determined by the average coupling strength of individual particles and are strongly influenced by the interaction of the porous system with water. A lot, if not most, of the information on the microstructure of hardened cement paste has been gained by means of sorption measurements.

The pioneering work of Powers and Brownyard (1947) can be looked upon as a real milestone in this area. Later Powers (1968) condensed his results in a model which is essentially based on the thermodynamics of adsorbed water. According to this model, creep and shrinkage are caused by the same mechanism, i.e. the squeezing out or re-entering of water at the wedge-shaped narrow gaps near points of contact in the xerogel. In case of creep the moisture migration is initiated by an external load and in case of shrinkage by a change of relative humidity. We will come back to this point while discussing real creep and shrinkage mechanisms.

Based on sorption and length-change measurements, Feldman and Sereda (1968) later developed a different model. They assume that hardened cement paste is built up by a layered irregular structure. But in contrast to Power's model

428   Mechanics of Geomaterials

according to Feldman and Sereda the main properties of the xerogel are influenced far more by the removal or re-entry of interlayer water than by adsorbed water. Interlayer water in a colloidal particle might then be squeezed out by an external load just as by a low water vapour pressure thus causing creep or shrinkage.

In an attempt to unify these diverging views, Kondo and Daimon (Taylor 1979) suggested another model for the microstructure of hardened cement paste. In this model there is a distinct difference between intergel particle pores, intercrystallite pores, and intracrystallite pores. A schematic representation of this model is shown in Figure 19.2. On the surface of gel particles within intercrystallite pores there is physically adsorbed water which may influence the properties of the xerogel in such a way as suggested by Powers (1968). Moreover, the interlayer water between layers of crystallites may be removed or it may re-enter as predicted by Feldman and Sereda (1968).

Taylor (1979) and Taylor and Roy (1980) combined recent results of different structural investigations such as trimethylsilylation (TMS) and suggested

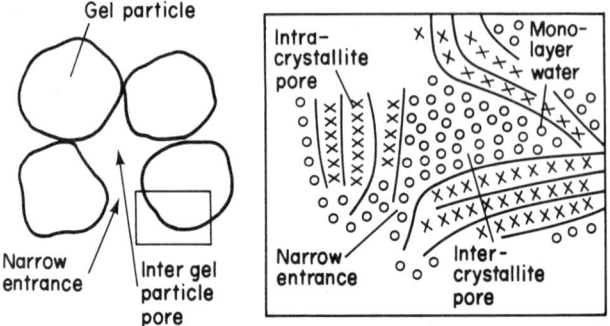

Figure 19.2   Simplified model for the microstructure of hardened cement paste as suggested by Kondo and Daimon

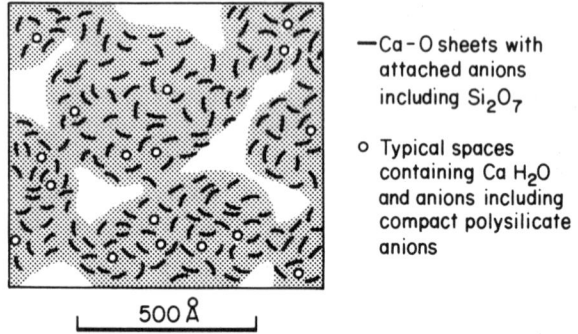

— Ca-O sheets with attached anions including Si_2O_7

○ Typical spaces containing Ca H_2O and anions including compact polysilicate anions

500 Å

Figure 19.3   Suggested structure of type III C-S-H according to Taylor (1979)

another model of the microstructure of type III C-S-H. The schematic representation of Taylor's model is shown in Figure 19.3. It is suggested that subcrystalline order can be confined to the immediate neighbourhood of small fragments of Ca-O layers which are distributed at random and only rarely parallel to each other. The intervening spaces could then be totally amorphous.

Chemical and mineralogical aspects of the microstructure have been recently compiled and treated in detail by J. F. Young (1982). We shall concentrate here, however, more on the mechanical behaviour of the material. Therefore we are going to introduce basic elements of another model: the Munich model for the microstructure of hardened cement paste. This model is up to now the only basis for a quantitative description of creep and shrinkage mechanisms.

The Munich model has been developed on the basis of a number of different approaches such as the direct observation of van der Waals interaction at short distances (Splittgerber and Wittmann, 1974) the study of the coupling of gel particles within a xerogel by means of the Mössbauer effect (Uberlback and Wittmann, 1976) or the determination of the complex permittivity in the range of microwave frequencies (Schlude and Wittmann, 1974) and in the range of Hertzian spectroscopy (Zech and Wittmann, 1974). It has been described in detail in literature (Wittmann, 1976, 1977). Therefore we have to recall relevant aspects only.

It is well known that a liquid droplet having a radius r is under a hydrostatic pressure P:

$$P = 2\gamma/r \qquad (19.1)$$

where γ represents the surface tension of the liquid. In liquids surface tension and surface energy are numerically equal. In solids these two values are at least in the same order of magnitude. In a colloidal system nonspherical particles may occur. Flood (1961) has shown that the mean pressure in solid particles created by surface tension in such a system can be estimated with the help of the following equation:

$$P = 2\gamma S/3 \qquad (19.2)$$

S represents the specific surface area and in this connection has to be expressed, i.e. as cm^2/cm^3. If the specific surface area is introduced instead of the radius, the actual particle size distribution is neglected or rather replaced by a means value. In C-S-H there are particles which are large enough to ensure that no appreciable internal pressure will be created by surface tension. Other particles in the same system will experience comparatively high pressures. The overall response of a system with active and inactive particles has been calculated by Krasilnikov and coworkers (1974). Their paper points out that expansion of gel particles in heterogeneous systems is not linearly related to the expansion of the total system but a geometrical magnification factor has to be taken into consideration.

Well aware of the simplifications implied, we may go back to Equation (19.1).

Now r has to be looked upon as a characteristic value of a given xerogel and P as a mean internal pressure. It is obvious that the resulting internal pressure changes as the surface energy is changed. The surface energy, or rather the interfacial energy of a colloidal system, may be changed by absorption of gases or vapours. If a film of thickness Γ is adsorbed at a given vapour pressure p the interfacial energy measured *in vacuo* decreases by $\Delta\gamma$ (Gibbs, 1975):

$$\Delta\gamma = \gamma_0 - \gamma = RT \int_0^p \Gamma \, d(\ln p) \tag{19.3}$$

If γ of Equation (19.3) is inserted into Equation (19.1) the change of internal pressure caused by a changing surface energy can be calculated. Each individual gel particle expands as the internal pressure is reduced. Bangham and Fakhoury (1931) showed that within certain limits a linear relation exists between the change of interfacial energy and the resulting length change:

$$\Delta l/l = \lambda \Delta\gamma \tag{19.4}$$

Later Hiller (Wittmann, 1974) expressed λ in terms of properties of the colloidal system. It is assumed that in the range of low RH the hygral length change can be described semi-phenomenologically by utilizing Equation (19.4). A more quantitative application of Equation (19.4) is impossible as decisive factors such as the particle size distribution are not sufficiently well known.

As the relative humidity is raised above 50 per cent some surfaces will be separated by the disjoining pressure. This leads to additional expansion of the colloidal system. Such length change is not caused by a corresponding change of surface energy. Therefore Equation (19.4) cannot be applied in this range. Simultaneously the total structure is weakened by the action of disjoining pressure.

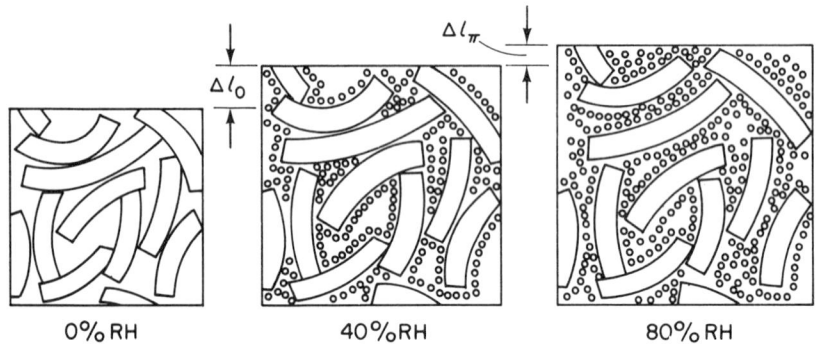

Figure 19.4 Schematic representation of three different stages of the xerogel within hardened cement paste (Munich model): (a) dry state, all particles are compressed by surface tension (see Equation (19.1)); (b) surface free energy is reduced by adsorbed water films and therefore the system expands by Δl_0 (see Equation (19.3)); and (c) additional swelling (Δl_π) is caused by disjoining pressure and some points of contact are interrupted

A simplifying sketch of the essential aspects of the Munich model is shown in Figure 19.4. Later we shall come back to this model.

It has to be pointed out, however, that the xerogel in hardened cement paste is an unstable phase. There is a strong tendency to reach a state of lower energy. Drying and heating both reduce the internal surface thus creating a more stable coarser structure with less energy. A rough estimate shows that if the internal surface is reduced from $200\,\text{m}^2/\text{g}$ to $100\,\text{m}^2/\text{g}$ an amount of energy of $40\,\text{J/g}$ is liberated.

19.2.3 Mesolevel: pores, inclusions, cracks, and internal stresses

In the preceding section we have discussed some details of the microstructure. Those aspects which can be linked with creep and shrinkage mechanisms have been dealt with in particular. So far we have, however, neglected the heterogeneous structure of the material and the influence of the geometry of a specimen on time-dependent deformation. We have to introduce pores, inclusions, and cracks as the main characteristic features of the mesolevel. In addition the internal state of stress created by the drying process has to be considered.

Hardened cement paste is a porous material with a wide range of pore sizes. The total porosity as well as the pore size distribution depends on the water/cement ratio and the degree of hydration. In Figures 19.5 and 19.6 some

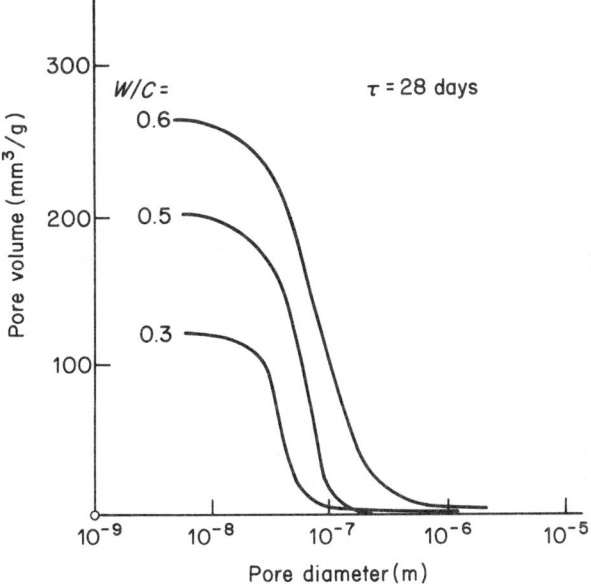

Figure 19.5 Influence of water/cement ratio on pore size distribution of hardened cement paste

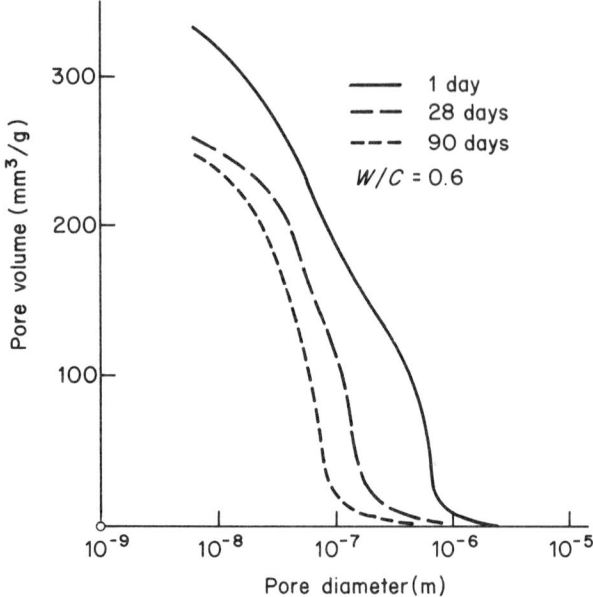

Figure 19.6 Influence of duration of hydration on pore size distribution of hardened cement paste

typical examples are given. The drying process and, as a consequence, the time-dependence of shrinkage of a porous material are governed essentially by the pore size distribution.

Porosity has a strong influence on creep, too. This is mainly due to increasing stress concentrations in the load-bearing solid skeleton as porosity increases. In concrete, in addition to the pores in hardened cement paste, we have compaction pores.

Most aggregates in concrete can be considered as reacting in a linear elastic way and undergoing negligible shrinkage. This is one major reason for the difficulty of relating observed creep and shrinkage strains of concrete with materials mechanisms. The aggregates do not participate in time-dependent deformation but modify the observed behaviour and in particular the internal stress distribution.

If we introduce viscosity η as a measure for creep velocity we can apply existing formulae which were derived to describe the flow behaviour of two-phase materials. For low volume concentrations C_v of aggregates the classical Einstein equation can be used:

$$\eta/\eta_0 = 1 + 2.5C_v + 4.4C_v^2 \tag{19.5}$$

The limits as well as possible extensions of this approach have been pointed out and discussed earlier (Wittmann, 1974).

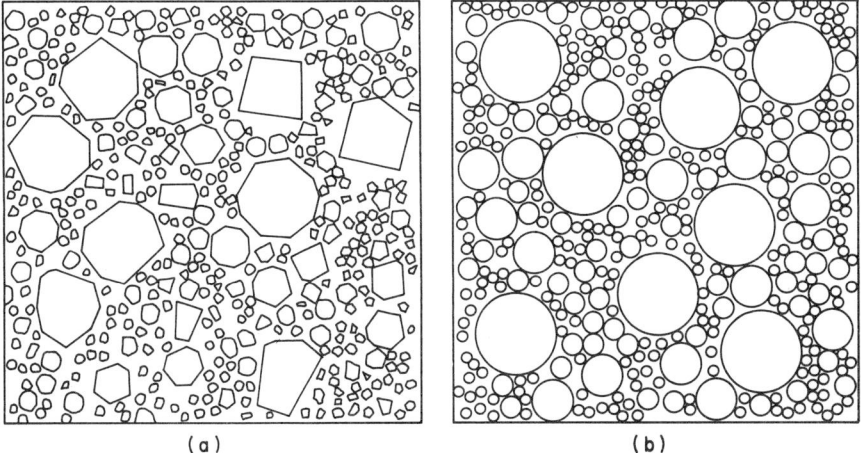

Figure 19.7 A computer simulated composite structure with polygonal (a) and with spherical (b) inclusions

Nowadays we can simulate composite structures of materials by means of computer programs. In this way grain size distribution, cement content, quality of the paste, etc. can be taken into consideration realistically. In Figure 19.7 two examples are given. One is a simulated structure with polygonal inclusions, the other with spherical ones. It is possible to simulate two- and three-dimensional composite structures.

If such a material is loaded elastic energy is gradually being transferred to the higher modulus inclusions. Drying process creates usually both matrix and interfacial cracks. Therefore it is obvious that the material response cannot directly be linked to the mechanisms discussed at the microlevel.

As drying of concrete is an extremely slow process there exists in any structural member a moisture distribution for a very long time. This moisture distribution provokes an internal state of stress which significantly influences the total deformation of a drying and loaded specimen.

Thus we have seen that porosity, inclusions, crack formation, and internal stresses cannot be neglected if we want to understand the behaviour of a porous composite material under load really.

19.2.4 Macrolevel: material laws

Research on creep and shrinkage of concrete has to be oriented towards the application in structural engineering. In order to achieve this aim, all information available has to be condensed in simplified but reliable material laws. On the macrolevel the material is considered to be homogeneous and all heterogeneities and structural defects described on the mesolevel are considered to be smeared out over the total volume.

So far empirical material laws serve to predict materials behaviour under service conditions and examples are relevant national codes or international recommendations. A rather detailed proposal typical for the macrolevel has been published by Bažant and Panula (1978–1979).

It may be mentioned that if one looks in greater detail even on the macrolevel, materials properties vary locally. In high columns the lower parts are denser, and show less creep, and reduced shrinkage due to hydration under increased hydrostatic pressure. In mass concrete the temperature rise due to hydration in the centre increases the degree of hydration and coarsens the structure and thus influences creep and shrinkage. Shrinkage and creep are higher in the surface zones if compared with bulk properties because of the increased content of viscoelastic hardened cement paste.

Although we have seen that, strictly speaking, concrete is inhomogeneous even on the macrolevel, for practical applications simplified material laws have to be used. In research, however, these effects cannot be neglected and further progress will depend on a balanced systematic study of the various structural details entering on the three different levels.

19.3 CREEP OF HARDENED CEMENT PASTE

19.3.1 Short-time creep

We will subdivide the discussion of creep mechanisms in those termed short-time creep and those termed long-time creep. Ruetz (1966) was probably the first to point out that a special short-time creep mechanism exists. Later Sellevold and Richards (1972) and Sellevold (1970) related low frequency internal friction with short-time creep of hardened cement paste. They determined the loss angle of a small vibrating cantilever of hardened cement paste with one free end. This method enables us to observe a distinct transition in the frequency range of about 1 Hz. By using Schwarzl's method (1969) this transition can be related directly with short-time creep. Thus we see that in this range hardened cement paste reacts like a linear viscoelastic solid.

Based on these results it can be concluded that short-time creep is caused by a stress induced redistribution of capillary water within the structure of hardened cement paste. Additional evidence for this hypothesis is provided by the measurements of the electromechanical effect (Wittmann, 1974). In this case an externally applied alternating electric field forces the capillary water due to the ζ-potential to move. In a small bar of hardened cement paste this water movement creates an alternating bending moment. If the frequency is too high the water cannot follow. As soon as the frequency approaches the characteristic frequency of the transition which corresponds to short-time creep the electrically induced bending moment increases and reaches a maximum value.

Thus we can conclude that the mechanism for short-time creep is water movement and redistribution in the porous structure. The thermodynamic approach developed by Powers (1968) seems to be adequate for the description of this type of creep.

19.3.2 Displacements, rate theory

For short-time creep we have identified one creep mechanism. For long-time creep the situation is far more complex. The xerogel forms a solid skeleton in hardened cement paste. This porous system interacts with adsorbed and capillary condensed water. In the unloaded state all gel particles are fixed to their surroundings partly by primary bonds and partly by secondary bonds. It is impossible to indicate quantitatively and in detail the bonding of an individual gel particle. But we can estimate the average coupling force (Uberlback and Wittmann, 1976).

If creep takes place the coupling force of a large number of gel particles has to be overcome so that these particles can leave their original position in the xerogel. This movement can be looked upon as playing the role of displacement of dislocation, and vacancies in crystalline materials.

Long-time creep in hardened cement paste is in fact the consequence of displacement of gel particles and to some extent creep within particles under high concentrated stress. We do not necessarily need to know all the different processes involved in order to describe the creep process realistically. If we know the average values well enough and if the number of particles and events is high enough we can apply rate theory. In a colloidal system such as hardened cement paste these two conditions are fulfilled. Therefore we will try to characterize the essential creep mechanisms by introducing basic elements of rate theory.

All gel particles are fixed to their position of rest by a characteristic coupling force. If one wants to separate a given gel particle from its surroundings so that it can be fixed in a new position, a certain energy Q is required. This energy is called activation energy of the creep process. The probability P that a particle is removed by thermal energy is given by:

$$P = P_0 \exp\left(-\frac{Q}{RT}\right) \qquad (19.6)$$

In Figure 19.8 the coupling of the colloidal particles is shown schematically by a potential trough. It can be seen that all particles which have at least the activation energy Q can leave the trough. Figure 19.8 is a two-dimensional representation of the real three-dimensional process. If there is no load applied the particles which are freed have equal probability to leave their original position to the right or to the left (in the two-dimensional model). That means no net deformation will be observed after many particles have changed their position.

If an external Force F is applied the potential trough becomes asymmetrical

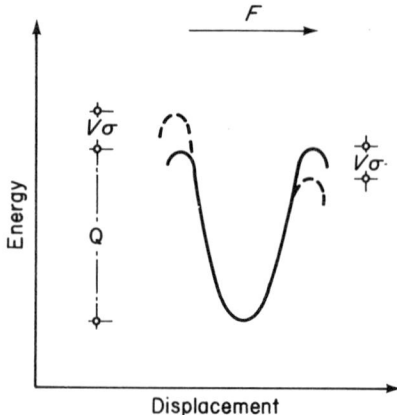

Figure 19.8 Schematic representation of the potential trough without (solid line) and with an external force F applied (dashed line). The activation energy Q corresponds to the average coupling energy of gel particles in the xerogel of hardened cement paste

and now the probability of a jump in the direction of the applied force is increased:

$$P' = P_0 \exp\left\{-\left(\frac{Q}{RT} - \frac{V\sigma}{RT}\right)\right\} \qquad (19.7)$$

and of course the probability of a jump in the opposite direction is decreased in the same way:

$$P'' = P_0 \exp\left\{-\left(\frac{Q}{RT} + \frac{V\sigma}{RT}\right)\right\} \qquad (19.8)$$

The asymmetrical trough is shown schematically by dashed lines in Figure 19.8. In the last two equations σ stands for the locally existing stress and V has the dimension of a volume. It is therefore called the activation volume of the creep process. If there is a well defined creep mechanism, the activation volume has the meaning of a volume involved in an elementary step. This can be interpreted by the cross-section of the moving particle multiplied by the distance of one jump. In a colloidal system Q and V have to be considered as average values representing a wide spectrum (see, for example, Niklas (1967)). Details of rate theory can be found in the book by Krausz and Eyring (1975).

By using Equations (19.7) and (19.8) the following relation for the creep rate can be deduced (Wittmann, 1974):

$$\dot{\varepsilon} = \dot{\varepsilon}_0 \exp\left(-\frac{Q}{RT}\right) \sinh\left(\frac{V}{RT}\sigma\right) \qquad (19.9)$$

If correct values for the activation energy and the activation volume are available, Equation (19.9) describes the influence of temperature and of stress level on creep of hardened cement paste. Corresponding values have been determined experimentally by several authors and are described in the literature (see, for example, Klug and Wittmann (1974), Straub and Wittmann (1976), Luijerink (1982).

If particles are moved in the xerogel they are normally fixed more strongly in their new positions. This can be explained by the fact that the weakest links are affected first. Then $\dot{\varepsilon}_0$ from Equation (19.9) can be written so as to describe the time dependence of creep (Wittmann, 1974):

$$\dot{\varepsilon}_0 = a_0(t - \tau)^{n-1} \tag{19.10}$$

where $(t - \tau)$ is the duration of load with t being the age of the specimen and τ the age of loading.

In this equation a_0 represents the density of creep centres at time $(t - \tau) = 1$, that means the volume concentration of volume elements which contribute to the creep process.

In Figure 19.9 some data taken from Hannant (1967) are replotted on a double logarithmic scale. The straight line indicates that the power function theoretically predicted by Equation (19.10) is a good approximation of the experimentally determined creep function. The power function has been found superior to other creep functions by extensive data fitting (Bažant and Thonguthai, 1976; Wittmann and Setzer, 1971).

From the temperature dependence of creep of concrete as shown in Figure 19.9 the activation energy can be determined. In this case a value of 6.7 kcal/mol has been found (Wittmann, 1971). Data obtained on hardened cement paste lead to similar results (Wittmann, 1974).

In concrete technology it is more usual to use creep deformation instead of

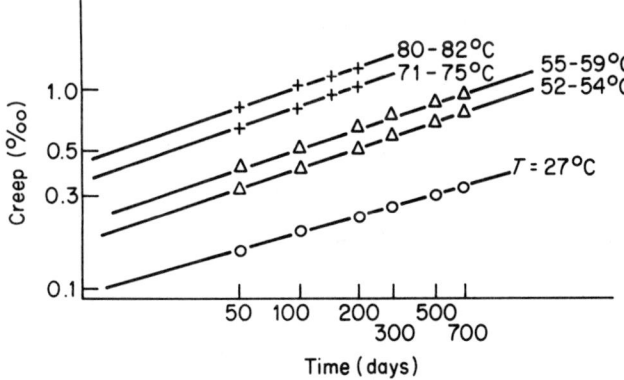

Figure 19.9 Creep of concrete as function of time. The parameter is the temperature (Hannant, 1967)

creep rate, therefore we rewrite Equation (19.10) in an integrated form:

$$\varepsilon_0 = a'(t - \tau)^n \qquad (19.11)$$

So far we have completely neglected the influence of hydration on creep. As the paste matures the xerogel becomes more stable and as a consequence the density of creep centres is reduced. It must be stated at this point that maturing of the xerogel can go on even if hydration has come to an end. In Figure 19.10 the creep deformation after a duration of load $(t - \tau) = 1$ day has been plotted on double logarithmic scale. For the interval chosen a power function represents this relation well:

$$a' = a\tau^{-m} \qquad (19.12)$$

By using Equations (19.11) and (19.12) we can rewrite Equation (19.9) in the integrated form:

$$\varepsilon = a\tau^{-m}(t - \tau)^n \exp\left(-\frac{Q}{RT}\right)\sinh\left(\frac{V}{RT}\sigma\right) \qquad (19.13)$$

This is the well known double-power law which is now widely used (see, for example, Bažant and Osman (1975). It is evident that all parameters a, m, n, Q, and V depend on the actual state of the xerogel.

We have already discussed the fact that the microstructure of hardened cement paste is an unstable phase and that drying especially changes its properties. Therefore it must be expected that a change of moisture content has a significant influence on creep. In Figure 19.11 the parameter a of Equation (19.13) is shown as function of relative humidity (Wittmann, 1974). Values of three test series which have been loaded with different levels are plotted in Figure 19.10. All specimens have been equilibrated with the corresponding relative humidity before loading and have been kept in the same relative humidity while under load.

It can be seen that there is no significant influence of RH on creep in the low humidity region. That means that a change of the interfacial energy does not have

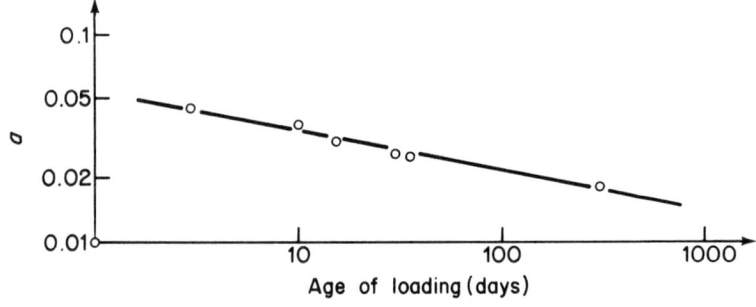

Figure 19.10 Creep of hardened cement paste At $(t - \tau) = 1$ (see Equation (19.11)) as a function of age of loading τ according to Ruetz (1966)

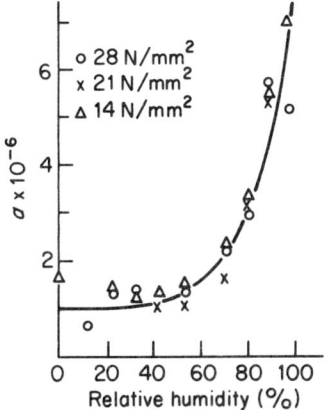

Figure 19.11 Factor a of Equation (19.13) as function of RH. At higher RH the action of disjoining pressure decreases the stability of the xerogel and hence increases the density of creep centres (Wittmann, 1974)

great influence on creep. If the disjoining pressure separates some of the gel particles, however, the creep rate increases sharply. This result can be interpreted by the long-time creep mechanism discussed in this section. Disjoining pressure causes loosening of the microstructure and hence increases the density of creep centres. A direct correlation between swelling due to disjoining pressure and change of creep has been pointed out by Setzer (1982).

Thus we have described creep of hardened cement paste as a thermally activated process which can be depicted by rate theory if average values are taken for the activation volume, activation energy, and density of creep centres. The influence of age of loading and moisture content on this creep mechanism can be taken into consideration in a reasonable way. In hardened cement paste next to the creep centres we find denser zones and unhydrous particles which store energy elastically. In this sense even hardend cement paste has to be looked upon as a composite material.

19.4 DRYING OF HARDENED CEMENT PASTE

The drying process of hardened cement paste can be adequately described by diffusion equations (Ruetz, 1966). In this context the relative humidity H with which the microporous system is in equilibrium is often used as variable. Formally the flux of the relative humidity J_H can then be expressed by the following equation:

$$J_H = -D \, \text{grad} \, H \qquad (19.14)$$

440 Mechanics of Geomaterials

In this equation D stands for the diffusion coefficient. Then the moisture change can be calculated by the following differential equation:

$$\frac{\partial H}{\partial t} = \text{div}(D \text{ grad } H) \tag{19.15}$$

It has been shown both experimentally and numerically (Roelfstra and Wittmann, 1983; Kiessl (1983); Bažant and Najjar, 1971, 1972) that D depends

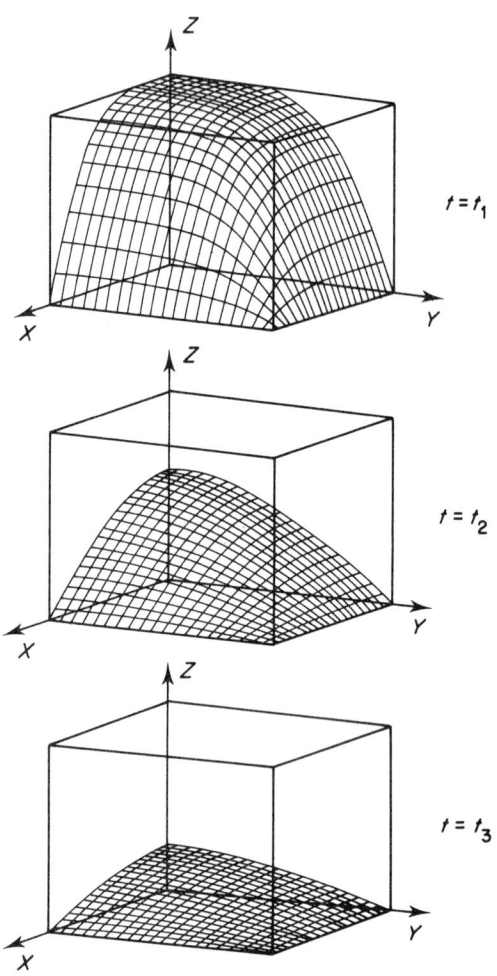

Figure 19.12 Calculated moisture distribution shown in one quarter of a drying prism. During the initial period the centre part of the specimen is still saturated

Deformation of Concrete at Variable Moisture Content 441

among other factors on humidity:

$$D = f(H) \qquad (19.16)$$

For this reason Equation (19.15) has to be solved numerically.

In Figure 19.12 an example for the moisture distribution in a drying prism with two sealed endfaces is shown. Three different durations of drying have been selected.

19.5 SHRINKAGE OF HARDENED CEMENT PASTE

19.5.1 Mechanisms and unrestrained shrinkage

Moisture loss of a microporous xerogel such as hardened cement paste is always accompanied by a corresponding volume change. In this contribution we will not deal explicitly with capillary or chemical shrinkage but concentrate on drying shrinkage.

If an infinite thin sheet dries out, instantaneous drying shrinkage is the consequence. This idealized material behaviour is called unrestrained shrinkage. It is a material property which depends on water/cement ratio, degree of hydration, type of cement, etc. In a real specimen with finite size, shrinkage of the outer drying shell will be hindered by the moist core. If the moisture distribution and the unrestrained shrinkage are known, we can calculate the internal state of stress and the total resulting deformation. This situation is schematically shown in Figure 19.13 taken from Bažant (1982).

Before dealing with the macroscopically observed drying shrinkage of hardened cement paste, we shall briefly discuss the main mechanisms involved. In Figure 19.14 swelling of the xerogel from 0 per cent RH to 100 per cent RH is shown.

In the low humidity region the surface energy of the xerogel is changed by

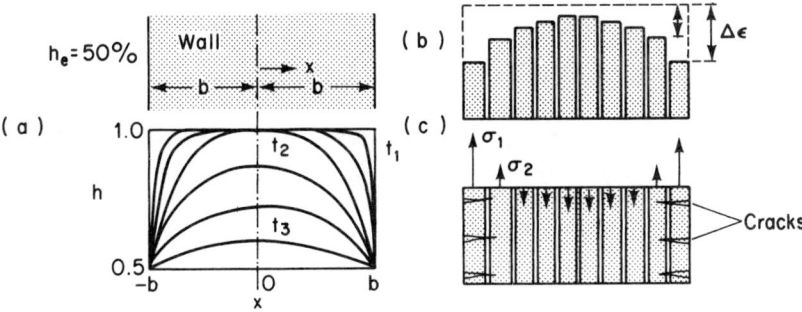

Figure 19.13 (a) Typical distributions of pore humidity at various times during drying; (b) free shrinkage and creep at various points of cross-section; (c) internal stresses (after Bažant, 1982)

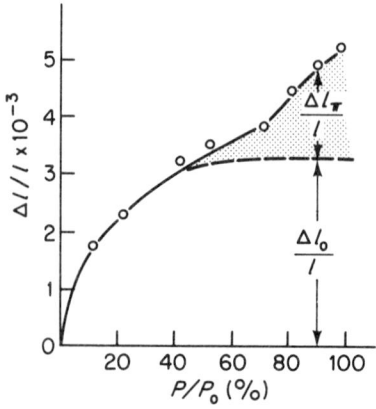

Figure 19.14 Swelling of the xerogel in hardened cement paste as a function of relative water vapour pressure (Wittmann, 1976)

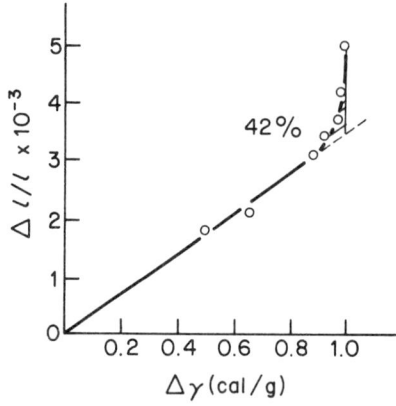

Figure 19.15 Measured swelling of hardened cement paste as shown in Figure 19.14 as function of change of interfacial energy

adsorbed water films. The Bangham equation (equation (19.4)) predicts a linear relationship between wetting swelling (or drying shrinkage) and the change of surface energy. If we replot the measured length change of Figure 19.15 as a function of the change of interfacial energy $\Delta\gamma$ we find the data shown in Figure 19.16. Up to a RH of about 50 per cent the Bangham equation is a good approximation and we can conclude that in the low humidity region change of surface energy due to adsorbed water films is the dominant mechanism of drying shrinkage. This result has been verified by many experimental data for different xerogels.

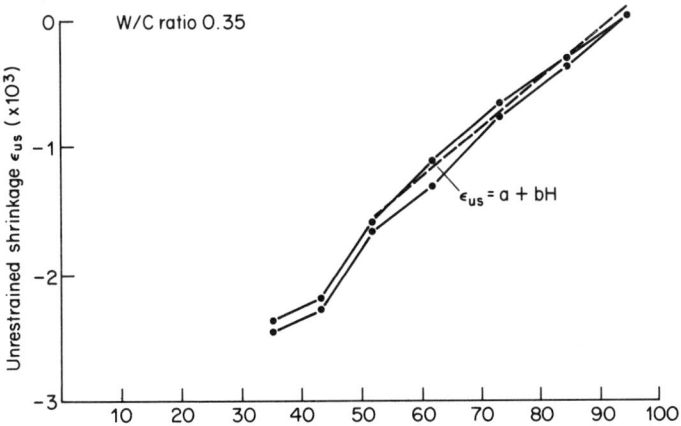

Figure 19.16 Unrestrained shrinkage as function of relative humidity

From the results shown in Figures 19.14 and 19.15 it is obvious that in the high humidity region an additional mechanism has to be taken into consideration. If capillary condensation takes place one ought to expect a contraction of the xerogel due to capillary pressure. This is evidently not correct, namely for the reason that water in fine capillaries differs in many respects from bulk water. The chemical potential is changed and the interaction with neighbouring solid surfaces imposes a certain structure on the mobile water molecules. In addition to this, an electrical double-layer is formed close to the surface. In narrow gaps the space charge (ζ-potential) creates a repulsive force.

Because of these peculiarities water in narrow gaps exerts a disjoining pressure on the pore walls (Stockhausen, 1981). At present it is not possible to predict quantitatively the action of disjoining pressure. Further experimental and theoretical investigations are needed. Therefore when we introduce disjoining pressure as the dominant mechanism of drying shrinkage in the high humidity region we have to keep in mind that we can explain the different components of disjoining pressure but do not yet fully understand the complex interaction of water with pore walls in fine capillaries.

We have used the Munich model to explain the two basic mechanisms of drying shrinkage. It should be noted at this point that quantitative interpretation of experimental data is hampered by the fact that in drying concrete several mechanisms have to be taken into consideration simultaneously.

We have shown above how moisture distribution can be numerically determined. Unrestrained shrinkage can be assumed to depend approximately linearly on relative humidity H:

$$\varepsilon_{us} = a + bH \tag{19.17}$$

Experimental values obtained on small cylinders of hardened cement paste with a water/cement ratio of 0.35 are shown in Figure 19.16. The dashed line in Figure 19.16 represents a reasonable approximation of the observed behaviour within the range of relative humidity between 50 and 100 per cent.

19.5.2 Shrinkage of a real specimen

A special finite element has been developed with the aim of calculating shrinkage deformation of a real specimen (Roelfstra, 1983). With the ring element shown in Figure 19.17 the total deformation as well as the curvature of the endface of a drying cylinder can be calculated. A typical result is shown in Figure 19.18.

It can be seen how the end-face is deformed during the shrinkage process and

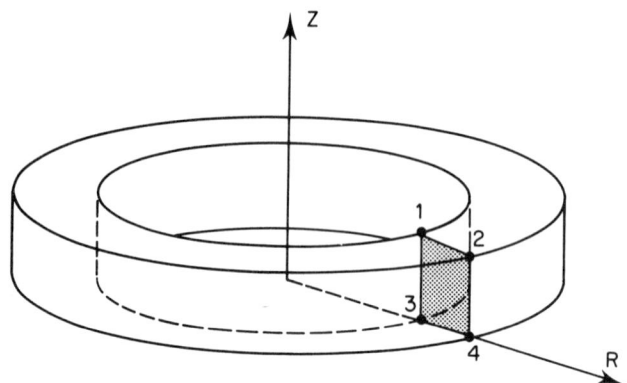

Figure 19.17 Finite element developed to calculate stress distribution in a cylinder

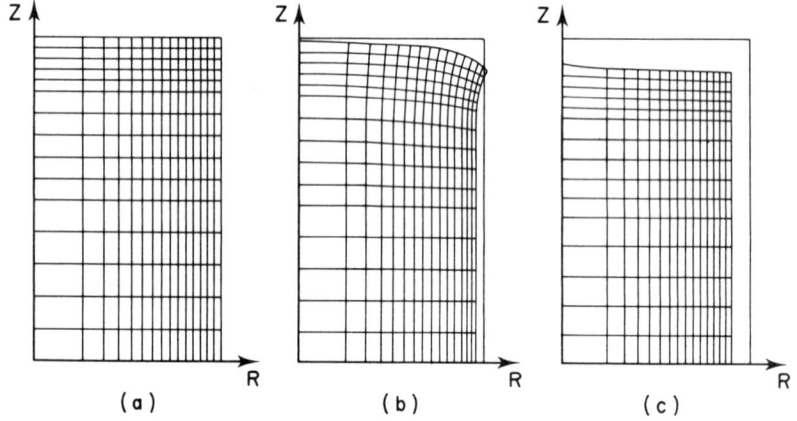

Figure 19.18 (a) The original shape of the cylinder; (b) and (c) the calculated deformations after two different durations of the drying process

that the deformation at the centre of the cylinder lags behind that of the outer shell.

To verify these numerically obtained results we measured the deformation of a drying cylinder of hardened cement paste directly by means of holographic interferometry. As an example the observed deformation after 1 h 23 min, 5 h 33 min and 8 h 32 min is shown in Figure 19.19. It can be stated that numerical predictions and experimental findings agree well.

More information can be gained, if both the numerical and the experimental methods are combined. It is possible for instance to calculate the initial strains which are necessary to create the observed curvature of the endface (Roelfstra, 1983). In Figure 19.20 one example is shown. In the upper part the measured total

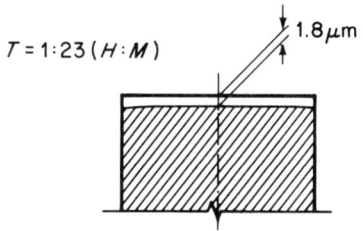

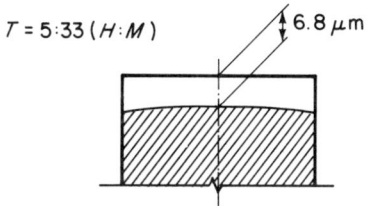

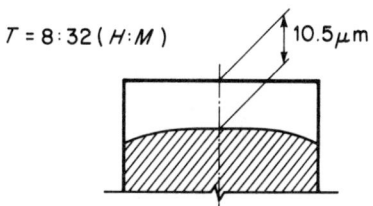

Figure 19.19 Measured curvature of the end-face and total deformation as function of time by means of holographic interferometry

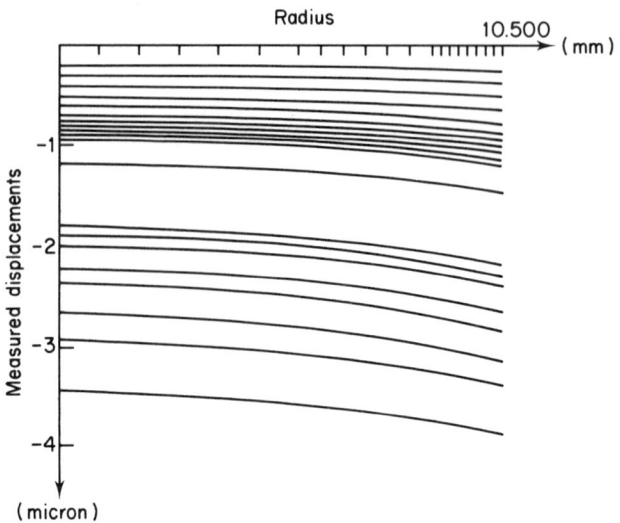

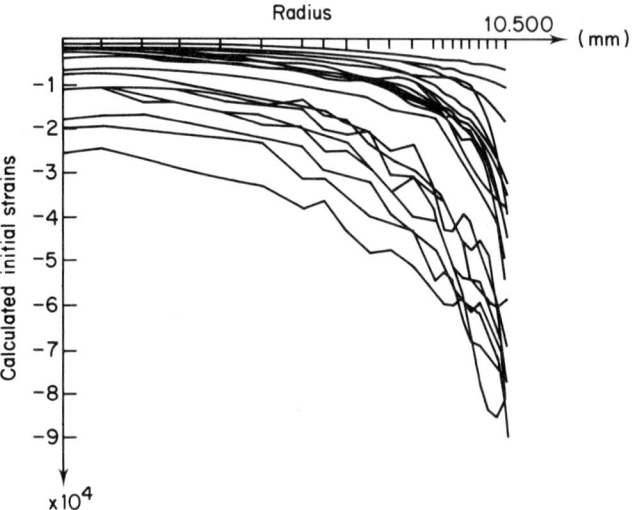

Figure 19.20 Measured total drying shrinkage and calculated initial strains of cylinder of hardened cement paste with a water/cement ratio of 0.35. Drying from 76 per cent RH to 55 per cent RH

deformation and the curvature of a drying cylinder of hardened cement paste with a water/cement ratio of 0.35 are plotted. The lower part shows the numerically obtained initial strains which correspond to the unrestrained shrinkage. In this case the cylinder was equilibrated with a relative humidity of 76 per cent and then dried in air with 55 per cent RH.

Deformation of Concrete at Variable Moisture Content 447

If we dry an identical specimen directly from 100 per cent HR to 55 per cent HR tensile stresses in the dry outer shell overcome tensile strength of the material. This can already be seen in Figure 19.21. Again drying shrinkage of a cylinder with water/cement ratio of 0.35 has been measured. Close to the circumference crack formation releases tensile stresses and therefore the curvature of the endface

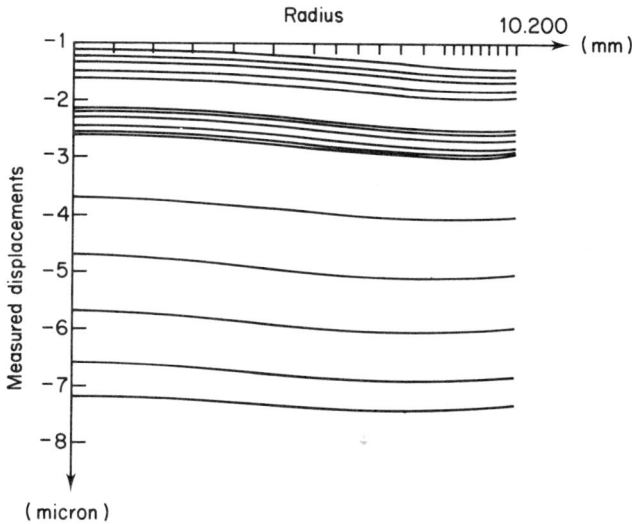

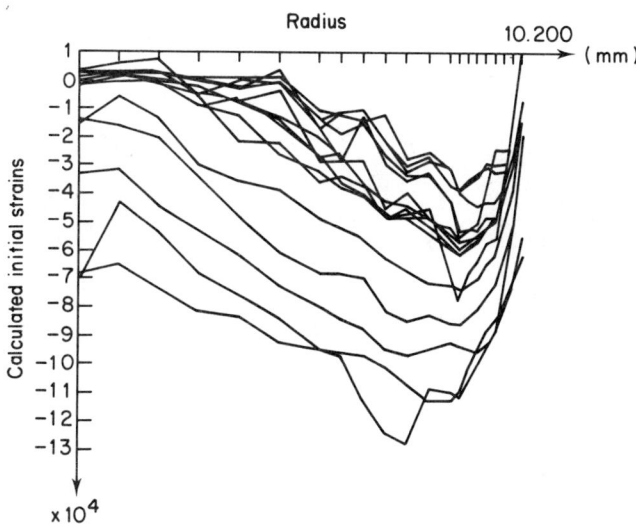

Figure 19.21 Measured total drying shrinkage and calculated initial strains of a cylinder of hardened cement paste with a water/cement ratio of 0.35. Drying from 100 per cent RH to 55 per cent RH

448 *Mechanics of Geomaterials*

bends upwards. Consequently the initial strains determined numerically decrease sharply towards the surface of the specimen. We can conclude that depending on the moisture gradient the time-dependent deformation of a drying specimen is more or less influenced by crack formation.

19.6 BEHAVIOUR OF THE COMPOSITE MATERIAL

19.6.1 Creep

In Figure 19.7 two computer simulated random composite structures are shown. If we know the elastic modulus of the aggregates and the viscoelastic properties of the matrix we can simulate time-dependent deformation under load. One possibility is to use rheological models. More realistic results are obtained by finite elements. In this case the volume concentration, the grading, and the geometry of the aggregates can be taken into consideration.

One example for a finite element generated structure is given in Figure 19.22. In this way the influence of the composition of concrete on creep can be studied.

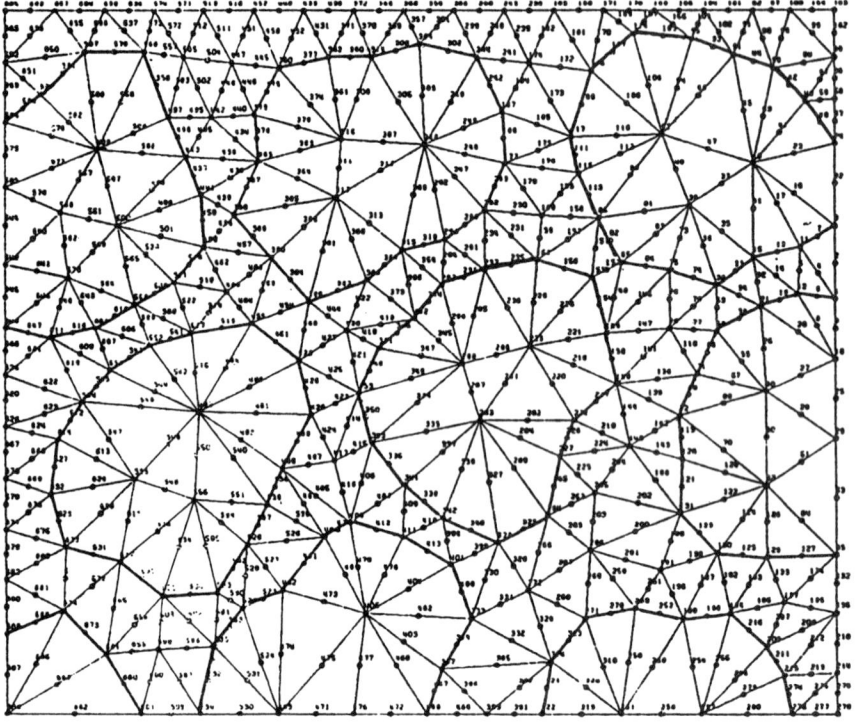

Figure 19.22   Finite element generated structure of a concrete-like two-phase material (Wittmann et al., 1983)

Deformation of Concrete at Variable Moisture Content

In paticular the storage of elastic energy in the aggregates and creep recovery are realistically represented.

19.6.2 Diffusion coefficient

We have pointed out that the diffusion coefficient of hardened cement paste depends among other parameters on water/cement ratio and on humidity content. Most aggregates in conventional concrete can be considered to be dense and thus do not contribute essentially to moisture movement in the composite material.

To simulate moisture diffusion in a composite material finite element method can be applied. In Figure 19.23 a computer generated random composite concrete-like structure is shown. The grading of the aggregates has been chosen to follow the Fuller-curve. In this case only large aggregates between 8 and 32 mm are represented. This composite structure can be composed by triangular finite elements combining the centres of three neighbouring aggregate particles. This 'super'-element is shown in Figure 19.24.

The effective diffusion coefficient of a composite structure can be calculated in this way if the diffusion coefficient of the matrix (mortar or hardened cement paste) is known (Roelfstra, 1983). In Figure 19.25 the influence of aggregate concentration on the effective diffusion coefficient is shown. Thus, the influence of grading and aggregate geometry on drying can be studied systematically.

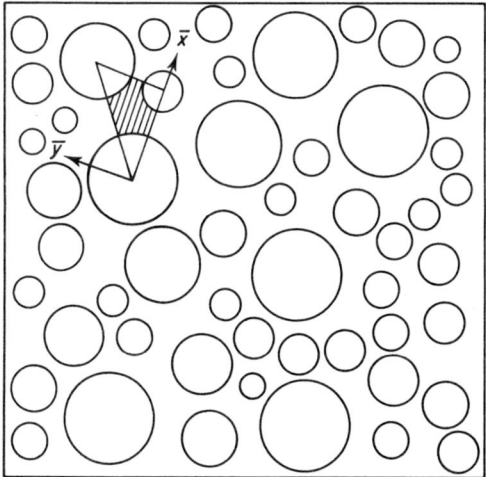

Figure 19.23 Computer-generated composite structure. Grading follows the Fuller-curve. Only the aggregates between dia. 8–32 mm are represented in this case

Figure 19.24 'Super'-element consisting of six isoparametric elements with each having twelve nodes

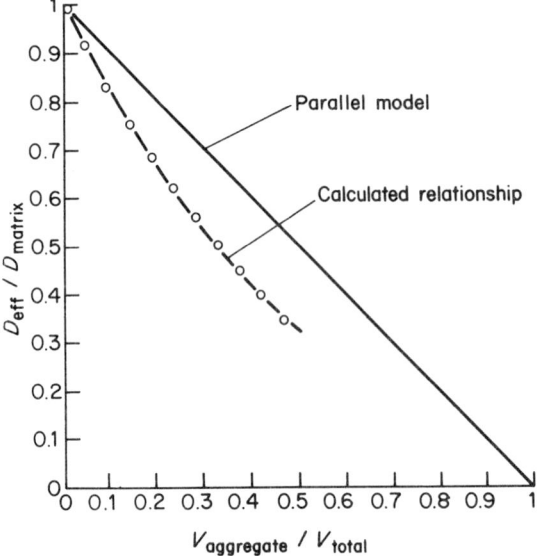

Figure 19.25 Influence of aggregate content on the diffusion coefficient

19.6.3 Crack formation

It has long been recognized that cracks are formed in the composite structure of concrete even at low loads. Different methods to detect crack formation are described in detail by Slate *et al.* (1983). These cracks evidently contribute to the total observed deformation. In normal concrete cracks frequently start from and run along interfaces. In Figure 19.26 computer-generated crack patterns for two different load levels are shown. These calculated crack patterns have been discussed and compared with experimental results in Zaitsev and Wittmann (1981) and Zaitsev (1983).

In a drying concrete specimen additional cracks are formed when the tensile stress in the outer zones overcomes tensile strength. These drying cracks have already been shown schematically in Figure 19.13. We shall deal with these cracks in particular in the next section because they influence the internal state of stress of a drying specimen significantly.

Finally the difference in volume change between the drying matrix and the inert aggregates creates internal stresses which can overcome tensile strength of the matrix and thus create cracks distributed at random in the composite structure. Podvalnyi has developed a model to calculate stresses and cracking in the composite structure during heating or cooling (1976). This model can be modified and adapted to cover the situation of drying concrete as well. Another

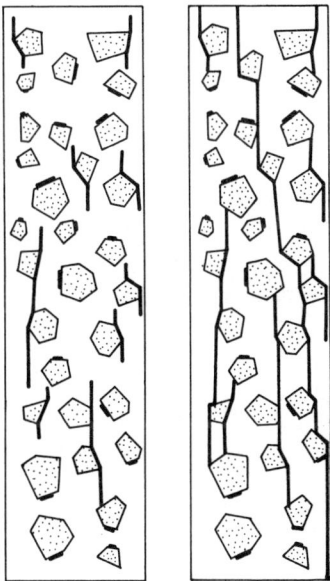

Figure 19.26 Crack pattern for two different load levels in normal concrete. Cracks are running around aggregates

452 Mechanics of Geomaterials

possibility to calculate stresses and cracking in a composite structure is given by finite element analysis.

19.7 SIMULTANEOUS CREEP AND SHRINKAGE

Now we come back to the initially formulated question if there are special mechanisms of accelerated creep under drying conditions or if the increased deformation marked with a shaded area in Figure 19.1 is caused by superposition of stress fields.

Drying of concrete maintains a stress distribution for a long time. In Figure 19.27

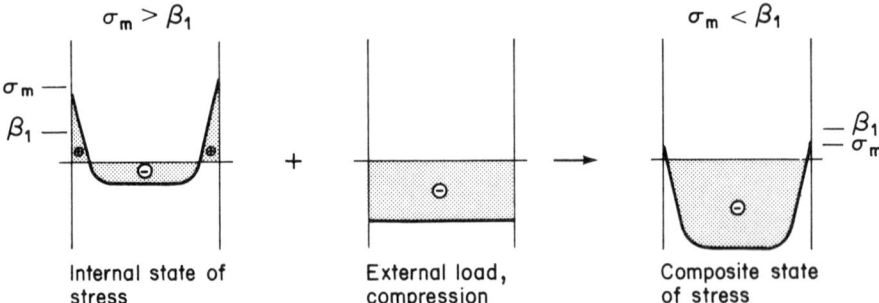

Figure 19.27 Internal state of stress due to a moisture gradient. In the example the estimated maximum tensile stress σ_m is greater than tensile strength β_t. If a compressive load is applied the tensile stress in the outer drying zoned is reduced and thus crack formation is prevented

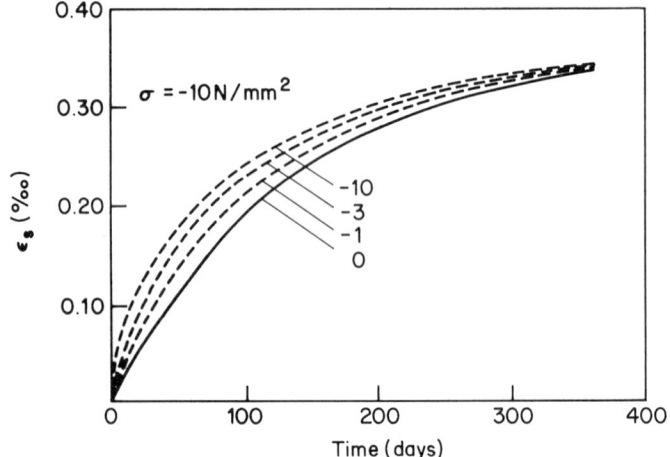

Figure 19.28 Shrinkage of concrete as calculated without an external load ($0\,N/mm^2$) and with external compressive loads of 1, 3, and $10\,N/mm^2$ (Wittmann and Roelfstra, 1980)

the stress field in a drying cylinder is schematically shown. If the moisture gradient is sufficiently large the resulting tensile stresses are bigger than tensile strength of the material. As a consequence the dry zones will be cracked. Above we have shown that drying shrinkage and the curvature of the end-face can be calculated if the moisture distribution and the unrestrained shrinkage are known. If a drying cylinder is loaded crack formation can be reduced or even prevented. In Figure 19.27 it is shown how an applied compressive stress modifies the resulting stress distribution in such a way that the remaining tensile stresses are below the tensile strength. In contrast if a tensile load is applied to a drying cylinder cracking is even more severe.

If we calculate shrinkage of an unloaded concrete specimen and another under different compressive loads we obtain results shown in Figure 19.28 (Wittmann and Roelfstra, 1980). It can be seen that the final value of shrinkage is hardly affected by crack formation because in the dry equilibrated state cracks are closed again. The time-dependence of shrinkage, however, is significantly influenced. The early shrinkage is apparently accelerated. This effect accounts for a large proportion if not for all of the deformation usually called drying creep. In (Wittmann and Roelfstra, 1980) it has been shown that experimental results can be explained without a special mechanism for drying creep. In a similar analysis Iding and Bresler have shown that tensile strength influences drying shrinkage of a concrete specimen (1982). Bažant suggests that this type of crack formation should rather be represented as strain softening than as an abrupt stress drop (Bažant and Raftshol, 1982).

As mentioned already an external tensile stress facilitates crack formation.

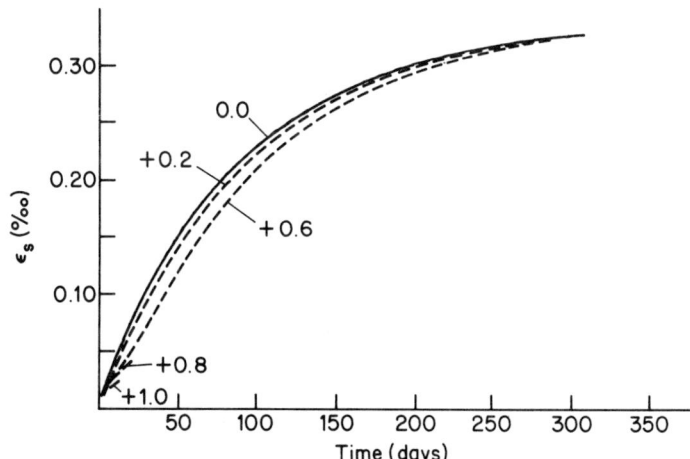

Figure 19.29 Shrinkage of an unloaded concrete specimen (solid line) and of companion specimens under a tensile load of 0.2 N/mm² and 0.6 N/mm². In the case of external tensile loads of 0.8 N/mm² and 1.0 N/mm² the specimen fails as the internal state of stress reaches a critical level

Results of numerical determination of shrinkage under constant tensile stress are shown in Figure 19.29. In this case modest stresses retard significantly early shrinkage and a stress of 0.8 N/mm² finally leads to failure after about 20 days of drying.

In this analysis creep has been neglected to demonstrate clearly the effect of an external load on shrinkage. Under the conditions chosen in this analysis the observed hygral length change may vary by a factor of two if only an external load is added to the internal stress field. This enormous increase is not related to a modified real shrinkage mechanism. From this analysis it follows that shrinkage mechanisms, i.e. processes in the microstructure, cannot be determined by simple observation of macroscopic deformation.

Furthermore it is evident that creep and shrinkage of a drying specimen under load cannot be separated into two components. The resulting total deformation does not depend on the internal state of stress and the external load separately but only on the composite state of stress. It follows that mechanisms which have been determined on the basis of this usual subdivision have no real meaning.

In a more general way we can say that macroscopically observed total deformation depends on the external state of stress, and the moisture and the temperature distribution. In addition the degree of hydration (effective age) may vary within a specimen. To find the contribution of real mechanisms to

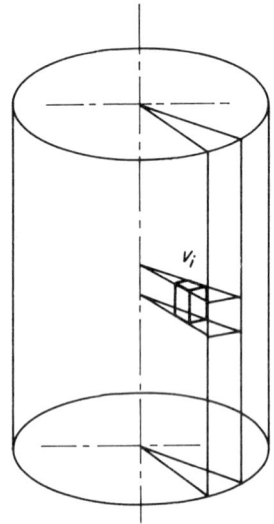

Figure 19.30 Creep deformation of a small volume element depends on stress, temperature, relative humidity, and degree of hydration. The total deformation of a macroscopic specimen is determined by the unrestrained deformations of all volume elements V_i

macroscopic deformation we have to subdivide the total volume into n elements. Each volume element V_i (see Figure 19.30) is chosen to be small enough so that for a given time interval Δt it may be looked upon to be in quasi-equilibrium. Then the deformation of this volume element V_i depends on the actual existing stress which is the sum of the locally observed part of the external load σ_e and of the internal stress field σ_i, the temperature T, the relative humidity RH, and the degree of hydration α:

$$\varepsilon_i(V_i) = f(\sigma_e, \sigma_i, T, RH, \alpha) \tag{19.18}$$

Within the small volume element we can characterize the time-dependent deformation by real mechanisms. Therefore we call $\varepsilon_i(V_i)$ the real creep function. In fact $\varepsilon_i(V_i)$ is a complementary value to the unrestrained shrinkage introduced earlier. If we know the real creep function and unrestrained shrinkage as function of RH we can calculate the total deformation.

From Figure 19.11 we learn that as the moisture content of hardened cement paste is lowered the specific creep function decreases. After having calculated the moisture distribution in a drying specimen we can predict that the dried outer zones will exhibit very little creep. As the drying process proceeds more and more volume elements will be characterized by low creep. This means that the stress field induced by the moisture distribution increases early apparent shrinkage by preventing crack formation. At a much later stage, however, creep rate of a drying specimen will be lower than that of sealed companion specimens. For small cylinders of hardened cement paste this transition can be observed after about 10 days of drying. This prediction is verified by experimental results (Wittmann, 1971, 1974) shown in Figure 19.31. In real concrete elements this transition occurs much later of course.

It has been shown that an increased total deformation of drying concrete under

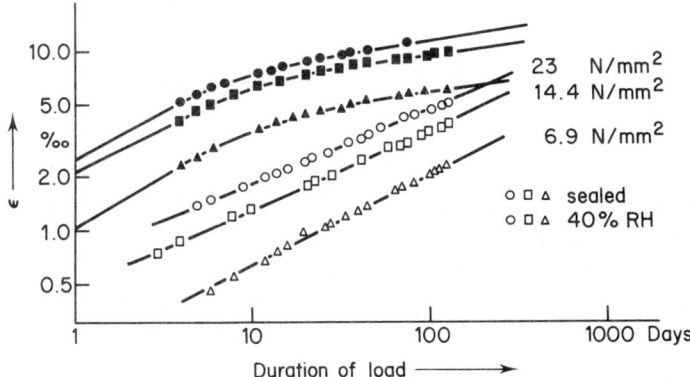

Figure 19.31 Total time-dependent deformation of sealed and drying (40 per cent RH) cylinders of hardened cement paste. Three different load levels have been applied at an age of 28 days (Wittmann, 1971, 1974)

load can be explained on the basis of a mechanical analysis. This part is independent of creep and shrinkage mechanisms. At present it cannot totally be excluded that under drying conditions accelerated mechanisms contribute to the observed increased total deformation. We can state, however, that if these mechanisms exist they are of minor importance.

While crack formation has a major influence on the early total deformation of drying and loaded specimens the influence of moisture content on material properties becomes dominant at a later stage.

19.8 CONCLUSIONS

1. There are two distinct creep mechanisms: short-time creep can be explained by redistribution of water in the microstructure, and long-time creep is caused by displacement of gel particles. Rate theory is a powerful tool to describe long-time creep.
2. Drying of hardened cement paste can be adequately described by nonlinear diffusion theory.
3. For drying concrete an effective diffusion coefficient can be introduced which takes volume concentration, grading, and geometry of aggregates into consideration.
4. Real shrinkage mechanisms have to be related to unrestrained shrinkage and real creep mechanisms to the real creep function. The observed time-dependent deformation cannot be linked directly with creep or shrinkage mechanisms. A number of apparent mechanisms are always involved and they modify the behaviour.
5. Separation of the total time-dependent deformation of a drying loaded concrete specimen in shrinkage, basic creep, and drying creep is meaningless.
6. Shrinkage and swelling are governed by the internal state of stress which is caused by the inhomogeneous moisture distribution.
7. Crack formation has a major influence on the early hygral length change. If crack formation is prevented by an external load the apparent shrinkage is accelerated.
8. At a later stage the reduced moisture content of a drying loaded specimen leads to a reduced creep rate.
9. Combination of experimental and numerical methods offer new and promising possibilities in material research.
10. For the study of the properties of a composite material such as concrete the introduction of three hierarchical structural levels has proved to be successful.

REFERENCES

Bangham, D. H., and Fakhoury, N. (1931) 'The swelling of charcoal, Part I, Preliminary experiments with water vapour, carbon dioxide, ammonia and sulphur dioxide', *Proc. R. Soc. A*, **130**, 81–89.

Bažant, Z. P. (1982) 'Mathematical models for creep and shrinkage in concrete', in (Z. P. Bažant and F. H. Wittmann, Eds.), *Creep and Shrinkage in Concrete Structures*, John Wiley & Sons.

Bažant, Z. P., and Najjar, L. J. (1971) 'Drying of concrete as a nonlinear diffusion problem', *Cem. Concr. Res.*, **1**, 461–473.

Bažant, Z. P., and Najjar, L. J. (1972) 'Nonlinear water diffusion in nonsaturated concrete', *Mat. and Struct.*, **5**, 3–20.

Bažant, Z. P., and Osman, E. (1975) 'On the choice of creep function for standard recommendations on practical analysis of structures', *Cem. Concr. Res.*, **5**, 129–138.

Bažant, Z. P., and Panula, L. (1978–1979) 'Practical prediction of time-dependent deformations of concrete', *Mater. Struct.*, **11**, 307–328, 415–434; **12**, 169–183.

Bažant, Z. P., and Raftshol, W. J. (1982) 'Effect of cracking in drying and shrinkage specimens', *Cem. Concr. Res.*, **12**, 209–226.

Bažant, Z. P., and Thonguthai, W. (1976) 'Optimization check of certain practical formulations for concrete creep', *Mater. Struct.*, **9**, 91–98.

Feldmann, R. F., and Sereda, P. J. (1968) 'A model for hydrated Portland cement paste as deduced from sorption-length change and mechanical properties', *Mater. Constr.*, **1**, 509–520.

Flood, E. A. (1961) 'Adsorption potentials, adsorbent self-potentials and thermodynamic equilibria, in solid surfaces and the gas–solid Interface', *Adv. Chem. Ser.*, No. 33, American Chemical Society, Washington, pp. 249–63.

Gibbs, J. W. (1957) *Collected Works*, Yale University Press, New Haven.

Hannant, D. J. (1967) 'Strain behavior of concrete up to 95°C under compressive stresses', *Proc. Conf. on Prestressed Concrete Pressure Vessels*, The Institution of Civil Engineers, London, pp. 177–91.

Iding, R., and Bresler, B. (1982) 'Prediction of shrinkage stresses and deformations in concrete', in *Fundamental Research on Creep and Shrinkage of Concrete* (Ed. F. H. Wittmann), Martinus Nijhoff, The Hague.

Kiessl, K., (1983) *Kapillarer und dampfförmiger Feuchtetransport in mehrschichtigen Bauteilen*, Dissertation, Essen University.

Klug, P., and Wittmann, F. H. (1974) 'Activation energy and activation volume of creep of hardened cement paste', *Mater. Sci. Ing.*, **15**, 63–66.

Krasilnikov, K. G., Podvalny, A. M., and Segalov, A. E. (1974) 'On the self-induced deformations in porous bodies', *Kolloidnyi. Zhurnal*, **36**, 266–271.

Krausz, A. S., and Eyring, H. (1975) *Deformation Kinetics*, Wiley, New York.

Luijerink, J. (1982) 'Deformation kinetics of concrete', in *Fundamental Research on Creep and Shrinkage of Concrete* (Ed. F. H. Wittmann), Martinus Nijhoff Publishers, The Hague, pp. 27–34.

Niklas (1967) 'Uber den Machanismus des Andrade-Kriechens von amorphen Polymeren im Glaszustand', *Z. angew. Physik*, **23**, 470–476.

Podvalnyi, A. M. (1976) 'Phenomenological aspect of concrete durability theory', *Mat. and Struct.*, **9**, 151–162.

Powers, T. C. (1968) 'The thermodynamics of volume change and creep', *Mater. Constr.*, **1**, 487–507.

Powers, T. C., and Brownyard, T. L. (1947) 'Studies of the physical properties of hardened Portland cement paste', *Portland Cement Association, Res. Bull.*, **22**. (Originally published in *J. Amer. Concr. Inst.*, Oct. 1946–Apr. 1947.)

Roelfstra, P. E., Ph.D. Thesis, Swiss Federal Institute of Technology Lausanne, Switzerland.

Roelfstra, P. E., and Wittmann, F. H. (1983) 'Numerical analysis of drying and shrinkage', in (F. H. Wittmann, Ed.), *Autoclaved Aerated Concrete*, Moisture and properties, Elsevier.

Ruetz, W. (1966) 'Das Kriechan des Zementsteins im Beton und seine Beeinflussung durch

gleichzeitiges Schwinden', *Dtsch. Ausschuss Stahlbeton, Schriftenr.*, Heft 183, Wilhelm Ernst & Sohn, Berlin.
Schlude, F., and Wittmann, F. H. (1974) 'Uber ein Verfahren zur raschen Bestimmung der Komplexen DK im Mikrowellenbereich', *Nachrichtentech. Ztg.*, **27**, 365–368.
Schneider, U. (1982) 'Creep effects under transient temperature conditions', in *Fundamental Research on Creep and Shrinkage of Concrete* (F. H. Wittmann, Ed.), Martinus Nijhoff Publishers, The Hague.
Schwarzl, F. R. (1969) 'The numerical calculation of storage and less compliance from creep data for linear viscoelastic materials', *Rheol. Acta*, **8**, 6–17.
Sellevold, E. J. (1970) 'Low frequency internal friction and short-time creep of hardened cement paste: an experimental correlation', *Proc. Conf. on Hydraulic Cement Pastes: their Structure and Properties*, Shefield, pp. 330–334.
Sellevold, E. J., and Richard, C. W. (1972) 'Short-time creep transition for hardened cement paste', *J. Am. Ceram. Soc.*, **55**, 284–289.
Setzer, M. J. (1982) 'A model of hardened cement paste for linking shrinkage and creep phenomena', in *Fundamental Research on Creep and Shrinkage of Concrete* (Ed. F. H. Wittmann), Martinus Nijhoff Publishers, The Hague, pp. 3–13.
Slate, F. O., et al. (1983) 'Experimental methods to detect crack formation', in Fracture Mechanics of Concrete (Ed. F. H. Wittmann), Elsevier (1983).
Splittgerber, H., and Wittmann, F. H. (1974) 'Einfluss absorbierter Wasserfilme auf die van der Waals Kraft zwischen Quarzglasoberflächen', *Surf. Sci.*, **41**, 504–514.
Stockhausen, N. (1981) 'Die Dilatation hochporöser Festkörper bei Wasseraufnahme und Eisbildung', Dissertation, Techn. Univ. Mü·nchen.
Straub, F., and Wittmann, F. H. (1976) 'Activation energy and activation volume of compressive and tensile creep of hardened cement paste', *Proc. Conf. on Hydraulic Cement Pastes: their Structure and Properties*, Sheffield, pp. 227–30.
Taylor, H. F. W. (1979) 'Cement hydration reactions: the silicate phases', *Proc. Eng. Found. Conf. on Cement Production and Use*, Franklin Pierce College, Eng. Found., New York, pp. 107–116 (1979).
Taylor, H. F. W., and Roy, D. M. (1980) 'Structure and composition of hydrates', *Proc. 7th Int. Congr. on Chemistry of Cement*, Paris, Vol. 1, Paper 11-2.
Uberlhack, H., and Wittmann, F. H. (1976) 'Coupling of colloidal particles and recoilless fraction', *J. Phys., Paris*, **37**, (C6), 269–271.
Wittmann, F. H. (1971) 'Discussion of some factors influencing creep of concrete', The State Institute for Technical Research, Finland, Fiedotus Sarja III–Rakennus 167.
Wittmann, F. H. (1974) 'Bestimmung physikalischer Eigenschaften des Zementsteins', *Dtsch. Ausschuss Stahlbeton, Schriftenr.*, Heft 232, Wilhelm Ernst & Sohn, Berlin, pp. 1–63.
Wittmann, F. H. (1976) 'The structure of hardened cement paste a basis for a better understanding of the materials properties', *Proc. Conf. on Hydraulic Cement Pastes: their Structure and Properties*, Sheffield, pp. 96–117.
Wittmann, F. H. (1977) 'Grundlagen eines Modells zur Beschreibung charakteristischer Eigenschaften des Betons', *Detsch. Ausschuss Stahlbeton Schriftener.*, Heft 290, pp. 43–101.
Wittmann, F. H. (1982a) 'Creep and shrinkage mechanisms', in *Creep and Shrinkage in Concrete Structures* (Eds. Z. P. Bažant and F. H. Wittmann), John Wiley & Sons.
Wittmann, F. H. (1982b) 'Modelling of concrete behaviour', *Proc. Conf. on Contemporary European Concrete Research*, Swedish Cement and Concrete Research Institute, Stockholm, pp. 171–189.
Wittmann, F. H., and Roelfstra, P. E. (1980) 'Total deformation of loaded drying concrete', *Cem. Concr. Res.*, **10**, 601–610.
Wittmann, F. H., and Setzer, M. J. (1971) 'Vergleich einiger Kriechfunktionen mit Versuchsergebnissen', *Cem. Concr. Res.*, **1**, 679–690.

Wittmann, F. H., Roelfstra, P. E., and Sadouki, H. (1983) 'Simulation and numerical analysis of composite materials' (to be published).

Young, J. F. (1982) 'The microstructure of hardened Portland cement paste', in *Creep and Shrinkage in Concrete Structures*, (Eds. Z. P. Bažant and F. H. Wittmann), John Wiley & Sons.

Zaitsev, Y. B., and Wittmann, F. H. (1981) 'Simulation of crack propagation and failure of concrete', and Struct., **14**, 357–365.

Zaitsev, Y. B. (1983) 'Crack propagation in a composite material', in *Fracture Mechanics of Concrete* (Ed. F. H. Wittmann), Elsevier.

Zech, B., and Wittmann, F. H. (1974) 'Studium des dielectrischen Verhaltens von dünnen adsorbierten Wasserfilmen', *Z. Phys. Chem., Neue Folge*, **92**, 45–62.

Mechanics of Geomaterials
Edited by Z. Bažant
© 1985 John Wiley & Sons Ltd

Chapter 20

Some Remarks on Constitutive Equations for Concrete and Geomaterials: Discusser's Report

K. S. Pister

20.1 INTRODUCTION

The chapters by Anderson and Wittmann in this volume, as well as discussion that developed at the Symposium have motivated this brief note, which consists of two parts: first, some general remarks on concepts underlying constitutive equations for materials, and their limitations when applied to complex materials such as concrete, rock, or soils are noted; then, suggestions pertinent to the modelling of a material such as concrete subjected to a mechanical, thermal, and hygral environment are presented.

20.2 REVIEW OF BASIC PRINCIPLES

Representation of material behaviour by means of constitutive equations should clearly not be an end in itself. Such equations comprise but one component of a simulation model typically constructed to predict the behaviour of a proposed engineering device or system. In the sense of mathematical physics, prediction of behaviour is reduced to the solution of field problems in which the dependent variables are functions of position and time, i.e. functions of points in a suitable space. This obvious fact has been the motivation (at least in part) for construction of a constitutive theory of material behaviour based on the abstract concept of a material body as a representation of actual materials (Truesdell and Noll, 1965). This abstraction contains the notion of a primitive called a 'particle' of the material, or a 'material point'; a set of particles comprises the material body. Such a body is accessible as a mapping onto a reference configuration for which particles are identified with points in the space in which the body is embedded. Changes in configuration of the body resulting from imposed actions can then be represented as further mappings from the reference configuration. A *constitutive equation* then defines the response of a particle *in a prescribed reference configuration* to exposure to a process (action) imposed on the particle. If the global response of the body requires a different constitutive description from particle to particle, the body is said to be non-homogeneous. The foregoing makes it clear that constitutive equations are point functions. To make the notion of a constitutive equation operative for a material body, additional postulates must be introduced. In the present context two are significant. First, it is typically asserted that stress (or other dependent field variables) in a body is determined by the history of the response of particles in the body to the prescribed process. In the case of a purely mechanical process, for example, stress is determined by the collection of histories of motion of all particles comprising the body. This unworkable principle is supplemented by a *localization* statement, i.e. in determining stress at a given particle x, the response outside an arbitrary neighborhood of x may be disregarded. Typically, this principle is made operational by taking the stress to be a function of the history of the deformation gradient of the motion (in the case of a mechanical process) evaluated at the particle. This 'homogenization' of the motion permits identification of the constitutive equation for a particle and a body from purely homogeneous deformation states, a very important concomitant for conduct of experiments.

This brief background is introduced to serve as a starting point for subsequent discussion of constitutive equations for real materials (not abstract material bodies!). Let us examine some of the limitations of the approach that has been introduced.

20.2.1 Composite materials

If a body is comprised of more than one kind of constituent material, it is obvious that the concept of particle must be extended. One possibility is to impute a structural arrangement to the heterogeneous set of particles. Such a step complicates the use of continuum principles and introduces scale effects in constitutive equations, however faithfully it reproduces the actual structure of the material.

Alternatively, there has developed a continuum theory of mixtures to address problems associated with representation of behaviour of composite and multi-phase materials. Theories of this type permit coexistence of different constituent material particles at a point in a body configuration, while not explicitly taking into account internal structure. As reflected by the spirited discussion at the Symposium, there is a wide range of views as to the adequacy of and need for the theory of mixtures to represent the behaviour of complex materials.

There have also emerged so-called multipolar and non-local theories of material bodies which are designed, *inter alia*, to reflect non-homogeneity and scale effects in materials. (See Truesdell and Noll, (1965) and Jaunzemis (1967) for background). Here, the traditional viewpoint embodied in the notion of a simple material (Truesdell and Noll, 1965) will be pursued to discuss the scale effect problem. The choice reflects the bias of the author, which simply stated, is that the complexity of current alternative approaches outweighs their practical utility.

20.2.2 Problem of scale effect

Representation of behaviour of composite materials, as well as nominally homogeneous materials, is confounded by the inevitable occurrence of structural defects and anomalies which are typically present in the virgin state of a material as well as developed by imposed processes. The dilemma thus produced results from the desire, rather the need, to utilize continuum concepts while simultaneously reflecting 'discontinuum' structure. One seldom is concerned with this dilemma in dealing with metals, where the scale of discontinuities renders them invisible to the naked eye (except when surface slip bands become visible). Non-homogeneity and structural complexity at this scale are comfortably compatible with Cauchy stress at a point, since neither is visible to the observer! On the other hand, for materials such as concrete and for geomaterials, heterogeneity, structure, and defects are all too apparent to the eye. Accordingly, resolution of the scale problem requires some care. To provide a focus for discussion in the sequel, consider a material volume of concrete. As noted by Wittmann (Chapter 19, this volume) in his three-level approach to concrete structure, one must reflect the fact that such a material volume comprises aggregate, cement gel, pores, and evaporable water in a structure containing cracks even in the absence of external forces. Recognition of this leads us to define a generic material element by selecting a cube of material whose side l is taken large enough to capture the relevant physics of behaviour associated with processes imposed on the material. Taking a mechanical process as an example, we must subject the cube to prescribed surface tractions and 'observe' the resulting deformation of the cube in order to start the process of developing mechanical constitutive equations for the material. By 'observe' one means to include statistical or finite element models designed to introduce an appropriate averaging process in the cause–effect relations reflected by stress and deformation measures in a constitutive equation. Inevitably, 'observe' must also include experiments on actual samples of the material, from which data constitutive model identification and parameter estimation may be extracted.

Implicit in this scheme is the recognition that the surface traction–deformation relations so obtained for the generic element can only be candidate constitutive equations for the material if a limiting process is involved, i.e. l must approach zero to permit one to utilize constitutive equations as point functions for use in solution of field problems. This self-evident observation leads to the very important result that one cannot hope or expect to explain any behavioural phenomenology in the material below the scale of l. Typically, insufficient attention has been given to this constraint in the literature.

20.3 AGEING, CREEP, AND SHRINKAGE IN CONCRETE

As a starting point for outlining a structure for modelling concrete behaviour, it will be helpful to state what is meant by the terms ageing, creep, and shrinkage in the current context:

ageing—change in constitutive properties resulting from processes to which a material is exposed. Ageing is only indirectly a function of clock time.
creep—time-dependent deformation occurring under constant applied stress.
shrinkage—deformation of a free material volume in the absence of surface tractions.

To these definitions is added the specification of the process for which constitutive equations are desired: here the process imposed on a material volume of concrete (suitably selected with reference to critical length, l) consists of prescribed surface tractions, surface heat, and water flux, all of which may be functions of time. Keeping in mind earlier comments concerning the scale effect problem, the detailed behaviour of the constituents comprising the generic element can be expected to be very complex. However, from the standpoint of constitutive theory the outcome of 'observation' of the response of the volume to the process must result in constitutive equations in which stress, strain, temperature, water concentration, heat, and water flux are point functions. Spatial gradients of these process measures (however important they may be *within* the material volume) do not appear in the constitutive equation for mechanical variables, which is our interest here. It may be noted that computation of the response of a *material body* to a process typically leads to solution of coupled field equations, which is beyond our focus here. We therefore introduce the following functions to define the mechanical constitutive equation at a *point*, assuming full kinematic linearization of the motion of the material element:

σ—stress
ε—strain
T—temperature
ω—evaporable water concentration

It may be noted that evaporable water concentration can be related to pore humidity, a choice of some researchers in the field. To these variables we add certain internal variables to define desired constitutive characteristics:

λ—maturity of the material
β—damage measure
ε^i—inelastic strain

Among the above variables, σ, ε, ε^i are symmetric, rank two tensors while others are scalars. All variables depend on time. Next we define a complementary free energy function

$$G = G(\sigma, T, \omega, \beta, \lambda, \varepsilon^i) \tag{20.1}$$

Constitutive Equations for Concrete and Geomaterials

It then follows from the entropy production inequality (Lubliner, 1972) that

$$\varepsilon = \frac{\partial G}{\partial \sigma} = G_{,\sigma} \tag{20.2}$$

Taking the time derivative of (20.2)

$$\dot{\varepsilon} = G_{,\sigma\sigma}\dot{\sigma} + G_{,\sigma T}\dot{T} + G_{,\sigma\omega}\dot{\omega} + G_{,\sigma\beta}\dot{\beta} + G_{,\sigma\lambda}\dot{\lambda} + G_{,\sigma\varepsilon^i}\dot{\varepsilon}^i \tag{20.3}$$

For convenience, in (20.3) set $G_{,\sigma\varepsilon^i} = 1$. Then we can rewrite (20.3) as a decomposition of strain-rates:

$$\dot{\varepsilon} = \dot{\varepsilon}^\sigma + \dot{\varepsilon}^T + \dot{\varepsilon}^\omega + \dot{\varepsilon}^\beta + \dot{\varepsilon}^\lambda + \dot{\varepsilon}^i \tag{20.4}$$

where

$$\dot{\varepsilon}^\sigma = G_{,\sigma\sigma}\dot{\sigma}$$
$$\dot{\varepsilon}^T = G_{,\sigma T}\dot{T}$$
$$\dot{\varepsilon}^\omega = G_{,\sigma\omega}\dot{\omega} \tag{20.5}$$
$$\dot{\varepsilon}^\beta = G_{,\sigma\beta}\dot{\beta}$$
$$\varepsilon^\lambda = G_{,\sigma\lambda}\dot{\lambda}$$

The functions $G_{,\sigma n}$ are rank four material compliance tensors that must be identified through experiments. It is often assumed that the mechanical compliance tensor $G_{,\sigma\sigma}$ is independent of inelastic strain, ε^i. In this instance (20.5)$_1$ can be identified as the elastic strain rate, complementary to the inelastic rate $\dot{\varepsilon}^i$. Strain-rates (20.5)$_2$–(20.5)$_5$ are respectively the thermal, hygral, damage, and autogeneous components of the total strain-rate. In general, all compliance tensors may depend on the same set of variables as the function G in (20.1), although of course they need out.

To complete the specification of constitutive equations for a material volume, we first recall that the stress, temperature, and evaporable water concentration are prescribed functions of time. We also note in passing that this requirement poses serious experimental difficulties, for most test specimens are so large and are exposed to mechanical and environmental surface effects such that the solutions of coupled boundary value problems are required to interpret experiments conducted on the specimens. Since the properties of the specimen are unknown, one is confronted with solution of a difficult inverse problem to determine constitutive properties.

If σ, T, ω are prescribed along with a structure for G (or the compliances in (20.5)), it remains to specify equations for maturity, damage, and inelastic strain.

20.3.1 Maturity

Maturity of concrete is primarily dependent on the degree of hydration of the cement paste, which in turn is influenced by temperature and water concentration. Since micro-cracking influences water migration through permeability and surface area effects, we postulate that maturity is determined by

$$\dot{\lambda} = \lambda(T, \omega, \lambda, \beta, \sigma) \tag{20.6}$$

For fixed environmental and loading conditions, one would expect λ to reach monotonically an asymptotic value.

20.3.2 Damage

A damage variable is introduced to account for degradation of the integrity of the material associated with micro-cracking. It might appear as an argument in any of the material compliance tensors, as well as produce strain in the material. The damage variable might also be expected to depend strongly on stress and weakly on other variables. Thus, we take

$$\dot{\beta} = \beta(\sigma, \beta, T, \omega, \lambda) \tag{20.7}$$

Further, it may be that β is a discontinuous function of σ, i.e. that damage occurs only after a threshold value of stress state is attained, and that damage may increase only under 'loading', however defined. For an isotropic damage model, β will depend on invariants of σ. For an anisotropic damage model, β must be a tensor-valued function of σ.

20.3.3 Inelastic strain

In the literature on behaviour of concrete, inelastic strain has commonly been modelled by selecting one or another of many types of linear viscoelastic models. At low stress levels reasonably good prediction of creep behaviour has been found, although there are shortcomings under stress reversal. At higher stress levels, sufficient internal damage occurs that standard linear viscoelastic models no longer suffice. In this note two approaches to the modelling problem will be briefly described.

20.3.4 Internal variables

The inelastic strain can be represented by a finite sum of strain contributions associated with internal variables q_n, i.e.

$$\varepsilon^i = \sum_n q_n \tag{20.8}$$

where the internal variables are determined by equations of the form

$$\dot{q}_n = f_n(\sigma, T, \omega, \beta, \lambda, q_i) \tag{20.9}$$

Constitutive Equations for Concrete and Geomaterials

If the functions f_n are chosen as a product of the form

$$f_n = F_n(\sigma, T, \omega, \beta, \lambda)[\sigma, q_n] \tag{20.10}$$

where the bracket denotes linearity in σ and q_n, it is possible to write (20.9) in the form

$$q'_n = g_n(\sigma, q_n) \tag{20.11}$$

where g_n is a linear function of σ, q_n and the superscript prime denotes differentiation with respect to a reduced variable y defined by

$$y(t) = \int_0^t F_n(\sigma, T, \omega, \beta, \lambda)(s)\,ds \tag{20.12}$$

This approach, originating in the treatment of so-called thermorheologically simple viscoelastic materials, has been used quite successfully by Schapery (1968), among others, to model nonlinear effects in materials associated with temperature and damage. This approach parallels and generalizes the familiar practice of employing linear rheological models noted by Anderson (Chapter 18, this volume). It will be observed that nonlinear phenomenology is captured by linear models through nonlinear mapping of the time scale. In connection with representation of ageing phenomena in viscoelastic materials a paper by Lubliner and Sackman (1966) deserves special mention.

20.3.5 Transformation of stress (or strain) measures

An alternative approach to reduction to linearity was introduced by Leaderman for viscoelastic materials (1943). For the present purposes one could write

$$\varepsilon^i = \int_0^t J(t-s)\frac{d}{ds}\left\{F[\sigma(s)]\right\}ds \tag{20.13}$$

In (20.13) the inelastic strain is a linear functional of the transformed stress, $F(\sigma)$, not the actual stress as in linear viscoelasticity. Experimental evidence is necessary to determine F as well as the creep compliance tensor J. Recent work by Browning, Gurten, and Williams (to appear), as noted by Anderson in this volume, incorporates Leaderman's idea of transformation of stress, but generalizes the model by adding concepts of loading, unloading, and reloading associated with rate-independent plasticity. More recently, Simo and Taylor (to appear) have utilized the same approach to develop a simple three-dimensional viscoelastic model which is capable of accounting for damage effects characteristic of highly filled elastomeric polymers. Continuation of such work for the modelling of concrete would appear to be a fruitful step at this time. In this context, whether or not the separate damage variable β previously introduced is still required, needs to be examined.

20.4 CONCLUDING REMARKS

This brief chapter has not been intended to provide an exhaustive review and analysis of the many issues surrounding modelling and computational mechanics for concrete and geomaterials, inasmuch as extensive reviews for concrete, at least, have appeared in recent years. For more detailed treatment of the internal variable formalism applied to ageing concrete, including computational aspects, Pister et al. (1978) may be consulted. Further discussion of materials in which coupling exists between instantaneous and rate-dependent inelastic strain response as well as shearing deformation and dilation can be found in Pister (1978).

REFERENCES

Browning, R. V., Gurten, M. E., and Williams, W. O. (to appear) 'A viscoplastic constitutive theory for filled polymers', *Int. J. Solids and Structures*.

Jaunzemis, W. (1967) *Continuum Mechanics*, Macmillan, New York.

Leaderman, H. (1943) *Elastic and Creep Properties of Filamentous Materials*, Textile Foundation, Washington, D.C.

Lubliner, J. (1972) 'On the thermodynamic foundations of nonlinear solid mechanics', *Int. J. non-Linear Mechanics*, **7**, 237–254.

Lubliner, J., and Sackman, J. L. (1966) 'On aging viscoelastic materials', *J. Mech. Physics Solids*, **14**, 25–32.

Pister, K. S. (1978) 'Constitutive equations for a class of inelastic materials', *Proc. of a Symposium on Appl. Mech. Honoring Henry L. Langhaar*, University of Illinois, Urbana, Illinois.

Pister, K. S., Argyris, J. H., and Willam, K. J. (1978) 'Creep and shrinkage of aging concrete', *Douglas McHenry International Symposium on Concrete and Concrete Structures*, Publication SP-55, Amer. Conc. Inst., Detroit, pp. 1–30.

Schapery, R. A. (1968) 'On a thermodynamic constitutive theory and its application to various nonlinear materials, *Proc. IUTAM Symposium East Kilbride*, Springer, pp. 259–285.

Simo, J. C., and Taylor, R. L. (1984) 'A simple viscoelastic model accounting for damage effects', Report No. UCB/SESM-84/06, Department of Civil Engineering, Univ. of Calif., Berkeley, Calif.

Truesdell, C. A. and Noll, W. (1965) 'The non-linear field theories of mechanics', in *Handbuch der Physik*, Vol. III3 (S. Flugge, Ed.), Springer-Verlag, Berlin.

PART VIII

NUMERICAL MODELLING

Mechanics of Geomaterials
Edited by Z. Bažant
© 1985 John Wiley & Sons Ltd

Chapter 21

Numerical Modelling and Geomechanics (Soil–Rock–Concrete)

O. C. Zienkiewicz

21.1 INTRODUCTION

In this symposium organized to commemorate William Prager and his inestimable contributions to the subject of constitutive modelling, I have been assigned the role of discussing the impact of numerical modelling on the field of geomechanics and concrete technology.

In many other chapters in this volume specific forms of constitutive relations and behaviour have been described. In this chapter we shall, perforce, deal with these only in a general manner. It is, however, an undisputed fact that the present interest and refinement of material modelling got its biggest impulse from the entry of the computer and numerical methods onto the scene of solving real physical situations. William Prager was one of the first to realize this even though many of his contributions predate the computer era.

Undoubtedly the biggest progress in the field is due to the collaboration of experimental 'modellers' and those engaged in the numerical solution business. Such a collaboration was envisaged by Roscoe of Cambridge, England, who was responsible for introducing into soil mechanics the first set of consistent plasticity models (critical state, or cap, models) and who established a collaboration with myself for solving realistic boundary value problems.

Though this collaboration was cut short by his untimely death, it was responsible for much of my own interest in Soil Mechanics which will transpire through this chapter.

'Collaboration' is not always the only way of achieving progress. The needs of numerical analysis have often not been met directly and necessitated the creation of 'ad hoc' models by those trying to solve problems. It is evident from the presenters at this conference that many of us entered the area of constitutive modelling via numerical analysis and that indeed many models have been established through their computational convenience—as we hope to demonstrate later.

This chapter will concern itself with three aspects of the problem.

(i) the communality of general behaviour of geomaterials—as two-phase media,
(ii) the present aspects of numerical solution of well posed problems, and
(iii) the nature of modelling effectively used to describe the materials.

I shall not separate static from dynamic behaviour, as the two are best dealt with together. Indeed the trend of present programs is to solve the former as a by-product of the latter formulation. In conclusion I shall show some practical, engineering, problems for which the numerical—finite element—solution presents the only line of attack. Here in particular the practical importance of step-by-step, nonlinear, analysis in the context of response of soil and concrete structures to earthquakes will be demonstrated.

21.2 BASIC MECHANICS AND FORMULATION

Soils, and to a lesser extent rock and concrete are porous media which when saturated exhibit a behaviour dependent on 'effective stress'. For all such materials Biot-type formulation (Biot, 1941, 1956, 1962a, 1962b) provides the description of behaviour which couples the dynamics of the solid and fluid phases. The essence of such behaviour has been investigated by many and a survey of the basic forms has recently been provided by the author (Zienkiewicz and Shiomi, 1984) summarizing the governing equations.

Below we quote the salient points.

1. In a porous material the total stress σ_{ij} is conveniently split into the 'effective stress'

$$\sigma'_{ij} = \sigma_{ij} - \delta_{ij} p \tag{21.1}$$

which is responsible for most of the deformation and failure behaviour of the material, and a hydrostatic component $\delta_{ij} p$, which together with the pore pressure results only in a uniform compression of the solid grains. Assuming isotropic, elastic, behaviour and a uniform strain rate this gives

$$\dot{\varepsilon}^p_{ii} = -\dot{p}/K_s \tag{21.2}$$

where K_s is the grain bulk modulus.

2. It is (relatively) easy to show Zienkiewicz et al. (1980a) that if a new effective stress is redefined as

$$\sigma''_{ij} = \sigma_{ij} - \alpha \delta_{ij} p \tag{21.3}$$

then the incremental constitutive relation can be written as

$$\dot{\sigma}''_{ij} = D_{ijkl} \dot{\varepsilon}_{kl} + \dot{\omega}_{ik} \sigma''_{kj} + \sigma''_{ik} \dot{\omega}_{jk} \tag{21.4}$$

where the last two terms ensure objectively in the Jaumann sense with $\dot{\varepsilon}_{ij}$ and $\dot{\omega}_{ij}$ being defined in terms of the skeleton displacement as

$$\dot{\varepsilon}_{ij} = (\dot{u}_{i,j} + \dot{u}_{j,i})/2; \quad \dot{\omega}_{ij} = (\dot{u}_{i,j} - \dot{u}_{j,i})/2 \tag{21.5}$$

The coefficient α is approximately given by

$$\alpha = \left(1 - \frac{K_T}{K_s}\right) \tag{21.6}$$

where K_T is the bulk modulus of the skeleton.

For soils $\alpha \approx 1$ and the difference between the two 'effective' stresses does not arise but for rocks and concrete the value of $\alpha \approx \frac{1}{3} - \frac{1}{2}$ is not uncommon and the difference is important.

In above the D matrix will in general depend on direction of straining, stress (σ' not σ'') and history in a manner specified by the appropriate constitutive law used.

21.2.1 Full equations

Defining U as being the mean displacement of the fluid in the pores the governing equation for the system can be written as

$$\sigma''_{ij} - (\alpha - n)p_{,i} + (1 - n)\rho_s b_i + (1 - n)\rho_s \ddot{u}_i - R_i = 0 \tag{21.7a}$$

for the momentum balance of the solid phase;

$$-np_{,i} + n\rho_f b_i - n\rho_f \ddot{U}_i + R_i = 0 \tag{21.7b}$$

for the momentum balance of the fluid phase; and

$$-\dot{U}_{i,i} n = (\alpha - n)\dot{\varepsilon}_{ii} + \frac{1}{Q}\dot{p} \tag{21.7c}$$

for the volumetric balance of fluid flow.
In the above

n is the porosity

ρ_s, ρ_f are the densities of the solid and fluid phases respectively giving an overall density as

$$\rho = (1 - n)\rho_s + n\rho_f$$

Q is the only new parameter introduced and corresponds to a fluid/solid compressibility

$$\frac{1}{Q} = \frac{n}{K_f} + \frac{\alpha - n}{K_s} \tag{21.8}$$

This often is considered to be zero but for numerical treatment is conveniently retained as a 'penalty' number.

Finally R_i represents the viscous drag of the fluid in motion which in terms of the well known permeability coefficient, k, can be written as

$$R_i = n^2 k^{-1}(\dot{U}_i - \dot{u}_i) \tag{21.9}$$

These equations together with the appropriate constitutive relation (21.4) and boundary conditions provide the basis for all geomechanical computation.

21.2.2 Approximate equations

It is convenient at times to use an approximate form of these equations in which we neglect the relative acceleration terms and replace $\ddot{U}_i$ by $\ddot{u}_i$. With this approximation it is possible to eliminate the variable U_i (and further neglecting the acceleration effects in the fluid flow equation) to write the system as

$$\sigma''_{ij,i} - \alpha p_{,i} + \rho b_i - \rho \ddot{u}_i = 0 \tag{21.10a}$$

for the overall momentum balance; and

$$[k(-p_{,i} + \rho_f b_i)]_{,i} + \alpha \dot{\varepsilon}_{ii} + \frac{1}{Q}\dot{p} = 0 \tag{21.10b}$$

which is the equivalent of the standard (consolidation), seepage, equation.

Again this system with boundary conditions and constitutive relations is properly posed.

21.2.3 Undrained conditions

If the permeability $k \to 0$ or if the time scale is sufficiently short undrained behaviour can be assumed. Now, as can easily be verified, the constitutive relation in the undrained form can be written in terms of total stress as

$$\dot{\sigma}_{ij} = \bar{D}_{ijkl}\dot{\varepsilon}_{kl} + \dot{\omega}_{ik}\sigma_{kj} + \sigma_{ik}\dot{\omega}_{jk} \tag{21.11a}$$

with

$$\bar{D}_{ijkl} = D_{ijkl} + \alpha^2 \delta_{ij}\delta_{kl}Q \tag{21.11b}$$

which relates uniquely the total to effective stress matrices.

Now only the overall momentum balance enters into the problem, i.e.

$$\sigma_{ij,j} + \rho b_i - \rho \ddot{u}_i = 0 \tag{21.12}$$

and this is the dynamic equation for a standard single-phase medium.

The range of applicability of the expressions has been studied and Figure 21.1 shows the limits of applicability in terms of two nondimensional parameters (Zienkiewicz et al., 1980a; Zienkiewicz and Bettess, 1980)

21.2.4 Discrete equations

The problem as stated in the full interation equations (Zienkiewicz and Bettess, 1980), the approximation (21.10) or in its undrained form (21.12) can be

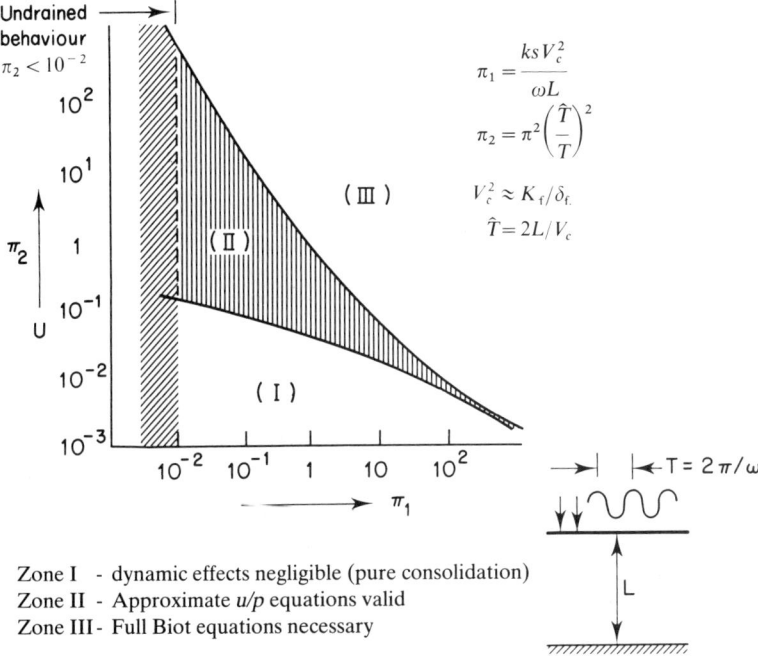

Figure 21.1 Limits of applicability of various assumptions in soil–fluid interaction

discretized using the finite element process (Zienkiewicz, 1977) and many alternatives of doing so have been discussed in the literature *extensively* (Zienkiewicz and Shiomi, 1984).

Below we give a list of typical coupled sets of equations for which in the space discretization linearity assumptions are used. We quote such forms not because of their practical value but to show the basic equation structure. In what follows the variables $\bar{\mathbf{u}}$ etc. stand for discrete (nodal) sets and symbols of standard matrices follow (Zienkiewicz, 1977) and Zienkiewicz, Shiomi (1984).

Thus proceeding to Equation (21.7) we can have with full (mixed) discretization

$$\begin{bmatrix} \mathbf{M}_s & 0 & 0 \\ 0 & 0 & 0 \\ 0 & 0 & \mathbf{M}_f \end{bmatrix} \begin{bmatrix} \ddot{\bar{\mathbf{u}}} \\ \ddot{\bar{\mathbf{p}}} \\ \ddot{\bar{\mathbf{U}}} \end{bmatrix} + \begin{bmatrix} \mathbf{C}_1 & 0 & -\mathbf{C}_2 \\ 0 & 0 & 0 \\ -\mathbf{C}_2^T & 0 & \mathbf{C}_3 \end{bmatrix} \begin{bmatrix} \dot{\bar{\mathbf{u}}} \\ \dot{\bar{\mathbf{p}}} \\ \dot{\bar{\mathbf{U}}} \end{bmatrix}$$
$$+ \begin{bmatrix} \mathbf{K} & \mathbf{G}_1 & 0 \\ \mathbf{G}_1^T & \mathbf{P} & \mathbf{G}_2^T \\ 0 & \mathbf{G}_2 & 0 \end{bmatrix} \begin{bmatrix} \bar{\mathbf{u}} \\ \bar{\mathbf{p}} \\ \bar{\mathbf{U}} \end{bmatrix} = \begin{bmatrix} \bar{\mathbf{f}}_s \\ \bar{\mathbf{f}}_p \\ \bar{\mathbf{f}}_f \end{bmatrix} \quad (21.13)$$

Or, if the value of $1/Q$ which is proportional to the matrix $\mathbf{P}$ is not zero (i.e. the fluid is compressible), the irreducible form can be obtained

$$\begin{bmatrix} \mathbf{M}_s & 0 \\ 0 & \mathbf{M}_f \end{bmatrix} \begin{bmatrix} \ddot{\mathbf{u}} \\ \ddot{\mathbf{U}} \end{bmatrix} + \begin{bmatrix} \mathbf{C}_1 & -\mathbf{C}_2 \\ -\mathbf{C}_2^T & \mathbf{C}_3 \end{bmatrix} \begin{bmatrix} \dot{\mathbf{u}} \\ \dot{\mathbf{U}} \end{bmatrix} + \begin{bmatrix} \mathbf{K}_1 + \mathbf{K} & \mathbf{K}_2 \\ \mathbf{K}_2^T & \mathbf{K}_3 \end{bmatrix} \begin{bmatrix} \mathbf{u} \\ \mathbf{U} \end{bmatrix} = \begin{bmatrix} \bar{\mathbf{f}}_u \\ \bar{\mathbf{f}}_U \end{bmatrix}$$
(21.14)

This form is similar (but not identical) to that used by Ghaboussi and Wilson (1972) in the first solutions of such problems by the finite element method.

If $1/Q = 0$, a penalty factor can be introduced by adding $(1/\lambda)p$ to Equation (21.7c) and now an alternative form appears in Zienkiewicz and Shiomi (1984); Prevost (1982); and van der Kogel (1977)

$$\begin{bmatrix} \mathbf{M}_s & 0 \\ 0 & \mathbf{M}_f \end{bmatrix} \begin{bmatrix} \ddot{\mathbf{u}} \\ \ddot{\mathbf{U}} \end{bmatrix} + \begin{bmatrix} \mathbf{C}_1 + \tilde{\mathbf{K}}_1 & -\mathbf{C}_2 + \tilde{\mathbf{K}}_2 \\ -\mathbf{C}_2^T + \tilde{\mathbf{K}}_2^T & \mathbf{C}_3 + \tilde{\mathbf{K}}_3 \end{bmatrix} \begin{bmatrix} \dot{\mathbf{u}} \\ \dot{\mathbf{U}} \end{bmatrix} + \begin{bmatrix} \mathbf{K} & 0 \\ 0 & 0 \end{bmatrix} \begin{bmatrix} \mathbf{u} \\ \mathbf{U} \end{bmatrix} = \begin{bmatrix} \bar{\mathbf{f}}_u \\ \bar{\mathbf{f}}_U \end{bmatrix}$$
(21.15)

A simpler form with a smaller number of variables derives from the approximation of Equations (21.10) (Zienkiewicz et al., 1979, 1980b, 1982).

$$\begin{bmatrix} \mathbf{M} & 0 \\ 0 & 0 \end{bmatrix} \begin{bmatrix} \ddot{\mathbf{u}} \\ \ddot{\mathbf{p}} \end{bmatrix} + \begin{bmatrix} \mathbf{C} & 0 \\ \mathbf{Q} & \mathbf{S} \end{bmatrix} \begin{bmatrix} \dot{\mathbf{u}} \\ \dot{\mathbf{p}} \end{bmatrix} + \begin{bmatrix} \mathbf{K} & -\mathbf{Q}^T \\ 0 & \mathbf{H} \end{bmatrix} \begin{bmatrix} \mathbf{u} \\ \mathbf{p} \end{bmatrix} = \begin{bmatrix} \bar{\mathbf{f}}_s \\ \bar{\mathbf{f}}_p \end{bmatrix} \quad (21.16)$$

Finally for undrained behaviour a single matrix form of

$$\mathbf{M}\ddot{\bar{\mathbf{u}}} + \mathbf{C}\dot{\bar{\mathbf{u}}} + \mathbf{K}\bar{\mathbf{u}} = \mathbf{f} \tag{21.17}$$

is obtained (Zienkiewicz, 1978).

All the equations are of a transient type with steady state arising as special cases.

21.3 NUMERICAL SOLUTION

The numerical solution of the, generally nonlinear, systems is a subject dealt with in many texts (viz. Zienkiewicz, 1977) and numerous papers.

Basically two alternatives are practised in the finite element applications.

(A) Explicit solution of the dynamic equation requiring in the case of lumped mass matrices *no equation solving*.

or

(B) Use of implicit methods which, just like steady state problems, result in a solution of an equation system in some parametric form

$$\psi(\mathbf{a}) = \mathbf{f} \tag{21.18}$$

Here the practice is to use one or other of Newton-like procedures in which the tangent matrix at its approximation is used and explicitly solved. The problem is reduced to a sequential solution

$$\mathbf{K}_n \Delta \mathbf{a}_n = g_{n-1} \tag{21.19}$$

and many variants are available (Zienkiewicz, 1977). Currently very popular are the so-called quasi-Newton (secant procedures) and much literature is devoted to this aspect.

I shall not dwell on this subject here as in some way it is possible that the finite element fraternity has chosen the wrong path in abandoning direct, iterative, solutions in favour of repeated equation solving characteristic of presently used processes. Many of us engaged in explicit dynamic solution procedures have found that the standard programs designed for dynamics can be conveniently used for static solutions resulting always in smaller storage requirements and simplicity and generally similar computer costs.

Today some new developments promise to allow an order to magnitude improvement of performance of such 'relaxation' methods (Zienkiewicz and Löhner, 1984). For instance if Equation (19.18) is complemented by a 'viscous' term (regularization) as

$$C\dot{a} + \psi(a) = f \qquad (21.20)$$

and its steady state solution sought by solving *explicitly* the time marching problem (Oden and Key, 1973) the so called 'viscous relaxation' process will be

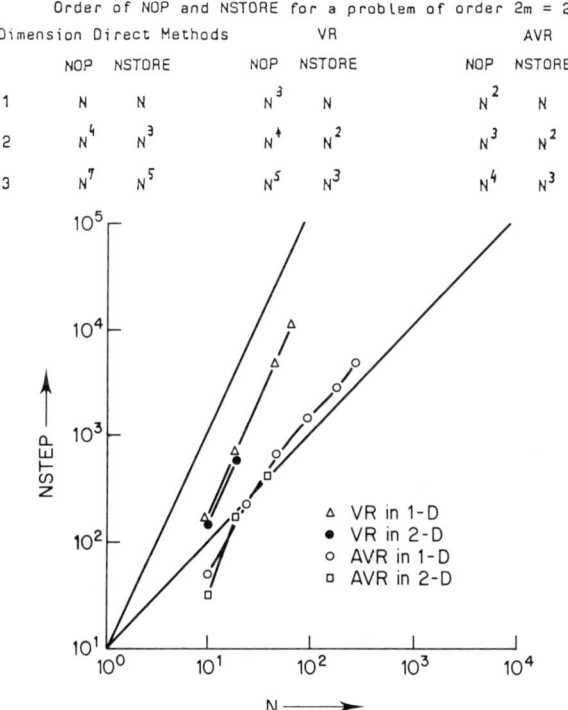

Figure 21.2 Comparison of operations (NOP) and storage (NSTORE) or time step (NSTEP) requirements for direct solution, viscous relaxation and accelerated viscous relaxation

478 *Mechanics of Geomaterials*

found with rather slow convergence. With a suitable acceleration (AVR) described in (Zienkiewicz and Löhner, 1984) computational times for solution of linear and nonlinear problems become almost identical.

Figure 21.2 shows a table of operation and storage counts for various procedures of solution in an N^i dimensional 'cube' ($i = 1-3$).

It is readily seen that for two dimensions the VR counts are identical to direct solvers as far as the number of operations to obtain a solution are concerned but that storage requirements are reduced; in the three dimensions the

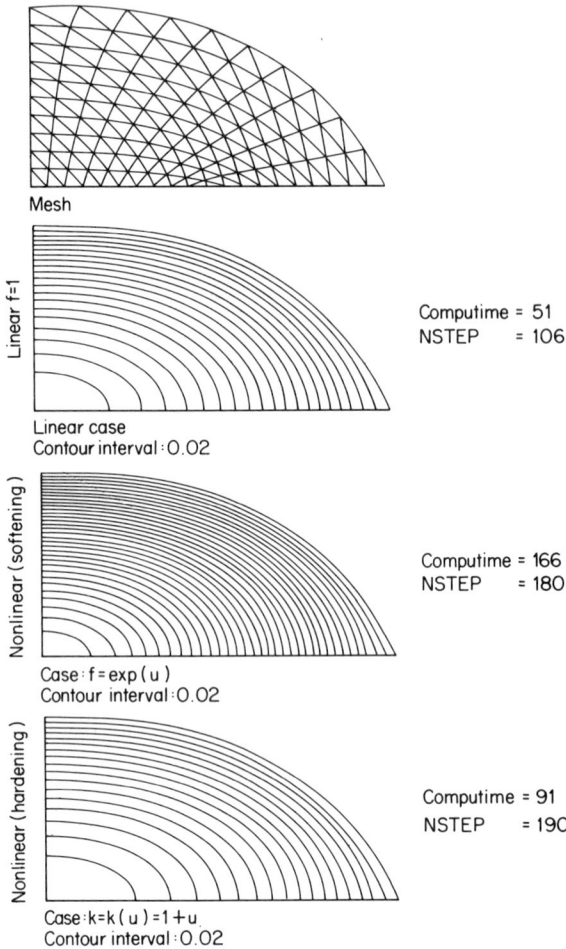

Figure 21.3 Comparison of computation times for linear or non-linear problem of heat conduction. $u = 0$ on boundary
$$\nabla(k\nabla u) + f = 0$$

improvement is achieved on all counts and when 'accelerated viscous relaxation' is used all comparisons favour the new process.

In a typical solution of a nonlinear (thermal) problem with some 120 DOF similar times were achieved for both linear and nonlinear problems (using some 180 steps) and for all the cost was less than that of a linear solution with a direct front solver. The cost effectiveness in larger problems and the possibility of using vector processes (or microcomputers) for such nonlinear solutions is very apparent.

21.4 THE MATERIAL MODELS

Here the problem should be posed in a manner which is 'as simple as possible but no simpler' to quote Einstein. Perhaps indeed in this statement the word 'possible' should be changed to 'required' as in many cases quite different forms of

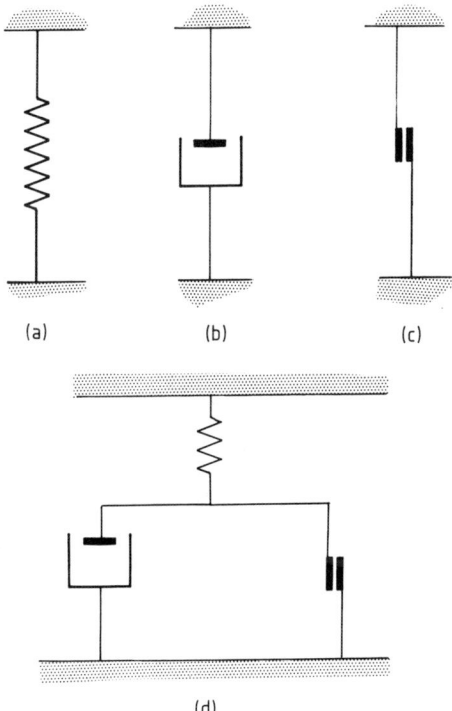

Figure 21.4 Basic components for material models. (a) Spring—reversible linear/nonlinear elasticity. (b) Dashpot—linear/nonlinear creep. (c) Slider—plastic resistance (strain dependent). (d) Possible elastic, viscoplastic assembly

constitutive models achieve the same prediction providing these contain the essential features of the problem.

Conceptually all material models can be built up for three simple ingredients illustrated in Figure 21.4—the spring, the dashpot, and the slider.

From such simple beginnings stem more complex forms of description using big words such as

> nonlinear (hyper) elasticity
> viscoelasticity
> viscoplasticity
> plasticity (hypoelasticity)
> endochronic theory, etc.

and these at times allow an improved characterization. Whatever the model chosen numerical procedures will allow it to be treated but occasionally the program efficiency dictates the preference. Thus for instance three types of models owe their emergence to the numerical procedures favoured.

(1) The overlay or sublayer models in which complex behaviour patterns are built up from such simple ingredients as ideally plastic components placed *in parallel*, viz. Besseling (1958), Zienkiewicz et al. (1972, 1974).
(2) The multi-laminate models where a series addition of components which include preferential (joint) direction sliding allow a physical introduction of joint direction (Zienkiewicz and Pande, 1977) or in a more general sense lead to a new category of models in which rotation of stresses influence yield (Pande and Sharma, 1980, 1981).
(3) The use of viscoplasticity as an artifice for solution of purely plastic problems—in a manner akin to the regularization method discussed in the previous section (Zienkiewicz and Cormeau, 1974; Zienkiewicz et al., 1975).

To some the assembly concept is indeed the only logical way of generating the behaviour patterns (Dougill, 1982) and gaining a physical insight. This is a very logical view and others prefer the use of direct, nonphysical descriptions. The author of this paper takes a pragmatic view, using whichever concept leads to the simplest and most direct computational success.

Space does not permit here a detailed description of the various models for geomaterials. However, we shall point out some typical classes of problems— and a few simple approaches which are common to many of such materials.

21.4.1 Tensile failure and cracking

All geomaterials have the common characteristic of being weak in tension. Soils have here a strength which is approximately zero while both rocks and concrete have a finite strength which is a fraction $1/10-1/5$ of their compressive limit.

Indeed in the design of plain concrete structures it has been often assumed that no reliance should be placed on such tensile strength and that a safe analysis would exclude it.

The first attempt to extend this concept to a nonlinear, finite element, analysis was made by Zienkiewicz et al. (1968) who introduced the 'no tension' material concept. To obtain more satisfactory modelling which included a finite tensile strength much further study has gone into this problem and a recent survey (Darwin et al. (1982)) summarized the state of the 'art'. Three basic approaches exist today and are shown in Figure 21.5. In these it can be assumed that

(a) cracking occurs in a distributed sense through the elements in a direction normal to the tensile principal stress;
(b) single cracks at element junctions develop for the same criteria as (a); or finally that
(c) cracking is governed by fracture mechanics criteria and results in element separation.

The first idea, similar to that used in no tension analysis is the simplest to implement computationally and has been used extensively. It was introduced initially by Rashid (1968) and developed further by others (Suidan and Schnobrich, 1973; Phillips and Zienkiewicz, 1976).

Bažant and Oh (1981a, 1981b) point out that a procedure based on smeared or distributed cracking is in fact consistent with energy criteria for fracture providing the total energy absorption in a cracked zone of a width w is given correctly. In this a softening zone is included and cracking deemed to start when *stress* reaches a limiting value corresponding to uniaxial strength (Figure 21.6).

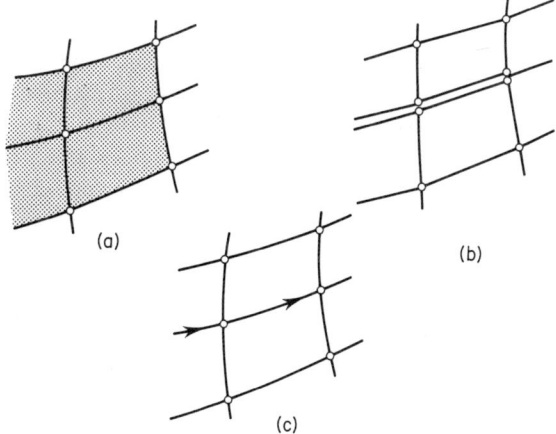

Figure 21.5 Alternative approaches to modelling of cracks. (a) Distributed cracking. (b) Crack on element interface (element separation). (c) Fracture mechanics—stress intensity factor approach

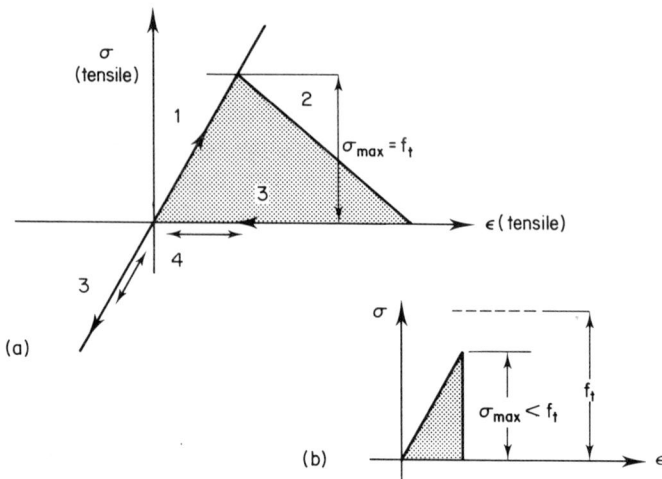

Figure 21.6 Behaviour in uniaxial tension. (a) (1) First tensile loading. (2) Strain softening. (3) Compressive reload. (4) Subsequent behaviour. Shaded area—energy dissipation per unit volume (E). Width of crack zone W. WE = constant. If W too large, reduce σ_{max} as shown in (b). f_t—uniaxial strength in tension

Only if the cracked zone is very wide is it necessary to reduce the strength—a logical conclusion including a size effect.

The importance of the softening zone is significant in dynamic analysis but for static computation it can be generally ignored.

Once cracking has developed in a certain direction the material acquires anisotropic properties maintaining zero tension while the total strain in the direction normal to the crack is positive but 'heals' once this has become compressive. This modelling is of extreme importance in dynamic problems when stress reversals occur (Zienkiewicz et al., 1983a; Pande, 1979).

The performance of distributed cracking models leads to excellent results when comparison with experiment are considered (Phillips and Zienkiewicz, 1976) Figure 21.7 shows a typical result.

21.4.2 Joints

Here when joints in rock or concrete are considered a discrete discontinuity surface exists here alternative approaches are possible. In the first a mixed formulation can be used in which the tractions across the joint are treated as problem variables and appropriate tensile/shear criteria applied to these. In the second the joint is considered as an element of a small but finite thickness in which the properties are again 'distributed' as in section 4.1 (Zienkiewicz et al., 1970; Ghaboussi et al., 1973; Zienkiewicz, 1977).

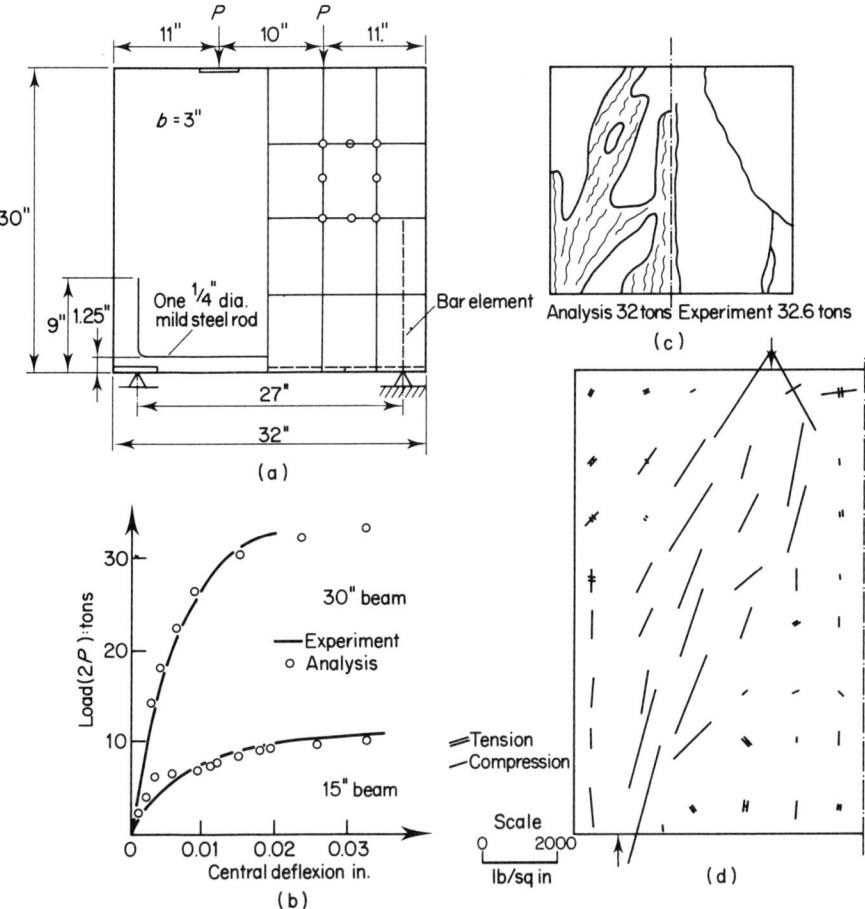

Figure 21.7 Results of analyses of deep beams under two-point loading. (a) Details and mesh for 30-in beam. (b) Load-displacement curves. (c) Crack patterns near failure. (d) Principal stresses in 30-in beam at 32 tons. (Phillips and Zienkiewicz, 1974 and 1976; experiments by Ramekrishnan and Anathanarayana, 1968)

21.4.3 Geomaterials under static loads

Soils and indeed most geomaterials exhibit 'failure' or yield characteristics with criteria which can be well described by Mohr–Coulomb type surface in the principal stress space showing an increase of strength with compressively increasing main stress.

Thus whatever type of model is adopted it is essential that this should show continuing deformation whenever the stresses reach the Mohr–Coulomb limit.

In Figure 21.8 we show stress space characterized by the first and second invariants (though the importance of the third invariant is considerable).

484   *Mechanics of Geomaterials*

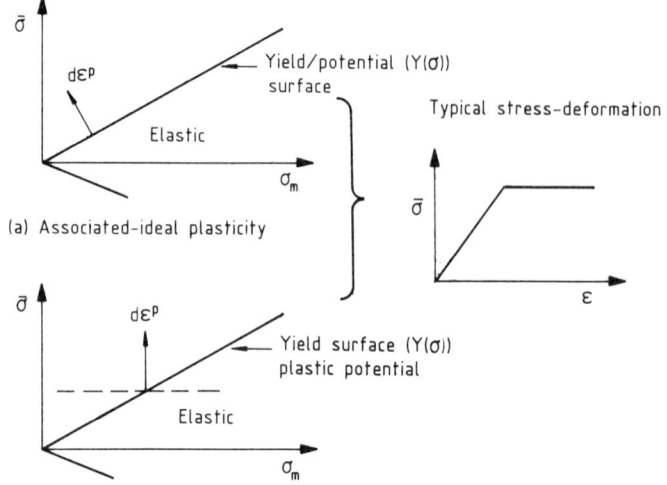

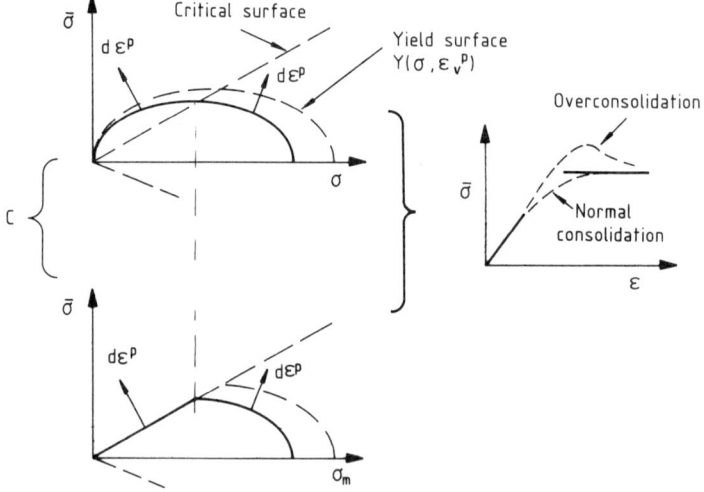

Figure 21.8 Typical elasto-plastic geomechanical models in deviatoric $\bar{\sigma}$/mean stress $\bar{\sigma}_m$/space (dependence on third invariant implied)

The most obvious approach to such modelling is to use ideal plasticity assumptions and to treat the Mohr–Coulomb surface as a yield-potential, Figure 21.8(a). This has shortcomings as the volumetric dilation is excessive and an alternative with non-associative plasticity can reduce this dilation to zero if desired as shown in Figure 21.8(b).

Finally, as suggested by Drucker *et al.* (1957) a strain-hardening surface

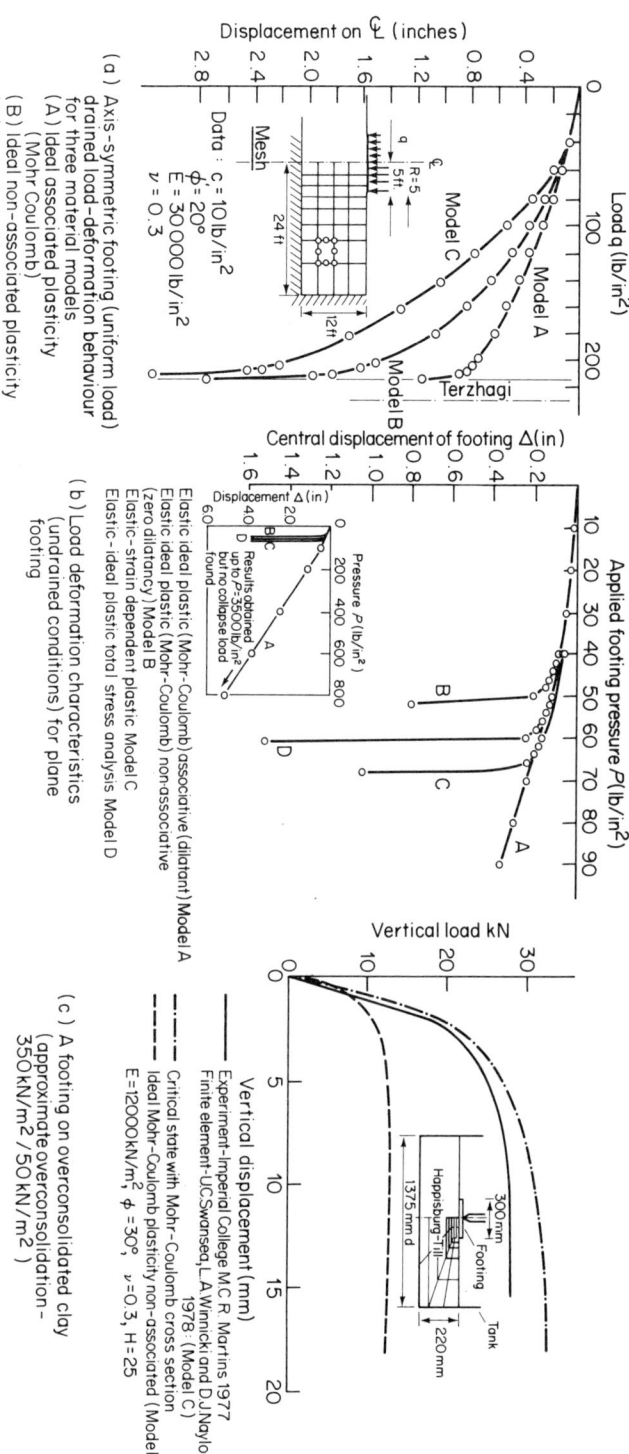

Figure 21.9 Drained/undrained behaviour of a footing normal consolidation and with overconsolidated conditions

dependent on the plastic volumetric strain in the hardening parameter can be used. This is indeed the critical state model adopted by Roscoe and his associates (Roscoe et al., 1958; Schofield and Wroth, 1968) later reintroduced as the cap model by Dimaggio and Sandler (1970).

Many variants and elaborations of such models have been introduced. We shall not discuss these here. The modelling is merely introduced to show that for many purposes (almost) identical behaviour can be obtained with considerably different assumptions providing these reproduce the essence of the behaviour. Thus in Figure 21.9(a) we see that each of the models predicts well the deformation and collapse behaviour of a drained material footing. However, when a volumetric strain constraint is introduced as occurs in an undrained situation only the non-associated of critical state models work as shown in Figure 21.9(b).

Finally, if a distinction is to be made between 'normally' or 'over'-consolidated materials only a strain-hardening model can produce reasonable answers as shown in Figure 21.9(c) (Zienkiewicz, 1979).

Thus at all times selectivity and sensitivity of modelling has to be considered and a 'horses for courses' policy adopted.

Similar behaviour patterns occur in concrete/rock under compression and similar models have successfully been used.

Although all geomaterials exhibit some creep form of behaviour which can (and must) on occasion be added by viscoplastic considerations (Zienkiewicz and Cormeau, 1974; Zienkiewicz et al., 1975) we shall not enter into this discussion now.

21.4.4 Transient or cyclic loads

While elasto-plastic models based in classical plasticity operate both in loading and unloading all fail to reproduce the essential feature of soil subject to such conditions in which a continuous pressure rise is observed due to a 'densification' of the material when subjected to a number of stress reversals (Figure 21.10(a)). This pressure rise leads to a progressive weakening of the material leading finally to liquefaction or to so-called cyclic mobility where large strain amplitudes occur for small stress cycles.

A very simple model for such materials can be obtained by 'augmenting' the classical plasticity forms by addition of a volumetric strain ε_v^0 dependent on a 'historical' variable linked to total strain path characteristics. This historical variable is indeed of a type associated with endochronic models by Valanis and Read (1982) and Bažant et al. (1982) but by separating the main effects achieves the model with very few additional parameters.

The whole subject of soil behaviour under cyclic loads is one of continuing interest and a recent symposium (Pande and Zienkiewicz, 1982) presents a picture of the different paths followed by various researchers.

Numerical Modelling and Geomechanics

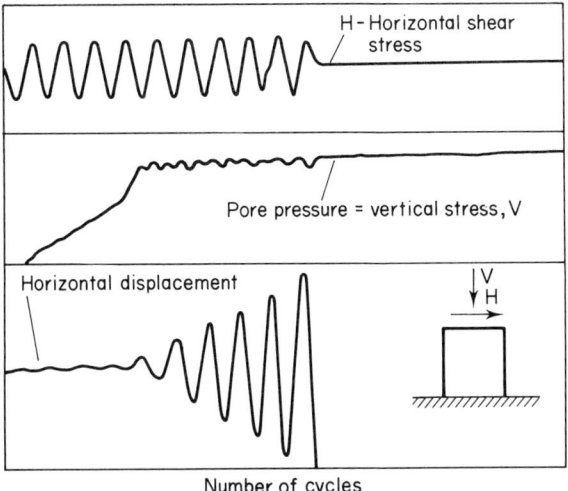

(a) Typical behaviour

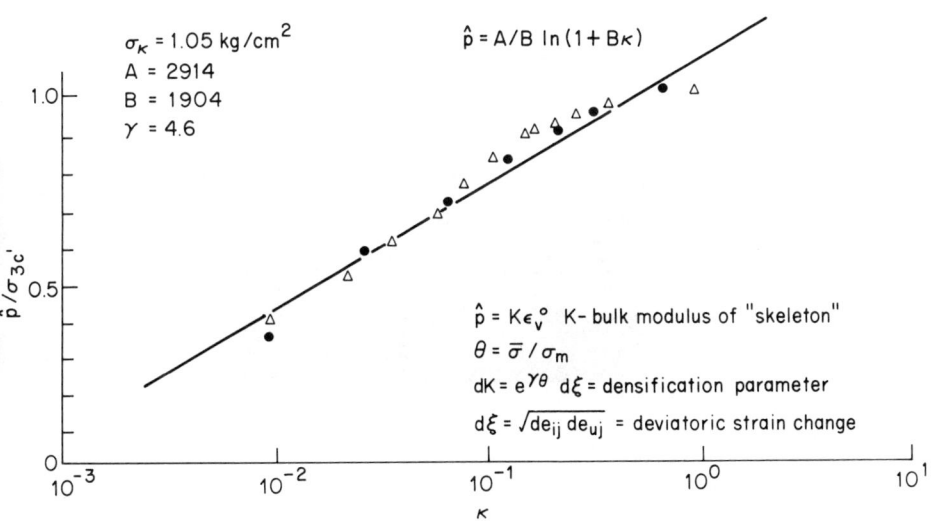

(b) Pressure changes or densification versus an endochronic parameter κ
(Lower San Fernando Dam – Experimental data by Seed et al 1980)

Figure 21.10 Densification of sands under cyclic loading

488 Mechanics of Geomaterials

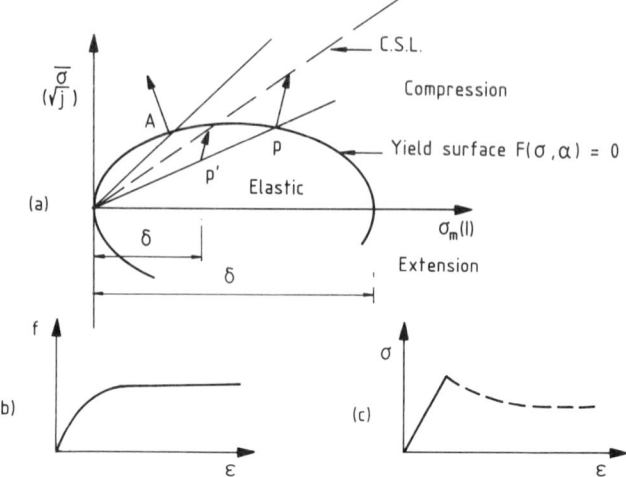

Figure 21.11 Bounding surface plasticity, critical surface model with volumetric and deviatoric hardening (Wilde, 1977). $\alpha = \varepsilon_{vol}^p + f(\varepsilon^{-p})$. (a) Yield surface in stress space. (b) Deviatoric hardening function. (c) Stress–strain behaviour in drained loading. Plastic modulus interpolation $K_{p'} = K_p \left(\dfrac{\delta_o}{\delta}\right) \gamma (\gamma = 1 - 10)$

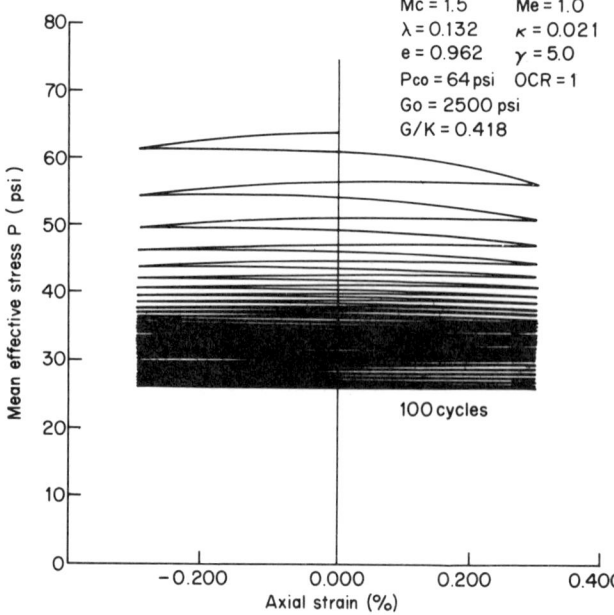

(a) Mean effective stress/axial strain

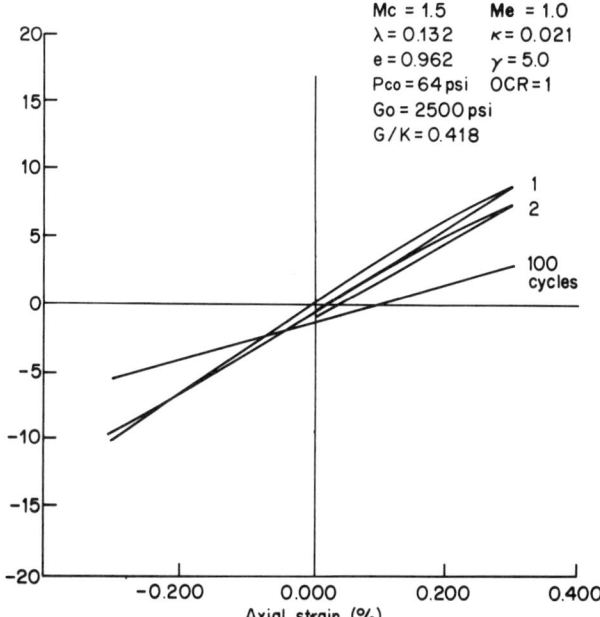

(b) Axial stress /axial strain

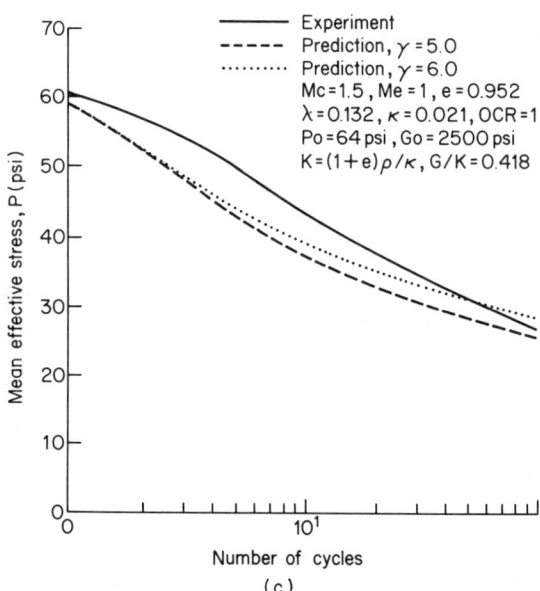

Figure 21.12 Performance of bounding surface model under cyclic strain in triaxial test (Taylor and Bacchus, 1969). Amplitude of strain ± 0.3%

One of the most successful paths is that of 'bounding surface' plasticity suggested by Dafalias and Herrmann, 1982 and followed by a 'generalized plasticity' definition (Zienkiewicz and Mroz, 1983). A recent model (Zienkiewicz and Mroz, 1983; Mroz and Zienkiewicz, 1983; Zienkiewicz et al. (in preparation) allows the whole densification process to be followed by the introduction of one additional parameter to the interpolation of the plastic modulus as shown in Figures 21.11 and 21.12.

This model is similar to one termed 'reflective' surface plasticity (Shiomi et al., 1982) and in Figure 21.13 we show a study of a sand layer and liquefaction prediction by the use of different models. Clearly both densification and reflective surface plasticity models give excellent results compared with the Carter et al. (1982) model.

The stage has now been reached where rational models of this kind can be used for the prediction of possible behaviour patterns under earthquake (or shock loading) conditions. In Figure 21.14 we show a study of the behaviour of the (lower) San Fernando dam which failed under the action of an earthquake on 9 February 1971 (Seed, 1979).

Here a full couples solution was used allowing for redistribution of pore pressures due to seepage in the manner discussed earlier.

It is of interest (an importance) to observe that now both positive and negative pressure can be generated showing that some parts of the structure are strengthened while others reach nearer liquefaction.

The satisfactory nature of the back calculation here should give the engineer confidence in applying such procedures in design in place of irrational 'linearized' analysis which in cost approximate to the same computer usage but are incapable of predicting permanent deformations.

21.4.5 Rate effects and deterioration

In soils it appears that rate of loading has little effect on the effective stress strength of the materials and that degradation of the materials in such terms is not an important factor. On the other hand in concrete (and presumably rock) such effects are of considerable importance as shown in the classical experiments of Hatano (Hatano, 1960; Hatano and Watanabe, 1971) shown in Figure 21.15.

Few attempts of modelling of such behaviour have been reported, but an effective way of doing so has been shown by Bicanic et al. (1978) and Bicanic and Zienkiewicz 1983 in which an adaptation of the viscoplastic model is used in the manner described in Figure 21.16.

In this model two essential features are introduced. The first is the concept of a monitoring 'failure' surface which degrades with the amount of plastic work (W^p) developed. When the stress reaches the failure surface a fairly abrupt drop of the yield surface F_D occurs to the residual material strength.

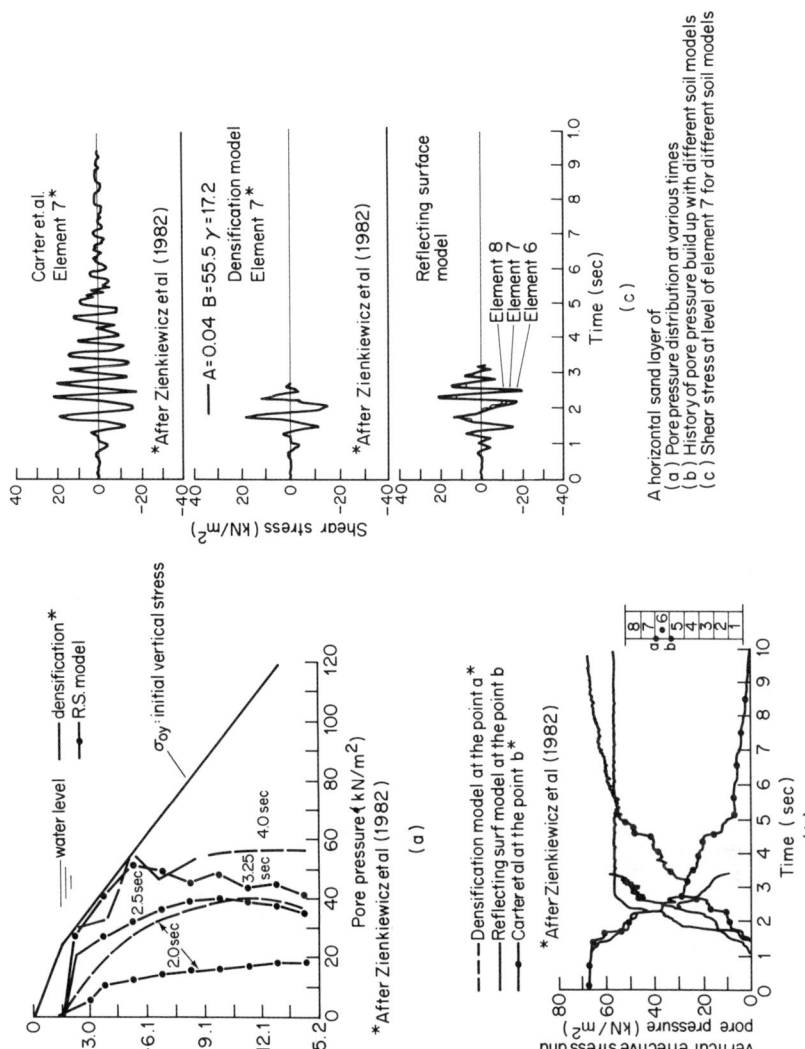

Figure 21.13 Comparative study of various models in prediction of liquefaction

492   Mechanics of Geomaterials

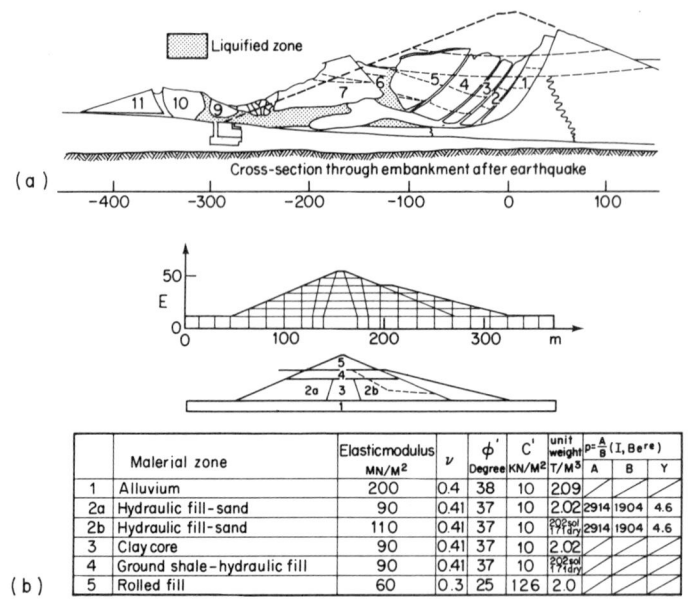

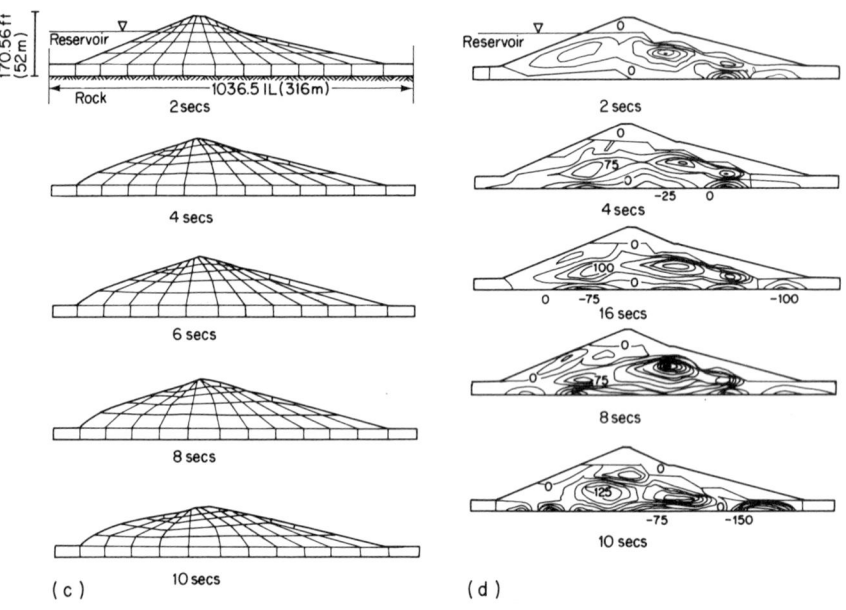

Figure 21.14 The failure of lower Sam Fernando dam; observation and analysis. (a) The failure (Seed *et al.*). (b) Finite element mesh and material properties. (c) Analysis results displacement × 5 (*y* node elements). (d) Analysis of excess pore pressure build. (Contour interval = 0.523 kips/ft^2 (25 kN/m^2))

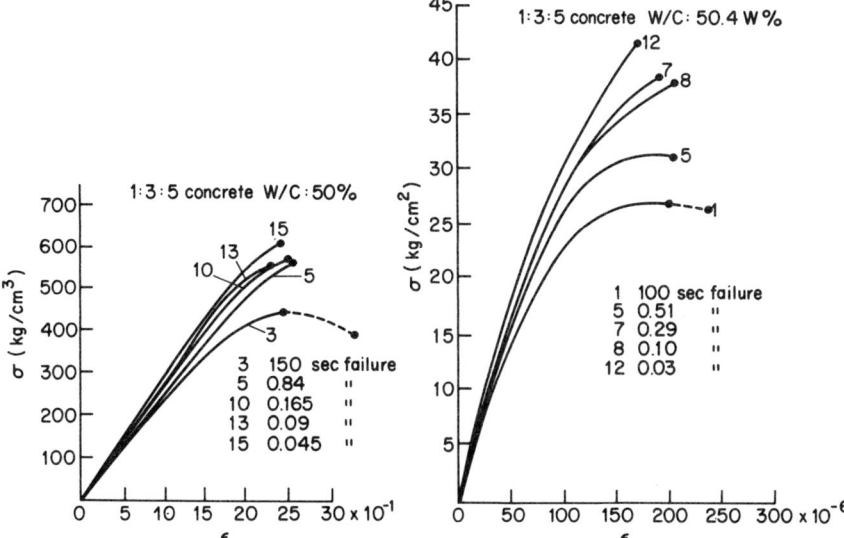

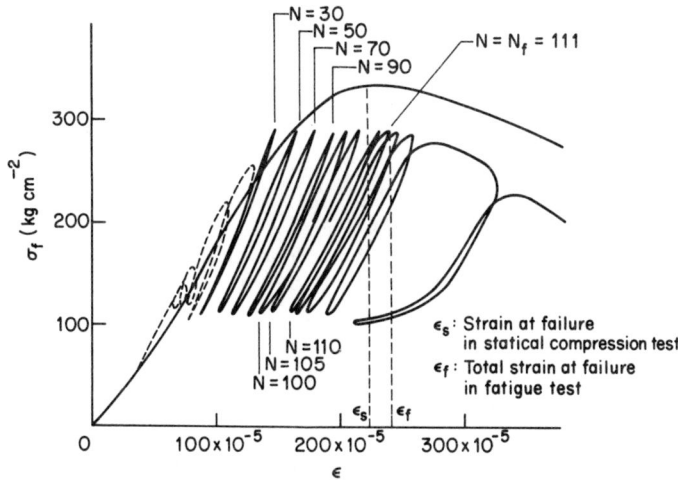

Figure 21.15

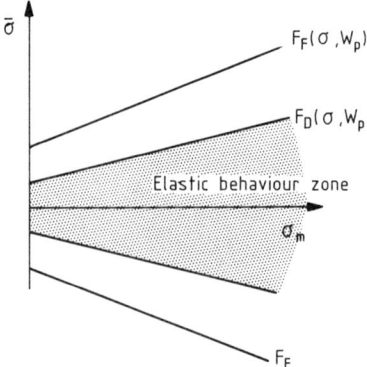

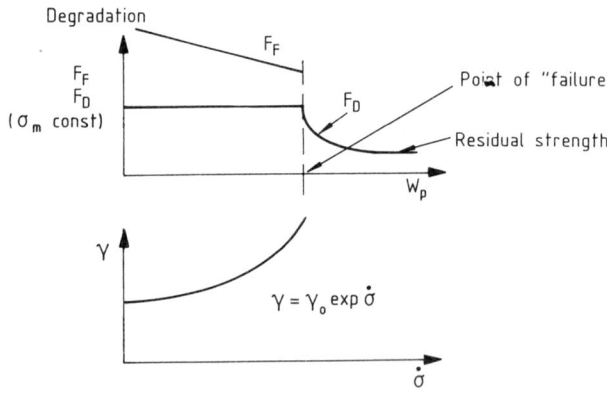

Figure 21.16 The visco-plastic-elastic model for rate effects and degradation of concrete (Bicanic et al., 1980)

The second essential feature is that of dependence of the fluidity parameter, γ, on the stress rate allowing a more rapid degradation at high stress rates.

Such models reproduce reasonably well behaviour at moderate or high strain rates when the yield surface is taken as the proportionality limit. Figures 21.17 and 21.18 show an analysis of the permanent cracking and displacements of the Koyna dam in India which suffered damage in the earthquake of 11 December 1967.

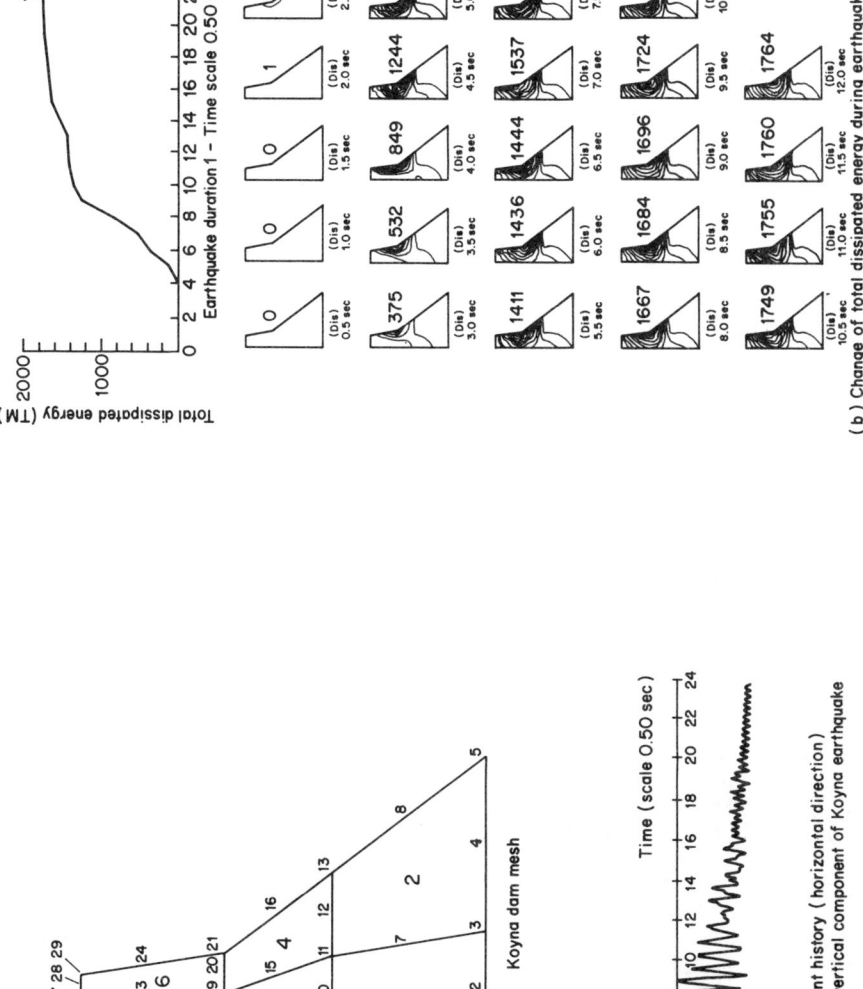

Figure 21.17 Koyna dam analysis

Once again it is shown that transient computation can be effectively used in such situations, though of course improved modelling should be considered.

21.5 CONCLUDING REMARKS

This survey of numerical computation and of some modelling aspects has demonstrated that

(a) both steady state and transient nonlinear analysis are today possible for a variety of problems at a cost comparable with that of linearized approximation;
(b) alternative modelling often results in identical predictions providing essential features of behaviour are included. Here perhaps, we should always consider a 'hierarchical' nature of modelling in which refinements are progressively introduced rather than attempting to model all aspects of behaviour in a single comprehensive package for which only a few parameters can be easily identified; and finally that
(c) progress can only be achieved through 'sensitivity' studies and a collaboration between those engaged in experimental and those attempting to model numerically the behaviour.

At all times it should be borne in mind *what* prediction of behaviour of a structure is sought rather than attempting a purely academic reproduction of all laboratory sample behaviour, noting in particular the variability of 'geomaterial' behaviour.

REFERENCES

Bažant, Z. P., and Oh, B. H. (1981a) 'Concrete fracture via stress-strain relations', Report 81-10/655C, Centre for Concrete and Geomaterials, Northwestern University, Evanston, Illinois.

Bažant, Z. P., and Oh, B. H. (1981b) 'Strain rate effect in rapid triaxial loading of concrete', *J. Engng. Mech. Div. ASCE*, **108**, EM5, 764–782.

Bažant, Z. P., Ansal, A. M., and Krizek, R. J. (1982) 'Endochronic models for soils', in *Soil Mechanics—Transient and Cyclic Loads* (Eds. G. N. Pande and O. C. Zienkiewicz), John Wiley, Chichester, Ch. 15, pp. 417–438.

Besseling, J. F. (1958). A theory of elastic, plastic and creep deformations of an initially isotropic material showing anisotropic strain hardening, creep recovery and secondary creep A.S.M.E. Applied Mechanics Division paper No. 58-APM-17, p. 1–8.

Bicanic, N., Hinton, E., Pande, G. N., and Zienkiewicz, O. C. (1978) 'Nonlinear seismic response of concrete gravity dams', *Proc. 6th European Conf. on Earthquake Engineering*, Sept., 18-22, Dubrovnik.

Bicanic N. and Zienkiewicz O. C. (1983) Constitutive model for concrete under dynamic loading, *Earthquake Eng. and Struct. Dyn.*, **11**, 689–710.

Biot, M. A. (1941) 'General theory of three dimensional consolidation', *J. Appl. Phys.*, **12**, 155–164.

Biot, M. A. (1956) 'Theory of propagation of elastic waves in a fluid saturated porous solid', *J. Acoust. Soc. of America*, **28**, 168–191.

Biot, M. A. (1962a) 'Mechanics of deformation and acoustic propagation in porous media', *J. Appl. Phys.*, **33**, 1482–1498.
Biot, M. A. (1962b) 'Generalized theory of acoustic propagation in porous dissipation media', *J. Acoust. Soc. of America*, **34**, 1254–1264.
Carter, J. P., Booker, J. R., and Wroth, C. P. (1982) 'A critical state model for cyclic loading', in *Soil Mechanics — Transient and Cyclic Loads* (Eds. G. N. Pande and O. C. Zienkiewicz), John Wiley, Chichester, Ch. 9, pp. 219–252.
Dafalias, Y. F., and Herrmann, L. R. (1982) 'Bounding surface formulation of soil plasticity', in *Soil Mechanics — Transient and Cyclic Loads* (Eds. G. N. Pande and O. C. Zienkiewicz), John Wiley, Chichester, Ch. 10, pp. 253–282.
Darwin, A. R., Ingraffea, A. R., Pecknold, D. A., and Schnobrich, W. C., 'Concrete cracking', Ch. 4.
Dimaggio, F. L., and Sandler, I. (1970) 'Material models for soils', Paul Weidlinger, Consulting Engineering, DASA-2521, New York, N.Y., April.
Dougill, J. W. (1982) 'Mechanics of concrete structures—current approaches to assessing material behaviour and some possible extensions,' in *Creep and Strength of Concrete Structures* (Eds. Z. P. Bažant and F. H. Wittman), John Wiley, Ch 2., pp. 23–50.
Drucker, D. C., Gibson, R. E., and Henkel, D. J. (1957) 'Soil mechanics and work hardening theories of plasticity', *Proc. ASCE*, **122**, 388.
Ghaboussi, J., and Wilson, E. L. (1972) 'Variational formulation of dynamics of fluid saturated porous elastic solids', *J. Engng. Mech. Div., ASCE*, **98**, EM4, 947–963.
Ghaboussi, J., Isenberg, J., and Wilson, E. L. (1973) 'Finite element for rock joints and interfaces', *J. of Soil Mech. and Found. Engng. Div., ASCE*, **99**, SM10.
Hatano, T. (1960) 'Dynamical behaviour of concrete under impulsive tensile loads', Report C-6002, Central Research Inst. of Electric Power Industry, Tokyo.
Hatano, T., and Watanabe, H. (1971) 'Fatigue failure of concrete under periodic compressive load', *Trans. Japan Soc. Civil Eng.*, **3**, 106–107.
Mroz, Z., and Zienkiewicz, O. C. (1983) 'Uniform formulation of constitutive equations for clays and sands', in· *Constitutive Laws for Engineering Materials*, John Wiley, Chichester.
Oden, J. T., and Key, J. E. (1973) 'Analysis of static, nonlinear response by explicit time integration', *Int. J. Num. Meth. Eng.*, **7**, 225–240.
Owen, D. R. J., Prakash, A., and Zienkiewicz, O. C. (1974) 'Finite element analysis of non-linear composite materials by use of overlay systems', *Computers and Structures*, **4**, 1251–1267.
Pande, G. N. (1979) 'Viscoplastic algorithm for modelling tensile nonlinearity in rock and concrete structures', *Proc. ASME Symp. on Mechanics of Bi-Modulus Material*, New York.
Pande, G. N., and Sharma, K. G. (1980) 'A micro-structural model for soils under cyclic loading', *Proc. Int. Symp. on Soils under Cyclic and Transient Loadings*, Swansea, Balkema Press, Rotterdam, Vol. 1, pp. 451–462.
Pande, G. N., and Sharma, K. G. (1981) 'Multi-laminate model of clays—a numerical study of the influence of rotation of the principal stress axes', in *Implementation of Computer Procedures and Stress-Strain Laws in Geotechnical Engineering* (Eds. Desai and Saxena), Vol. 2, Acorn Press, Durham, N. C., pp. 575–590.
Pande, G. N., and Zienkiewicz, O. C. Eds. (1982) *Soil Mechanics — Transient and Cyclic Loads*, John Wiley, Chichester.
Phillips, D. V. and Zienkiewicz, O. C. (1976) 'Finite element analysis of concrete structures', *Proc. Int. Civil Eng.*, **61**, March, 59–88. (See also D. V. Phillips, Ph.D. Thesis, University of Wales, Swansea, 1974).
Prevost, J. H. (1982) 'Nonlinear transient phenomena in saturated porous media', *Comp.*

Meth. in Appl. Mech. Eng., **20**, 3–8.

Rashid, Y. R. (1968) 'Ultimate strength analysis of prestressed concrete pressure vessels', Nuclear Engng. and Design, **7**, 334–344.

Roscoe, K. H., Schofield, A. N., and Wroth, C. P. (1958) 'On the yielding of soils', Geotechnique, **8**, 2253.

Schofield, A. N., and Wroth, C. P. (1968) Critical State Soil Mechanics, McGraw-Hill.

Seed, H. B. (1979) 'Considerations in the earthquake resistant design of earth and rock fill dams', Geotechnique, **29**, 215–263.

Shiomi, T., Pietruszczak, S., and Pande, G. N. (1982) 'A liquefaction study of sand layers using the reflecting surface model', Proc. Int. Symp. on Numerical Models in Geomechanics, Balkema Press, Rotterdam, pp. 411–418.

Suidan, M., and Schnobrich, W. C. (1973) 'Finite element analysis of reinforced concrete', J. Struct. Engng. Div., ASCE, **99**, ST10, 2109–2132.

Valanis, K. C., and Read, H. E. (1982) 'A new endochronic plasticity model for soilds', in Soil Mechanics—Transient and Cyclic Loads (Eds. G. N. Pande and O. C. Zienkiewicz), John Wiley, Chichester, Ch. 14, pp. 375–417.

Van Der Kogel, H. (1977) 'Wave propagation in saturated porous media', Ph.D. Thesis, Cal. Inst. Tech.

Zienkiewicz, O. C. (1977) The Finite Element Method, 3rd edn. McGraw-Hill.

Zienkiewicz, O. C. (1979) 'Constitutive laws and numerical analysis for soils foundations under static, transient or cyclic loads,' BOSS, 1979 (Proc. 2nd Int. Conf. on Behaviour of Offshore Struct.) pp. 391–406. (Also reprinted by Applied Ocean Research **2**, 23–31 (1980).)

Zienkiewicz, O. C., and Bettess, P. (1980) 'Soil and other saturated porous media under transient, Dynamic conditions. General formulation and the Validity of Various simplifying assumption', in Soil Mechanics—Transient Loads (Eds. G. N. Pande and O. C. Zienkiewicz), Chichester John Wiley, Ch. 1, pp. 1–16.

Zienkiewicz, O. C., and Cormeau, I. C. (1974) 'Visco-plasticity—plasticity and creep in elastic solids—a unified numerical solution approach', Int. J. Num. Meth. Engng., **8**, 821–845.

Zienkiewicz, O. C., and Leung, K. H. (in preparation) 'A simple model for cyclic and transient loadings for soils'.

Zienkiewicz, O. C., and Löhner, R. (1985) 'Accelerated "relaxation" or direct solution? Future prospects for FEM', Int. J. Num. Meth. Eng., **21**, 1–11.

Zienkiewicz, O. C., and Mróz, Z. (1983) 'Generalized plasticity formulation and application to geomechanics, in Contitutive Laws for Engineering Materials, John Wiley, Chichester.

Zienkiewicz, O. C., and Pande, G. N. (1977), 'Time-dependent multilaminate model of rocks—a numerical study of deformation and failure of rock masses', Int. J. Num and Anal. Meth. in Geomechanics, **1**, 219–247.

Zienkiewicz, O. C., and Shiomi, T. (1984) 'Dynamic behaviour of saturated porous media—the generalized Biot formulation and its numerical solution', Int. J. Num. Anal. Meth. in Geomechanics, **8**, 71–96.

Zienkiewicz, O. C. Valliappan, S., and King, I. P. (1968) 'Stress analysis of rock as a notension material', Geotechnique, **18**, 56–66.

Zienkiewicz, O. C., Best, B., Dullage, C., and Stagg, K. G. (1970) 'Analysis of nonlinear problems in rock mechanics with particular reference to jointed rock systems', 2nd Int. Congress on Rock Mechanics, Belgrade, 1–9, 1970.

Zienkiewicz, O. C., Nayak, G. C., and Owen, D. R. J. (1972) 'Composite and "overlay" models in numerical analysis of elasto-plastic continua', in Foundations of Plasticity, (Ed. A. Sawczuk), Nordhoff Press, pp. 107–112.

Zienkiewicz, O. C., Humpheson, C., and Lewis, R. W. (1975), 'Associated and non-associated visco-plasticity and plasticity in soil mechanics', *Geotechnique*, **25**, 617–689.

Zienkiewicz, O. C., Chang, C. T., and Hinton, E. (1978) 'Nonlinear seismic response and liquefaction', *Int. J. for Num. and Anal. Meth. in Geomechanics*, **2**, 381–404.

Zienkiewicz, O. C., Chang, C. T., Bicanic, N., and Hinton, E. (1979) 'Earthquake response of earth and concrete dams in the partial damage range', *Proc. 13th Int. Congress of Large Dam*, ICOLD, New Delhi, R. 14, 1033–1047.

Zienkiewicz, O. C., Chang, C. T., and Bettess, P. (1980a) 'Drained, undrained, consolidating and dynamic behaviour assumptions in soils. Limits of validity', *Geotechnique*, **30**, 385–395.

Zienkiewicz, O. C., Leung, K. H., Hinton, E., and Chang, C. T. (1980b) 'Earth dam analysis for earthquake—numerical solution and constitutive relations for nonlinear (damage) analysis', *Proc. Conf. on Design of Dams to Resist Earthquakes*, Institution of Civil Engineers, London, pp. 179–194.

Zienkiewicz, O. C., Leung, K. H., and Hinton, E. (1982) 'Earthquake response behaviour of soils with drainage', *Proc. 4th Int. Conf. of Numerical Methods in Geomechanics* (Ed. Z. Eisenstein), Edmonton, Vol. 3, pp. 983–1002.

Zienkiewicz, O. C., Bicanic, N., and Fejzo, R. (1983a) 'Experience in analyzing plain concrete structures using a rate sensitive model with crack monitoring capabilities', *Proc. Conf. on Constitutive Laws for Engineering Materials*, University of Arizona, Tucson, January 1983.

Mechanics of Geomaterials
Edited by Z. Bažant
© 1985 John Wiley & Sons Ltd

Chapter 22

Numerical Models for Dynamic Loading

I. Sandler and M. Baron

22.1 INTRODUCTION

The constitutive modelling of geological materials for applications involving dynamic loading has become a major area of interest in applied mechanics. The problem is complicated by the fact that the choice of model must depend on the application at hand: the appropriate model to be used depends on the type of material and the geometry of the problem as well as the loading rates, stress levels, and periodicity or repetitiveness of the loading. In this paper some of the considerations which affect the choice of such models will be examined. The range of problems considered here is indicated by the following list:

Ground shock resulting from explosions.
Soil–structure interaction under explosive loading.
Ground shock under repeated loadings (multiple bursts).
Soil–structure interaction under earthquake loading.

For each of the above classes of problems, the level of sophistication in the mathematical modelling of the geological materials has risen considerably in the last two decades. The applicability of the models ranges from essentially hydrodynamic fluid behaviour for the extremely high pressures (megabars) in the neighbourhood of a burst point through inelastic solid material in regions at intermediate pressures to nearly linear behaviour at sufficiently low pressure levels. Generally, no single model can be applied for all of the problems of interest.

Because real materials consist of variable amounts of mineral grain, air voids, and water, their mechanical behaviour can exhibit wide variation and can be quite complicated. One must, however, consider carefully the scale and scope of the problems of interest since, for practical purposes, a number of essentially microscopic effects can be averaged in many cases. For large-scale problems, which cover distance from tens to many thousands of feet, no practical attempt can be made to account for the interaction of the various individual constituents of materials, but averaged solid-material models which exhibit nonlinear behaviour (with appropriate hysteretic effects in both pressure and shear) can be developed. Isotropy, homogeneity, and rate dependence are usually assumed although experimental evidence generally indicates that the quantitative response of geological materials will be more complicated. The mathematical models are fitted, wherever possible, from experimental data corresponding to the applicable loading of the problem under consideration.

Parallel to the theoretical development of more realistic material models have been the increasingly difficult demands on experimentalists for tests which would reproduce the behaviour of the material throughout the entire range of pressures. These tests, which are an essential tool in model development, serve two major purposes: (1) to give an indication of the behaviour at appropriate pressure ranges; and (2) to provide data for the evaluation of the various material constants which appear in the mathematical models. Generally, static and 'dynamic' laboratory tests on small samples are the major source of material property data. (Both of these tests usually are, in fact, 'quasi-static' insofar as wave-propagation effects in the specimen are unimportant.) Different tests are generally available for soil and rock materials. For those tests which are common to both materials, e.g. triaxial compression and proportional loading (occasionally available), different phenomena are of importance for soils than those for rocks. For example, a major effect in a soil is an irreversible volume decrease during compression (compaction), although this is generally not of major importance for rocks.

In addition to laboratory tests, there are sometimes large-scale field events which are adequately instrumented. Such tests may be viewed as material property tests, but they also serve as check results for both the modelling and the calculational procedures. The relation between the properties obtained from tests in small laboratory samples and the properties of the *in situ* material are also of considerable importance, but this is a question which can be appropriately answered only by carefully controlled and coordinated suites of large *in situ* and small-scale experiments which are rarely, if ever, performed. Such questions are touched on later in this paper.

22.2 GENERAL CONSIDERATIONS

As noted in the introduction, a model developed for a particular class of problem may not be directly applicable to problems of an entirely different nature. For example, models developed for ground shock in which at most a few cycles of predominantly P-wave motion occur, have to be modified for use in seismic problems in which many cycles of shearing motion predominate.

It is desirable that a particular form of model be able to fit a wide class of geological materials. In this manner, the same form of model could be used for the different materials or layers found in a single problem, with only the various parameters changed. A special case of this occurs when setting certain parameters to special values (such as zero) reduces the more complex model to a simpler one.

The model should satisfy the theoretical requirements needed to prove existence, uniqueness, and stability of solutions. This is necessary in order to be confident that any numerical solution is an approximation (in some sense) of the physical problem and not nonsense that will vary widely with computer accuracy or choice of algorithm. Further, in any particular analysis dependence or unknown (or not readily available) variables must be avoided if calculations are to have any real value.

Numerous models are currently used for both explosively induced ground shock and seismic analysis. The simplest of these is the elastic model (most often linearly elastic) which also serves as the most common starting point from which other models are developed. Moreover, this simple model continues to be useful because most analytical solutions can be found only for linear problem formulations. In addition, elastic models are often used to represent hard competent basement rock layers even when more complex models are used for the near-surface materials.

Viscoelastic models are often used in seismic analysis. The hysteresis found in cyclic loading is approximately by linear viscoelasticity. The resulting linear problem may then be solved in the frequency domain. Nonlinear (strain-dependent) hysteresis is approximated by an iterative scheme. During each of the successive iterations, the amount of damping in every element is adjusted to correspond to the strain amplitude found in the previous iteration for that element.

In addition, viscoelasticity can be combined with other types of models which are used to represent the plastic behaviour exhibited by geological materials. These include variable moduli, cap, and endochronic models, some of which will be considered in detail in subsequent sections of this chapter.

22.3 SIMPLE PLASTICITY MODELS

The early plasticity models were elastic–ideally plastic (EIP), i.e. there is a fixed yield condition

$$F(\sigma_{ij}) = 0 \tag{22.1}$$

which restricts the magnitude of the stress. If the material is isotropic, the yield condition can depend only on the principal stresses, or alternatively on the stress invariants

$$F(J_1, J_2, J_3) = 0 \tag{22.2}$$

Within the yield surface the material is elastic, i.e.

$$d\varepsilon_{ij}^E = \frac{1}{2G} ds_{ij} + \frac{1}{9K} \delta_{ij} dJ_1 \tag{22.3}$$

where $d\varepsilon_{ij}^E$ is the strain increment, ds_{ij} the increment in the stress deviator and δ_{ij} the Kronecker delta. The bulk and shear moduli are K and G, respectively.

Stresses outside the yield surface are not possible. On the yield surface, the strain increment will consist of elastic and plastic parts

$$d\varepsilon_{ij} = d\varepsilon_{ij}^E + d\varepsilon_{ij}^P \tag{22.4}$$

The elastic part $d\varepsilon_{ij}^E$ is given by Equation (22.3), while the plastic part is

$$d\varepsilon_{ij}^P = \lambda \frac{\partial \phi}{\partial \sigma_{ij}} \tag{22.5}$$

where λ is a non-negative scaler function, and ϕ is the plastic potential. When the plastic potential and the yield condition are the same,

$$\phi \equiv F \tag{22.6}$$

in which case Equation (22.5) becomes the associated flow rule

$$d\varepsilon_{ij}^P = \lambda \frac{\partial F}{\partial \sigma_{ij}} \tag{22.7}$$

and the plastic strain increment is normal to the yield surface at the current stress point. Furthermore, when F is convex, unique solutions are assured. When the non-associated flow rule (Equation (22.5) with $\phi \neq F$) is used, one cannot in general prove uniqueness. The simplest form of the yield condition is the von Mises condition

$$\sqrt{J_2'} = k \tag{22.8}$$

where J_2' is the second invariant of the stress deviator

$$\sqrt{J_2'} = \tfrac{1}{2} s_{ij} s_{ij} \tag{22.9}$$

and k is a constant. The von Mises condition is actually a good representation of

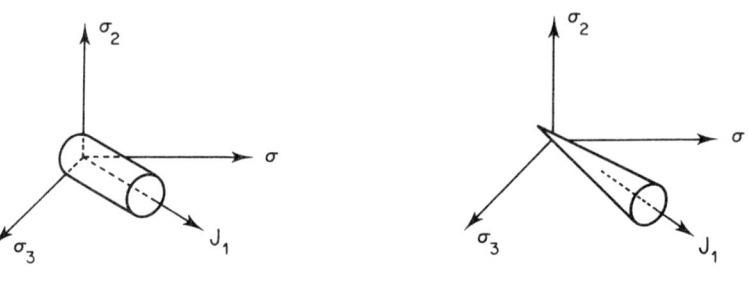

(a) Von Mises (cylinder)   (b) Drucker – Prager (cone)

Figure 22.1 Simple yield surfaces

Numerical Models for Dynamic Loading

the failure surface of many saturated clays. The von Mises yield surface is a cylinder in principal stress space, Figure 22.1(a).

A more realistic yield condition for granular material

$$J'_2 = k + \alpha J_1 \tag{22.10}$$

was suggested by Drucker and Prager (1952). It is shown as the cone of Figure 22.1(b), opening towards the positive (compressive) J_1 axis. When the cohesion k is zero, Equation (22.10) reduces to the Mohr–Coulomb relation.

It is instructive to examine the behaviour of the two simple models in uniaxial strain. This is illustrated in Figure 22.2 for constant elastic moduli. The stress strain curve starts with the elastic constrained modulus $K + \frac{4}{3}G$ as its slope. The stress path s_1 versus p likewise starts with its elastic slope $4G/3K$. At point A the yield surface is reached, and in the von Mises case (Figure 22.2(b)), the axial stress

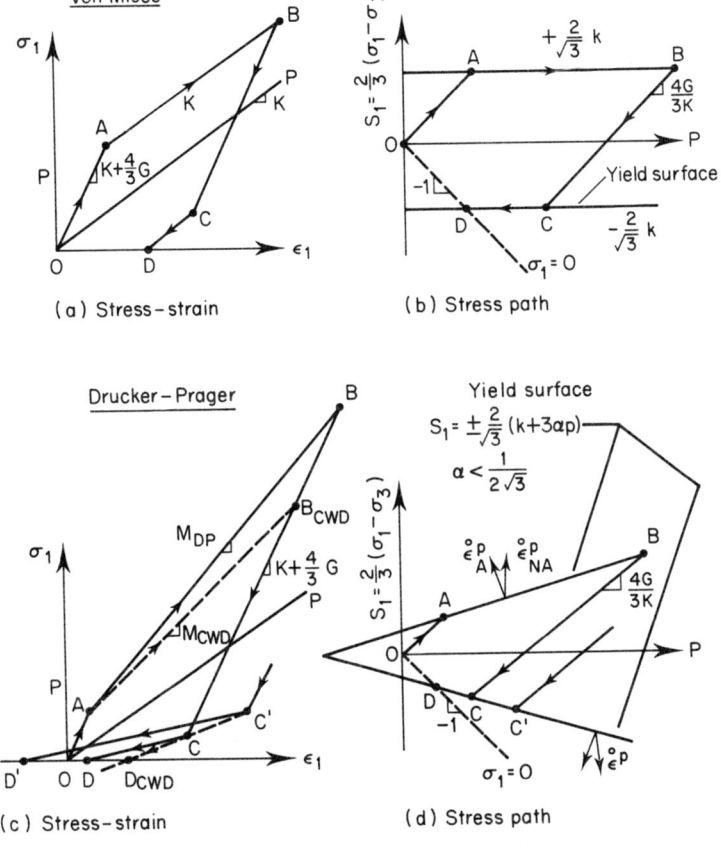

Figure 22.2 Uniaxial strain behaviour of simple elastic ideally plastic models

deviator cannot increase further, beyond the limiting value $2/\sqrt{3}k$. Any further loading, say to point B, can result only in an increase in the mean stress $p = \frac{1}{3}J$; this means that AB, the corresponding segment of the stress–strain curve in Figure 22.2(a) is parallel to the hydrostat, i.e. it has a slope K. Upon unloading at B, the stress point moves away from the yield surface, and the material behaves incrementally elastically, i.e. the lines BC and OA are parallel in both Figures 22.2(a) and 22.2(b). At point C, the reverse side of the yield condition is reached, and s_1 is again limited, this time to $-2/\sqrt{3}k$. Thus, any further unloading will be along the yield surface and the slope in the stress–strain plot again will be K. The cycle is complete at point D at which $\sigma_1 = s_1 + p = 0$.

In the case of the Drucker–Prager yield condition, Figures 22.2(c) and 22.2(d), the behaviour is similar until yield is first reached at point A. As loading continues to point B, the value of s_1 increases according to the yield condition.

$$s_1 = \frac{2}{\sqrt{3}}(k + 3\alpha p) \qquad (22.11)$$

If this value of s_1 were added to hydrostat, the slope of the resulting stress–strain curve would be

$$M_{CWD} = K(1 \pm 2\sqrt{3}\alpha) \qquad (22.12)$$

(with the upper sign). Equation (22.12) is represented by line AB_{cwd} in Figure 22.2(c). This is not the solution for the Drucker–Prager material, which is shown as the solid line slope

$$M_{DP} = K\frac{(1 \pm 2\sqrt{3}\alpha)^2}{1 + 9\alpha^2 K/G} \qquad (22.13)$$

(again with the upper sign). The difference is explained in Figure 1(d). With the associated flow rule, Equation (22.7), the plastic strain rate vector along AB, $\dot{\varepsilon}_A^P$, has a negative (tensile) volumetric component. Therefore, a greater stress is required (Figure 22.2(c)) to reach the same strain as compared to the case of a non-associated flow rule (one without any plastic volume change). The CWD subscript used in Equation (22.12) and Figure 22.2 refers to 'Coulomb without dilatancy' and results when a von Mises plastic potential is used in conjunction with the Drucker–Prager yield condition.

Unloading from point B is again elastic with slope of $K + \frac{4}{3}G$ and $4G/3K$ in Figures 22.2(c) and 22.2(d), respectively. At point C the far side of the yield condition is again reached. The stress point moves along the yield surface until $\sigma_1 = 0$ at point D in Figure 22.2(d). In the case of the CWD material, the negative value of s_1 is added to the hydrostat and the resultant slope of segment CD_{CWD} in the stress–strain curve is given by Equation (22.12) with the lower sign. In the case of the associated flow rule, there is again a volume expansion component to the plastic strain rate vector, and the resulting modulus from Equation (22.13), with the lower sign, is necessarily softer than that of the CWD material. While unloading from point B has resulted in a permanent compaction of the material at $\sigma_1 = 0$ (point D), unloading from some higher stress would have reached the far side of the yield surface at C', and resulted in net volume expansion, (dilatancy) at the end of the cycle at D'. This dilatancy is not observed in experiments and was the primary motivation which led to the development of more advanced EIP models. Nevertheless, to relatively low stress levels such as to those depicted in Figure 22.2(c), the Drucker–Prager model stress–strain leads to a uniaxial strain response which has many of the qualitative features observed experimentally. The model continues to be frequently used to represent 'frictional' materials.

22.4 MORE DETAILED PLASTICITY MODELS

In order to represent soil behaviour, in which permanent compaction is usually a dominant characteristic, the simple plasticity model was expanded to include a nonlinear pressure–volume relationship involving two different bulk moduli, one for initial loading, and a second for unloading/reloading. Mathematically,

$$dp = K\, d\varepsilon_{kk} \qquad (22.14)$$

where

$$K = K_L(p) \quad \text{for initial loading} \qquad (22.15a)$$

and

$$K = K_U(p) \quad \text{for unloading/reloading} \qquad (22.15b)$$

Requiring

$$K_U(p) \geq K_L(p) \qquad (22.16)$$

ensures volumetric hysteresis even in infinitesimal initial loading–unloading cycles. Another modification was the generalization of the yield condition to

$$J'_2 = f(J_1) \qquad (22.17)$$

which is sketched in Figure 22.3. The function $f(J_1)$ is chosen such that it is

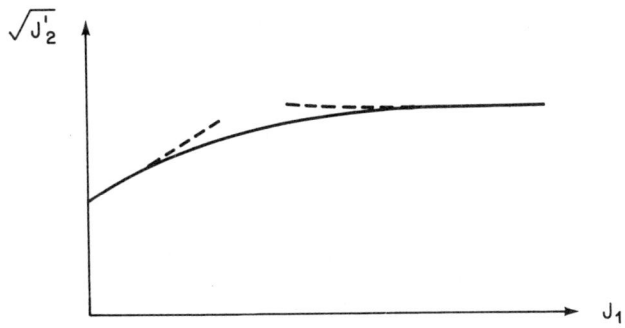

Figure 22.3 Yield condition for the advanced elastic–ideally plastic model

approximated by the Drucker–Prager condition, Equation (22.10), at low pressures, and by the von Mises condition, Equation (22.8), at high pressures.
The final modification involves the treatment of the deviatoric stresses within the yield surface. The incremental relation

$$ds_{ij} = 2G\, de_{ij} \qquad (22.18)$$

is used with a variety of specifications of the shear modulus. The simplest assumption is to consider the shear modulus G to be constant. While this model (with an associated flow rule) satisfies all theoretical requirements, the computed stress path in uniaxial strain is unrealistic when $K_U \gg K_L$. If K_L increases with increasing pressure, the stress path is shown by the dashed line in Figure 22.4. The path would gradually approach, and finally reach, the upper yield surface. On unloading, at least in the case of soils where the unloading bulk modulus is appreciably larger than the initial loading modulus, the unloading path would not generally intersect the lower yield surface. In fact, for moderate peak pressures (and when $K_U \gg K_L$), the unloading stress path would cross the loading path and intersect the upper yield surface. This behaviour is not experimentally observed in most cases where stress path data is available.

An alternative simple assumption which has been used is that of a constant ratio of K/G or a constant Poisson's ratio v. For this case, the stress path is shown by the solid line in Figure 22.4. It starts with a upper branch of the yield surface until a load reversal occurs. When unloading begins, the subsequent stress path is parallel to the initial loading path until the lower yield surface is encountered. The segment along the lower yield surface, until the line $s_1 + p = \sigma_1 = 0$ is reached, is associated with the tail at the bottom of the unloading segment of the stress–strain curve. A variant of this approach but one which matches ratio, v_L and v_U. The choice could depend whether K_L or K_U were being used, or on the sign of the

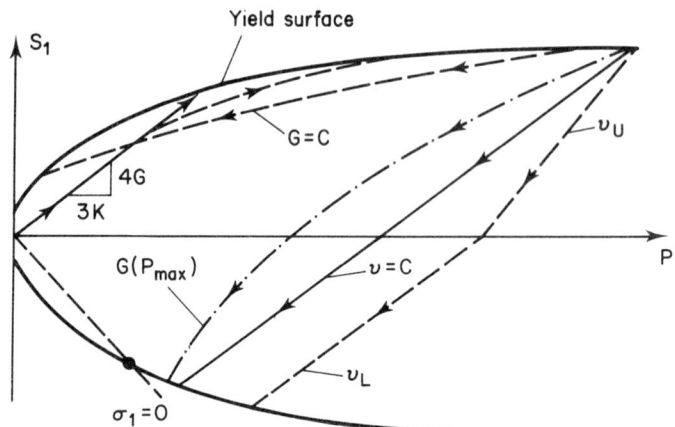

Figure 22.4 Loading paths for elastic ideally–plastic models

Numerical Models for Dynamic Loading

incremental deviatoric work, $s_{ij}\, de_{ij}$. The corresponding stress path is illustrated as a broken dotted line in Figure 22.4. In either case, the stress paths agree, at least qualitatively, with experimental observations. The difference between the two simple assumptions of constant G or constant v in loading is not of real importance. This is not true in unloading.

Unfortunately, it is possible to introduce stress paths for which a constant Poisson's ratio model (with K_U increasing with pressure) will produce energy for each load–unload cycle. The same criticism may be applied to other models which have been used in which v was piecewise constant or in which G was a more general function of pressure.

Different constant values of G in loading and unloading have also been used. Models of this type are not subject to the above criticism. However, they fail to satisfy a continuity condition which will be discussed in conjunction with the variable modulus model in the next section.

In addition to the constant G model, another possibility which also satisfies theoretical requirements and moreover, leads to stress paths which are closer to those observed experimentally, is the use of a shear modulus G which depends only upon maximum previous pressure. During initial loading, the shear modulus increases as the pressure increases. On unloading or reloading to the maximum previous pressure p_{max}, the value of $G(p_{max})$ remains constant. The uniaxial strain–stress path for such a model is shown in Figure 22.4 for the case when K_U increases with increasing pressure (dash–dot curve).

The models described above have been used extensively for ground shock in both soil and rock. In rock, the pressure–volume hysteresis is much smaller than in soils, and in competent rocks, it is often neglected entirely, i.e. a single bulk modulus is used. The use of an associated flow rule in conjunction with a pressure dependent yield condition produced dilatancy, which is generally observed in most rocks. Therefore the Drucker–Prager model is a better approximation to rock than to soil behaviour at low stresses.

It is of interest to consider the triaxial behaviour of advanced elastic–ideally plastic material models of the general type discussed in this section. This is shown in Figure 22.5, which represents the results in the form of a $(\sigma_1 - \sigma_3)$ versus ε_1 diagram. Because of the variable bulk modulus $K_L(p)$, the initial loading curve first softens and then hardens. As the yield surface is reached (at different levels for different values of the lateral stress σ_3), the material 'fails' and 'flows' as represented by horizontal lines shown parallel to the strain axis. It is clear that none of the simple treatments of the shear modulus described previously can mirror the usual stress–strain curve in a triaxial compression test; these are the dashed lines in Figure 22.5.

In order to overcome this defect and to match laboratory data from all available tests, the 'variable moduli' models were introduced, Nelson and Baron (1971). In these models both the bulk and shear moduli are taken as nonlinear functions of the stress and/or strain tensor invariants. Different functions are used in loading and unloading. The variable moduli material has no unique stress–strain relation, even in initial loading, but rather is defined in terms of stress rate–strain rate (incremental) relations, Equations (22.14) and (22.18). No explicit yields condition is specified. However, the behaviour of the variable moduli models corresponds in many respects to that of elastic–plastic models.

The pressure volumetric strain relation is similar to that used in the advanced EIP models. Namely, one bulk modulus function of pressure (or volumetric strain) is used in initial loading ($\dot{p} > 0$ and $p = p_{max}$), and another is used in unloading ($\dot{p} < p$) or reloading ($\dot{p} > 0$, but $p < p_{max}$). There are also different shear moduli used in deviatoric (initial) loading and unloading, i.e.

$$G = G_{LD}(p, J'_2) \qquad \dot{J}'_2 > 0 \qquad (22.19)$$

$$G = G_{UN}(p, J'_2) \qquad \dot{J}'_2 \geqslant 0 \qquad (22.20)$$

The condition

$$0 < G_{LD}(p, J'_2) \leqslant G_{UN}(p, J'_2) \qquad (22.21)$$

insures deviatoric energy dissipation in any incremental load–unload cycle.

Although there is no explicit yield surface, at any constant pressure, the (initial) loading shear moduli functions get smaller as J'_2 increases. The limiting condition

$$G_{LD}(p, J'_2) = 0 \qquad (22.22)$$

corresponds to an implicit yield condition.

Such a model was fitted to a rather complete set of laboratory data for a particular soil material and good agreement for uniaxial strain, triaxial compression, and proportional loading tests was obtained. Moreover, it would

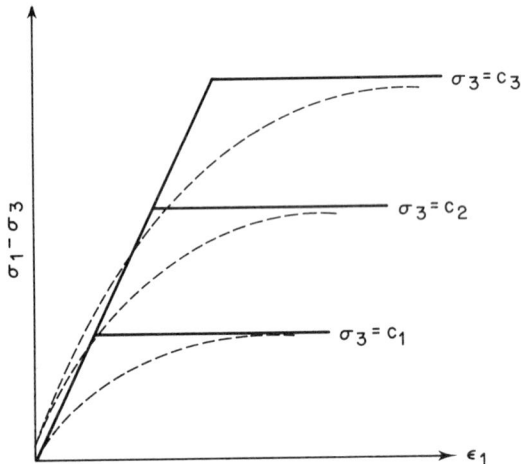

Figure 22.5 Triaxial behaviour of elastic–ideally plastic models compared to real soils

Numerical Models for Dynamic Loading

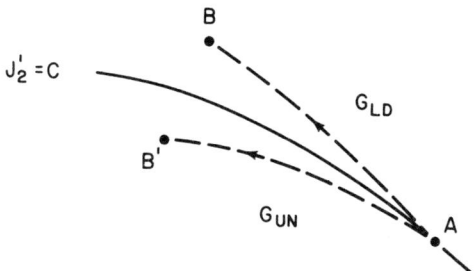

Figure 22.6 Example of discontinuous behaviour in variable moduli model

appear that the variable moduli approach would be ideal for implicit finite element codes in which the current values of K and G in every element would be available to construct the stiffness matrix.

The major problem with the variable moduli formulation, is qualitatively illustrated in Figure 22.6, where the point A lies on a surface $J'_2 = $ constant. Consider two paths, AB and AB', arbitrarily close to each other, but on opposite sides of the surface. On the outer path AB, the shear modulus G_{LD} would apply, while on AB', G_{UN} would be used. Consequently, there will be a definite difference in strain, even when the points B and B' are infinitesimally close to one another.

The violation of continuity in neutral loading ($J'_2 = $ constant) or near-neutral loading is similar to that discussed by Handelman, Lin, and Prager (1947), in their studies on deformation theories of plasticity. In fact, deformation theory may be considered a special case in which the rate equations of the variable moduli formulation are integrable.

In the case of proportional loading in shear, i.e. when there is a single independent stress deviator, the variable moduli approach can be made to satisfy all theoretical requirements, including continuity. This is true for planar, or spherically symmetric problems, as well as for all of the common laboratory tests. In other problems, in which the stress history is close to proportional loading, the present theory, like deformation theory, is probably satisfactory. For problems in which the stress history involves an appreciable amount of neutral (or near-neutral loading, use of the variable moduli formulation is not recommended.

22.5 THEORETICAL CONSIDERATIONS

Continuum models intended for use in dynamic multidimensional problems should satisfy certain theoretical requirements. These requirements ensure that the initial and boundary value problems involving the constitutive model, together with the equations of continuum mechanics (e.g. conservation of mass, momentum, energy), be properly posed, i.e. that such problems have solutions which exist, are unique, and depend continuously on the initial and boundary conditions. These seemingly abstract requirements are of considerable practical importance in this age of numerical solutions. In particular, one should avoid attempting computer solutions for problems without solutions and, in addition, avoid problems with several solutions for fear that a non-physical, though mathematically correct, solution would be obtained. Further, because all numerical solutions are subject to several kinds of error (due to truncation or round-off, the order of accuracy associated with the chosen numerical scheme, and errors in specification of initial and boundary conditions), any solution which is unduly dependent on such errors is highly suspect. Therefore, the continuous dependence of solutions on the input data is directly related to the confidence with which numerical solutions can be obtained for real problems.

(Continuous dependence on the data implies that a small change in the surface loading will lead to a correspondingly small change in the solution. A simple example of a discontinuous material is an elastic string with tensile strength σ_T. For loads $P < \sigma_T A$, the uniform stress will be $\sigma = P/A$ and the corresponding strain $\varepsilon = P/EA$. For loads just greater than $\rho_T A$, the string will break and the 'solution' will be entirely different. The soil and rock models discussed up to this point have avoided any mention of the tensile stress region. Clearly natural materials, especially rocks, fail suddenly. Yet, to allow the nature of a numerical solution to depend entirely on such artificial parameters as mesh box size, time step, or numerical algorithm is completely unsatisfactory).

The continuity problem discussed in connection with the variable moduli model constitutes a lack of continuous dependence on the input data. Slightly different input, or slightly different numerical techniques, could result in a significantly different computed results.

A major contribution to the continuum models was the introduction of Drucker's stability postulate, Drucker (1956). Non-negative work must be done by an external agent in any excursion from equilibrium. In particular, for any stress cycle, where σ_{ij}^0 is the stress at the equilibrium state,

$$\int (\sigma_{ij} - \sigma_{ij}^0) d\varepsilon_{ij} \geqslant 0 \qquad (22.23)$$

The equal sign applies only for elastic or reversible paths. Satisfying Drucker's postulate is sufficient (but not necessary) to ensure unique solutions, and continuous dependence on the data.

A geometric interpretation of Drucker's postulate is shown in Figure 22.7. By eliminating the elastic or reversible strains and by choosing σ_{ij}^0 on the yield surface, one obtains the condition for stability in the 'small' for elastic–plastic models, i.e.

$$d\sigma_{ij} d\varepsilon_{ij}^P \geqslant 0 \qquad (22.24)$$

A consequence of Equation (22.24) is that the yield condition can only move outward (or not move) at a stress point, i.e. work softening or strain softening is not

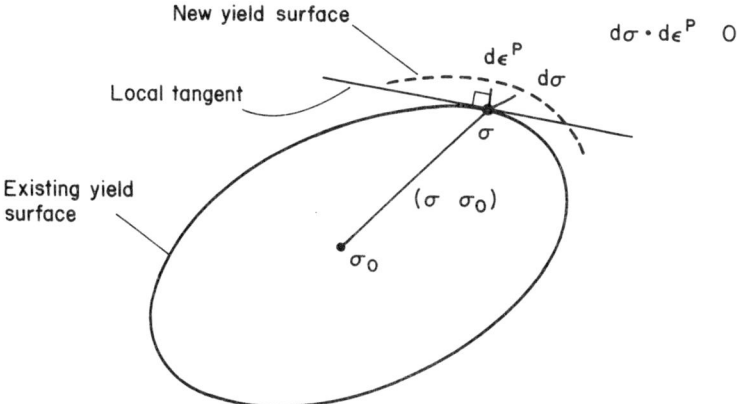

Figure 22.7 Geometric interpretation of Drucker's stability postulate

permitted. One may also obtain the condition for stability in the 'large' for elastic–plastic materials i.e.

$$(\sigma_{ij} - \sigma_{ij}^0)d\varepsilon_{ij}^P \geq 0 \qquad (22.25)$$

Equation (22.25) requires the normality of the plastic strain rate vector, and the convexity of the yield condition.

22.6 CAP MODELS

In order to satisfy uniqueness continuity and stability requirements, the cap model was developed. It is a continuum material model which is based on the classical incremental theory of plasticity. It has been used to represent both the high and low pressure mechanical behaviour of a number of geological materials, including sands, clays, and various types of rock. The model has been used primarily for computational studies of ground shock and structure medium interaction effect arising from nuclear or chemical explosions. In addition, it can be used for the study of earthquake effects; some work has already been done to apply the model to seismic problems.

The discrepancy between the observed compaction of soils and the dilatancy resulting from normality to any yield surface similar to that of Drucker–Prager disturbed investigators for many years. Drucker, Gibson, and Henkel (1957) first added a movable cap to the yield surface to eliminate this discrepancy. In the 1960s Roscoe and his coworkers at Cambridge University in England introduced the critical state model (Schofield and Wroth, 1968). It had many similarities to the earlier model and to the current cap model. Finally, a group at MIT also worked on a similar model, e.g. Christian (1966).

A cap model was proposed by Dimaggio and Sandler (1971) for the representation of granular soils, and similar models have been used for many ground shock calculations. The yield surface (Figure 22.8), is composed of a fixed failure envelope, Equation (22.17), and a movable cap which crosses the p axis. The combined yield surface is everywhere convex, and the associated flow rule is used throughout, i.e. the components of the plastic strain rate form a vector in stress space which is normal to the yield surface at the stress point is outwardly directed, as shown in Figure 22.8. Three different modes of behaviour are possible for the model: elastic, failure, and cap. Elastic behaviour occurs when the stress is within the failure envelope and stress changes result in recoverable deformations. Although various types of nonlinearly elastic behaviour can be modelled in complex cases, the model considered here (for isotropic materials) uses a constant bulk modulus, K, and a constant shear modulus, G. During the postulated elastic behaviour the volumetric and deviatoric components of stress and strain are decoupled, i.e. a purely volumetric change in strain does not affect the deviatoric stress components and a purely deviatoric strain increment produces no change in pressure.

During the failure mode of behaviour, the stress point lies on the failure envelope represented by

$$\sqrt{J_2'} = A - C\exp(3B_p) \qquad (22.26)$$

where A, B, and C are material constants. As shown in Figure 22.8 the associated flow rule requires that the plastic strain rate vector be directed upward to the left. Therefore, the plastic strain during failure is composed of a deviatoric, or shear, component together with a volumetric, or dilatant, component.

The cap mode of behaviour occurs when the stress point lies on the movable cap and pushed it outward. The motion of the cap is related to the plastic strain by means of a hardening rule. Although considerable leeway exists in the choice of

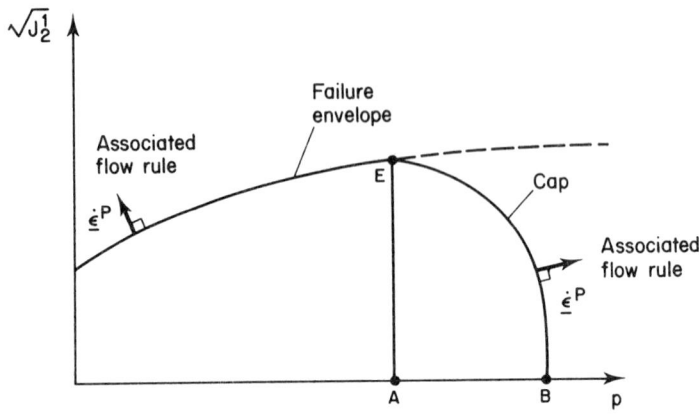

Figure 22.8 Yield surface in the cap model

the cap, an elliptical surface of the form

$$(p - p_A)^2 + \frac{1}{9}R^2 J_2' = (p_B - p_A)^2 \tag{22.27}$$

is found acceptable for a wide range of geologic materials. In Equation (22.27), p_A and p_B represent the values of p at points A and B in Figure 22.8, while R can be function of the position of the cap. In the simple eight-parameter version of the model discussed here, R is assumed to be constant. The pressures p_A and p_B which define the extent of the cap are not independent. From Figure 22.8, it is clear that, because point E lies on both the failure envelope and the cap

$$\sqrt{J_{2E}'} = A - C \exp(-3 B p_E) \tag{22.28}$$

and

$$(p_E - p_A)^2 + \frac{1}{9}R^2 J_{2E}' = (p_B - p_A)^2 \tag{22.29}$$

Further, since $p_E = p_A$, Equations (22.28) and (22.29) lead to

$$p_B - p_A = \frac{1}{3}R[A - C \exp(-3 B p_A)] \tag{22.30}$$

for the relation between p_A and p_B. Therefore, the specification of either p_A or p_B is sufficient to describe the position of the cap.

The cap position is related to the plastic strain history of the material through a hardening rule, which, again, can be chosen with considerable leeway. For the eight-parameter model discussed here, the hardening rule is assumed to be

$$\bar{\varepsilon}_v^p = W[1 - \exp(-D p_B)] \tag{22.31}$$

in which W and D are material constants and $\bar{\varepsilon}_v^p$ is related to the plastic strain history. The hardening parameter $\bar{\varepsilon}_v^p$ depends upon the history of the dilatant as well as compactive plastic strain for soils, while for rocks it depends only on the plastic strain which has been produced by cap action, i.e. compaction.

As shown in Figure 22.8, the associated flow rule requires that during cap action the plastic strain rate vector be directed upward and to the right. This implies that the plastic strain rate produces an irreversible decrease in volume in conjunction with the irreversible shear strain. This reduction in volume is referred to as compaction and represents the volumetric hysteresis observed during compression of most geologic materials.

As the cap action proceeds, the compaction resulting from the associated flow rule leads to an increase in the cap parameter $\bar{\varepsilon}_v^p$ which, through Equation (22.31), leads in turn to an increase of p_B. Therefore the cap moves to the right in Figure 22.8, increasing the extent of the elastic region inside the new yield surface. Either p or J_2' (or both) must increase in such a way as to keep stress point on the cap in order to maintain this mode of behaviour.

516 Mechanics of Geomaterials

The cap does not move during purely elastic deformation. The behaviour of the cap when the stress point lies on the failure envelope alone, however, depends upon the amount of dilatancy, or plastic volumetric expansion. This dilatancy leads to a decrease in $\bar{\varepsilon}_v^p$, resulting in the leftward movement of the cap. The cap movement is limited if and when the cap reaches the stress point (so that the stress point lies at the corner of the yield surface).

For most soils and for weak rocks, the low stress behaviour of the model can be simplified. In the range designated as 'small cap' (Figure 22.9), the centre of the elliptical cap is at the origin. The transition point is reached when the cap is large enough to intersect the failure surface, as it is shown in the figure.

The soil cap model described above was developed primarily for use in computations for explosions which usually involve much higher stress levels than are involved in earthquake-induced ground response and which are generally characterized by a single peak earth-compressive stress followed by smaller stresses. An exception is the case of outrunning ground motion in layered soil media, which involves cyclic, low amplitude response signals similar to those of earthquakes. As a rule, however, hysteresis in cyclic loading subsequent to an initial pulse is generally viewed as having secondary importance. Hysteresis becomes quite important, however, for earthquake-induced loadings where cyclic shear is the predominant effect.

The adaptations of the cap model described below make the basic model suitable for the seismic environment. These extensions of the model represent hysteresis in cyclic shear loading and also include pore water effects in wet media.

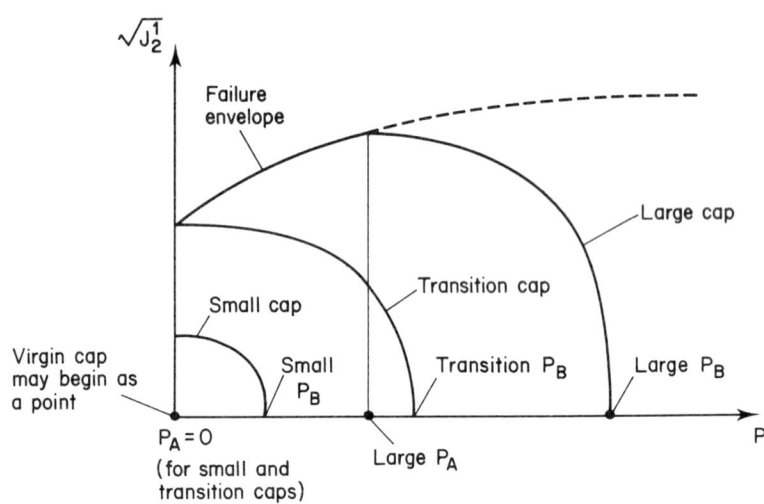

Figure 22.9 Some details of the cap models

Numerical Models for Dynamic Loading

22.7 CAP MODEL FOR CYCLIC LOADING

A version of the cap model was developed to represent materials which exhibit viscoelastic behaviour during loading. It has also been used to represent rate independent cyclic hysteresis over a limited range of loading rates.

In the viscous cap model linear viscous damping is introduced into the previously elastic portion of the cap model. For example, a standard solid was chosen to govern shear behaviour within the yield surface. The parameters which define the nonplastic portion of the model are an instantaneous modulus G_F, a long-term modulus $G_s < G_F$, and a relaxation rate τ. The model is shown schematically in Figure 22.10. The parameters G_s and τ are related to those in the figure via

$$G_s = \frac{G_F G_V}{G_F + G_V} \quad (22.32)$$

and

$$\tau = \mu(G_F - G_S)/G_F^2 \quad (22.33)$$

The deviatoric stress–strain relation for the model is

$$\frac{ds_{ij}}{dt} = 2G_F \frac{de_{ij}^v}{dt} + (2G_S e_{ij}^v - s_{ij})/\tau \quad (22.34)$$

where e_{ij}^v is the viscoelastic deviatoric strain, i.e. the total strain minus the plastic strain.

The shear response which may be obtained from the model is illustrated by the solid line in Figure 22.11 where stress difference versus computed strain difference is plotted for a simulated stress-controlled triaxial compression test. The figure illustrates the qualitative behaviour possible with this model. The large permanent strains in the figure during initial loading result from plastic

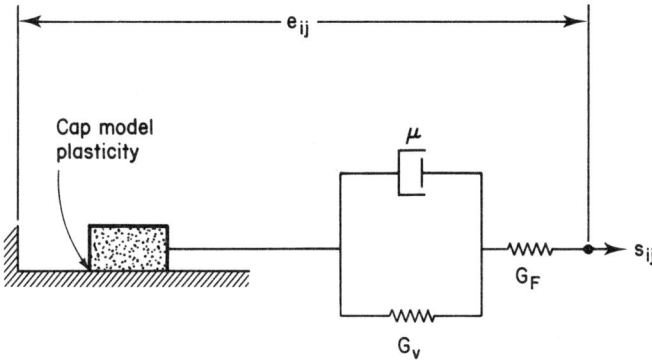

Figure 22.10 Schematic representation of viscous cap model

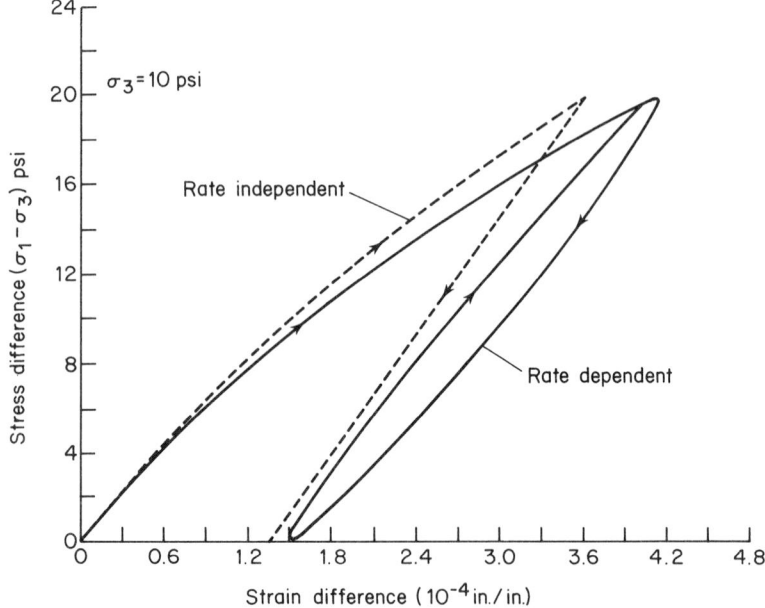

Figure 22.11 Comparison of standard and viscous cap models

deformation associated with the expanding cap. The small loop results from the viscoelastic behaviour within the yield surface. For comparison, the response of the standard (nonviscous) cap is shown by the dashed curve. There are two effects of the modified model which are apparent in the figure. The first is the creep relative to the instantaneous cap response which increases slightly as the loading frequency decreases. The second is the hysteresis loop for each successive cycle. With appropriate parameters, both the average slope and the area of the loops are relatively insensitive to the frequency of loading (about some 'centring' frequency).

The pressure-volumetric strain relation within the plastic yield surface may be chosen as elastic, or as a similar standard solid with constants K_F, K_S, and τ_v. It should be noted that the instantaneous response for any level of stress and/or strain is given by the current cap model with elastic constants K_F and G_F. Therefore, implementation of this type of model in various dynamic codes becomes a relatively straightforward modification of the 'elastic portion' of the subroutine.

For situations in which experimental data suggest that the amount of hysteresis is independent of strain rate, a cap model can be constructed using kinematic hardening (Prager, 1966). This extended model is obtained by replacing the stress tensor ρ_{ij} by the quantity $(\rho_{ij} - \alpha_{ij})$, where α_{ij} is a tensor whose components are memory parameters defining the translation of the yield surface

Numerical Models for Dynamic Loading

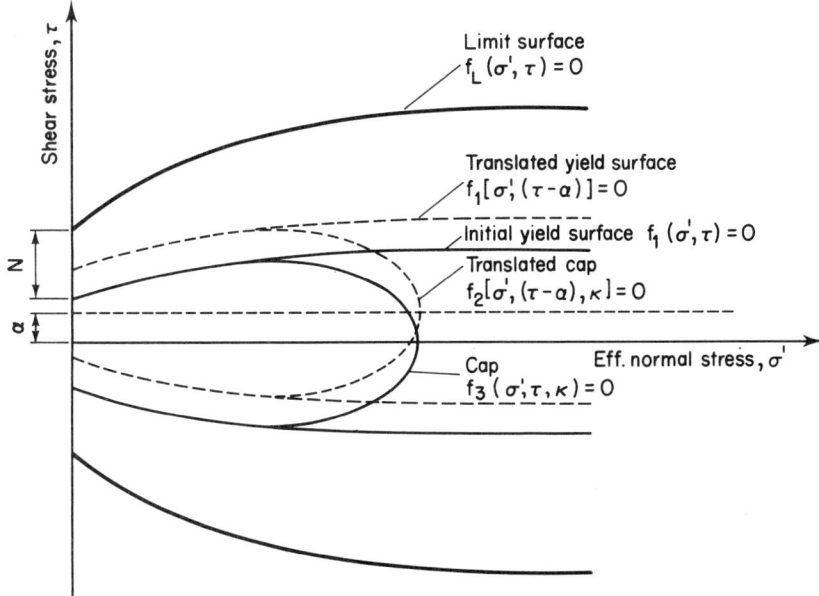

Figure 22.12 Kinematic hardening cap model

in stress space. Because the kinematic hardening is assumed to occur in shear only, only five of the six α_{ij} are independent, and

$$\alpha_{11} + \alpha_{22} + \alpha_{33} = 0 \tag{22.35}$$

This type of hardening is illustrated qualitatively in Figure 22.12 for the one-dimensional case in which only a single normal stress component ρ' and a shear stress component τ are considered. The entire yield surface translates along the τ-axis by the amount α.

To complete the specification of the model, an evolutionary equation which governs the value of the memory parameter α_{ij} is required. This is the kinematic hardening rule which may be expressed as

$$\dot{\alpha}_{ij} = f_{ijkl}(\sigma_{ij}, \alpha_{ij}, \kappa, \varepsilon_{ij}) \dot{e}^p_{kl} \tag{22.36}$$

where $\dot{e}^p_{kl}$ are the deviatoric components of plastic strain and the usual summation convention is implied.

Obtaining the function f_{ijkl} so as to fit the relevant available data is the key to constructing a model for a specific type of soil. Figure 22.13 illustrates the behaviour of kinematic cap model with a nonlinear choice of function f_{ijkl}.

The stress path for a cyclic triaxial stress test is shown in Figure 22.13(a) while the corresponding stress–strain curve is shown in Figure 22.13(b). The material

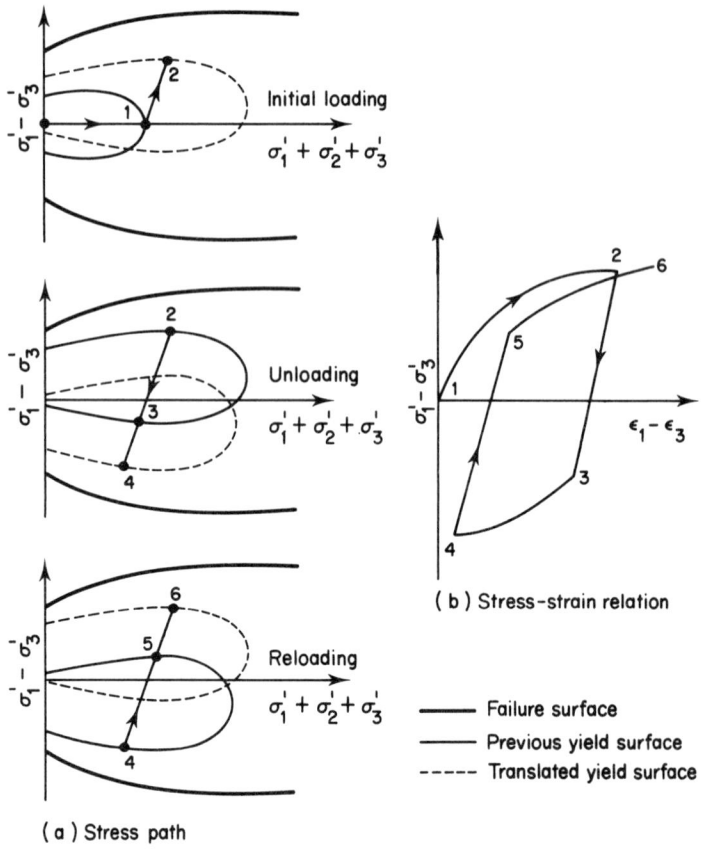

(a) Stress path

Figure 22.13 Triaxial behaviour of the kinematically hardening cap model

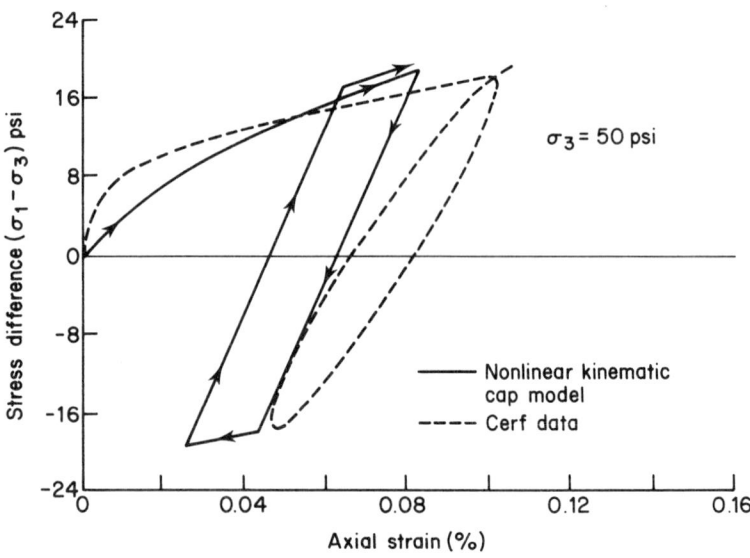

Figure 22.14 Fit of a kinematically hardening cap model (low shear)

behaviour illustrated in the latter figure is qualitatively similar to much of the laboratory data obtained for dry sands.

As more experimental data become available, various nonlinear hardening rules such as those proposed in Mroz, Shrwastava, and Dubey (1976) or

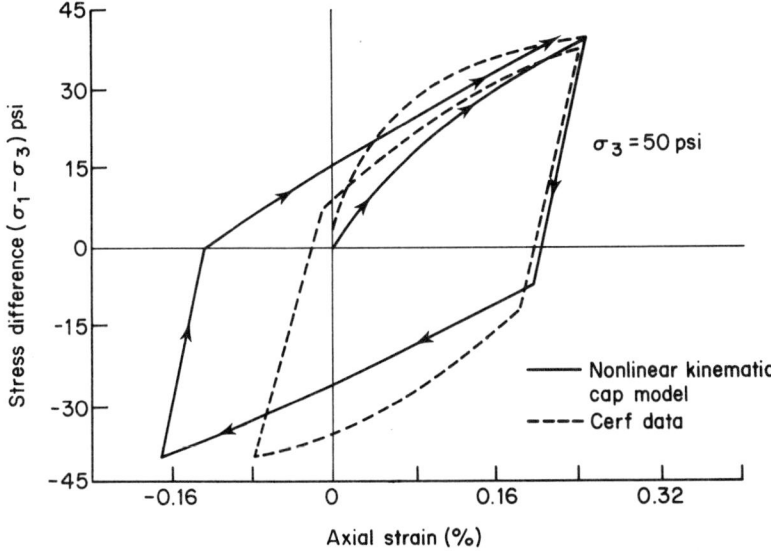

Figure 22.15 Fit of a kinematically hardening cap model (moderate shear)

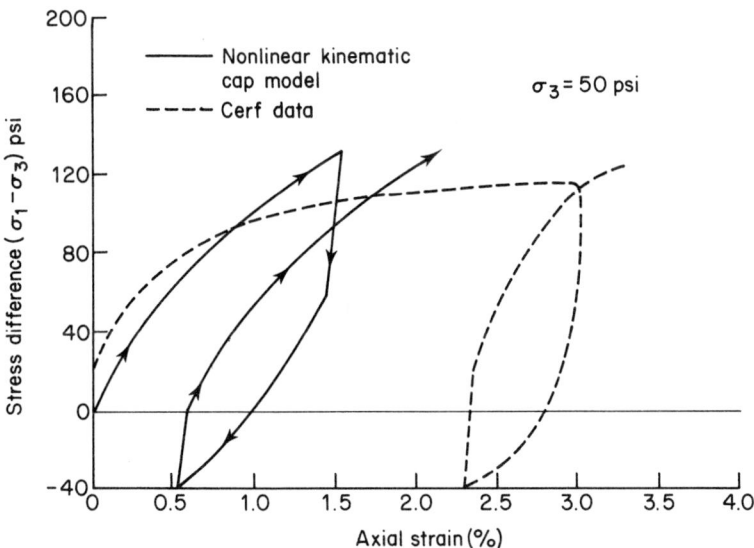

Figure 22.16 Fit of a kinematically hardening cap model (high shear)

522   *Mechanics of Geomaterials*

Ghaboussi and Karshenas (1977) may be required to fit the behaviour of any single material. Comparisons of a particular model with test data are shown in Figures 22.14 to 22.16. The model was used in computations of a structure–medium interaction experiment involving an artificially generated seismic environment.

22.8 THE MODELLING OF SATURATED MEDIA

The introduction of substantial amounts of water into the solid matrix of soils and rocks may lead to a level of complexity above and beyond what has been discussed so far in this paper. If the migration of pore fluid is a significant factor a multiphase analysis is required; each phase must be modelled separately, and a system of individual phase conservation laws must be applied.

For excitations resulting from explosions and for short-lived seismic motions, it may be possible to neglect the migration of pore water. In such cases a single phase analysis is possible by combining the solid matrix constitutive equation with a Terzaghi (1936) type of pore-pressure approach.

One such approach is described in Sandler, Dimaggio, and Baron (1983) in which a laboratory remodelled kaolinite wet clay was modelled. Ordinarily such clay would be expected to be nearly isotropic. However, because the clay was K_0 consolidated prior to triaxial testing, and because the subsequent triaxial tests

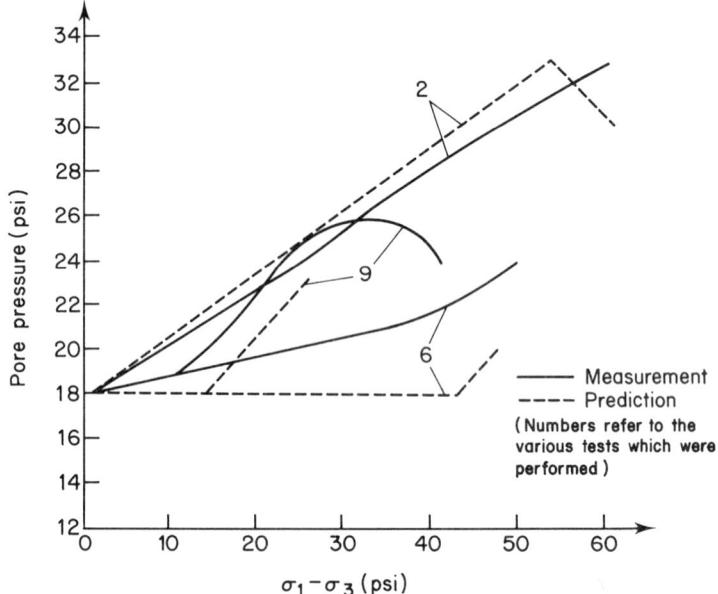

Figure 22.17 Comparison of predicted and observed pore pressure for kaolinite

were known to exhibit substantial anisotropy, the model with a kinematically and isotropically hardening cap (where the cap has an offset in the direction of vertical compression) was utilized. In addition, the material was assumed to be incompressible under the undrained test conditions, i.e. the elastic and plastic volumetric strains were taken equal to each other in magnitude but of opposite sign.

Since all the laboratory tests—those for which data was provided and those to be predicted—were performed in a narrow range of mean pressure (in the vicinity of 40 psi), a simple nine-parameter model was used with linear failure and hardening. This made it possible to fit and exercise the undrained triaxial tests on horizontal and vertical specimens together with consolidation data. The model was exercised to predict behaviour of inclined specimens under a series of stress paths. (These predictions were made before corresponding laboratory data were available for purposes of comparison.) Very good agreement between predictions and experiments was obtained. A typical example of the pore pressure predictions of the model is shown in Figure 22.17. Further details may be found in Sandler, Dimaggio, and Baron (1983).

22.9 ROLE OF *IN SITU* AND LARGE-SCALE TESTS

This would not be complete without a discussion of the role of the material property data on the models. In order to illustrate some points of interest let us consider the problem of validation of procedures used to analyse and predict explosively induced ground shock.

There were a number of large-scale field tests in the USA in the early 1970s to test the validity of ground shock calculations. Two of the major series of events were MIDDLE GUST and MIXED COMPANY. The procedure in force at the time consisted of a site investigation and a core sampling program. Intact 'undisturbed' samples would be brought to the laboratory and tested in uniaxial strain and triaxial compression. Representative property data would be chosen for each of the distinct materials or layers present at the site. The recommended properties would then be fitted with a constitutive model and a finite difference calculation would be run to simulate the field test.

An example of the 'agreement' found between the calculational results and the field measurements is shown in Figure 22.18 for MIDDLE GUST—Event II, 100-ton spherical shot of TNT. Both the small displacement LAYER code, and LAYER II (which included transport terms) code results are very different from that measured in the field, despite the fact that the cap model was used in the calculations.

The discrepancies found in MIDDLE GUST and MIXED COMPANY were the subject of a great deal of investigation and soul-searching on the part of the ground shock community. Much of the work in the area, in recent years, has been directed towards resolving these questions.

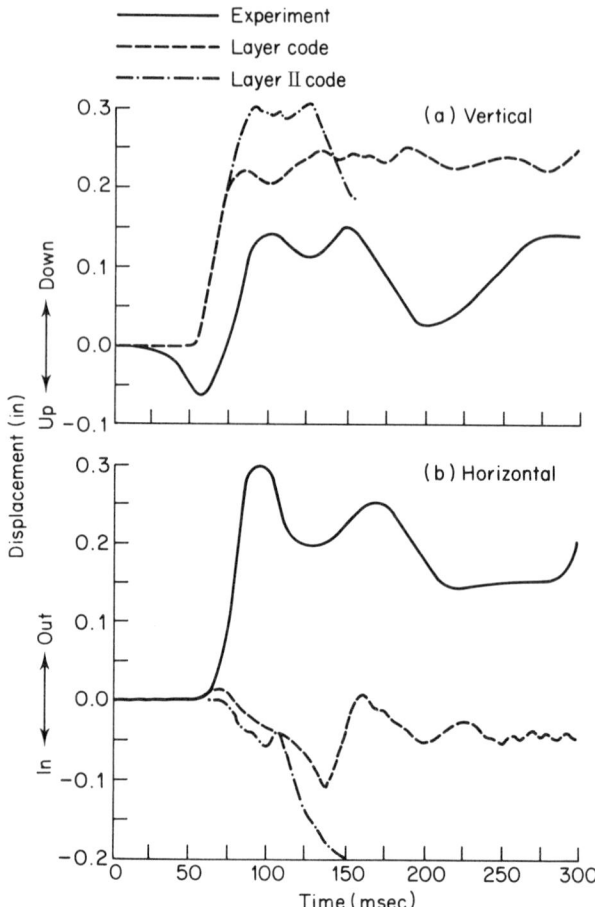

Figure 22.18 Comparison of measured and calculated displacements for MIDDLE GUST II

The main conclusions of these studies is that material model development cannot be based solely on laboratory tests but must also be based on the *in situ* behaviour of the material. In many cases, the behaviour of small disturbed samples of material in the laboratory does not adequately represent the *in situ* behaviour of the material. There are several reasons for this. For certain materials it is almost impossible to obtain truly undisturbed samples for laboratory testing. In addition, large-scale inhomogeneities may make adequate sampling of the site impracticable. The presence of anisotropy may lead to unrepresentative behaviour in the laboratory specimens (which are almost always obtained from vertical cores). The release of lithostatic stresses of unknown magnitude, which may be significant at rock site, may alter laboratory behaviour. Finally the rates at which some laboratory tests are performed may differ from those observed in ground response in explosive events.

Because of the above considerations and because material models should be based on as wide a range of loading paths as possible, a number of *in situ* testing procedures to supplement the laboratory tests have been developed.

In addition to the field seismic survey, these *in situ* tests include wave propagation tests with planar, cylindrical, and spherical symmetry. All of *in situ* tests give response data which can be interpreted only if some assumptions are made with respect to material behaviour. For the relatively complicated material behaviour that real geological materials exhibit, it is difficult if not impossible to obtain mathematical models based solely on the *in situ* tests. Consequently a preliminary mathematical model is fitted to the comprehensive data from laboratory tests as well as any *in situ* material wave speed test that is available. This model is then used to compute the *in situ* test configurations and amended so that the computed *in situ* response of the media mirrors the measured response. The final computational material model is thus determined by means of an iterative procedure in which both the laboratory data and the *in situ* data are utilized. The extremely important (and often major role that the *in situ* data plays in the development of mathematical models for ground shock calculations is now widely recognized, and this iterative procedure is currently being used whenever possible.

22.10 CONCLUSIONS

An overview of some of the considerations involved in the construction of numerical models for dynamic loading of geological materials has been given. Various models appropriate for a number of physical situations of interest are presented and brief descriptions of their behaviour are given.

The role of laboratory and *in situ* experimental data in material modelling is discussed in order to indicate how the phenomenological and practical aspects of modelling interwine. These aspects, together with the theoretical requirements, then constitute the salient points of model development.

ACKNOWLEDGEMENTS

Much of the work described herein was funded by US Government agencies, in particular the Defence Nuclear Agency.

The authors would like to thank their colleagues at Weidlinger Associates especially Dr Ivan Nelson, for their many contributions to the work described in this paper. Thanks are also due to Indy Parker for preparing the manuscript.

REFERENCES AND BIBLIOGRAPHY

Baron, M. L., McCormick, J. M., and Nelson, I. (1969) 'Investigation of ground shock effects in nonlinear hysteretic media', Symposium on Computational Approaches in

Applied Mechanics, American Society of Mechanical Engineers (also available in *Mechanics* 1970, American Academy of Mechanics, University Park, Pennsylvania).

Baron, M. L., Nelson, I., and Sandler, I. S. (1973) 'Influence of constitutive models on ground motion predictions', *Journal of the Engineering Mechanics Division, American Society of Civil Engineers*, **99**, EM6, 1181–1200.

Christian, J. T. (1966) 'Plain strain deformation analysis of soil', Contract Report 3–129, Report 3, Contract DA-22-079-eng-471, Department of Civil Engineering, Massachusetts Institute of Technology.

Dimaggio, F. L., and Sandler, I. S. (1971) 'Material model for granular soils', *Journal of Engineering Mechanics Division, American Society of Civil Engineers*, **97**, EM3, 935–950.

Drucker, D. C. (1956) 'On uniqueness in the theory of plasticity', *Quarterly of Applied Mathematics*, **14**, 35–42.

Drucker, D. C., and Prager, W. (1952) 'Soil mechanics and plastic analysis or limit design', *Quarterly of Applied Mathematics*, **10**, 157–175.

Drucker, D. C., Gibson, R. E., and Henkel, D. J. (1957) 'Soil mechanics and work-hardening theories of plasticity', *Transactions American Society of Civil Engineers*, **122**, 238–346.

Ghaboussi, J. G., and Karshenas, M. (1977) 'On the finite element analysis of certain material nonlinearities in geomechanics', *Intl. Conf. on Finite Elements in Nonlinear Solids and Structural Mechanics*, Geilo Norway, Tapir Publishing Co., Trondheim, Norway.

Handelman, G. H., Lin, C. C., and Prager, W. (1947) 'On the mechanical behavior of metals in the strain-hardening range', *Quarterly of Applied Mathematics*, **4**, 397–407.

Mroz, Z., Shrwastava, H. P., and Dubey R. N. (1976) 'A nonlinear hardening model and its application to cyclic loading', *Acta Mechanica*, **25**, 51–61.

Nelson, I. (1978) 'Constitutive models for use in numerical computations', *Dynamic Methods in Soil and Rock Mechanics, Proceedings*, Karlsruhe 1977, Vol. 2: *Plastic and Long-term Effects in Soils*, Balkema, Rotterdam.

Nelson, I., and Baron, M. L. (1971) 'Application of variable Moduli models of soils behavior', *International Journal of Solids and structures*, **7**, 399–417.

Nelson, I., Baron, M. L., and Sandler, I. S. (1971) 'Mathematical models for geological materials for wave propagation studies', *Shock Waves and the Mechanical Properties of Solids*, Syracuse University Press, Syracuse, New York.

Prager W. (1966) 'Models of plastic behavior', *Proc. Fifth US Cong. Applied Mechanics*, pp. 435–450.

Sandler, I. S. (1976a) 'The cap model for static and dynamic problems', *Proceedings of the 17th US Symposium on Rock Mechanics*, Snowbird, Utah.

Sandler, I. S. (1976b) 'Material modeling based on CIST test and laboratory data', Final Report DNA 3970F, Constract DNA001-75-C0239, prepared by Weidlinger Associates for Defense Nuclear Agency.

Sandler, I. S., Dimaggio F., and Baron, M. (1983) 'An extension of the cap model for the inclusion of pore pressure effects and kinematic hardening in a cap model representation of an anisotropic wet clay', *Mechanics of Engineering Materials*, John Wiley.

Schofield, A., and Wroth, P. (1968) *Critical State Soil Mechanics*, McGraw-Hill, London.

Terzaghi, K. (1936) 'The shearing resistance of saturated soils and the angle between the Planes of shear', *International Conference on Soil Mechanics and Foundation Engineering*, Cambridge, Mass., Vol. 1.

Mechanics of Geomaterials
Edited by Z. Bažant
© 1985 John Wiley & Sons Ltd

Chapter 23
Numerical Aspects in Modelling of Inelastic Materials: Discusser's Report

P. G. Bergan

23.1 GENERAL COMMENTS

The importance of the numerical aspects in constitutive modelling of materials has become increasingly more evident in recent years. Yet the fundamental concepts and principles behind the constitutive theories that we use today were established at a time when the word computer was hardly known. In the 1940s and 1950s fore-sighted researchers, like Prager and others, developed mathematical theories for inelastic materials with such a degree of complexity that these theories, at the time, hardly could be used for anything more than solving highly simplified academic problems. Today, the same theories have been adapted and implemented in a very large number of computer programs, and they are used in a routine manner to solve large and important practical problems.

The basic principles of modern constitutive modelling are much the same as before. However, it is also clear that the emphasis is moving away from pure analytical formulations to material parameterization and suitability for numerical implementations. Today, constitutive modelling and numerical modelling of materials are almost synonymous concepts.

The modelling of materials may be viewed as one aspect of the more extensive process of developing simulation models for mechanical systems, see Figure 23.1. Given an external input in terms of forces, temperature, etc., the real, physical systems give an immediate response, most often in agreement with what we believe to be the laws of nature. The numerical simulation models, on the other hand, have to be constructed step by step. First, they require a material behaviour, often denoted the 'material law'. These materials models must also be defined in terms of a parameterization describing the unique behaviour of the real material to be analysed.

It is also necessary that the simulation model is expressed in terms of a finite number of degrees of freedom with which the computer can work. This spatial description of the field functions is normally done by means of finite differences or finite elements. The material model and the spatial discretization is then utilized in a variational principle, usually the principle of virtual work. The resulting

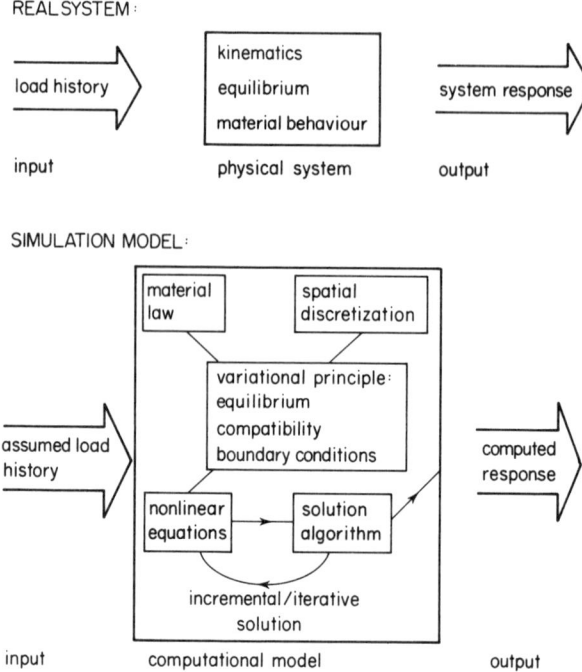

Figure 23.1 Relationship between real system and computational model

variational formulation ensures satisfaction of balance (internal and external equilibrium), kinematic compatibility, and boundary conditions. These conditions are normally not satisfied in a pointwise manner throughout the entire system, but only at selected points or in an integral manner.

In the general case the variational formulation leads to nonlinear equations in the primary variables. These equations may be expressed in many different ways, such as potential energy, force balance equations, incremental equilibrium equations, etc. A very large number of solution techniques is available for solving nonlinear equations, see for instance Taylor (1980), Wunderlich (1981), and Hinton (1982).

It must be recognized that the different constituents of the simulation model introduce various forms of errors. The degree of accuracy of the material model should thus be viewed in light of errors due to spatial discretization and introduced by solution algorithms. Hence, the description of the material behaviour does not have to be 'exact'. Another great uncertainty in the simulation process is simply the input of the load history. For complex problems it is often very difficult to prescribe a load history that is fully relevant for the problem at hand. One typical example of such a predicament is selection of earthquake loading.

It is most important to be aware of the direct interaction between material model, spatial discretization, and the technique used for solving the nonlinear equations. This interdependency will be discussed in some detail in the following sections, and the various implications that has for formulation of the constitutive equations will be pointed out.

23.2 SOME EFFECTS OF THE SPATIAL DISCRETIZATION

The inelastic deformations of geomaterials may to a large extent be attributed to accumulated micro-cracking, localized shear slip, and formation of large cracks. Another important effect is the volumetric changes that take place, such as compaction of sand during stress cycling and dilatation of rock and concrete during shear straining.

It is characteristic for many inelastic effects that they are highly local material phenomena, involving slip or shear bands, discrete cracks, etc. It is a major concern for many researchers to develop material models that can simulate these phenomena in the best possible manner. Much progress has been made towards this end. At the same time, it is important to recognize the rather limited capabilities that most discretization techniques offer in expressing very local deformations.

Let the case of a shear band development in a finite element be considered. Figure 23.2(a) shows a situation where a shear band of limited width cuts through a rectangular element in an arbitrary way. The deformational capabilities of the element are determined by the interpolation functions for the displacements (shape functions). These functions are always continuous, smooth functions incapable of expressing abrupt bands of local deformations. Thus, even if the material model is capable of expressing local slip mechanisms in the material, the basic assumptions of the spatial discretization techniques may prevent it from happening.

Use of highly refined element meshes (small elements) may improve the situation somewhat. Bands of local deformations can be modelled by finite elements provided the elements are of the same size of smaller than the width of the band. The situation is best when the bands follow the natural directions of the

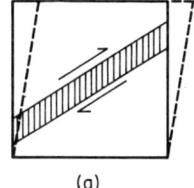
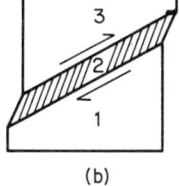

(a) (b)

Figure 23.2 Shear band formation through rectangular element

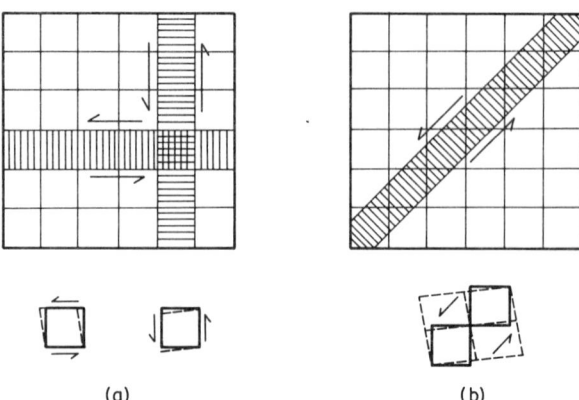

Figure 23.3 Shear band formation through rectangular element meshes

element mesh, such as the orthogonal directions of the rectangular mesh in Figure 23.3(a). However, locking constraints at the boundaries may prevent certain deformational mechanisms from developing.

The capabilities of an element mesh to express bands of local deformations in arbitrary, skew directions are normally not good. As an example, the shear band following a diagonal direction of the mesh in Figure 23.3(b) is in poor agreement with the deformational capabilities of the elements. When the elements on the main diagonal are sheared, it is also necessary that the neighbouring elements are deformed in a similar manner in order to maintain compatibility. The width of the shear band will therefore tend to spread and involve a zone the size of several elements.

The situation may be improved somewhat through use of element meshes that not only suit the geometry and physics of the problem, but are also selected with the purpose of expressing the anticipated mode of failure. The local band of deformations in Figure 23.2(a) can be expressed quite well by subdividing the rectangular element into three smaller ones, as shown in Figure 23.2(b). In modern finite analysis such mesh adjustments may be performed through use of interactive computer graphics. As an example, earthfill dams may be investigated for several modes of failure by orienting elements along the periphery of alternative slip circles.

Another implication of the spatial discretization has to do with the fact that element forces and element stiffnesses for inelastic problems are always calculated by means of numerical integration. The overall behaviour of an element reflects the material response at all contributing integration points within the element. The result of this is a smoothing effect, and the effect becomes more pronounced the higher is the number of integration points. This means that the element becomes relatively insensitive towards local material effects, such as local 'strain softening' or cracking. For instance, having zero or negative stiffness at some integration points will normally not result in an overall zero or negative element stiffness because positive stiffness at intact integration points dominates. Even an entirely failed element with negative stiffness does not necessarily mean that the system equations will lose positive definiteness. The stiffness at a node is assembled from all adjacent elements, and stiffness and strength at partly intact elements will often dominate over one that has failed. A global system instability requires that an instability mechanism has developed, i.e. there must be formed a kind of door that the energy may escape through in an uncontrolled manner.

It is thus clear that the spatial discretization has an important smoothing on the material behaviour and material instability in particular. Even though this may be detrimental to the actual modelling special material effects, it also has a positive stabilizing influence on the solution of the nonlinear equations. The smoothing influence is to a large extent dependent on the actual element sizes.

23.3 INTERACTION WITH THE SOLUTION ALGORITHM

The material model must be viewed and formulated under consideration of the solution algorithm with which it is going to interact. In most cases solution of the nonlinear equations is based on two ingredients, namely incrementation of external loads $\{R\}$ and iteration for equilibrium between these loads and the resulting internal forces $\{R_{\text{int}}\}$. Normally, simple one-step incrementation is employed

$$[K_{I,i+\alpha}]\{\Delta r_{i+1}\} = \{\Delta R_{i+1}\} \quad (23.1)$$

Index i indicates the number of the load step. The increment of applied loads is

$$\{\Delta R_{i+1}\} = \{R_{i+1}\} - \{R_i\} \quad (23.2)$$

The linearized form of Equation (23.1) may be solved with respect to the incremental displacements $\{\Delta r_{i+1}\}$, hence, the new displacement state is found to be

$$\{r_{i+1}\} = \{r_i\} + \{\Delta r_{i+1}\} \quad (23.3)$$

The accuracy and stability of the incremental scheme depends very much on the properties of the incremental stiffness $[K_{I,i+\alpha}]$. For each new load step it is customary to set $[K_{I,i+\alpha}]$ equal to $[K_{T,i}]$, which is the tangent stiffness at the last obtained equilibrium state 'i'. The system tangent stiffness is an integral expression for the tangential constitutive relations throughout the material. A tangential, forward stiffness expression is not representative for a finite step, and it is possible to recalculate the incremental relation (23.1) with an improved incremental stiffness

$$[K_{1,i+\alpha}] = (1-\alpha)[K_{T,i}] + \alpha[K^*_{T,i+1}] \tag{23.4}$$

Here, $[K^*_{T,i+1}]$ is the tangent stiffness for the currently best estimate of state '$i+1$', for instance using the state obtained by one simple forward integration step. α is a weighting parameter. $\alpha = 0$ corresponds to a forward algorithm while $\alpha = 1$ gives a pure backward algorithm. $\alpha = \frac{1}{2}$ normally gives the most stable algorithm. However, $\alpha = 0$ is most used for the incremental equations because it gives a cheap one-step procedure.

Assuming that the inelastic material law is available on a rate form, a similar calculation of incremental stresses takes place at all integration points

$$\{\Delta\sigma^*_{i+1}\} = [C_{1,i+\alpha}]\{\Delta\varepsilon_{i+1}\} \tag{23.5}$$

This gives a new trial state of stress

$$\{\Delta\sigma^*_{i+1}\} = \{\sigma_i\} + \{\Delta\sigma^*_{i+1}\} \tag{23.6}$$

The stress $\{\sigma^*_{i+1}\}$ has to be modified in order to obtain a stress $\{\sigma_{i+1}\}$ which lies on the yield surface. This is normal done through scaling of stress components or by finding the closest stress projection on the yield surface. It is advisable to carry out weighting also for the incremental material law

$$[C_{1,i+\alpha}] = (1-\alpha)[C_{T,i}] + \alpha[C_{T,i+1}] \tag{23.7}$$

where $[C_{T,i+1}]$ is the current tangent stiffness at load level '$i+1$'. Typically, increasingly improved values of this tangent stiffness is obtained during equilibrium iterations.

The weighting factor α plays an important role in stabilization of the solution. A value of $\alpha = \frac{1}{2}$ is normally preferable because it tends to stabilize the effects of discontinuous material behaviour. However, it has also been found that some problems are best solved with $\alpha = 1$. It should be noted that α, which in fact is a guiding parameter for the solution algorithm, also is an integrated part of the material subroutine in a computer program.

The process of scaling stresses back to the yield surface, discontinuous material behaviour, and possible geometric nonlinearities imply that equilibrium iterations have to be used. The iteration equation for the total system most often takes the form

$$\{r^{j+1}\} = \{r^j\} + \omega[K_T^j]^{-1}(\{R^{j+1}\} - \{R^j_{int}\}) \tag{23.8}$$

j is an index counting the number of the iteration cycle. The expression in the parentheses represents the difference between applied loads and the current internal reaction forces. This quantity is often denoted unbalanced, or residual, forces. $[K_T^j]$ is the gradient stiffness used for the iterative correction; this is normally taken to be the last available tangent stiffness. ω indicates the possibility of using an over-relaxation factor for the displacement corrections.

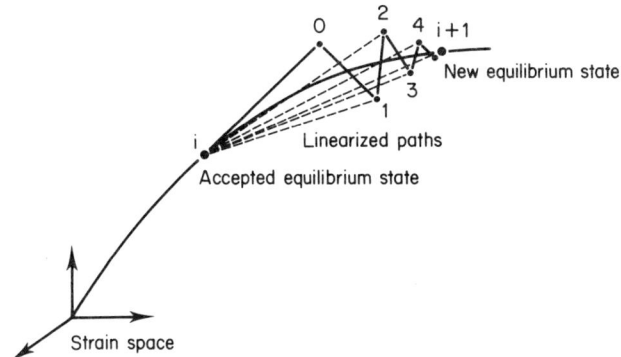

Figure 23.4 Zigzagging strain history during incrementation/iteration

The scheme defined in Equation (23.8) is called true Newton–Raphson iteration provided $[K_T]$ is updated for every cycle. The method is usually called modified Newton–Raphson when the gradient is updated only at intervals. The more primitive method of employing the linear, elastic stiffness corresponds to what is called an initial stress or initial strain technique. Some form of modified Newton–Raphson iteration is normally the most efficient for geomaterials like concrete and rock.

The convergence of an iteration process is quite often of highly oscillatory type, particularly when cracking and non-smooth material behaviour is involved. This type of behaviour requires special precautions in updating of the material data. Figure 23.4 illustrates a typical strain history at a material point. Point 'i' in the diagram represents the accepted equilibrium state at load level 'i'. A load increment in accordance with Equations (23.1), (23.2), and (23.3) produces the strain state '0'. From here equilibrium iterations are performed, giving the path 0–1–2–3–4 etc. at load level '$i+1$'. It is most important that this history of deformation is interpreted in a proper way. The zigzagging path leading to the convergence point '$i+1$' is in fact a product of the solution sequence, and it does not by any means represent the true history. The best approach for updating the material quantities is to assume a direct, straight path from equilibrium state 'i' to the current iteration state 'j'. After a new iterative correction the strains are changed by $\{\Delta\varepsilon^{j+1}\}$. The linearized strain increment to be applied for the material should be taken as

$$\{\Delta\varepsilon_{i+1}^{j+1}\} = \{\varepsilon_{i+1}^{j+1}\} - \{\varepsilon_i\} \tag{23.9}$$

where the current state of strain is

$$\{\varepsilon_{i+1}^{j+1}\} = \{\varepsilon_{i+1}^{j}\} + \{\Delta\varepsilon^{j+1}\} \tag{23.10}$$

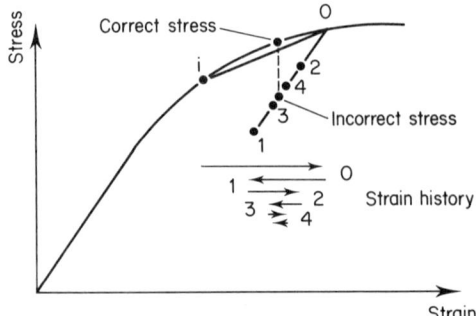

Figure 23.5 False unloading for elasto-plastic material

All intermediate states along the iteration path are thus 'forgotten', and the straightest incremental step $\{\Delta\varepsilon_{i+1}^{j+1}\}$ is always used.

The consequences of what may seem to be a computational detail can in fact be quite serious. Using the zigzagging history rather than the linearized history may lead to a phenomenon for elasto-plastic materials that may be called 'false unloading'. Figure 23.5 illustrates an oscillatory convergence giving the strain history i–0–1–2–3–4 resulting in unloading, this is primarily because the strain at '1' is smaller than the strain at '0'. But in fact, the new strain is larger than at the last equilibrium state 'i', and no real unloading takes place. Use of Equations (23.9) and (23.10) prevents the effect of false unloading.

The situation is similar for a cracking material with strain-softening as shown in Figure 23.6. A possible strain history during iterations is indicated below the σ–ε diagram. The state '0' results in full cracking, a process that is truly irreversible. However, the final state '$i+1$' should in fact render the material intact and elastic. A proper interpretation of the intermediate states of

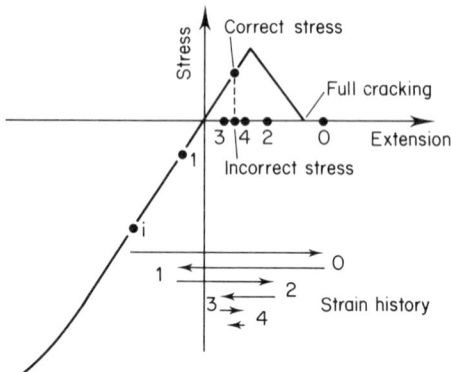

Figure 23.6 False cracking for brittle material

deformation would have prevented cracking in the simulation model. This phenomenon, which may be called 'false cracking', is particularly nasty because it tends to result in a 'domino effect' of progressive cracking, poor convergence, and incorrect results.

A proper interpretation of the strain history is equally important when using accumulative type internal variables, such as equivalent plastic strain (hardening measure), endochronic time, and internal damage parameters. Much improved convergence and accuracy has been experienced using the linearized path interpretation.

It is quite easy to carry through these measures in a computer program. First of all, the material subroutine must be told whether an incremental step is caused by a load increment or it is only an iterative correction. Secondly, it is necessary that the state of stress, internal variables, constitutive matrix, etc. from the last accepted equilibrium state is made available to the material subroutine. It is hence necessary that this information has been stored. Similar information from the last state of iteration must also be given in addition to the new strain increment. The problem of 'forgetting' previous iteration states is conveniently solved by overwriting with the latest obtained results. New computer storage is used only when a new load step is applied. This kind of strategy has successfully been used in the computer program FENRIS (Bergan, 1983; Bergan and Arnesen, 1983).

23.4 CONCLUSIONS

The field of material modelling is much more than just curve-fitting of experimentally observed results. The simulation of material behaviour is really a broad synthesis of many disciplines, such as mathematics, thermodynamics, continuum mechanics, experimental techniques, numerical techniques, discretization methods, and computer technology. In particular, this author would like to emphasize the importance of the numerical aspects. Proper formulation and numerical implementation of the material equations may yield very high rewards in terms of applicability of the theory, economy in the computations, and reliability of the results. It is only through a good comprehension of all the associated disciplines of material modelling that it can be ensured that the sophisticated new material models will turn out to be useful in solving practical problems.

REFERENCES

Bergan, P. G. (1983) 'Nonlinear finite element analysis by use of FENRIS', *Proc. 7th Int. Seminar on Computational Aspects of the Finite Element Method* (CAFEM-7), Chicago, Aug. 29–30, 1983, (Ed. J. Gloudeman).

Bergan, P. G., and Arnesen, A. (1983) 'FENRIS—a general purpose non-linear finite element program', *Proc. 4th Conf. on Finite Element Systems*, Southampton, July 6–8, 1983, (Ed. C. Brebbia).

Hinton, E. (Ed.) (1982) *Recent Advances in Nonlinear Computational Mechanics*, Pineridge Press, Swansea.

Taylor, C. (Ed.) (1980) *Numerical Methods for Non-linear Problems*, Pineridge Press, Swansea.

Wunderlich, W. (Ed.) (1981) *Nonlinear Finite Element Analysis in Structural Mechanics*, Springer Verlag, Berlin.

PART IX

MODERN TRENDS AND NEW DIRECTIONS

Mechanics of Geomaterials
Edited by Z. Bažant
© 1985 John Wiley & Sons Ltd

Chapter 24

Current Problems and New Directions in Mechanics of Geomaterials

Z. Mróz

24.1 INTRODUCTION

The present chapter is aimed at general discussion of some major problems encountered in inelastic analysis of geomaterials such as rocks, soils, and their interaction with structures. Since the elastic solutions have only a limited range of applications, a reasonable use should be made of inelastic analysis. However, in order to develop a sufficiently reliable and simple system of inelastic analysis, numerous new problems have to be solved. These problems are associated with the formulation of constitutive models, their implementation in solving boundary-value problems, development of numerical procedures, and finally proper identification of material parameters both in the laboratory and *in situ*. Whereas for researchers these problems constitute a challenging field of their activity, the application of inelastic analysis to design problems is so far rather limited not only due to its insufficient development but also due to lack of confidence in its accuracy and reliability. The insufficient knowledge of inelastic material response for a variety of loading conditions, and of the relevant material parameters to be incorporated into the constitutive models, contributes also to the difficulty in systematic application of inelastic analysis to actual engineering problems.

In this chapter, some essential elements of the system of inelastic analysis will be discussed, indicating types of problems actually investigated or to be solved in the future. It is believed that such a general discussion may prove useful in identifying areas for future investigation and also in appreciating the need for a coordinated research.

In section 24.2, the concept of a system of inelastic analysis will be discussed and in section 24.3 the applicability of constitutive models to solve problems in both stable and unstable domains will be considered. In section 24.4, the problem of stability and post-critical behaviour will be introduced together with the concept of static and dynamic failure modes. The importance of developing a general failure theory accounting for dynamic effects will be emphasized in the last section where rock burst phenomena will be briefly discussed.

24.2 THE SYSTEM OF INELASTIC ANALYSIS

The general concept of the system is illustrated in Figure 24.1. The system is composed of several interacting blocks pertaining to different classes of problems of inelastic analysis. The block I on 'Constitutive models' contains various formulations of constitutive equations accounting for plasticity, fracture, and degradation of materials for both monotonic and variable loading histories. The block II on 'Identification and sensitivity analysis' contains methods of identifying relevant material parameters from both laboratory and *in situ* measurement data (triaxial tests, penetrometer, pressuremeter data, etc.), and also provides methods of sensitivity analysis of model predictions with respect to variation of material parameters. The block III containing 'Numerical procedures' provides a set of algorithms and programs for solutions of boundary and initial-value problems. Benchmark problems are selected in order to verify the accuracy of numerical procedures and also accuracy of model predictions and therefore analytical solutions or experimental data are needed form such problems.

The fifth most important block on 'Design methods' contains codes of design and calculus used in practical engineering design. This block is supposed to assimilate developments of previous blocks and apply in design problems. However, this process is not straightforward and therefore a vertical semi-permeable wall is placed between the first four blocks containing mostly developments of researchers and the last block containing recommended methods of design calculus. Therefore, the information flux emanating from the system of four blocks does not penetrate totally to the right-hand side of the wall and usually drops to the information storage (libraries, research or conference

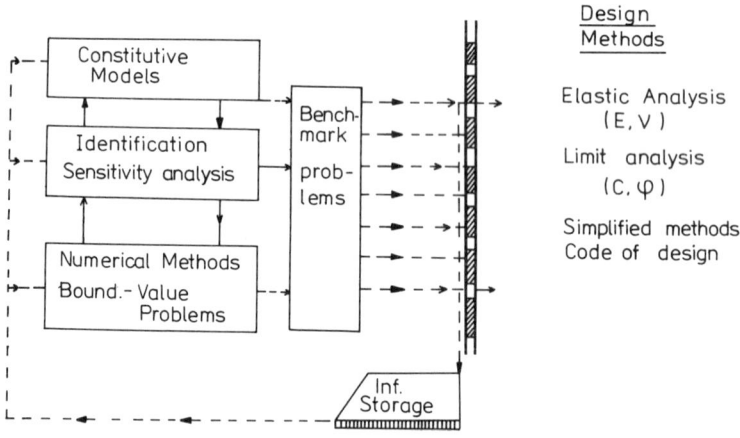

Figure 24.1   System of inelastic analysis

Current Problems and New Directions 541

reports). If this information is next recirculated in order to improve performance of any of the blocks, it is still regarded as useful information. If it does not recirculate and is only stored, it should be regarded as useless information constituting a loss to the society. Only a small portion of information penetrates to the other side of wall and is directly used in design methods.

This situation does not reflect conservatism of designers but rather follows from different goals and philosophies of two groups of people. Whereas the major goal of researchers is to provide the most accurate description of material behaviour for a variety of loading conditions, and the most effective numerical tools to apply developed constitutive models to boundary-value problems, the design as a decision making process is usually based on a single rule: *identify the potential failure modes and provide the rational design preventing these modes* (by using proper safety factors for particular modes). In many cases the failure modes and the associated critical loads can be evaluated by applying the rigid-plastic model assumptions and some simplified versions of limit analysis, for instance, equilibrium method associated with a selected failure mechanism or static and kinematic bounds to limit load based on theorems of limit analysis. Only two material parameters are then needed (cohesion c and friction angle φ) and they are easily identified from simple tests. Slope stability and bearing capacity of foundations provide good examples of such problems. Similarly, the linear elastic analysis, though of limited range of applicability, provides in some cases a useful information on the initial stress and displacement fields. The concept of applying as simple model as possible to identify the failure mode leads to reluctance in accepting more sophisticated descriptions involving numerous material parameters and more complex mathematical structure. However, there are many problems in geotechnical engineering, for instance, those associated with cyclic loading, earthquake processes, or dynamic loading, when simple models cannot provide a meaningful information and a more sophisticated description of material behaviour is necessary. Therefore, the system of inelastic analysis using more complex models will become a useful tool in the design process though it must be realized that the transfer of information from research to application is not direct but involves recycling and consecutive improvements of particular elements.

Let us now discuss some most important blocks of the system presented in Figure 24.1.

24.2.1 Constitutive models

The application of linear elasticity in soil and rock mechanics provides only first approximation valid for stress states lying far below the limit state. The failure condition based on local stress then provides very conservative assessment of limit load since the linear elastic analysis does not describe the considerable stress redistribution occurring in the nonlinear range before ultimate failure. On the other hand, the limit analysis is based on the assumption of unlimited material ductility and hence on complete stress redistribution corresponding to kinematically admissible failure mode. This approach therefore usually overestimates the real carrying capacity and therefore may provide non-conservative evaluation of the limit load. In many cases there is a need to trace the whole deformation history and specify the limit load by incremental analysis.

Considering *incremental* or *rate* description of time-independent material response, the following classes of models can be mentioned:

(i) non linear elasticity and hypoelasticity formulations,
(ii) perfectly plastic models with associated or non-associated flow rules,
(iii) isotropic hardening models with density (volumetric and/or deviatoric hardening),
(iv) anisotropic hardening models involving multiple loading surfaces,
(v) endochronic or incrementally nonlinear formulations avoiding loading–unloading conditions,
(vi) combined elasto-plastic and damage models accounting for elastic stiffness degradation.

For time-dependent material behaviour, the following models are applied:

(i) viscous or viscoplastic models using yield conditions and hardening rules formulated for time-independent plasticity,
(ii) viscoplastic degradation models describing creep and rate-dependent stress–strain behaviour.

For monotonic loading, the simple elasto-plastic or nonlinear elastic material models can be used to simulate the deformational response of a material with sufficient accuracy. For a perfectly-plastic model both hardening and softening phenomena are neglected and the material is assumed to exhibit unlimited ductility. For an isotropic hardening model, the irreversible void ratio or density are usually assumed as state variables. Both hardening and softening phenomena are usually predicted by this class of models together with the critical state surface corresponding to vanishing hardening modulus. The associated or non-associated flow rules are used depending on the accuracy of description material dilatancy. For more complex loading programs, in particular, for cyclic loading, more complex hardening rules should be introduced in order to describe hysteresis effects and cyclic material degradation resulting in eventual failure or liquefaction for undrained deformation. Proper account of material memory is then obtained by introducing multiple loading surfaces and defining particular loading events.

In the classical theory of plasticity, it is usually assumed that elastic behaviour occurs for stress states represented by the interior of the yield surface. However, for geological materials, such as clay, sand, or rock, there is no distinct domain or elastic behaviour and the yield surface, when defined by a small offset value

Current Problems and New Directions

usually encloses very small domain lying in the vicinity of the loading point. Therefore a more accurate formulation should allow for nullifying the elastic domain and for defining the inelastic unloading or reloading events. Such possibility is offered in multiple-surface anisotropic hardening rules or in incrementally nonlinear formulations.

In the present paper, we will not discuss all particular formulations of constitutive models. An extensive discussion of such formulations was presented earlier, cf. Mróz and Norris (1982) and Mróz and Zienkiewicz (1983). To supplement this discussion, we shall concentrate here on restrictive *admissibility conditions* following from the convexity requirement imposed on incremental relations within the stable range of material behaviour. These conditions specify also the limit surface for stress-controlled processes, whereas other limit surfaces can be specified for strain-controlled or mixed-controlled deformation processes. The idea to impose some restrictive admissibility conditions on constitutive relations follows from Prager's early work (1949) and subsequently this idea was elaborated in numerous works.

Considering the rate-independent behaviour of the material, the constitutive relations can be expressed as follows

$$\dot{\varepsilon} = \mathbf{C} \overset{\triangledown}{\boldsymbol{\sigma}} \quad \text{or} \quad \overset{\triangledown}{\boldsymbol{\sigma}} = \mathbf{D} \dot{\varepsilon} \tag{24.1}$$

where the matrices $\mathbf{D}$ and $\mathbf{C}$ depend on stress, strain, hardening, and memory parameters. For incrementally linear formulation, $\mathbf{D}$ or $\mathbf{C}$ do not depend on stress or strain rates. Here $\dot{\varepsilon}$ denotes the strain rate and $\overset{\triangledown}{\boldsymbol{\sigma}}$ is the objective stress rate, for instance, specified by the relation

$$\overset{\triangledown}{\boldsymbol{\sigma}} = \dot{\boldsymbol{\sigma}} + \boldsymbol{\sigma} \operatorname{div}(\mathbf{v}) - \omega\boldsymbol{\sigma} + \boldsymbol{\sigma}\omega \tag{24.2}$$

where ω is the spin of the element, $\mathbf{v}$ denotes the velocity vector and $\dot{\boldsymbol{\sigma}}$ is the material stress rate. In what follows, our discussion will be mainly confined to small strain theory for which configuration changes are neglected in defining the stress rate or increment. Let there exist at each stage of the deformation process a yield or loading surface in the stress or strain space separating domains of loading and unloading (not necessarily elastic), that is

$$f(\boldsymbol{\sigma}, \boldsymbol{\alpha}, \mathbf{p}) = 0, \quad \text{or} \quad \phi(\varepsilon, \varepsilon^p, \boldsymbol{\alpha}, \mathbf{p}) = 0 \tag{24.3}$$

where $\boldsymbol{\alpha}$ denotes collectively the state variables and $\mathbf{p}$ are the memory parameters which are consecutively stored or erased from the material memory. The usual decomposition of the strain tensor ε into elastic and plastic portions is assumed to apply, $\varepsilon = \varepsilon^e + \varepsilon^p$. As different rate relations occur active loading and unloading trajectories, we can write for strain-controlled processes

$$\overset{\triangledown}{\boldsymbol{\sigma}} = \mathbf{D}_2 \dot{\varepsilon} \quad \text{for} \quad \phi = 0, \dot{\phi}_\varepsilon = \frac{\partial \phi}{\partial \varepsilon} \cdot \dot{\varepsilon} > 0,$$

$$\overset{\triangledown}{\boldsymbol{\sigma}} = \mathbf{D}_1 \dot{\varepsilon} \quad \text{for} \quad \phi < 0, \quad \text{or} \quad \dot{\phi}_\varepsilon < 0, \phi = 0, \tag{24.4}$$

and similar inverse relations

$$\dot{\varepsilon} = \mathbf{C}_2 \overset{\triangledown}{\sigma} \quad \text{for} \quad f = 0, \quad \dot{f}_\sigma = \frac{\partial f}{\partial \sigma} \cdot \overset{\triangledown}{\sigma} > 0$$

$$\dot{\varepsilon} = \mathbf{C}_1 \overset{\triangledown}{\sigma} \quad \text{for} \quad f < 0, \quad \text{or} \quad \dot{f}_\sigma < 0, f = 0$$

(24.5)

where dot between two symbols denotes the scalar product. Assume that the continuity condition is satisfied for stress or strain rates directed tangentially to yield surfaces

$$\mathbf{D}_2 \dot{\varepsilon} = \mathbf{D}_1 \dot{\varepsilon} \quad \text{for} \quad \phi = \dot{\phi}_\varepsilon = 0$$

$$\mathbf{C}_2 \dot{\sigma} = \mathbf{C}_1 \dot{\sigma} \quad \text{for} \quad f = \dot{f}_\sigma = 0$$

(24.6)

For regular loading surfaces these continuity conditions impose essential constraint on constitutive matrices, namely

$$\mathbf{D}_2 = \mathbf{D}_1 + \mathbf{h}\frac{\partial \phi}{\partial \varepsilon}, \quad \mathbf{C}_2 = \mathbf{C}_1 + \mathbf{g}\frac{\partial f}{\partial \sigma}$$

(24.7)

where $\mathbf{h}$ and $\mathbf{g}$ are arbitrary tensors. We have therefore

$$\dot{\sigma}_2 = \mathbf{D}_2 \dot{\varepsilon} = \mathbf{D}_1 \dot{\varepsilon} + \mathbf{h}\boldsymbol{\phi}_n \cdot \dot{\varepsilon} = \overset{\triangledown}{\sigma}_1 - \overset{\triangledown}{\sigma}^r$$

$$\dot{\varepsilon}_2 = \mathbf{C}_2 \overset{\triangledown}{\sigma} = \mathbf{C}_1 \overset{\triangledown}{\sigma} + \mathbf{g}\mathbf{f}_n \cdot \overset{\triangledown}{\sigma} = \dot{\varepsilon}_1 + \dot{\varepsilon}^p$$

(24.8)

where $\boldsymbol{\phi}_n = \partial \phi / \partial \varepsilon$ and $\mathbf{f}_n = \partial f / \partial \sigma$.

The relations (24.8) can be interpreted as the familiar elasto-plastic relations. Identifying the matrices $\mathbf{D}_1$ and $\mathbf{C}_1$ with the elastic stiffness and compliance matrices $\mathbf{C}^e$ and $\mathbf{D}^e$, there is $\dot{\varepsilon}^e = \dot{\varepsilon}_1 = \mathbf{C}_1 \overset{\triangledown}{\sigma}$ and $\overset{\triangledown}{\sigma}^e = \overset{\triangledown}{\sigma}_1 = \mathbf{D}_1 \dot{\varepsilon}$. Now $\dot{\varepsilon}^p$ and $\overset{\triangledown}{\sigma}^r$ represent the plastic strain rate and the relaxation stress rate superposed upon the elastic stress rate. Denoting

$$\mathbf{h} = -M\mathbf{n}_h \frac{1}{|\boldsymbol{\phi}_n|}, \quad \mathbf{g} = \frac{1}{K}\mathbf{n}_g \frac{1/|\mathbf{f}_n|}{}$$

(24.9)

where $\mathbf{n}_h$ and $\mathbf{n}_g$ are normalized tensors, $\mathbf{n}_h \cdot \mathbf{n}_h = 1$, $\mathbf{n}_g \cdot \mathbf{n}_g = 1$, and M, K are scalar functions, the relations (24.8) take the form

$$\dot{\varepsilon} = \dot{\varepsilon}^e + \dot{\varepsilon}^p = \mathbf{C}^e \overset{\triangledown}{\sigma} + \frac{1}{K}\mathbf{n}_g \mathbf{n}_f \overset{\triangledown}{\sigma}, \quad f = 0, \quad \dot{f}_\sigma > 0$$

$$\overset{\triangledown}{\sigma} = \overset{\triangledown}{\sigma}^e - \overset{\triangledown}{\sigma}^r = \mathbf{D}^e \dot{\varepsilon} - M\mathbf{n}_h \mathbf{n}_\phi \cdot \dot{\varepsilon}, \quad \phi = 0, \quad \dot{\phi}_\varepsilon > 0$$

(24.10)

where

$$n_\phi = \frac{\partial \phi / \partial \varepsilon}{|\boldsymbol{\phi}_n|}, \quad n_f = \frac{\partial f / \partial \sigma}{|\mathbf{f}_n|}$$

(24.11)

are normalized gradient tensors. The scalar functions K and M are called respectively the hardening and relaxation moduli and they occur in the flow or relaxation rules

$$\dot{\varepsilon}^p = \frac{1}{K}\mathbf{n}_g(\mathbf{n}_f \cdot \overset{\triangledown}{\boldsymbol{\sigma}}), \quad \overset{\triangledown}{\boldsymbol{\sigma}}^r = M\mathbf{n}_h(\mathbf{n}_\phi \cdot \dot{\varepsilon}) \qquad (24.12)$$

and

$$K = \frac{\mathbf{n}_f \cdot \overset{\triangledown}{\boldsymbol{\sigma}}}{|\dot{\varepsilon}^p|}, \quad M = \frac{|\overset{\triangledown}{\boldsymbol{\sigma}}^r|}{(\mathbf{n}_\phi \cdot \dot{\varepsilon})} \qquad (24.13)$$

that is the hardening modulus K is obtained by projecting the stress rate onto the normal vector $\mathbf{n}_f$ and dividing by the modulus of $\dot{\varepsilon}^p$. The relaxation modulus M is defined as the ratio of the modulus of $\overset{\triangledown}{\boldsymbol{\sigma}}^r$ to the projection of $\dot{\varepsilon}$ onto the normal $\mathbf{n}_\phi$ to the loading surface in the strain space.

In the case of non-associated flow or relaxation rules there is $\mathbf{n}_g = \mathbf{n}_f$ and $\mathbf{n}_\phi = \mathbf{n}_h$. The plastic strain rate vector $\dot{\varepsilon}^p$ then departs from the normal vector $\mathbf{n}_f$ in the stress space and the direction of $\overset{\triangledown}{\boldsymbol{\sigma}}^r$ departs from $\mathbf{n}_\phi$ in the strain space. For the case of associated flow and relaxation rules there is $\mathbf{n}_g = \mathbf{n}_f$ and $\mathbf{n}_h = \mathbf{n}_\phi$, so that (24.12) becomes

$$\dot{\varepsilon}^p = \frac{1}{K}\mathbf{n}_f(\mathbf{n}_f \cdot \overset{\triangledown}{\boldsymbol{\sigma}}), \quad \overset{\triangledown}{\boldsymbol{\sigma}}^r = M\mathbf{n}_\phi(\mathbf{n}_\phi \cdot \dot{\varepsilon}) \qquad (24.14)$$

The hardening and relaxation moduli can be interrelated by inverting (24.10), namely

$$\overset{\triangledown}{\boldsymbol{\sigma}} = \left[\mathbf{D}^e - \frac{\mathbf{D}^e \mathbf{n}_g \mathbf{n}_f \cdot \mathbf{D}^e}{K + \mathbf{n}_f \cdot \mathbf{D}^e \mathbf{n}_g} \right] \dot{\varepsilon} = \overset{\triangledown}{\boldsymbol{\sigma}} - \overset{\triangledown}{\boldsymbol{\sigma}}^r \qquad (24.15)$$

and noting that

$$\frac{\partial \phi}{\partial \varepsilon} = \mathbf{D}^e \frac{\partial f}{\partial \boldsymbol{\sigma}} \quad \mathbf{n}_\phi = \frac{\mathbf{D}^e \mathbf{n}_f}{|\mathbf{D}^e \mathbf{n}_f|} \qquad (24.16)$$

$$\mathbf{h} = \mathbf{D}^e \mathbf{g} \quad \mathbf{n}_h = \frac{\mathbf{D}^e \mathbf{n}_g}{|\mathbf{D}^e \mathbf{n}_g|}$$

we obtain

$$\frac{1}{M} = K' + \frac{\mathbf{n}_f \cdot \mathbf{n}_h}{|\mathbf{D}^e \mathbf{n}_f|} \quad K' = \frac{K}{|\mathbf{D}^e \mathbf{n}_f||\mathbf{D}^e \mathbf{n}_g|} \qquad (24.17)$$

The relations (24.10) or (24.15) are most general forms of linear equations associated with two domains of loading and unloading. This general formulation should be complemented by the evolution rules for state variables α. The scalar memory parameters $p_1, p_2, \ldots$ can be used in order to specify these rules for particular loading events, for instance

$$\dot{\alpha} = \mathbf{A}_1(\boldsymbol{\sigma}, \boldsymbol{\alpha})\dot{\varepsilon}^p, \dot{p}_1 > 0 \qquad \text{—loading}$$
$$\dot{\alpha} \mathbf{A}_2(\boldsymbol{\sigma}, \boldsymbol{\alpha})\dot{\varepsilon}^p, p_2 < p_{1m}, \dot{p}_2 > 0 \qquad \text{—unloading} \qquad (24.18)$$
$$\dot{\alpha} = \mathbf{A}_3(\boldsymbol{\sigma}, \boldsymbol{\alpha})\dot{\varepsilon}^p, p_3 < p_{1m}, p_3 < p_{2m}, \dot{p}_3 > 0 \qquad \text{—reloading}$$

where p_{1m} is the maximal value reached during the loading process for which $\dot{p}_1 > 0$. The unloading process starts when $\dot{p}_1 < 0$. The value p_{1m} is stored in the material memory and a new parameter p_2, $\dot{p}_2 > 0$ specifies the unloading event. Similarly, when $\dot{p}_2 < 0$, the two values p_{1m} and p_{2m} are stored in the material memory and the third loading event commences for which $\dot{p}_3 > 0$. Such memory rules were discussed in detail in the papers by Mróz and Norris (1982) and Dougill (1976) and exemplified by using multiple loading and stress reversal surfaces for which scalar memory parameters are represented by diameters of these surfaces.

For stress-controlled processes, the limit states are usually attained when the hardening modulus vanishes, that is $K = 0$. This means that the stress is stationary for growing strain. Similarly, for strain-controlled processes the limit state corresponds to stationary strain for varying stress, and the relaxation modulus then tends to infinity, thus $K + \mathbf{n}_f \cdot \mathbf{D}^e \mathbf{n}_g = 0$. However, to examine more carefully the limit states, let us impose the *convexity* or *uniqueness* requirement on constitutive relations, that is

$$J = \Delta \overset{\triangledown}{\boldsymbol{\sigma}} \cdot \Delta \dot{\boldsymbol{\varepsilon}} > 0 \qquad (24.19)$$

for any two pairs $\overset{\triangledown}{\boldsymbol{\sigma}}{}^{(1)}, \overset{\triangledown}{\boldsymbol{\sigma}}{}^{(2)}$ and $\dot{\boldsymbol{\varepsilon}}^{(1)}, \dot{\boldsymbol{\varepsilon}}^{(2)}$ interrelated by the constitutive equations, where $\Delta \overset{\triangledown}{\boldsymbol{\sigma}} = \overset{\triangledown}{\boldsymbol{\sigma}}{}^{(2)} - \overset{\triangledown}{\boldsymbol{\sigma}}{}^{(1)}$ and $\Delta \boldsymbol{\varepsilon} = \dot{\boldsymbol{\varepsilon}}^{(2)} - \dot{\boldsymbol{\varepsilon}}^{(1)}$. Let us note that for the small strain formulation, the condition (24.19) is sufficient for the uniqueness of a boundary-value problem, since

$$\int \Delta \overset{\triangledown}{\boldsymbol{\sigma}} \cdot \Delta \boldsymbol{\sigma} \, \mathrm{d}V = \int J \, \mathrm{d}V = 0 \qquad (24.20)$$

for any two solutions $\overset{\triangledown}{\boldsymbol{\sigma}}{}^{(2)}, \dot{\boldsymbol{\varepsilon}}^{(2)}$ and $\overset{\triangledown}{\boldsymbol{\sigma}}{}^{(1)}, \dot{\boldsymbol{\varepsilon}}^{(1)}$ satisfying the boundary conditions and both the equilibrium and the kinematic relations. The condition (24.19) was listed in early Prager's work as most important in formulating the constitutive equations. A general discussion of uniqueness and stability conditions with account for configuration changes was presented by Hill (1958).

The properties of convex transformations were examined in the papers by Mróz (1963, 1966) and here only the most important implications are listed, namely

(i) The convexity condition $J > 0$ implies the continuity of transformation (24.6) for stress or strain paths tangential to loading surfaces, thus

$$\mathbf{D}_2 \dot{\boldsymbol{\varepsilon}} = \mathbf{D}_1 \dot{\boldsymbol{\varepsilon}} \text{ for } \mathbf{n}_\phi \cdot \dot{\boldsymbol{\varepsilon}} = 0 \text{ if } \Delta \overset{\triangledown}{\boldsymbol{\sigma}} \cdot \Delta \dot{\boldsymbol{\varepsilon}} > 0 \qquad (24.21)$$

(ii) When the continuity condition (24.6) is satisfied, the convexity condition $J > 0$ occurs for positive definite operators $\mathbf{D}_1$, $\mathbf{D}_2$, or $\mathbf{C}_1$, $\mathbf{C}_2$, that is

$$\Delta \overset{\triangledown}{\boldsymbol{\sigma}} \cdot \Delta \dot{\boldsymbol{\varepsilon}} = (\mathbf{D}_2 \dot{\boldsymbol{\varepsilon}}_2 - \mathbf{D}_1 \dot{\boldsymbol{\varepsilon}}_1) \cdot (\dot{\boldsymbol{\varepsilon}}_2 - \dot{\boldsymbol{\varepsilon}}_1) > 0 \quad \text{if} \quad \mathbf{D}_1 \dot{\boldsymbol{\varepsilon}}_1 \cdot \dot{\boldsymbol{\varepsilon}}_1 > 0, \, \mathbf{D}_2 \dot{\boldsymbol{\varepsilon}}_2 \cdot \dot{\boldsymbol{\varepsilon}}_2 > 0 \quad (24.22)$$

(iii) When the continuity condition (24.6) is violated, then the convexity condition (24.19) specifies conical domains of applicability of constitutive equations for positive-definite constitutive matrices $\mathbf{D}_1$ and $\mathbf{D}_2$ (or $\mathbf{C}_1$ and $\mathbf{C}_2$).

The second property implies that in order to satisfy the convexity condition, it suffices to demonstrate that both $\mathbf{D}_1$ and $\mathbf{D}_2$ are positive-definite. Since $\mathbf{D}_1$ is identified with the elastic matrix which is positive, definite, then

$$\Delta \overset{\triangledown}{\boldsymbol{\sigma}} \cdot \Delta \dot{\boldsymbol{\varepsilon}} > 0 \quad \text{if} \quad \overset{\triangledown}{\boldsymbol{\sigma}} \cdot \dot{\boldsymbol{\varepsilon}} = \mathbf{D}_2 \dot{\boldsymbol{\varepsilon}} \cdot \dot{\boldsymbol{\varepsilon}} > 0 \quad \text{for} \quad \mathbf{n}_\phi \cdot \dot{\boldsymbol{\varepsilon}} > 0 \quad (24.23)$$

Consider now the implication of the inequality (24.23) on the critical value of the hardening modulus for the non-associated flow or relaxation rules. This question was discussed in detail in earlier papers by Mróz (1963, 1966) and also recently in a more general context by Maier and Hueckel (1979) and Raniecki (1979). Requiring that

$$\mathbf{D}^e \dot{\boldsymbol{\varepsilon}} \cdot \dot{\boldsymbol{\varepsilon}} - \frac{(\mathbf{D}^e \mathbf{n}_g \cdot \dot{\boldsymbol{\varepsilon}})(\mathbf{n}_f \cdot \mathbf{D}^e \dot{\boldsymbol{\varepsilon}})}{K + \mathbf{n}_f \cdot \mathbf{D}^e \mathbf{n}_g} > 0 \quad (24.24)$$

the following condition is obtained

$$2K > 2K_{cr} = \sqrt{\mathbf{n}_g \cdot \mathbf{D}^e \mathbf{n}_g} \sqrt{\mathbf{n}_f \cdot \mathbf{D}^e \mathbf{n}_f} - \mathbf{n}_g \cdot \mathbf{D}^e \mathbf{n}_f \quad (24.25)$$

In particular, when $\mathbf{n}_g = \mathbf{n}_f$, the inequality (24.25) provides

$$K > K_{cr} = 0 \quad (24.26)$$

where K_{cr} is the critical value of hardening modulus.

The inequalities (24.25) and (24.26) provide the definitions of limit or critical states on the loading path. For the associated flow rule the limit state occurs at the stationary stress. In fact, for the strain rate coaxial with $\mathbf{n}_f$, that is $\dot{\boldsymbol{\varepsilon}} = c\mathbf{n}_f$ from (24.15) it follows that $\overset{\triangledown}{\boldsymbol{\sigma}} = 0$ when $K = 0$. This strain is therefore the eigenvalue of the constitutive matrix. However, for the non-associated flow rule, the bifurcation may occurs at the critical value of hardening modulus $K = K_{cr} > 0$ and the bifurcation mode is specified $\Delta \dot{\boldsymbol{\varepsilon}} = c_1(\mathbf{n}_g + \mathbf{n}_f)$, $\Delta \overset{\triangledown}{\boldsymbol{\sigma}} = c_2 \mathbf{D}^e(\mathbf{n}_g - \mathbf{n}_f)$, cf. Raniecki (1979) where c_1 and c_2 are constants. When $K = 0$ for the non-associated flow rule, the eigenstrain is $\dot{\boldsymbol{\varepsilon}} = c\mathbf{n}_g$ and then $\overset{\triangledown}{\boldsymbol{\sigma}} = 0$.

It is seen that there are two kinds of critical points. For the non-associated flow rule, the *bifurcation surface* is first reached for which $K = K_{cr} > 0$ whereas the *limit surface* corresponds to $K = 0$ that is the stationary stress on the loading path. For the associated flow rules these two surfaces coincide. Figure 24.2 presents schematically the limit surfaces in the stress and strain spaces. The limit surface $F_1 = 0$ in the stress space corresponds to $K = 0$ and its image in the strain space is $F_1^\varepsilon = 0$. The limit surface $\phi_1 = 0$ for strain-controlled processes corresponds to stationary strain, thus $1/M = K + \mathbf{n}_f \cdot \mathbf{D}^e \mathbf{n}_g = 0$ on $\phi_1 = 0$. In Figure 24.3(a) the limit surface for the critical state model with the associated flow

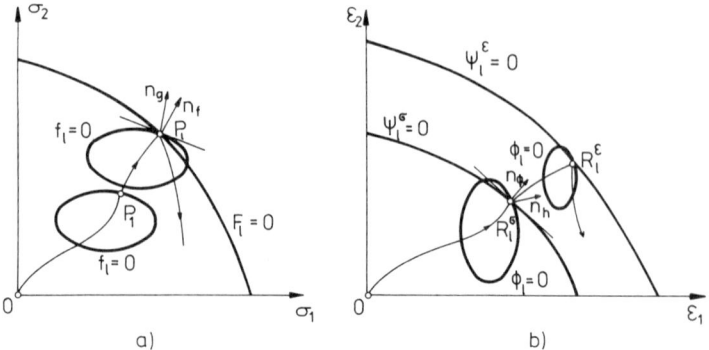

Figure 24.2 Yield and limit surfaces in (a) stress and (b) strain spaces

rule is shown. It is composed of a portion of the yield surface in the softening domain and by the critical state surface in the hardening domain. In Figure 24.3(b) both the bifurcation and the limit surfaces are shown for the hardening Coulomb material with the non-associated flow rule.

The localization in the shear band studied by Rudnicki and Rice (1975) and Rice (1976) is a particular bifurcation mechanism and therefore the critical values of hardening moduli for localization should lie below the values specified by (24.25). For plane strain, the inequality (24.25) provides close upper bound to critical modulus for localization.

The convexity condition can also be imposed on a more general structure of constitutive laws when the elastic degradation effects due to micro-cracks are accounted for, cf. Dougill (1976), Dragon and Mróz (1979a), Bažant and Kim (1979). The total strain and stress rates are now decomposed into three portions

$$\dot{\varepsilon} = \dot{\varepsilon}^e + \dot{\varepsilon}^p + \dot{\varepsilon}^f, \quad \overset{\triangledown}{\sigma} = \overset{\triangledown}{\sigma}^e - \overset{\triangledown}{\sigma}^r - \overset{\triangledown}{\sigma}^f \qquad (24.27)$$

where $\dot{\varepsilon}^f$ represents the fracture strain rate due to variation of elastic moduli and

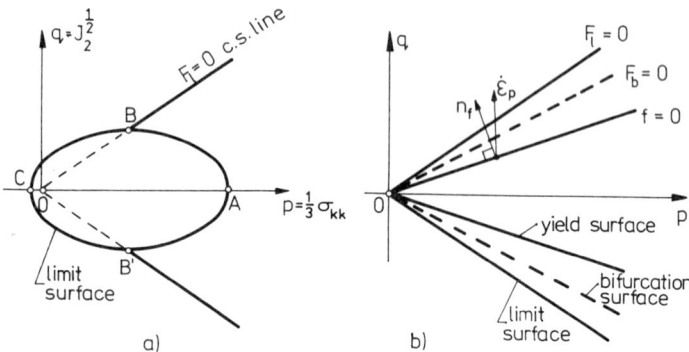

Figure 24.3 (a) The limit surface for the critical state model, (b) yield, bifurcation, and limit surfaces for the hardening Coulomb material

$\overset{v}{\sigma}{}^f$ is the corresponding relaxation stress rate. Besides the yield or loading surfaces for plastic deformation, let us introduce the fracture surface specifying the domain of progressive fracture

$$f^f(\boldsymbol{\sigma}, \boldsymbol{\varphi}) = 0, \quad \text{or} \quad \phi^f(\boldsymbol{\varepsilon}, \boldsymbol{\varphi}) = 0 \qquad (24.28)$$

where φ denotes the damage tensor. The constitutive relations for $\dot{\boldsymbol{\varepsilon}}^f$ and $\dot{\boldsymbol{\sigma}}^f$ now are

$$\dot{\boldsymbol{\varepsilon}}^f = \dot{\mathbf{C}}^e \boldsymbol{\sigma} = \frac{1}{L} \mathbf{s}(\mathbf{m}_f \cdot \overset{v}{\boldsymbol{\sigma}}), \quad f^f = 0, \quad \mathbf{m}_f \cdot \overset{v}{\boldsymbol{\sigma}} > 0$$

$$\overset{v}{\boldsymbol{\sigma}}{}^f = \dot{\mathbf{D}}^e \boldsymbol{\varepsilon} = N \mathbf{t}(\mathbf{m}_\phi \cdot \dot{\boldsymbol{\varepsilon}}), \quad \phi_f = 0, \quad \mathbf{m}_\phi \cdot \dot{\boldsymbol{\varepsilon}} > 0 \qquad (24.29)$$

and are analogous to flow and relaxation rules in plasticity. Here **s** and **t** are the unit vectors spacifying the directions of $\dot{\boldsymbol{\varepsilon}}^f$ and $\overset{v}{\boldsymbol{\sigma}}{}^f$ and $\mathbf{m}_f$, $\mathbf{m}_\phi$ are the unit normal vectors to fracture surfaces in the stress and strain spaces. When $\mathbf{s} = \mathbf{m}_f$, $\mathbf{t} = \mathbf{m}_\phi$, the associated rules for rates of fractures strain and stress occur. Similarly as in the previous analysis, the critical values for the moduli L and N can be established.

24.2.2 Identification and sensitivity analysis

The rate equations (24.8) and (24.18) can be briefly rewritten as follows

$$\overset{v}{\boldsymbol{\sigma}} = \mathbf{D}(\boldsymbol{\sigma}, \boldsymbol{\alpha}, \mathbf{p}, \mathbf{a}) \dot{\boldsymbol{\varepsilon}}$$

$$\dot{\boldsymbol{\alpha}} = \mathbf{A}(\boldsymbol{\sigma}, \boldsymbol{\alpha}, \mathbf{p}, \mathbf{a}) \dot{\boldsymbol{\varepsilon}}^p \qquad (24.30)$$

where $\mathbf{a}(a_1, a_2, \ldots, a_k)$ is the set of parameters entering into the description. Carrying out tests for uniform stress, the model prediction can be compared with experimental results and the material parameters **a** can be identified. The *identification set* of tests should contain the smallest number of simple experiments, usually represented by straight paths in the stress or strain spaces. The *verification set* of tests specifies the domain of applicability of the model and should contain more complex loading programs. This set is used to improve accuracy of identification and also to verify the basic assumptions of model formulation.

The identification procedures have been extensively studied in system theory, cf. Eykhoff (1974), Pister (1975) and usually are concerned with minimization of distance between experimental and predicted response curves. Considering stress–strain curves obtained for strain-controlled loading programs, it can required, for instance

$$\min I(a) = \sum_l (\boldsymbol{\sigma}^{\exp} - \boldsymbol{\sigma}) \cdot (\boldsymbol{\sigma}^{\exp} - \boldsymbol{\sigma})_l \qquad (24.31)$$

or

$$\min J(a) = \int_0^\varepsilon (\boldsymbol{\sigma}^{\exp} - \boldsymbol{\sigma}) \cdot d\varepsilon \qquad (24.32)$$

where σ^{exp} and σ are the experimental and predicted stress components for the same value of strain and the same loading program, whereas $l = 1, 2, 3, \ldots$ denotes the set of discrete points on the response curve. The direct search or gradient methods can be applied in the optimization procedure. The sensitivity equation

$$\frac{\partial \dot{\sigma}}{\partial \mathbf{a}} = \left(\frac{\partial \mathbf{D}}{\partial \sigma} \cdot \frac{\partial \sigma}{\partial \mathbf{a}} + \frac{\partial \mathbf{D}}{\partial \alpha} \cdot \frac{\partial \alpha}{\partial \mathbf{a}} + \frac{\partial \mathbf{D}}{\partial \mathbf{a}} \right) \dot{\varepsilon} \qquad (24.33)$$

coupled with the second equation (24.30) provides now the information on variation of the stress rate with respect to variation of the parameter vector **a**.

The full-scale tests provide the possibility of identifying the parameters from the solution of a boundary value problem by comparing prediction and measurements at a selected set of points, and minimizing the distance norm between measured and predicted values of displacement or strain. Such *in situ* identification will usually provide different values of parameters than those obtained from laboratory tests on small specimens. This difference can be ascribed to presence of defects, such as cracks, faults, inhomogeneities, not accounted for in model formulations. Some of these defects imply also *size-dependence* of the macro-response and hence uniform stress laboratory experiments or model tests cannot provide complete information on material parameters.

Consider the finite-element formulation of a boundary-value problem for the material specified by (24.30), namely,

$$\mathbf{K}(\sigma, \alpha, \mathbf{p}, \mathbf{a}, \mathbf{f}) \dot{\delta} = \dot{\mathbf{F}} \qquad (24.34)$$

and the distance norm expressed in terms of nodal displacements at the measurement points l

$$J(\mathbf{a}, \mathbf{f}) = \sum_l (\delta^{\text{exp}} - \delta) \cdot (\delta^{\text{exp}} - \delta)_l \qquad (24.35)$$

where **K** denotes the global stiffness matrix, δ and **F** are the nodal displacements and forces. The symbol **f** denotes collectively the imperfection parameters not accounted for in local formulation. The identification problem is now not only concerned with the parameter set **a**, but also with the imperfection set **f**. As the imperfection set is statistically distributed, the more extensive empirical data and filtering techniques are necessary in parameter estimation.

The full-scale tests provide also the possibility of performing the sensitivity analysis of the solution with respect to model parameters **a**, imperfections **f**, and also with respect to selected class of models. It may turn out that a simpler model with smaller parameter set **a** provides more accurate description in view of identification error associated with more complex formulation.

24.2.3 Benchmark problems

These problems can be selected (i) in order to verify the accuracy of numerical procedures, (ii) to verify constitutive model predictions, and (iii) to verify both numerical techniques combined with the assumed model and its parameters. The centrifuge model tests provide a possibility of specifying such benchmark problems for which scaling factors are preserved. However, in view of the size effect of post-critical softening behaviour and damage progression, full-scale tests are more representative as benchmark problems.

24.3 SOLUTION OF BOUNDARY-VALUE PROBLEMS IN POST-CRITICAL RANGE

In section 24.2, we discussed briefly the basic structure of constitutive models and the determination of local limit surfaces separating stable and unstable ranges of deformation. However, in practical cases the solution should proceed beyond the states specified by the limit surface. The analysis should usually determine the failure surface for the whole structure with the associated failure mode which locally implies progression beyond the peak strength. Unfortunately, the standard finite element procedure suffers from numerical instability and sensitivity to mesh size. In fact, the deformation tends to localize along concentrated shear bands whose width is sensitive to the size of selected mesh of finite elements. To avoid this sensitivity, several modifications in formulations are necessary, for instance

(i) assume a prescribed width of the shear band and produce the solution within the band and in its exterior by elements of different sizes,
(ii) reformulate the problem by introducing the effect of existing shear band into the stiffness matrix of the element,
(iii) assume contact constitutive relations along the shear band relating normal and tangential tractions to conjugate displacement discontinuities. The finite element formulation should allow for such discontinuities at the nodes.

Let us discuss one of possible modifications, cf. Pietruszczak and Mróz (1981), by assuming the representative element to be composed of two sub-elements connected in series, Figure 24.4. In the sub-element (a) the deformation is concentrated in a shear band of width d inclined at an angle α to the $O - x$ axis. The remaining portion of this sub-element is assumed to be rigid. The sub-element (b) is assumed to behave elastically.

Consider first the small strain formulation. The plastic flow within the shear band is described by the contact yield condition and the flow rule

$$f(\sigma_n, \tau, \beta) = 0 \qquad (24.36)$$

$$\dot{\varepsilon}_n^P = \lambda \frac{\partial f}{\partial \sigma_n}, \quad \dot{\gamma}^P = \lambda \frac{\partial f}{\partial \tau}, \quad \dot{\varepsilon}_t^P = 0 \qquad (24.37)$$

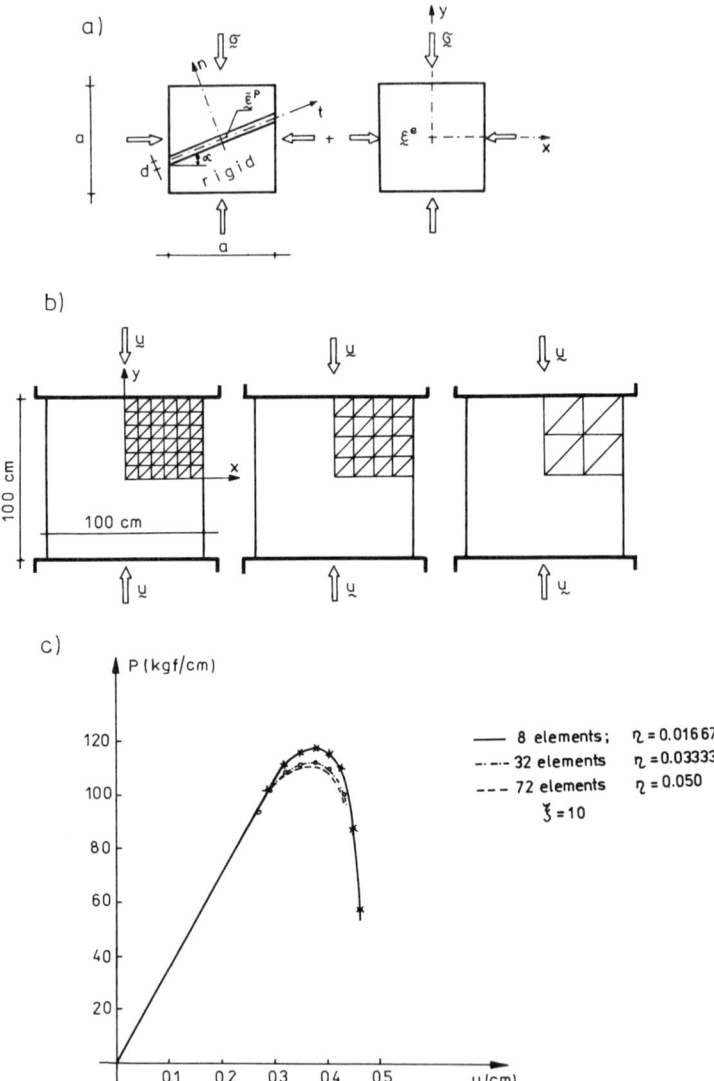

Figure 24.4 Solution for a softening material: (a) element containing shear band, (b) various mesh sizes, (c) force–displacement curves

where $\dot{\varepsilon}_n^P, \dot{\gamma}^P, \dot{\varepsilon}_t^P$ are the strain rates within the shear band referred to the local reference system (n, t). Here σ_n, τ denote the normal and shear stresses and β is the softening variable. The flow rule can also be written as follows

$$\dot{\varepsilon}^P = \frac{1}{H}[\mathbf{C}^P]\dot{\boldsymbol{\sigma}}, \quad [\mathbf{C}^P] = \left(\frac{\partial f}{\partial \boldsymbol{\sigma}}\right)^T \frac{\partial f}{\partial \boldsymbol{\sigma}} \qquad (24.38)$$

The average plastic strain in the element is expressed as follows

$$\dot{\varepsilon}^P = \frac{F_b}{F_e}\dot{\varepsilon}^P = \frac{d}{a\cos\alpha}\frac{1}{}\dot{\varepsilon}^P = \frac{d}{a\cos\alpha}\frac{1}{H}[\mathbf{C}]\dot{\boldsymbol{\sigma}} \quad (24.39)$$

where F_b and F_e denote the areas of the shear band and of the element. The angle α varies within the interval $0 \leqslant \alpha \leqslant \pi/4$. The effect of position of the shear band within the element is neglected.

The constitutive relations for the sub-element (b) averaged for the whole element take the form

$$\dot{\varepsilon}^e = [\mathbf{C}^e]\dot{\boldsymbol{\sigma}} \quad (24.40)$$

Adding elastic and plastic strain rates, we have

$$\dot{\varepsilon} = \left\{\frac{d}{a}\frac{1}{\cos\alpha}\frac{1}{H}[\mathbf{C}^P] + \frac{1}{E}[\mathbf{C}^e]\right\}\dot{\boldsymbol{\sigma}} \quad (24.41)$$

or

$$\boldsymbol{\sigma} = [\mathbf{D}]\dot{\varepsilon} \quad \text{where} \quad [\mathbf{D}] = \left\{\frac{d}{a\cos\alpha}\frac{1}{H}[\mathbf{C}^P] + [\mathbf{D}^e]\right\}^{-1} \quad (24.42)$$

and **D** is the tangential stiffness matrix referred to the coordinates (n, t). Transforming this matrix to the global coordinate system, we obtain

$$\dot{\boldsymbol{\sigma}} = [\mathbf{T}]^T\left\{\frac{F_b}{F_e}\frac{1}{H}[\mathbf{C}^P] + [\mathbf{C}^e]\right\}^{-1}[\mathbf{T}] \quad (24.43)$$

where $F_b/F_e = \eta/\cos\alpha$ and $\eta = d/\sqrt{F_e}$; **T** denotes the rotation matrix.

It is seen that the stiffness of the element now depends on the ratio d/a or F_b/F_e. The softening rate will increase with the size of the element and above the critical size a_c the displacement element loses its stability.

The familiar size effect is therefore naturally incorporated into the element formulation.

Consider for simplicity the Coulomb yield condition in the form

$$f = \tau + \sigma_n \tan\varphi - c(\beta) = 0 \quad (24.44)$$

where φ is the angle of internal friction and c is a variable cohesion of the material. Assume that

$$\dot{\beta} = \dot{\gamma}^P, \quad c = c_0 - c_1\beta \quad (24.45)$$

where c_0 is the initial cohesion and c_1 denotes the softening modulus. Thus the cohesion is assumed to decrease in the course of shearing and the ultimate state corresponds to the cohesionless material. The flow rule now takes the form

$$\dot{\varepsilon}^P = -\frac{1}{\xi c_0}[\mathbf{C}^P]\dot{\boldsymbol{\sigma}}, \quad [\mathbf{C}^P] = \begin{bmatrix} \tan^2\varphi & 0 & \tan\varphi \\ 0 & 0 & 0 \\ \tan\varphi & 0 & 1 \end{bmatrix} \quad (24.46)$$

where $\xi = c_1/c_0$. The inclination of the shear band to the maximal principal stress equals

$$\theta = \tfrac{1}{2}\tan^{-1}(2\tau_{xy}/(\sigma_y - \sigma_x)) \tag{24.47}$$

For more complex yield condition the determination of the angle θ should be carried out at each incremental step.

Figure 24.4(b) presents the results of the sensitivity study to mesh size. Considering an elasto-plastic strip of Coulomb-softening material compressed between two rigid plates in plane strain condition, it is found that for different mesh sizes the post-critical load–displacement curve is practically the same whereas the standard formulation exhibited great sensitivity to the size of finite element mesh, Figure 24.4(c).

This concept of a smeared shear band is similar to that considered recently in blunt crack propagation analysis by Bažant and Cedolin (1979). An extensive discussion of solutions of boundary-value problems accounting for concentrated shear zones was presented by Cleary (1978). The uniqueness problem in softening range was discussed by Prevost and Hoeg (1975).

Whereas the overall element strain and rotation is assumed to be small, the shear band strain and rotation may be large. To account for finite configuration changes within the band, the analysis presented can be reformulated. The spin within the band equals

$$\bar{\omega}_{(xy)} = \frac{1}{2}\dot{\gamma}^P = \frac{1}{2}\frac{F_e}{F_b}\dot{\gamma}^P = \mathcal{H}^T \dot{\varepsilon}^P \frac{F_e}{F_b} \tag{24.48}$$

where

$$\mathcal{H}^T = \{-sc, sc, \tfrac{1}{2}(c^2 - s^2)\}, s = \sin\alpha, c = \cos\alpha \tag{24.49}$$

The plastic strain rate within the band is expressed as follows

$$\dot{\varepsilon}^P = \frac{1}{H}[\mathbf{T}]^T[\mathbf{C}][\mathbf{T}]\overset{\triangledown}{\boldsymbol{\sigma}} = \frac{1}{H}[\mathbf{C}'](\dot{\boldsymbol{\sigma}} + \bar{\omega}_{(xy)}\mathbf{S}) = \frac{1}{H}[\mathbf{C}'](\dot{\boldsymbol{\sigma}} + \mathcal{H}^T\dot{\varepsilon}^P\frac{F_e}{F_b}\mathbf{S}) \tag{24.50}$$

where

$$[\mathbf{C}'] = [\mathbf{T}^T][\mathbf{C}][\mathbf{T}], \quad \mathbf{S} = [-2\tau_{xy}, 2\tau_{xy}, \sigma_x - \sigma_y]^T \tag{24.51}$$

and $\mathbf{T}$ is the rotation matrix. The average plastic strain rate equals

$$\dot{\varepsilon}^P = \frac{F_b}{F_e}\frac{1}{H}[\mathbf{C}']\dot{\boldsymbol{\sigma}} + \frac{1}{H}[\mathbf{C}']\mathbf{S}\mathcal{H}^T\dot{\varepsilon}^P \tag{24.52}$$

so that finally we have

$$\dot{\varepsilon}^P = \frac{F_b}{F_e}\frac{1}{H}\left([\mathbf{I}] - \frac{1}{H}[\mathbf{C}']\mathbf{S}\mathcal{H}^T\right)^{-1}[\mathbf{C}']\dot{\boldsymbol{\sigma}} = [\mathbf{C}^P]\dot{\boldsymbol{\sigma}} \tag{24.53}$$

$$\dot{\varepsilon}^e = [\mathbf{C}^e]\dot{\boldsymbol{\sigma}}$$

Further analysis proceeds similarly as before. This approach provides also the possibility to account for structural changes within the shear band for large strains, for instance, cracking developed along parallel planes which are subsequently rotated due to progressing shear.

The problem of effective solution procedures with account for softening and localization is still open for further research. Much effort has been spent on determining conditions of initiation of shear bands but much less on providing solution with progression of such bands. The analytical study of shear band propagation was initiated by Palmer and Rice (1973).

24.4 INSTABILITY CONDITIONS AND MODES OF FAILURE

One of most important problems from the practical viewpoint is the prediction of maximal load at failure and the associated failure mode. For a perfectly plastic material model the limit load is a well defined concept as it corresponds to failure mechanism developing under stationary load. The value of limit load does not depend on the initial stress state nor on the deformation history. On the other hand, for materials which exhibit softening or limited ductility with subsequent rupture, this simple concept does not hold. The value of maximal load depends on the initial stress, rate of softening and on elastic material properties. Moreover, the structure may possess several load maxima and once element of structure has failed, the ensuing dynamic process may result in consecutive failure of remaining elements, thus leading to failure mechanism through dynamic mode.

To illustrate these phenomena, consider a simple one-degree of freedom system shown in Figure 24.5(a). Two elements 1 and 2 are connected in series with the mass m attached at their connection. The element 1 is linearly elastic, Figure 24.5(b), whereas the element 2 is nonlinear and inelastic, possessing softening and hardening characteristics shown in Figure 24.5(c). Denoting by u_1, P_1, and u_2, P_2 displacements and forces acting in elements 1 and 2, for the static equilibrium we have

$$P_1 = P_2 = P, \quad u = u_1 + u_2 \tag{24.54}$$

The force–displacement relation between the external force and displacement is

$$P = \frac{K_1 K_2^s}{K_1 + K_2^s} u \tag{24.55}$$

where K_1 is the constant modulus of the element 1 and K_2^s is the secant modulus of the characteristics $P = P(u_2)$ for the element 2. For increments ΔP and Δu from an arbitrary state P_i, u_i, we have

$$\Delta P = \frac{K_1 K_2}{K_1 + K_2} \Delta u \tag{24.56}$$

where K_2 is the tangent modulus of the curve $P = P(u_2)$ of Figure 24.5(b),

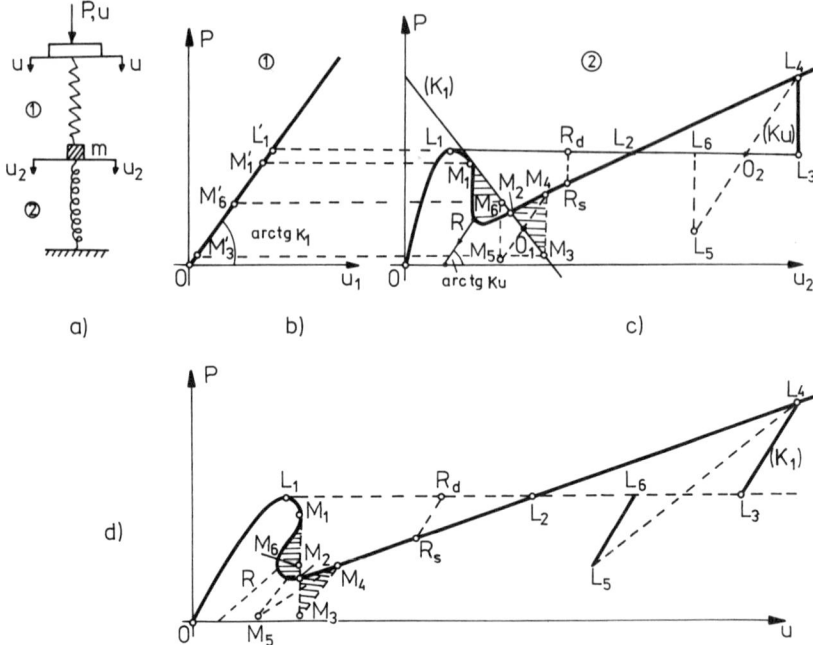

Figure 24.5 System composed of linear and nonlinear softening springs (a) force–deflection curves (b, c), and the resulting P–u curve (d)

at the initial point u_i. It is seen that there is a limit point L_1 for force control where $K_2 = 0$ and $\Delta P/\Delta u = 0$, and the limit point M_1 for displacement control for which $K_2 = -K_1, \Delta P/\Delta u = \infty$. The resulting equilibrium force–displacement diagram for the whole system is shown in Figure 24.5(d). It is seen that at M_1 the static load-controlled process terminates and the dynamic motion commences for constant P with the dynamic path $L_1 - L_2 - L_3$. Similarly, the static displacement controlled process terminates at M_1 and the subsequent dynamic path is $M_1 - M_2 - M_3$.

The dynamic equation of motion takes the form

$$\left(\Delta P - \frac{K_1 K_2}{K_1 + K_2}\Delta u\right) + \frac{m}{K_1 + K_2}(\Delta \ddot{P} - K_1 \Delta \ddot{u}) = 0 \qquad (24.57)$$

where $\Delta \ddot{P}$ and $\Delta \ddot{u}$ are the second time derivatives. For the specified load there is $\Delta P = 0$ and (24.57) provides

$$m\Delta \ddot{u} = -K_2 \Delta u = P - P_2 \qquad (24.58)$$

where P_2 is the force in the element 2 during dynamic motion starting from L_1. Denoting $\Delta \ddot{u} = \dot{v}$ one obtains

$$m\Delta \ddot{u} = m\dot{v} = m\frac{dv}{du_2}\frac{du_2}{dt} = m\frac{dv}{du_2}v = P - P_2 \qquad (24.59)$$

since $\Delta u_1 = 0$, $\Delta u = \Delta u_2$. Integrating (24.59) leads to

$$\frac{mv^2}{2} - \frac{mv_0^2}{2} = \int (P - P_2) du_2 = \int_0^{u_k} (P - P_2) du \qquad (24.60)$$

This equation allows for simple geometric interpretation. Since the right-hand side of (24.60) represents the area between the line L_1–L_2–L_3 and the static curve of Figure 24.5(c), it can be stated that the kinetic energy acquired on $L_1 L_2$ is equal to this area. When $v_0 = 0$, the motion terminates at L_3 for which the areas $L_1 M_2 L_2$ and $L_2 L_3 L_4$ are equal. In Figure 24.5(d) the corresponding static and dynamic points R_s and R_d lie on the line parallel to the $P - u_1$ line in Figure 24.5(b). At L_3 the reverse motion begins. Assuming the unloading stiffness of 2 as equal to K_u, the mass m will execute harmonic oscillations between L_3 and L_6 so that areas $O_2 L_3 L_4$ and $O_2 L_5 L_6$ are equal.

A similar situation occurs for displacement controlled deformation of the system. For fixed displacement at the limit point M_1 there is $\Delta u = \Delta u_1 + \Delta u_2 = 0$ and from (24.57) it follows that

$$m\Delta\ddot{P} + \Delta P(K_1 + K_2) = 0 \qquad (24.61)$$

Equation (24.61) specifies the relaxation process at constant external displacement. This process can be studied in Figure 25.5(c) and (d). In Figure 24.5(c) the deformation of 2 proceeds along the line M_1–M_2–M_3 inclined at $-\tan^{-1}(K_1)$ to the u_2-axis. The motion terminates at M_3 when the areas $M_1 M_2 R$ and $M_2 M_3 M_4$ are equal. The subsequent oscillatory motion occurs between M_3 and M_6. In Figure 24.5(d) the relaxation process proceeds along the vertical line M_1–M_2–M_3.

Figure 24.6 presents the parallel system of two elements of specified piecewise-linear characteristics shown in Figure 24.6(b), (c). Since $u_1 = u_2 = u$, $P = P_1 + P_2$, the static curve is $P = (K_1 + K_2)u$. The behaviour of the system depends on eight parameters $K_1^e, K_2^e, K_1^s, K_2^s, P_1^m, P_2^m$ and P_1^r, P_2^r. Figure 24.6(d), (e),(f) show three static load–deflection curves that are possible for this system. In Figure 24.6(d), after reaching the point 1, the dynamic motion occurs under constant load and the acquired kinetic energy equals the area 1–2–2′. If $E_{k1} > E_{k2}$, the system passes through the segment 2′–3′ where the static equilibrium is possible and produces the failure of the second element. From 3′ the accelerated motion continues. In Figure 24.5(e), the system is characterized by the inequality $E_{k1} < E_{k2}$ and the dynamic path terminates at 3′ with subsequent oscillatory motion around this point. In Figure 24.5(f) the accelerated motion starts at 1 and there is no equilibrium domain on the dynamic path.

It is thus seen that the behaviour of the system depends much on the kinetic energy acquired during the first period of motion and the failure energy of consecutive elements. Though statically higher load P_3 can be obtained in Figure 24.6(d), the ultimate failure occurs at the load level $P = P_1$. Both *single* and *sequential failures* of elements may occur in a dynamic mode.

These examples serve as prototypes for more complex problem of instability

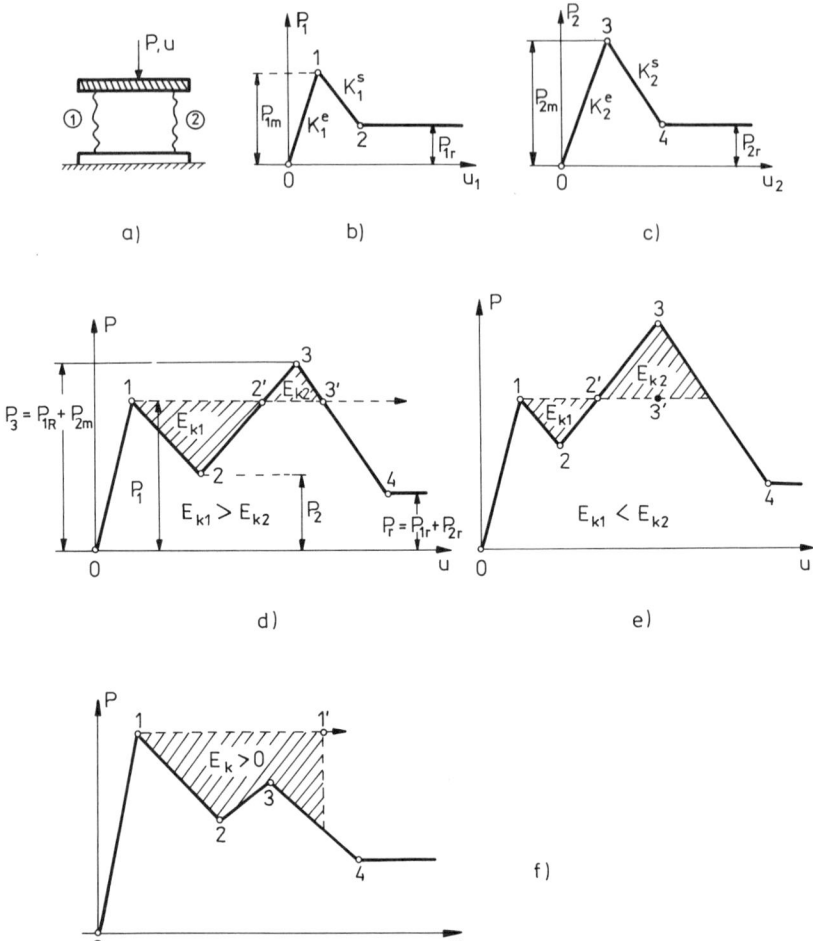

Figure 24.6 System of two elements connected in parallel (a), of hardening and softening characteristics (b, c), and three possible P–u diagrams (d, e, f)

and failure mode within a rock mass surrounding the excavated domain. Consider the domain V occupied by rock and two tunnels, Figure 24.7, with surface tractions and displacements specified on its boundary. Consider the subdomains V_1 and V_2, where the subdomain V_1 is assumed to correspond to stable elastic or elasto-plastic behaviour and the subdomain V_2 is in the post-critical softening state, so that

$$d\boldsymbol{\sigma} \cdot d\boldsymbol{\varepsilon} = d\boldsymbol{\sigma} \cdot d\boldsymbol{\varepsilon}^e + d\boldsymbol{\sigma} \cdot d\boldsymbol{\varepsilon}^p > 0 \quad \text{—stability}$$

$$d\boldsymbol{\sigma} \cdot d\boldsymbol{\varepsilon} = d\boldsymbol{\sigma} \cdot d\boldsymbol{\varepsilon}^e + d\boldsymbol{\sigma} \cdot d\boldsymbol{\varepsilon}^p < 0 \quad \text{—instability}$$

(24.62)

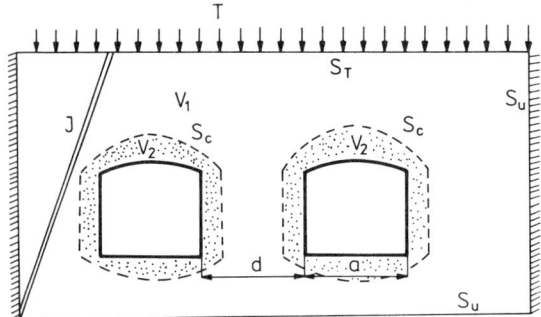

Figure 24.7 Rock domain with two tunnels

and for arbitrary departures from the considered state σ_s there is

$$U(\sigma_T - \sigma_s) = \int_{\varepsilon_s}^{\varepsilon_T} (\sigma - \sigma_s) \cdot d\varepsilon > 0 \text{—stability}$$

$$U(\sigma_T - \sigma_s) = \int_{\varepsilon_s}^{\varepsilon_T} (\sigma - \sigma_s) \cdot d\varepsilon \leqslant 0 \text{—instability}$$

(24.63)

Since $d\sigma \cdot d\varepsilon^e > 0$ in two domains, the following inequalities define hardening and softening

$$\text{hardening:} \quad d\boldsymbol{\sigma} \cdot d\varepsilon^P > 0, \quad \text{softening:} \quad d\boldsymbol{\sigma} \cdot d\varepsilon^P < 0 \quad (24.64)$$

The state presented in Figure 24.7 was attained in a quasi-static process of excavation resulting in stress and strain redistribution from the initial state. This process is here controlled not by external loads or displacements but by geometric parameters, such as a, d, specifying the configuration and size of excavation.

In order to examine stability of a particular equilibrium configuration, let us introduce the disturbance into the system. For instance, the disturbance can be thought of as an additional impulse imposing the velocity field within the domain so that boundary conditions are satisfied. The disturbance can also be conceived as an additional field of body forces, surface tractions, boundary velocities acting over specified period of time or as a sudden removal of the stressed material element. With each disturbance, the amount of energy supplied to the system can be associated and the stability of subsequent motion will usually depend on this energy.

Consider the disturbance in a form of the dynamic impulse superimposed at $t = t_0$ upon the equilibrium state, so that $\dot{\mathbf{u}}^0(x) = \mathbf{v}(x)$ within V at $t = t_0$ and the kinetic energy of this impulse equals

$$\Delta K_0 = \int \tfrac{1}{2}\rho \mathbf{V}^0 \cdot \mathbf{V}^0 dV \qquad (24.65)$$

where ρ denotes the material density. Denote the stress, strain, and displacement fields at $t = t_0$ by σ^0, ε^0, u^0. The subsequent values at $t > t_0$ are

$$\sigma = \sigma^0 + \int_{t_0}^t \dot{\sigma}\,dt, \quad \varepsilon = \varepsilon^0 + \int_{t_0}^t \dot{\varepsilon}\,dt, \quad u = u^0 + \int_{t_0}^t \dot{u}\,d\tau \qquad (24.66)$$

and the subsequent motion is compatible with the boundary conditions on S_T and S_u. The increment of the elastic potential energy $\Delta\Pi_e$ and of the dissipated work ΔD equals

$$\Delta\Pi_t = \Delta\Pi_e + \Delta D = \int\int_{\varepsilon_0}^{\varepsilon} (\sigma - \sigma^0)\cdot d\varepsilon\,dV \qquad (24.67)$$

In view of the principle of conservation of energy, there is

$$\Delta K_0 = \Delta\Pi_t + \Delta K(t) \qquad (24.68)$$

where $\Delta K(t)$ denotes the kinetic energy at any subsequent instant.

The divergence instability occurs when the kinetic energy of imposed motion monotonically grows. To prevent this instability, the sufficient condition is

$$\Delta\Pi_t = \int\int_{\varepsilon_0}^{\varepsilon} (\sigma - \sigma^0)\cdot d\varepsilon\,dV > 0 \qquad (24.69)$$

for any kinematically admissible deformation path issuing from the equilibrium position. In the incremental form (24.69) becomes

$$d\Pi_t = \tfrac{1}{2}\int d\sigma\cdot d\varepsilon^e\,dV + \tfrac{1}{2}\int d\sigma\cdot d\varepsilon^p\,dV > 0 \qquad (24.70)$$

and is obviously satisfied for a stable portion of the stress–strain curve.

Consider now the cyclic motion for which elastic unloading is followed by plastic deformation during successive motion reversals. For the kth reversal, instead of (24.70), we have

$$\Delta\Pi_t^{(k)} = \Delta\Pi_e^{(k)} + \sum_{l=1}^{k} \Delta D^{(l)} = \int U^{(k)}(d\varepsilon^e)\,dV + \tfrac{1}{2}\sum_{l=1}^{k}(d\sigma\cdot d\varepsilon^p)^{(l)}\,dV \qquad (24.71)$$

where $U(d\varepsilon^e) = (\tfrac{1}{2}\mathbf{D}^e\,d\varepsilon^e\cdot d\varepsilon^e)$ is the incremental elastic energy which may grow or decrease during the cyclic motion, whereas the second term represents the accumulated incremental plastic work. Assume that $d\sigma\cdot d\varepsilon^p < 0$ that is the softening response occurs for each incremental plastic flow. Since the maximal value of $\Delta\Pi_t$ equals ΔK_0 then in view of (24.71), the amplitude of elastic strain must grow in consecutive portions of motion, that is $\Delta\Pi_e^{(k-1)} > \Delta\Pi_e^{(k)}$.

Thus there exists the possibility of *cyclic instability* for which the amplitude of superposed motion will grow due to repetitive plastic flow during each cycle. Figures 24.8(a)–(c) present schematically the instability mechanisms on the phase plane $(u, \dot{u})$ where u denotes the displacement at a typical point. It can also be

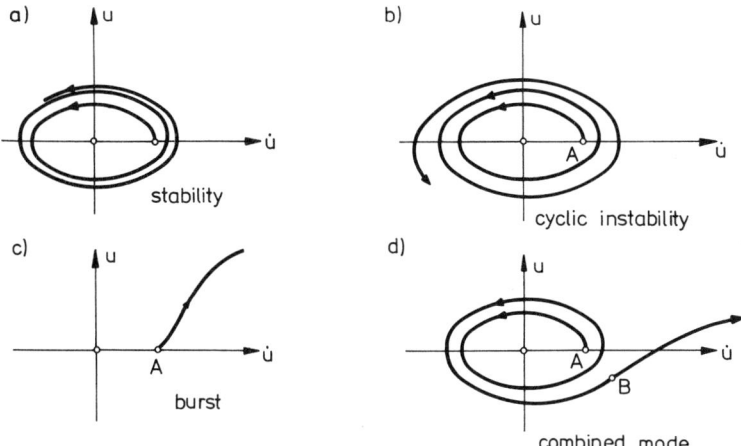

Figure 24.8 Possible stability and instability modes

conceived that a combined mode of instability may occur when after several cycles the condition (24.70) is violated and the subsequent motion ensues with the unbounded kinetic energy.

When joint or fault planes exist within the domain V, the stability condition (24.70) takes the form

$$\tfrac{1}{2}\int d\boldsymbol{\sigma}\cdot d\boldsymbol{\varepsilon}\, dV + \tfrac{1}{2}\int d\tilde{\mathbf{T}}\cdot d\tilde{\mathbf{u}}\, dS_j > 0 \qquad (24.72)$$

where $d\tilde{\mathbf{T}}$, $d\tilde{\mathbf{u}}$ are increments of contact tractions and displacement discontinuity on joints.

Figure 24.9 presents schematically the proposed approach to this class of problems. As the instability point A is difficult to determine, consider the equilibrium state before the limit point A, and impose the dynamic impulse of the kinetic energy $\Delta K_0 > E_1$. Starting from 1 the system will pass dynamically to the instability point 2 with subsequent growth of kinetic energy (Figure 24.9(a)). In Figure 24.9(b) the dynamic motion is initiated by the applied step loading.

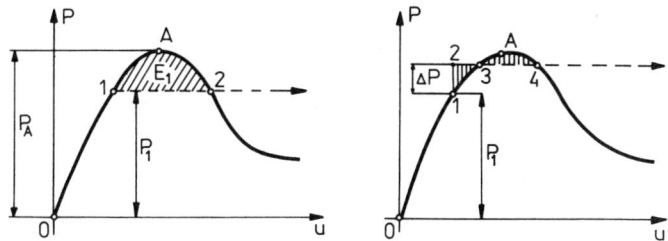

Figure 24.9 Static equilibrium at 1 with superposed impulse or step loading

562   Mechanics of Geomaterials

Assuming this dynamic approach, several examples were solved by Zubelewicz and Mróz (1983). Consider, for example, a horizontal rectangular tunnel within the inhomogeneous rock. The stiff rock is separated from soft rock by a horizontal layer of intermediate stiffness Figure 24.10. Consider the solution domain shown in Figure 24.10 where the initial static state corresponds to non-

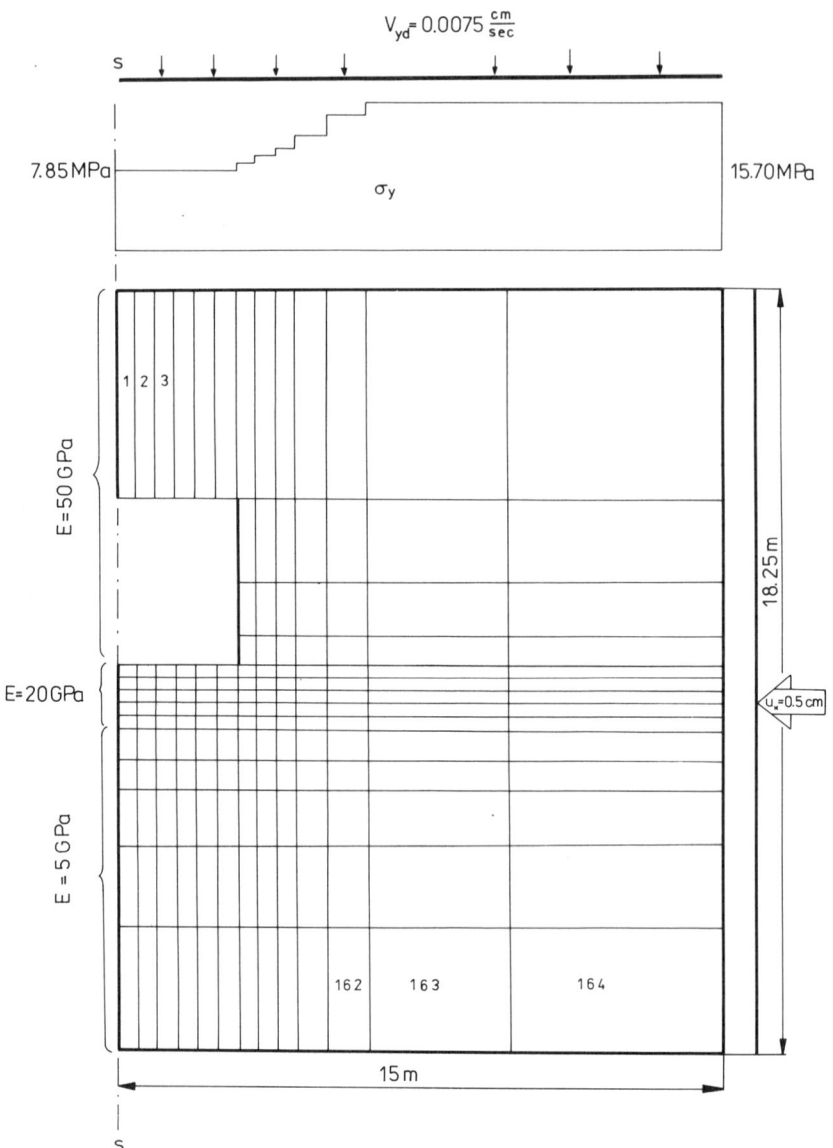

Figure 24.10 Tunnel within inhomogeneous rock

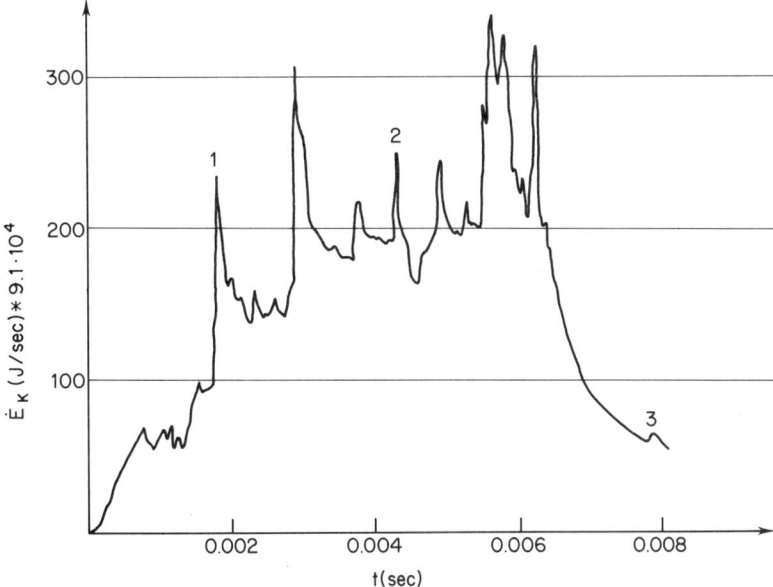

Figure 24.11 Rate of kinetic energy variation in time after imposed impulse

uniform vertical stress σ_y applied at the top boundary and horizontal displacement $u_x = 0.5$ cm at the right vertical boundary, with the bottom boundary constrained. The subsequent dynamic solution exhibits the rupture mode. Figure 24.11 shows the variation of the rate of growth of kinetic energy and Figure 24.12 shows the progression of damaged zones at instants 1, 2, 3 shown in

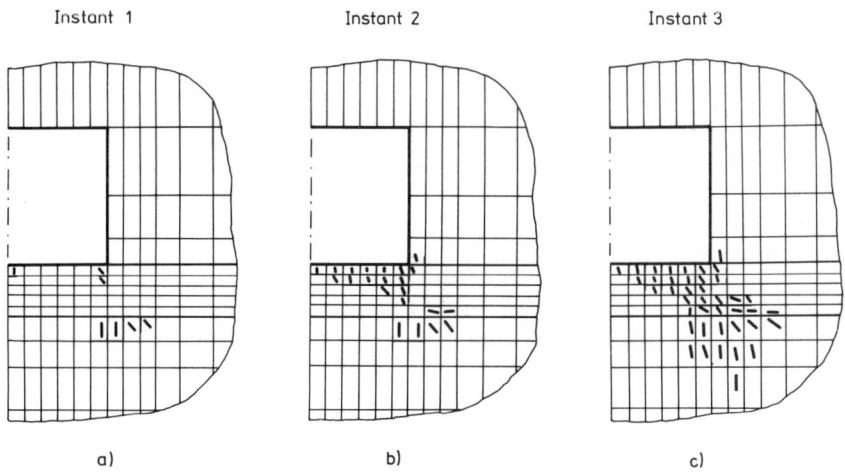

Figure 24.12 Growth of damaged zones near the tunnel

564 Mechanics of Geomaterials

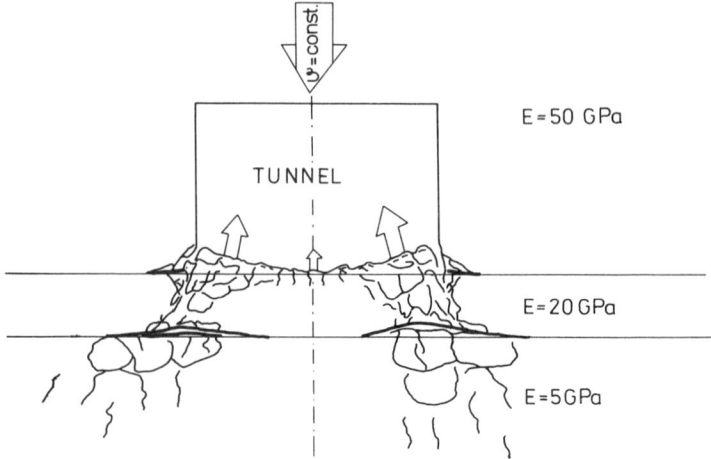

Figure 24.13 Failure mode schematically shown

Figure 24.12. The failure mode is schematically shown in Figure 24.13. During to tensile action, the delamination between the layer and the soft rock occurs with the associated energy release and rupture of the layer. The damaged material is pushed into the tunnel exhibiting the familiar effect of rock burst. The static stability conditions for the excavation were also discussed by Petukchov and Linkov (1979) by considering virtual motion of surface separating the subdomains V_1 and V_2.

24.5 CONCLUDING REMARKS

The current and new problems in geomechanics follow from the scheme of inelastic analysis system shown in Figure 24.1. The future developments should proceed on both sides of the wall, that is in research and in design methods. However, it is clearly seen from Figure 24.1 that the current research effort is not evenly distributed between particular blocks and therefore the applicability of inelastic analysis much depends on further developments in domains not discussed during the present symposium. Let us briefly mention major problems to be solved from the point of view of system performance.

(i) Uniform formulation of constitutive models should be achieved including plasticity, damage, and fracture effects. Both stable and unstable ranges of deformation should be described with account for localization phenomena. As the concept of dynamic failure modes becomes essential, the constitutive description should account for rate dependent deformation and degradation.

(ii) The identification and sensitivity analysis should provide not only model parameters but also assessment of imperfections of actual full-scale systems and sensitivity of model predictions with respect to parameter and imperfection variation. Domain of applicability of the model should be specified by means of laboratory specimen testing and by applying the model to properly selected benchmark problems.

(iii) The fundamental understanding of failure modes in geotechnical systems should be attained through application of inelastic analysis to various problems including static and variable loading with the associated liquefaction effects in soils or dynamic instability effects in rock systems. A new limit state or failure theory is needed in which the limited ductility and softening rate will be accounted for in determining failure modes and critical load factors.

REFERENCES

Bažant, Z. P. (1976) 'Instability, ductility, and size effect in strain softening concrete', *J. Eng. Mech. Div., ASCE*, **102**, EM2, 331–344.

Bažant, Z. P., and Cedolin, L. (1979) 'Blunt crack band propagation in finite element analysis', *J. Eng. Mech. Div., ASCE*, **105**, EM2, 297–315.

Bažant, Z. P., and Kim, S. S. (1979) 'Plastic-fracturing theory for concrete', *J. Eng. Mech. Div., Proc. ASCE*, **105**, 429–446.

Chaboche, J. L., Lemaitre, J., Marquis, D., and Savalle, S. (1981) 'Discussion on problems of models identification,' *Proc. IUTAM Symp. on 'Physical non-linearities in structural analysis'* (Eds. Hult, J. and Lemaitre, J.), Springer Verlag, pp. 37–51.

Cleary, M. P. (1978) 'Fracture discontinuities and structural analysis in resource recovery endeavors', *J. Press. Vess. Petr. Techn.*, **100**, 1–11.

Dougill, J. W. (1976) 'On stable progressively fracturing solids', *Zeitschr. Ang. Math. Phys.*, **27**, 423–437.

Dragon, A., and Mróz, Z. (1979a) 'A continuum model for plastic-brittle behaviour of rock and concrete', *Int. J. Eng. Sci.*, **17**, 121–137.

Dragon, A., and Mróz, Z. (1979b) 'A model for plastic creep of rock-like materials accounting for the kinetics of fracture', *Int. J. Rock Mech. Min. Sci. and Geom. Abstr.*, **16**, 253–259.

Drucker, D. C. (1951) 'A more fundamental approach to plastic stress–strain relations', *Proc. 1st US Nat. Congr. Appl. Mech.*, pp. 487–491.

Eykhoff, P. (1974) *'System Identification, Parameter and State Estimation'*, John Wiley.

Hill, R. (1958) 'A general theory of uniqueness and stability in elastic–plastic solids', *J. Mech. Phys. Solids*, **6**, 236–249.

Ilyushin, A. A. (1964) *Plasticity* (in Russian), Moscow.

Maier, G., and Hueckel, T. (1979) 'Non-associated and coupled flow rules of elastoplasticity', *Int. J. Rock. Mech. Min. Sci. and Geom. Abstr.*, **16**, 77–92.

Mróz, Z. (1963) 'Non-associated flow laws in plasticity', *J. de Mecanique*, **2**, 21–42.

Mróz, Z. (1966) 'On forms of constitutive laws for elastic–plastic solids', *Arch. Mech. Stos.*, **18**, 3–35.

Mróz, Z. (1980) 'On hypoelasticity and plasticity approaches to constitutive modelling of inelastic behaviour of soils', *Int. J. Num. Anal. Meth. Geom.*, **4**, 45–55.

Mróz, Z., and Angelillo, M. (1982) 'Rate-dependent degradation model for concrete and rock', *Proc. Intern. Symp. 'Numerical Models in Geomechanics'* (Ed. Dungar, R. *et al.*), A. A. Balkema Publ., pp. 208–217.

Mróz, Z., and Norris, V. A. (1982) 'Elastoplastic and viscoplastic constitutive models for soils with application to cyclic loading', in *Soil Mechanics–Transient and Cyclic loads* (Eds. Pande, G. N., and Zienkiewicz, O. C.), John Wiley pp. 173–217.

Mróz, Z., and Zienkiewicz, O. C. (1983) 'Uniform formulation of constitutive equations for clays and sands', in *Constitutive Models for Engineering Materials* (Ed. Desai, C. S.), John Wiley.

Palmer, A. C., and Rice, J. R. (1973) 'The growth of slip surfaces in the progressive failure of overconsolidated clay', *Proc. Roy. Soc., Ser. A*, **332**, 527–548.

Petukchov, I. M., and Linkov, A. M. (1979) 'The theory of post-failure deformations and the problem of stability in rock mechanics', *Int. J. Rock Mech. Min. Sci. Geom. Abstr.*, **16**, 57–76.

Pietruszczak, S., and Mróz, Z. (1981) 'Finite element analysis of deformation of strain softening materials', *Int. J. Num. Meth. Eng.*, **17**, 327–334.

Pietruszczak, S., and Mróz, Ž. (1983) 'On hardening anisotropy of K_0-consolidated clays', *Int. J. Num. Anal. Meth. in Geom.*, **7**, 19–38.

Pister, K. S. (1975) 'Some thoughts on the material identification problem,' *Proc. 'Workshop on Applied Thermoviscoplasticity'*, (Ed. S. Nemat-Nasser), Northwestern University Press, Evanston, Illinois.

Prager, W. (1949) 'Recent developments in the mathematical theory of plasticity', *J. Appl. Phys.*, **20**, 235–241.

Prevost, J. H., and Hoeg, K. (1975) 'Soil mechanics and plasticity analysis of strain softening', *Geotechnique*, **25**, 279–297.

Raniecki, B. (1979) 'Uniqueness criteria in solids with non-associated plastic flow laws at finite deformations', *Bull. Ac. Pol. Sci. Ser. Sci. Techn.*, **27**, 391–399.

Rice, J. R. (1976) 'The localization of plastic deformation', *Proc. 14th IUTAM cong. on Theoret. Appl. Mech.* (Ed. Koiter, W. T.), North-Holland Amsterdam, pp. 207–220.

Rudnicki, J. W., and Rice, J. R. (1975) 'Conditions for the localization of deformation in pressure-sensitive dilatant materials', *J. Mech. Phys. Sol.*, **23**, 371–394.

Vermeer, P. A. (1980) 'Formulation and analysis of sand deformation problems', Delft University Techn. Rep. No. 195, Geotechn. Lab.

Zubelewicz, O.C., and Mróz, Z. (1983) 'Numerical simulation of rock-burst processes treated as problems of dynamic instability', *Rock. Mech.* (in press).

APPENDIX

SUMMARIES OF DISCUSSIONS FROM THE SYMPOSIUM

Mechanics of Geomaterials
Edited by Z. Bažant
© 1985 John Wiley & Sons Ltd

A. Summary of Discussions on Constitutive Modelling of Nonlinear Triaxial Behaviour (Part II)

Y. F. Dafalias

Following the discusser's report, Professor E. Juarez-Badillo presented a general compressibility equation under isotropic stresses and showed its applicability to soils and concrete. The equation requires one or two parameters, depending on the choice of a reference volume. For the one-parameter case the value of the parameter may be estimated by an empirical correlation with the liquid limit of a soil medium.

Subsequently, Dr V. Cuellar summarized the endochronic model for soils utilizing Bažant's concept of 'jump' kinematic hardening, which in his version requires thirteen model constants and exhibits a cone type limit condition. By properly changing the values of the model constants he showed an improved fitting of experimental data for soils obtained from the 1982 Grenoble Workshop on 'Soil Constitute Modeling'.

Professor M. Carroll showed some experimental data on the deformation response of rock-type porous materials, where the coupling between deviatoric and hydrostatic stress changes with volumetric and deviatoric strain changes was strongly pronounced. Professor A. Schofield briefly discussed the preceding experimental data within the framework of critical state soil mechanics plasticity. Professor K. Gerstle observed that similar data can be obtained for concrete. Dr W. Wawersik discussed a particular behaviour observed in the post-failure stress–strain curve for porous media, where a creep or relaxation type behaviour appears to be superposed to the inelastic instantaneous response.

Professor J. Rice asked Professor J. W. Dougill if the concrete sample in the softening region presents localized deformations, and to which extent such deformations influence the sample's homogeneity. Professor Dougill answered that the localized deformations do not appear to be strongly pronounced in compression, but the size effect is important and suggested that one should account for the possible localized zones by a continuum description. Professor A. Sawczuk posed the question if an elastic–plastic description without a damage law is sufficient to represent realistically the concrete response, and if isotropic hardening alone is an appropriate assumption since the cracks in concrete are directional and, therefore, induce anisotropy which must be reflected in the stress–strain relations. Professor Dougill answered that damage can be accounted for by changing the stiffness moduli via elastoplastic coupling, and this change must indicate anisotropy.

Professor Z. Bažant brought attention again to the point raised by Professor Rice, that is the localization of deformation shown by concrete around aggregates, and emphasized the necessity to consider more carefully the microstructural features. Professor Z. Mróz briefly presented a scheme where two surfaces are used in the constitutive description, are controlling the fracture process due to crack growth, and the other the plasticity due to frictional slip. Hence, the normality condition is not necessary.

Professor Y. Dafalias commented on the simple fact that if one uses for a macroscopic formulation tensorial structure variables of rank higher than zero in the state functions (yield surface, elastic potential, etc.), the material is anisotropic and the evolution of these variables define the evolution of the anisotropy. If, however, the structure variables (or internal variables) are scalar valued, the material remains istropic with respect to its fabric, but still one can describe an isotropically induced elasto-plastic coupling or damage. He also commented on a necessary modification of the loading–unloading criterion in stress space presented by Professor Dougill, in order to account for a proper description in the softening regime.

Following the short contributions by H. Okamura, H. C. Wu, and F. Darve, Professor Sawczuk asked Professor Wu if the endochronic version he presented exhibits nonzero strain at zero stress and the answer was that zero strain is obtained for zero stress. Professor Dafalias observed that the model presented by Dr Darve cannot account for structural anisotropy because, upon rotation of the loaded configuration, the same strain increment is obtained for the same stress increment as before rotation, since no tensorial variable exists which can account for structural anisotropy. Dr Darve answered that this is true for the simplified version he presented, but in a more general version where stress and strain components are included, such anisotropy can be accounted for. Professor Rice observed that vertex models can accomplish similar description of material response characteristics as the ones embodied in Dr Darve's model. Professor G. Gudehus observed that vertex discontinuities are not good for numerical convergence purposes, but Professor Bažant disagreed, noting that such discontinuities can still be handled is such a way as to obtain convergence. Professor Sawczuk closed the session by observing that the necessary homogeneity of order zero for rate independence imposes certain relations on the state variables.

Mechanics of Geomaterials
Edited by Z. Bažant
© 1985 John Wiley & Sons Ltd

B. Summary of Discussions on Behaviour of Solids with a System of Cracks (Part III)

D. J. Holcomb

Two principal lectures were to be presented in the session on 'Behaviour of Solids with a System of Cracks'. Owing to difficulties with travel from the USSR, Y. V. Zaitsev was unable to attend and present the paper entitled 'Inelastic properties of solids with random cracks'. As scheduled, the paper entitled 'The mechanics of fracture under high-rate stress loading' was presented by D. Grady.

One of the emerging concepts in the study of dynamic bulk fracture is that a measure of the dynamic fracture energy must be included in a successful theory. B. Atkinson asked what method was preferred for measuring the fracture energy. Grady replied that one commonly used method, the Hopkinson bar, was not completely satisfactory because of geometry problems. He suggested that results derived from spalling of plates were better, but not perfect.

In reply to a question from J. Rice, Grady stated that the initial flaw distribution was unimportant in determining the final fragment size distribution. Most materials have so many potential crack sites, that there are always enough to be activated by the loading stress.

There was a lengthy and inconclusive discussion on the number of parameters that it was appropriate and useful to include in a constitutive model. The questioner seemed to imply that there was some fixed number of parameters that could be defined in principle. As pointed out by D. Drucker, even metals which are simpler than geomaterials require new parameters for each new type of test. Several people made the point that for practical applications only a very few parameters can or will be used. Thus there is a responsibility on the part of those developing complex general models to derive the simple special cases that are likely to be of widespread use. In the opening address, Professor Drucker described Prager's approach to presenting new ideas. Instead of presenting the full complexity, a simple, practical problem would be solved, with more complexity introduced as audience understanding grew. The value of this approach was pointed out in the discussion of the paper by R. B. Stout dealing with the thermodynamics of crack propagation. A very general theory was presented, but the generality made it difficult to compare with existing work. In response to a question from J. Rice, Stout stated that the theory had not been applied to the case of a single crack in tension, which would make comparison with other work quite simple.

A brief presentation was made by K. Gerstle and H. Y. Ko, to announce the availability of their report on multiaxial stress testing of concrete following various load paths. In particular, paths were followed where stress invariants were held constant. The report gives stress and strain data in graph and tabular form. It should be useful to anyone investigating path dependent effects. The report may be obtained by writing to them at the address given for Gerstle in the list of participants.

In describing brittle geomaterials where deformation and failure are due to the growth and coalescence of cracks, the concept of damage is useful. Damage is a rather vague concept in most formulations, including such ideas as energy loss during cyclic loading, degradation of elastic moduli, crack density, crack size, and possibly orientation. J. Dougill was asked by D. Krajcinovic if it was possible to distinguish between strain that was plastic and strain due to damage. Dougill replied that the division was arbitrary, an opinion with which Z. P. Bažant concurred. D. Krajcinovic suggested that if a cyclic test were performed and the unloading was an elastic process with unrecovered strain, then the strain could be considered plastic. In general a cyclic test results in hysteresis and unrecovered strain with no clear way to attribute strain to damage or plasticity. In his presentation, J. Dougill suggested a measure of strain energy for quantifying damage. Damage is produced only during the first cycle along a given stress path. Thus the energy absorbed during the first cycle includes energy required to produce damage. Energy absorbed in subsequent cycles to the same stress state is due to some other non-damaging process. Thus Dougill proposed that the difference between energy absorbed in the first cycle and energy absorbed in the second cycle was a measure of the damage produced in the first cycle.

A fundamental question, first raised at this meeting by L. Margolin, is whether the formalism of plasticity is really appropriate to the study of brittle geomaterials. In general these materials do not exhibit a yield surface and there is usually no stress regime where they are elastic. An alternative is to change from a strain-based formalism to one based on damage. In order to begin such a program the current ideas of damage must be made clearer. As a first step, methods of quantifying damage and detecting when damage is being generated are necessary. The energy method proposed by Dougill is suitable for quantifying damage. For detecting damage, one possibility is the use of acoustic emissions, which (in rock and concrete) are the bursts of high frequency elastic waves emitted by a local failure such as a micro-crack propagating or pores collapsing. In many metals, it is observed that acoustic emissions are detected only during the first loading to a given stress state, a phenomenon known as the Kaiser effect. A similar effect occurs in rock and concrete under compressional loading (Kurita and Fujii, 1979; Holcomb, 1983; Spooner and Dougill, 1975). If the previously applied stress state is exceeded, then acoustic emission events are again observed. Because acoustic emissions are associated with cracking and collapse, it is intuitively satisfying to identify states that produce acoustic emissions with states where damage is increasing.

As a guide in constructing a constitutive theory based on damage, it may be enlightening to make an analogy between plasticity and damage. Plastic strain is replaced by a measure of damage such as the energy absorbed. Yield is replaced by the production of acoustic emissions instead of plastic strain. Yield surfaces are replaced by damage surfaces which are defined as the locus of stress states where acoustic emissions begin to be produced. The location of the damage surface for a given sample depends on the history of the sample. Within the damage surface, material behaviour is not elastic but no further damage is produced as evidenced by the lack of acoustic emissions and the repeatable hysteresis loops generated. Such damage surfaces have been mapped for stress states attainable in a triaxial test frame (Holcomb, 1983). For the simple stress paths available, clearly defined damage surfaces could be located using acoustic emissions as an indicator.

REFERENCES

Kurita, K. and Fujii, N. (1979) 'Stress memory of crystalline rock in acoustic emission', *Geophys. Res. Lett.*, **6**, 9–12.

Spooner, D. C., and Dougill, J. W. (1975) 'A quantitative assessment of damage sustained in concrete during compressive loading', *Magazine of Concrete Research*, **27**, 92, 151–160.

Holcomb, D. J. (1983) 'Using acoustic emissions to determine in-situ stress: problems and promise', *Geomechanics* (Ed. S. Nemat–Nasser) AMD-Vol. 57, Pub. ASME.

C. Summary of Discussions on Shear Localization, Faulting, and Frictional Slip (Part IV)

L. W. Teufel

Two principal lectures were presented in the session on 'Shear localization, faulting, and frictional slip'. A. L. Ruina presented a paper on 'Constitutive relations for frictional slip' and T. F. Wong presented a paper co-authored by B. Evans on 'Shear localization in rocks induced by tectonic deformation'. These papers were discussed in a formal presentation by J. R. Rice.

Ruina was asked if machine stiffness affected the velocity dependence of rock friction or the observation that the frictional force changes its value over a characteristic distance when the steady state slip velocity is changed from one velocity to another. Ruina briefly described his servo-controlled experimental apparatus, noting that in his experiments the displacements and velocity were measured directly on the specimen and that the displacement signals were used for control, and therefore, stiffness depended only on the efficiency of the servo-control system. Accordingly, in his experiments the stiffness of the machine did not have an effect on the velocity dependence of friction or the displacement required for the frictional force to change during incremental velocity changes. This characteristic displacement has been observed to be independent of the magnitude of the incremental velocity change and the normal stress, but was found to be dependent on the roughness of the sliding surface. Experimental observations by Dieterich (1978) have shown that this characteristic displacement increases with increasing surface roughness, suggesting that it reflects the distance required to change one asperity contact population to a new population. Using the concept of a characteristic displacement and time-dependent friction Dieterich has attempted to explain the relationship between dynamic and static friction.

For friction experiments designed to investigate instability of quasi-static slip motion, machine stiffness, coupled with the effect of the applied normal stress across the sliding surface, are critical parameters which influence the transition from stable to unstable sliding. An analysis of a one-dimensional spring and slider system by Dieterich (1978) supports experimental observations establishing that the transition from stable to unstable sliding is a consequence of time-dependent friction and is a function of normal stress, stiffness, and surface finish of the sliding surfaces. Stable sliding is favoured by conditions of high stiffness and low normal stress. Observations of frictional instabilities in rock appear to be analogous with those in metals.

Rice was queried by Bažant about the application of shear localization calculations to earthquake prediction studies. Bažant noted that in concrete structure work there is a need to determine the displacement distribution within the fracture or softening zone. Rice pointed out that in the context of large-scale tectonic faulting there is also interest in the localized zone where softening might take place. Work by Rudnicki (1977) on strain weaking zones during fault initiation is an example. In this study Rudnicki developed a model for analysing earthquake instabilities, in which a limited zone of rock is assumed to be stressed into the inelastic and, ultimately, strain weakening regime while the surrounding rock mass remains elastic and is subjected to steadily increasing remote stress. The concept of a localized strain weakening zone developing during the earthquake process is supported by field observations which show the occurrence of small magnitude seismicity along a fault preceding and following some major earthquakes.

The influence of grain size and fabric anisotropy on rock deformation was raised by several participants. Wong was asked if there was experimental observations available on the effect of grain size and anisotropy on shear band formation and if it was possible to incorporate these parameters into a corresponding constitutive law. Wong indicated that he had not studied the effect of grain size nor anisotropic rocks. However, he noted that it is well known that deformation behaviour of rock is highly depended on the grain size and the orientation of the fabric anisotropy with respect to the principal stress directions. The influence grain size and anisotropy on onset of yield, failure strength and elastic compliance has been briefly reviewed by Paterson (1978).

With regard to including the effect of anisotropy into constitutive laws Rudnicki commented that he had studied an incrementally-linear, transversely isotropic solid and made calculations similar to Rice. His studies were done to simulate observations of compression experiments which show that cracks grow vertically, so even if one has an anistropic structure, it is essentially axisymmetric. Predictions from this model correlate better than isotropic models with experimental observations, but he concluded that this study was somewhat inconclusive because of a lack of incremental parameters for the models that were well determined in the laboratory.

REFERENCES

Dieterich, J. H. (1978) 'Time-dependent friction and the mechanics of stick–slip', *Pure and Applied Geophysics*, **116**, 790–806.

Paterson, M. S. (1978) *Experimental Rock Deformation: The Brittle Field*, Springer-Verlag, New York, 254 pages.

Rudnicki, J. W. (1977) 'The inception of faulting in a rock mass with a weakened zone', *Jour. Geophys. Res.*, **82**, 844–854.

Mechanics of Geomaterials
Edited by Z. Bažant
© 1985 John Wiley & Sons Ltd

D. Summary of Discussions on Fracture Propagation and Fracture Energy (Part V)

F. Ouchterlony

The presentations in this session were largely concerned with various crack models and their usefulness in the fracture propagation prediction in structures of rock and concrete. So was all of the discussion. The complexity of an appropriate crack model depends on the relative size of the structure e.g. expressed as the relation of crack length a to a characteristic length l_{ch} of the material. Since l_{ch} roughly is of the order 0.01–0.1 m for rock and 0.1–1.0 m or even more for concrete, a rock structure of given absolute size may need a different crack model than one of concrete.

Sharp crack models (SCM) are useful when $a \gg l_{ch}$. This is the realm of linear elastic fracture mechanics (LEFM) as well as anisotropic and global nonlinear behaviour, see A. Ingraffea's and M. Cleary's presentations. For LEFM of metals ASTM sets the lower bound $a_{min} \geq 2.5\, l_{ch}$, for concrete it may be as much as $10\, l_{ch}$ (Hillerborg, 1983). The SCMs have been embedded in numerical codes based on finite element methods (FEM) or boundary element methods (BEM), enhanced by interactive graphics and remeshing to incorporate mixed mode and curved fracture propagation (see Ingraffea), or coupled hybrid methods with the same capacity that obviate the remeshing (see Cleary). Both approaches are now being extended to three-dimensional problems.

A SCM typically has a single parameter that determines the onset and propagation of fractures, such as the fracture toughness in LEFM or a critical J-integral value. Though reported on for rock by Ingraffea, fracture toughness testing was not discussed directly during this session.

When $a \simeq l_{ch}$ multi-parameter crack models become necessary to explain R-curve effects etc. that result when the nonlinearities contained in the fracture process zone have an appreciable influence. Several such two-dimensional models for mode I propagation were presented during this session, primarily the fictitious crack model (FCM) by A. Hillerborg (Hillerborg et al., 1976) and the blunt crack band model (BCBM) by Z. Bažant (Bažant and Cedolin, 1979).

The BCBM has much in common with the continuum damage modelling (CDM) briefly touched on in an earlier session by A. Ehrlacher (Bui and Ehrlacher, 1981). All of these are cohesive or softening zone models with either gradual or abrupt decohesion, alternatively brutal damaging. The FCM is a discrete inter-element crack model, in the other two the cracking or damage is smeared out over a finite width.

The main discussion grew into a thorough comparison of the FCM with the BCBM. Apart from elastic properties such as the modulus E and Poisson's ratio v, the FCM contains two fracture parameters, the fracture energy G_F and either the tensile strength f_t or the maximum crack opening displacement w_1 when a linear decohesion law $\sigma(w) = f_t(1 - w/w_1)$ is assumed. The length of the fracture process zone becomes a result of the calculation. In the FCM l_{ch} is defined as

$$l_{ch} = EG_F/f_t^2$$

According to Hillerborg it has no direct physical meaning but may of course be interpreted as a qualitative measure of the process zone length.

The BCBM contains three fracture parameters, the fracture energy G_f, f_t, and the modulus E_t of the linear softening law. This defines the width of the crack band as

$$w_c = 2EG_f/[f_t^2(1 - E/E_t)]$$

which thus is a material parameter too. Likewise in CDM a mode I damage zone has the characteristic width

$$w_c = 2EG/[f_t^2(1 - v^2)]$$

in which G is the fracture energy rate dissipated through the damage front.

The question about the physical meaning of w_c was answered by Bažant who related it to the mesostructure of the material. For concrete w_c is 1.5 to 6 times the maximum aggregate size with a flat optimum around three. For rock optimum w_c lies around five times the grain size. The BCBM does not work as well with mortar though.

For computational work both models are implemented in FEM codes. Other questions were how to obtain mesh independent results. With the SCM and the FCM this is achieved by a comparison with results from increasingly finer mesh sizes.

In using the BCBM, an element size of w_c may be impracticably small. Mesh independent results are then obtained by the artifice of relating the fracture work density W_f in the crack band to the actual mesh size h so that the material parameter $G_f = h \cdot W_f$ is constant. Since $W_f = \int \sigma(\varepsilon) d\varepsilon$ the area under the actual uniaxial stress–strain curve must be scaled in proportion to w_c/h. Usually this is done by scaling $f_t \propto 1/\sqrt{h}$ and keeping E_t constant, i.e. scaling size but not shape of the $\sigma(\varepsilon)$ curve. The converse, scaling E_t and keeping f_t constant has also been tried with insignificant differences in the results.

According to Bažant, the FCM and the BCBM yield identical results for all practical purposes. J. Rice stated that their general consistency can also be established by a more fundamental approach. It is well known that for the FCM, like for any cohesive line zone model, the J-integral yields the potential energy release rate—dU/da and a measure of the cohesive forces as well. When the evaluation path is shrunk onto the process zone only the traction term contributes. It yields $J_{tip} = G_f$, the area under the decohesion curve $\sigma(w)$.

The J-integral also yields $-dU/da$ between two infinitesimally neighbouring states of stress free and smooth notches (Rice, 1968). Regarding notch growth as a change of state rather than material removal, this describes fracture propagation in the BCBM with abrupt softening and in the CDM with brutal damaging. When the evaluation path in shrunk onto the process zone this time, the traction term in J vanishes. The remaining strain energy density terms yields $J_{tip} = G_f = w_c \cdot W_f$ with W_f being the area under a $\sigma(\varepsilon)$ curve with abrupt vertical stress drop once f_t is reached ($E_t \to \infty$).

The corresponding discrete formulation for ΔU is entirely equivalent of course. Rice suggested that its volume term could be used to demonstrate that W_f equals the total area under a $\sigma(\varepsilon)$ curve with gradual softening but that then the energy from stresses parallel to the crack band would have to be negligible. The most general BCBM description does, however, include such a residual stress. A recent generalized J-integral analysis of fracture propagation in dilatant transforming materials (Budiansky *et al.*, 1983) might perhaps be extended to cover this case.

Thus the FCM and the BCBM indeed seem to be generally consistent if G_F equals G_f, but they differ somewhat in their residual strains. Both rationalize $\sigma(w)$ and the softening part of $\sigma(\varepsilon)$ respectively from the uniaxial tensile stress test, taking into account stability conditions for strain localization, machine stiffness, and specimen length. Both have been modified successfully to account for fibre reinforcement in concrete. The FCM essentially requires mesh colinear fracture propagation at present. The BCBM does not; it can describe skew cracking as zigzag growth. That the FCM requires no assumption about crack bandwidth is of minor importance since w_c and l_{ch} obviously are equivalent.

Both the FMC and the BCBM can predict R-curves previously measured on concrete test specimens. Since these curves are specimen type and size dependent they are not material properties. L. Keer presented some new applications of a recent FCM version (Wecharatana and Shah, 1983) in which K_I at the process zone tip is not necessarily zero. S. Shah stated that they use the optically observed crack width in the post-peak region of the uniaxial tensile test as the inelastic deformation in the decohesion law and that hence it includes the unloading of the side material.

Fracture propagation prediction under mixed mode loading when $a \simeq l_{ch}$ was also discussed. The microplane model was offered as a practical alternative by Bažant. It is relatively easy to evaluate and it is not in conflict with tensorial invariance restrictions.

The final contribution to the discussion was made by M. Braestrup. In it he challenged the audience to come up with models that allow multiple crack systems to develop so that a realistic description can be reached of the ultimate state in an orthogonally reinforced concrete panel subjected to shear, for example. The details are given in a separate contribution.

REFERENCES

Bažant, Z. P., and Cedolin, L. (1979) 'Blunt crack band propagation in finite element analysis', *J. Engng. Mechs. Div. ASCE*, **105**, EM2, 297–315.
Budiansky, B., Hutchinson, J. W., and Lambropoulos, J. C. (1983) 'Continuum theory of dilatant transformation toughening in ceramics', *Int. J. Solids Structures*, **19**, 337–355.
Bui, H. D., and Erlacher, A. (1981) 'Propagation of damage in elastic and plastic solids', in *Advances in Fracture Research*, Vol. 2 (Proc. 5th Intnl. Conf. Fracture, ICF5, Eds. D. Francois *et al.*), Pergamon, Oxford, pp. 533–551.
Hillerborg, A., Modéer, M. and Peterson, P. E. (1976) 'Analysis of crack formation and crack growth in concrete by means of fracture mechanics and finite elements', *Cement and Concrete Research*, **6**, 773–782.
Hillerborg, A. (1983) 'Numerical methods to simulate softening and fracture of concrete', in *Fracture Mechanics Applied to Concrete Structures* (Ed. G. C. Sih), Martinus Nijhoff Publishers, The Hague (in press).
Rice, J. R. (1968) 'Mathematical analysis in the mechanics of fracture', in *Fracture, An Advanced Treatise*, Vol. 2 (Ed. H. Liebowitz), Academic Press, New York, pp. 191–311.
Wecharatana, M., and Shah, S. P. (1983) 'Predictions of nonlinear fracture process zone in concrete', to appear in *Proc. ASCE Engng. Mechs. Div.*

Mechanics of Geomaterials
Edited by Z. Bažant
© 1985 John Wiley & Sons Ltd

E. Summary of Discussions on Fluid-Infiltrated Geomaterials (Part VI)

S. C. Cowin

The two principal lectures were delivered, but the discussers' report was not because V. N. Nikolaevsky had difficulty in travelling from the USSR. Also, the announced chairman, B. Budianski, could not attend and O. C. Zienkiewicz served as chairman of the session.

The discussion session opened with a brief presentation by M. P. Cleary. A summary of this presentation, prepared after the meeting by Cleary, is quoted below:

'To crystallize and focus discussion on this long-standing issue, I would like to suggest for debate and long-term testing the following
THEOREM: There is not any physically verifiable or irrefutable phenomenology which can be predicted by formal mixture theory of multiple interacting continua and which is not implied by rational extensions of a physically based formulation (e.g. that of Biot)'.

Although this theorem may seem controversial, it is almost a truism. It may be supported by noting the following central points:

1. Prediction of any physical phenomenon should always be achievable with a sound physically based theory; any new observations should be easily incorporated by a natural extension of the theory, otherwise the whole formulation must be fundamentally in error.

2. Comparison of theory to experiment requires use of physically measurable quantities in the formulation; Biot theory may have to be rephrased (e.g. *à la* Rice and Cleary) to achieve this, but mixture theories often prove impossible to rationalize in such a way.

3. Mixture theory is purely a mathematical formalism, expressing all possible relations between measures defined on the basis of a continuum theory for each component—even though the individual phases will probably not act physically as such isolated/interacting continua. Even after imposition of 'thermodynamic' and form/invariant/symmetry arguments, this theory may still seem to predict more general results than the simplest physical theory suggests. However, there are only two natural resulting possibilities:

—the predictions involved are not relevant, the case of 'zero coefficients', or
—the physical theory must be extended and should have been realized in this

form from the outset, if the corresponding thought experiment had been employed in its development.

Thus, as an analogy, mixture theory is to multiphase materials what functions of several variables are to classical thermodynamics: both provide a guide as to the structure that the theory can take, but only the physics allow a real determination of the relevant variables and forms. There is nothing wrong with a formal mixture theory approach, provided it is phrased in terms of measurable variables and subjected to all the known physical laws; actual measurements then allow determination of the coefficients, as against some vague suggestions that certain unobserved behaviour may be possible, and the overall result could equivalently be extracted by an intelligent development of a physically based theory.

This presentation generated a discussion which involved M. M. Carroll, M. P. Cleary, S. C. Cowin, K. S. Pister, J. H. Prevost and J. R. Rice. There was general agreement that mixture theory had not demonstrated any advantage over the Biot theory for fluid saturated porous linearly elastic solids. However, it was pointed out that mixture theory had as its objective the formulation of a general framework for developing models of multiphase mixtures and the fact that it could not improve upon the first prototype of such a theory, namely the Biot theory, did not mean that further development of mixture theory should be abandoned. Perhaps the proper approach to the criticism of mixture theory is to evaluate and assess the axioms upon which it is based.

C. A. Anderson asked about the state of numerical calculations for the Biot theories and J. W. Rudnicki indicated that they were still in an evolutionary stage because certain physical constants had not yet been measured.

Mechanics of Geomaterials
Edited by Z. Bažant
© 1985 John Wiley & Sons Ltd

F. Summary of Discussions on Creep, Shrinkage, and Ageing (Part VII)

L. Cedolin

Two principal lectures were presented in the session on 'Creep, shrinkage, and ageing'. The lecture given by C. A. Anderson, entitled 'Creep and thermal effects in ageing solids', treated topics ranging from the creep analysis of ageing concrete structures to the description of a numerical model of the interaction of the mantle convention with the Earth's crust. During the discussion, D. C. Drucker asked if the dimensions of the convection cells adopted in the numerical study had been chosen after a preliminary study of a model of the whole earth globe, and O. C. Zienkiewicz asked if the viscoelastic properties of the earth crust had been taken into account. Anderson, after pointing out the complexity of the problem, stated that these aspects were part of planned future developments of the work.

The lecture given by F. H. Wittman, entitled 'Deformation of concrete at variable moisture content', illustrated the role of the various mechanisms which are responsible for the time-dependent behaviour of concrete. During the discussion, Wittman emphasized the fact that in the deformation of concrete specimens the separate effects of creep and shrinkage cannot be added, because shrinkage alone is accompanied by cracking due to differential strains, while this cracking may be eliminated by the presence of a compressive stress due to external loading. This stress, however, does not have any direct influence on unrestrained shrinkage, with the exception of a short-time reaction (30 to 60 s) due to the mechanical squeezing out of the water near the specimen's walls. Wittman stated that the stress level considered is such that extensive cracking is not caused by the load, i.e. the stress is less than 50 per cent of ultimate.

In reply to a question from M. P. Cleary, Wittman confirmed that there are experimental investigations on the stress distribution due to shrinkage, which utilize instrumented artificial aggregates or steel bars. J. R. Rice asked if shrinkage induces plastic deformations. Wittman answered that if one considers the process of drying of a specimen from 100 per cent to 50 per cent of relative humidity, cracking occurs both in the matrix and at the interfaces with aggregate. In reply to a question by Drucker, Wittman stated that the experimental evidence available so far indicates that unrestrained shrinkage of the cement paste is not influenced by the stress state. O. Buyukozturk pointed out the need for three-dimensional numerical calculations of the stress state due to shrinkage; Rice asked if some parallelism may be found between shrinkage stresses in concrete and a mechanism which exists in steel, where the nucleation of carbides produces a greater volume and consequently a smaller stress. Wittman replied that stresses in concrete are due to the geometry of the specimen and not to a migration process.

Z. P. Bažant essentially agrees with the remarks of Wittman on the effect of drying creep, which were adopted also in a previous work (Bažant and Wu, 1974). He observes, however, that in order for this explanation to work over a range of time delays (1 day to 10,000 days) it is necessary to consider gradual strain softening instead of a sudden stress drop, and, even more importantly, to take into account the irreversibility of deformation after strain softening. H. W. Reinhardt pointed out that the same difference in behaviour as that produced by moisture gradient in loaded and unloaded specimen may be also produced by thermal gradient. Wittman agreed, noting the advantage of a much quicker evolution of the process toward equilibrium, and he mentioned the fact that he is in the process of actually performing such test.

In his discusser's report, K. S. Pister analysed the concepts of shrinkage, creep, and ageing from the point of view of continuum mechanics. He pointed out that the process of determining the constitutive relation of concrete involves the definition of the characteristic length of an element of material capable of representing the physical behaviour, and it consists in an averaging process and in a passage to the limit in order to obtain a point function. This remark on the problem of the size effect was followed by an extended discussion. Rice pointed out that for concrete the dimensions of the material element may be comparable with structural dimensions, so that the definition of test specimen may be invalid. I. Sandler focused on the difference between material inhomogeneity and effect of high gradients of deformation. A given material, although inhomogeneous at a certain scale, may be analysed as homogeneous on a larger scale. Strain-softening, however, is a situation in which any homogeneous approach, even if a size effect is included, cannot any more control the real behaviour. At peak stress inhomogeneities increase their effect, and they play the same role as perturbations in instability. One cannot distinguish any longer between nonhomogeneities due to the material or due to gradient of deformation. Bažant pointed out the complexity of this approach. Sandler explained that the type of constitutive relation adopted in a certain application depends on many factors, and principally the objectivity requirement, capability of description, information needed, resources available, etc. Carrol expressed the opinion that the representative volume element should be the laboratory sample and that one should not use micropolar analysis or solve boundary value problems in order to interpret experiments. S. Sture pointed out that this is not always possible and cited the case of granular loose soils, for which the presence of the body force in large specimens requires the solution of the inverse problem. K. Willam observed that in the computational procedure the kinematic relations are deduced from the continuum theory and are separated from the constitutive relations. The latter, however, must be spread over a certain zone, even if in reality the system may be highly discontinuous. Sandler remarked that this is possible only if one has previously characterized the disturbances, since strain-localization is produced by inhomogeneity due to material properties and boundary conditions.

The session ended with two short contributions. S. C. Cowin presented a paper entitled 'Mechanical modeling of the shrinking and swelling of porous solids' in which he used the theory of elastic materials with voids. In reply to a question by Rice, Cowin explained that he did not discuss in the results presented the effect of the coupling with the diffusion of moisture through the medium. H. W. Reinhardt presented a paper entitled 'Plain concrete in uniaxial post-peak cyclic tensile-compressive loading' in which the crack propagation in concrete is assumed as basis for studying the fatigue behaviour. Bažant pointed out the importance of including the consideration of creep and of the increase of the size of the softening zone produced by the successive cycles.

REFERENCES

Bažant, Z. P., and Wu, S. T. (1974) 'Creep and shrinkage law for concrete at variable humidity', *Journal of the Engineering Mechanics Division, ASCE*, **100**, 1183–1209.

Mechanics of Geomaterials
Edited by Z. Bažant
© 1985 John Wiley & Sons Ltd

G. Summary of Discussions on Numerical Modelling (Part VIII)

J. H. Prevost

The discussions are summarized in the form of abbreviated principal questions and answers by symposium participants.

DISCUSSIONS OF 'NUMERICAL MODELLING AND GEOMECHANICS' BY O. C. ZIENKIEWICZ (CHAPTER 21)

K. Pister. Questioned whether it is wise to only implement and consider simple (but not simpler) models. He pointed out that retrospective predictions allow tuning of a model but also exhaust data. Emphasized that it is better to start with a complex model and simplify it through a 'consistent reduction of theory'. His advice is that one should not use a model so simple that one cannot get any predictive capabilities from it.

O. C. Zienkiewicz. Agreed very much with K. Pister but pointed out that 'cheating' or 'tuning' of a model cannot usually be done since it is too expensive. He pointed out that the dam calculations reported in his presentation had only been done once. Also, he pointed out that SWANSEA has been working on complex models through its collaboration with Z. Mróz and others.

J. Rice. Asked whether the finite element model used for the dam (San Fernando) could capture localization of deformations and subsequent failure mechanisms as observed in the dam failure.

O. C. Zienkiewicz. Yes, answered Dr Zienkiewicz, and referred to a figure in his latest book which illustrates the failure of a slope.

DISCUSSIONS OF 'NUMERICAL MODELS FOR DYNAMIC LOADING' BY I. SANDLER, AND M. BARON (CHAPTER 22)

Z. Mróz. Remarked on the continuity requirement. Pointed out that some models do not satisfy that continuity requirement and pointed out that this is dangerous since the solution may drift away. Followed earlier comments on model selection. Pointed out that for some problems, all models will give approximately the same answer. However, for other problems, the models will give different answers. Stressed that it is important to find, for each model, the range of its applicability.

Z. P. Bažant. Re-emphasized continuity requirement. However, he pointed out that there are physical cases (evidences) where continuity is not satisfied; e.g. failure mechanisms, fracture, etc. So, one cannot always insist on continuity requirement.

I. Sandler. Yes! But modelling those situations is tricky!

Z. P. Bažant. Agree, but they are important!

I. Sandler. However, the computer does not know how to deal with a lack of continuity! Analysis is too sensitive, and one cannot ascertain the reliability of the calculations in that case.

DISCUSSION FOLLOWING DISCUSSER'S REPORT BY P. G. BERGAN (CHAPTER 23)

O. C. Zienkiewicz. Commented on localization—showed a slide of a slope stability calculation with a crude mesh. Claimed good representation of slip. Mentioned other attempts to capture localization phenomena using special slip elements and stiff elements.

J. R. Rice. Emphasized that a stress versus displacement relation on the slip surface is more appropriate. Pointed out to some potential difficulties in strain-softening systems.

J. H. Prevost. Emphasized that 'reasonable' meshes do capture localization adequately.

P. Bergan. How about post-localization and resulting failure mechanisms?

J. H. Prevost. After localization, the stress–strain equations may not be appropriate in the band (Ref. Rice, previous comment). Furthermore, equations changed regime (elliptic to hyperbolic), so solution alogrithm is no more appropriate, in general.

R. Galloway. Made a short presentation on a practical example.

GENERAL DISCUSSION

K. Willam. Emphasized Sandler's remark that a sound constitutive model is essential, and pointed out that implementation details also play an essential role in successful calculations.

Z. P. Bažant. Brought back the issue of continuity requirements, and pointed out that successful calculations have been reported with endocronic models which violate continuity.

R. B. Stout. Short presentation on dislocation model.

J. R. Rice. Criticized and strongly objected to Dr Stout's arguments.

Appendix

O. C. Zienkiewicz. Pointed out that automatic mesh refinement procedures may help in capturing discontinuous (localized) solutions.

C. S. Desai. Pointed out the great difficulties encountered in modelling accurately interface conditions. Also mentioned that numerical modelling of well-defined experiments (boundary value problems) may help in developing better constitutive models.

DISCUSSIONS OF J. BLAAUWENDRAAD'S SHORT CONTRIBUTION ON THE USE OF THE ROUGH CRACK MODEL OF WALRAVEN AND THE FICTITIOUS CRACK MODEL OF HILLERBORG IN FINITE ELEMENT ANALYSIS

Z. P. Bažant. Commented that Budianski slip model can provide a unified approach to cracks (rough and fictitious models).

H. Madsen. Asked whether the aggregates or the reinforcement (or both) were modelled to scale in the experiments.

J. Blaauwendraad. The concrete was the same, only the reinforcement was scaled.

S. G. Lee. Asked whether it is true that the model used a shear stress vs. opening relation monotonically increasing. If so, he pointed out that might explain why the computed stiffness was too high.

J. Blaauwendraad. No horizontal cut-off was used in the stiffness calculations.

From the floor. Were interlock effects accounted for?

J. Blaauwendraad. Yes. Pointed out that one needs release and slip in the crack model to get good agreement with experiments.

Mechanics of Geomaterials
Edited by Z. Bažant
© 1985 John Wiley & Sons Ltd

H. Summary of Discussions and Short Contributions on Modern Trends and New Directions (Part IX)

S. Sture

The discussions are summarized in the form of abbreviated principal questions and answers by symposium participants.

DISCUSSION OF 'CURRENT PROBLEMS AND NEW DIRECTIONS IN MECHANISMS OF GEOMATERIALS' BY Z. MRÓZ (CHAPTER 24)

A. Sawczuk emphasized that Mróz had given a very detailed presentation of issues related to parameter identification, theory, and experiments for solution of specific boundaries value problems. How should one proceed in the case of an unspecified boundary value problem? We need to identify analysis techniques in conjunction with experimental procedures and parameters also for non-specified problems.

Z. P. Bažant asked Mróz whether the formulation he had presented was for problems subjected to force or displacement control loading, and he argued that the given technique would be unstable for force control conditions.

Z. Mróz answered that the model is capable of describing both conditions. However, he emphasized that limit point processes cannot proceed in a purely static manner and that we are in reality considering a mixed boundary process. We need to consider motion in the course of characterizing material as well as structural degradation. Rate dependent damage models appear to be good candidates for assessing post-peak strength or strain-softening processes in structures, and we need new failure theories and methodologies for searching for limit points in static as well as dynamic analyses.

C. A. Anderson asked Mróz what the current state-of-the-art is for computing limits states.

Z. Mróz answered that the limit theorems in plasticity still seem to be the only feasible methodology. Yet when he discussed limit points for structures and materials, he really had bifurcation and instability phenomena in mind for both the local and global levels.

K. Willam asked Mróz if he has a preference between the so-called smeared approach or the discrete approach for analysing fracture phenomena in concrete and geomaterials.

Z. Mróz answered that he has not reached any conclusion whether to adopt one or the other approach. Their use is a matter of convenience, since he cannot see any fundamental arguments for adopting one before the other.

A. N. Schofield asked Mróz if pore water diffusion and generation in geomaterials posed any problem in the formulation of the theory he had presented.

Z. Mróz answered that he could not see any practical or analytical problems in incorporating classical Biot-type coupled pore water pressure–displacement analysis procedures as part of an incremental nonlinear material formulation. The generation and diffusion of excess pore fluid pressure within a solid skeleton would in fact provide a good tool for describing destruction on the material as well as structural level especially during dynamic processes.

A GENERAL DISCUSSION RELATED TO NUMERICAL MODELLING OF CAUSE–EFFECT RELATIONS IN CONCRETE AND GEOMATERIALS

A. W. Jenike stated that apparent viscous behaviour of flowing dry particles in hoppers, etc. is often exhibited because of varying pressure gradients in the air trapped in the voids between the grains. He then asked if anybody working in the areas of concrete and geomaterials has considered including transient gas pressures in the solid voids in their computations or if anybody has considered the solid to behave as a viscoelastic material in conjunction with Biot-analysis procedures.

Y. F. Dafalias responded that incorporating viscoelastic or plastic behaviour for the solid skeleton in Biot-formulation diffusion problems do not pose any problem. In fact many clays behave in this manner, and it would be entirely appropriate to include rate dependent models for the solid skeleton.

C. S. Desai posed the general question how to make an initially isotropic soil or particle assembly sample for laboratory testing. How do we evaluate the degree of isotropy or anisotropy in a laboratory specimen, and what are the minimum number of tests required to determine such features?

A. W. Jenike responded that perhaps Gudehus's triaxial device could be used for such purposes.

D. Krajcinovic stated that we should focus on the causes for nonlinearity in materials rather than readily adopt conventional plasticity procedures and formulations. He maintained that parameters such as porosity, etc. would be relevant internal variables for describing damage, compaction, and dilation. He also emphasized that we may be trying to get too much mileage out of plasticity theory, where such models in reality are not valid.

Y. F. Dafalias injected that the term plasticity derives from the Greek word for clay.

J. Uitterbogaard posed the general question how to select or develop the constitutive model that is relevant for a given boundary value problem. Would it be possible to device a procedure involving preliminary analyses of the problem where stress and strain paths are mapped out for certain regions or elements from which laboratory testing programs such subsequent parameters and models could be selected?

S. Sture responded that such procedures are currently in use in conventional geotechnical engineering under the name 'stress path methods'. However, these procedures are rather crude and involve linear or very simple nonlinear analyses. The objective is mainly to define mean stress level and straight line stress paths for conventional triaxial testing. Secant moduli are often extracted from such tests. Modern stress and strain path procedures have recently been adopted in finite element analyses in conjunction with laboratory experiment programs for defining parameters in elasto-plastic models as part of iterative schedules for refining the solution. The technique is as follows: a first nonlinear finite element analysis including a workable constitutive model and a preliminary assessment of parameters and load history is performed. The evolution of stresses and strains for all elements of importance during the load history is reviewed, and the responses are compared to available laboratory information. The model input information is updated for the elements that have 'strayed' too far away from experimental findings. The predictor–corrector procedure has 'converged' when the stress–strain responses in all elements in the analysis correspond to or are in close agreement with experimental information. The constitutive model will in this procedure act more like a response function rather than a general model.

C. S. Desai asked Krajcinovic how porosity alone can be used as a damage, etc. variable or measure. Would it not be more appropriate to have a tensorial or vector quantity representation? Porosity relates to strain?

D. Krajcinovic responded that the porosity certainly can change independently of strain and vice versa. Porosity can be an independent kinematic variable and it can be a tensor quantity.

L. Margolin wanted to reinforce the comments made by Krajcinovic. We need to consider the physical damaging processes in greater detail, such as grain slippage, dislocation, etc. When we consider dynamic processes we should also be aware that uniqueness may disappear. The problem becomes hyperbolic and not elliptical. On the material level we still do not understand how concrete behaves at high strain rates.

I. Sandler commented that rate considerations very often removes uniqueness. If strain-softening behaviour is considered as a rate independent phenomenon, we may still be considering an ultra-hyperbolic problem and not an elliptical problem. He asserted that for rate independent softening problems there are not the remotest resemblance of uniqueness available.

A. Sawczuk concluded the general discussion by saying that we need to examine more closely the continuum assumptions we make with regard to material model definition. We should not only consider plasticity theories but also more general damage or fracture theories.

SUMMARY AND DISCUSSION OF A. SAWCZUK'S SHORT CONTRIBUTION ON MODELLING OF NONLINEAR MECHANICAL RESPONSE AND FAILURE OF SOLID

A. Sawczuk presented a constitutive theory based on polynomial representations of tensor-valued tensor functions and their applications to modelling creep, plasticity, oriented damage, and fissurization in materials. He reviewed earlier work involving istropic formulations, and he discussed how an isotropic tensor-valued tensor function of one tensor variable has three basic invariants and three so-called tensor generators. Any isotropic elastic or inelastic constitutive law can be developed by this procedure. He then went on to derive anisotropic formulations and requirements, where vectors or tensors characterizing cause–effect features with preferred orientation in the material are incorporated. The number of tensor generators increases with the complexity of the problem and mixed or co-invariants between stresses and strains evolve from the formulation. The representations are in this case quite involved. He specified the problem of inherent material transverse anisotropy and subsequent yielding behaviour controlled by the initial anisotropy and not stress induced anisotropy. The angle of inclination between the principal stresses and the preferred material fabric was included in the theory. It was also demonstrated that the plastic flow law is not necessarily associated with a yield condition and that the general forms of inelastic behaviour identified in the constitutive equations did not incorporate the concept of yield surfaces. However, anisotropic strain-hardening rules need to be incorporated which are representative of the specific loading case. It is not clear how this feature interacts with the stress and strain joint invariant formulation in the absence of a general kinematic rule. He also demonstrated how the theory could be used to characterize anisotropic creep behaviour. The general theory is restricted to small deformation theory. Porous materials with channels were also considered, and permeabilities were formulated based on orthogonal and simple tetrahedral channel structures. The variables in this coupled case and the equations resemble those of classical Biot-mixture theory. Sawczuk discussed how the general tensor function representation theory could be used to chracterize coupled thermal, stress, deformation, and fluid pressure phenomena in concrete and geomaterials. He concluded that tensorial functions are ideally suited to account for internal structure and nonlinearities in materials.

I. Sandler asked Sawczuk how stress and strain invariants can be used to characterize anisotropy in materials. How can such scalar representations reflect oriented effects?

C. S. Desai stated that he had also incorporated stress invariants, plastic strain invariants and co-invariants to describe inherent as well as stress-induced anisotropies in geomaterials.

S. Sture asserted that joint or co-invariant constitutive formulations may work for inherently anisotropic cases, but there is no directional feature such as a kinematic rule incorporated in such models that can properly control stress induced anisotropic features.

A. Sawczuk responded that he could not see why there would be any problem using such joint invariant expressions for both cases of anisotropy.

SUMMARY AND DISCUSSION OF P. MARTI'S SHORT CONTRIBUTION ON PLASTIC ANALYSIS OF STRUCTURAL CONCRETE

P. Marti discussed issues related to limit analysis methods and elasto-plastic finite element procedures for design and analysis of reinforced and prestressed concrete structures and structural components. Present knowledge and the state-of-the-art analysis procedures were reviewed, and it was stated that not very much progress has been made in the use of inelastic methods in engineering practice. There is definitely a need to improve our capability of analysing the behaviour of reinforced concrete and specifically interactions between concrete, reinforcement, ties, etc. The load transfer mechanisms and fracture behaviour of reinforced concrete components such as deep beams are far from fully understood. He also questioned the validity of applying conventional continuum mechanics concepts even outside the fracture zones in concrete and even between bond-slip regions. The relevance of using limit plasticity procedures for both under- and over-reinforced concrete members was discussed, and it was argued that these methods are simple to use in practice because of their transparency and adaptability. It was emphasized that the upper-bound method is best suited for analysing existing designs, and it is a very good tool for interpreting experimental findings. For elastic–plastic modelling of concrete-reinforcing composites four groups of constitutive relations are needed for characterizing concrete, reinforcement, aggregate interlock between concrete and concrete bonded to the reinforcing system, and aggregate interlock within fractured concrete. It was maintained that reliable constitutive formulations exist only for the reinforcing steel at this point. Much further analytical and experimental work needs to be carried out before we can do appropriate elastic–plastic finite element analyses of reinforced or prestressed concrete components. Continuum and fracture mechanics concepts are not yet 'married' into a unified constitutive theory, and he questioned if we ever will be able to join them. Shall we ever be able to 'smear' the constitutive behaviour of structural concrete with confidence?

Z. P. Bažant expressed total agreement with Marti. He pointed out that Marti did not consider size effects in his discussion. The issue of no size effect vs. wrong size effect was raised. The example of punching shear mechanisms in concrete was introduced to demonstrate that conventional elasto-plastic procedures only give appropriate solutions when f'_t is approximately equal to or less than 0.5 per cent of f'_c which is totally unrealistic and unacceptable relative to code description. Size effects would be quite significant in many analyses; beams have one size and PCRVs have another size and plasticity theory may yet not apply unless steel definitely governs failure.

P. Marti expressed himself in disagreement if Bažant meant to include limit theorem plasticity as part of the size effect issue. He agreed that elastic–plastic constitutive formulations at this stage have limited applicability in current design and analysis work.

SUMMARY AND DISCUSSION OF M. W. BRAESTRUP'S SHORT CONTRIBUTION ON PLASTIC ANALYSIS OF STRUCTURAL CONCRETE

M. W. Braestrup reviewed very conventional plasticity concepts including the classical Mohr–Coulomb condition as it applies to plane stress, plane strain, and fully three-dimensional concrete structure components. He discussed the apparent need to depart from associated plasticity considerations in order to prevent excessive dilatation or contraction. The argument for a curved ultimate strength locus for the Mohr–Coulomb condition was also presented.

Z. P. Bažant, M. W. Braestrup, J. W. Dougill, and *J. Warner* then discussed the influence of stirrups on concrete ductility as well as on post-peak response. It was agreed that strain-softening in tension is a relatively 'slow' process, whereas strain-softening in shear is quite unpredictable and 'sudden'.

List of Symposium Participants

Anderson, C. A., Los Alamos National Laboratory, MS-J576, Los Alamos, New Mexico 87545.
Astill, C. J., National Science Foundation, MEAM/Solid Mechanics Program, Rm. 1108, 1800 G. Street, N. W., Washington, D.C. 20550.
Atkinson, B., Imperial College of Science and Technology, Geology Department, London SW7 2BP, England.
Badillo, E. Juarez, Graduate School of Engineering, National University of Mexico, Tepanco 32, Coyacan, 04030 Mexico, D.F., Mexico.
Ballesteros, P., Privada de Los Cedros 98, Mexico 01720, D. F., Mexico.
Bažant, Z. J., Visitor, Northwestern University; Permanent Address: Prof. Emeritus, Tech. Univ. (ČVUT), Thákurova 7, 16629 Praha 6, Czechoslovakia.
Bažant, Z. P., Department of Civil Engineering, Northwestern University, Evanston, Illinois 60201 (Director, Center for Concrete and Geomaterials).
Babendreier, C. A., Geotechnical Division, Civil Engineering Program, National Science Foundation, 1800 G. Street, N.W., Washington, D.C. 20550.
Belytschko, T. B., Department of Civil Engineering, Northwestern University, Evanston, Illinois 60201.
Bergan, P. G., Division of Structural Mechanics, The Norwegian Institute of Technology, Institutt for Statikk, N-7034 Trondheim-NTH, Norway.
Blaauwendraad, J., Rijkswaterstaat Structural Research, Postorder Box 20000, 3502 La Utrecht, The Netherlands.
Boley, B. A., Dean, The Technological Institute, Northwestern University, Evanston, Illinois 60201.
Braestrup, M. W., Department of Structural Engineering, Technical University of Denmark, Building M8, DK-2800 Copenhagen Lyngby, Denmark.
Buyukozturk, O., Department of Civil Engineering, Massachusetts Institute of Technology, 77 Massachusetts Avenue, Cambridge, Massachusetts 02139.
Cao, Zhiping, Designing Institute of Yellow River Water Conservancy Commission, Zhengzhou, China.
Carroll, M., Department of Mechanical Engineering, University of California, Berkeley, California 94720.
Cedolin, L., Politecnico di Milano, Departimento di Ingegneria Strutturale, Piazza Leonardo da Vinci 32, 20133 Milano, Italy.
Červenka, V., Na Hřebenkách 55, 1500 Praha 5, Czechoslovakia.
Chen, W. F., School of Civil Engineering, Purdue University, West Lafayette, Indiana 47907.

Chiorino, M. A., Politecnico di Torino, Scienza-Construzioni, Facolta Architettura, Viale Mattioli 30, 1-10126 Torino, Italy.

Cleary, M. P., Massachusetts Institute of Technology, 77 Massachusetts Avenue, Cambridge, Massachusetts 02139.

Cowin, S. C., Department of Biomedical Engineering, Tulane University, New Orleans, Louisiana 70118.

Crawford, A. M., Department of Civil Engineering, University of Toronto, 35 St. George Street, Toronto, Ontario, Canada, M5S 1A4.

Cuellar, V., Laboratorio del Transporte, Mechanica del Suelo, Alfonso XII, 3 Madrid–7, Spain.

Dafalias, Y. F., Department of Civil Engineering, University of California at Davis, Davis, California 95616.

Darve, F., Institut de Mécanique de Grenoble, B.P. 68, 38402 Saint Martin D'Heres Cedex, France.

Darwin, D., Department of Civil Engineering, University of Kansas, Lawrence, Kansas 66045.

Desai, C. S., Department of Civil Engineering and Engineering Mechanics, University of Arizona, Tucson, Arizona 85721.

Dougill, J. W., Department of Civil Engineering, Imperial College of Science and Technology, Imperial College Road, London SW7 2BU, United Kingdom.

Dowding, C. H., Department of Civil Engineering, Northwestern University Evanston, Illinois 60201.

Drescher, A., Institute of Fundamental, Technological Research, Warsaw, Poland, Visiting Professor, University of Minnesota, Minneapolis, Minnesota.

Drucker, D. C., College of Engineering, University of Illinois, 106 Engineering Hall, 1308 W. Green Street, Urbana, Illinois 61801.

Dundurs, J., Department of Civil Engineering, Northwestern University, Evanston, Illinois 60201.

Ehrlacher, A., Laboratoire de Mécanique des Solides, Ecole Polytéchnique, 91128, Palaiseau Cedex, France.

Galloway, R. G., Air Force Weapons Laboratory, Air Force Systems Command, Kirtland Air Force Base, New Mexico 87117.

Gambarova, P., Politecnico di Milano, Departimento di Ingegneria Strutturale, Piazza Leonardo da Vinci 32, 20133 Milano, Italy.

Gerstle, K. H., Department of Civil Engineering, Campus Box 428, University of Colorado, Boulder, Colorado 80309.

Ghaboussi, J., Department of Civil Engineering, University of Illinois at Urbana-Champaign, 2129 Civil Engineering Bldg., Urbana, Illinois 61801.

Ghosh, S. K., Portland Cement Association, 5420 Old Orchard Road, Skokie, Illinois 60077.

Goldberg, J., Program Director, Department of Civil Engineering, Rm. 1130, National Science Foundation, 1800 G. Street, N. W., Washington, D.C. 20550.

Grady, D., Sandia National Laboratories, Division 1534, P.O. Box 5800, Albuquerque, New Mexico 87185.
Gudehus, G., Institut für Bodenmechanik und Felsmechanik, D7500 Karlsruhe, 1, Richard-Willstatter-Allee, Postfach NR 6380, West Germany.
Higgs, N. G., Amoco Production Company, P.O. Box 591, Tulsa, Oklahoma 74102.
Hillerborg, A., Division of Building Materials, Lund Institute of Technology, Box 725, S-22007 Lund, Sweden.
Hodge, P. G., Department of Mechanics, University of Minnesota, Minneapolis, Minnesota 55455.
Holcomb, D. J., Sandia National Laboratories, Division 1542, Albuquerque, New Mexico 87185.
Horrigmoe, G., Sivilingenior Ravlo Multiconsult A.S., Dronningens GT51, P.B. 381, 8501 Narvik, Norway.
Ingraffea, A. R., School of Civil Engineering, Hollister Hall, Cornell University, Ithaca, New York 14853.
Jaeger, Z., Israel Atomic Energy Commission, Soreq Nuclear Research Center, Yavne 70600, Israel.
Jenike, A. W., Jenike and Johanson, Inc., 2 Executive Park Drive, North Billerica, Maine 01862.
Jeter, J. W., Theoretical Analysis Division, New Mexico Engineering Research Institute, Campus P.O. Box 25, The University of New Mexico, Albuquerque, New Mexico 87131.
Jonasson, Jan-Erik, Swedish Cement and Concrete Research Institute, S-10044, Stockholm, Sweden.
Katsube, N., Dept. of Engineering Mechanics, Boyd Laboratory, Ohio State University, 155 W. Woodruff Avenue, Columbus, Ohio 43210.
Keer, L. M., Department of Civil Engineering, Northwestern University, Evanston, Illinois 60201.
Krajcinovic, D., Department of Civil Engineering & Mechanics, University of Illinois at Chicago, Box 4348, Chicago, Illinois 60680.
Krizek, R. J., Chairman, Department of Civil Engineering, Northwestern University, Evanston, Illinois 60201.
Kusters, G., TNO–IBBC, P.O. Box 49, 2600 AA Delft, The Netherlands.
Lazić, J. D., Gradjevinski Fakultet Postanski fah 895, Bulevar revolucije 73/1, 11000 Beograd, Yugoslavia.
Lee, Soo-Gon, Department of Architecture, College of Engineering, Chonnam National University, Gwangju, Korea.
Levine, H., Weidlinger Associates, Suite 155, Building 4, 3000 Sand Hill Road, Menlo Park, California 94025.
Li, V., Department of Civil Engineering, Massachusetts Institute of Technology, Cambridge, Massachusetts 02139.

Lippmann, H., Lehrstuhl für Mechanik, Technische Universität, Arcisstrasse 21 D-8000 Munchen 2, West Germany.

Madsen, H. O., Det Norske Veritas, P.O. Box 300, N1322 Høvik, Oslo, Norway.

Maekawa, K., The Technological University of Nagaoka, Nagamine 1630-1, Kamitomioka, Nagaoka, Niigata, Japan.

Malvern, L., Department of Engineering Sciences, University of Florida, Gainsville, Florida 32601.

Marchertas, A., RAS (Division of Reactor Analysis and Safety), Building 208, Argonne National Laboratory, 9700 S. Cass Avenue, Argonne, Illinois 64039.

Margolin, L., Los Alamos National Laboratory, P.O. Box 1663, Los Alamos, New Mexico 87545.

Marti, P., Department of Civil Engineering, University of Toronto, Toronto, Ontario M5S 1A4, Canada.

Masur, E., Chairman, Department of Civil Engineering, Mechanics & Metallurgy, University of Illinois at Chicago, Box 4348, Chicago, Illinois 60680.

Meyer, C., Department of Civil Engineering, Columbia University, New York, New York 10027.

Mróz, Z., Institute of Fundamental Technological Research, Polish Academy of Sciences, Swietokrzyska 21, 00-049 Warsaw, Poland.

Mura, T., Department of Civil Engineering, Northwestern University, Evanston, Illinois 60201.

Murray, D. W., Department of Civil Engineering, University of Alberta, Edmonton, Alberta, Canada.

Nilsson, L., Division of Computer Aided Analysis and Design, Luela Technical University, S-95187 Lulea, Sweden.

Okamura, H., Department of Civil Engineering, University of Tokyo, 7-3-1 Hongo, Bunkyo-ku, Tokyo 113, Japan.

Orkisz, J. Institut Mechaniki Budowli, Politechnika Krakowska, Warszawaka 24, 31155 Krakow, Poland.

Ottosen, N. S., Risø National University, P.O. Box 49, DK-4000, Roskilde, Denmark.

Ouchterlony, F., Swedish Detonic Research Foundation, P.O. Box 32058, S-12611 Stockholm, Sweden.

Pan, Xuewen, China National New Building Materials Corporation, Zi Zhu Yuan Road, Xi Jieo, Beijing, China.

Papadopoulos, J., Department of Mechanical Engineering, Massachusetts Institute of Technology, 77 Massachusetts Avenue, Cambridge, Massachusetts, 02139

Pister, K. S., College of Engineering, University of California, Berkeley, California 94720.

Powell, G., Division of Structural Engineering and Structural Mechanics, University of California, Berkeley, California 94720.

Appendix 601

Prevost, J. H., Department of Civil Engineering, School of Engineering/Applied Science, Princeton University, Princeton, New Jersey 08544.

Reinhardt, H. W., Department of Civil Engineering, Delft University of Technology, Stevinweg 4, NL-2628 CN, Delft, The Netherlands.

Rice, J. R., Division of Applied Sciences, Pierce 224, Harvard University, 29 Oxford Street, Cambridge, Massachusetts 02138.

Rudnicki, J. W., Department of Civil Engineering, Northwestern University, Evanston, Illinois 60201.

Ruina, A., Theoretical and Applied Mechanics, Thurston Hall, Cornell University, Ithaca, New York 14853.

Saada, A. S., Department of Civil Engineering, Case Western Reserve University, University Circle, Cleveland, Ohio 44106.

Sandler, I., Weidlinger Associates, 333 Seventh Avenue, New York, New York 10001.

Sawczuk, A., Laboratoire de Mécanique et d'Acoustique 31, Chemin Joseph-Aiquir, B.P. 71, 13277 Marseille, Cedex 9, France.

Schofield, A., University of Cambridge, Department of Engineering, Trumpington Street, Cambridge CB2 1PZ, United Kingdom.

Shah, S. P., Department of Civil Engineering, Northwestern University, Evanston, Illinois 60201.

Stout, R. B., Earth Sciences Department, Lawrence Livermore National Laboratory, P.O. Box 808, L-200, Livermore, California 94550.

Sture, S., Department of Civil Engineering, University of Colorado, Campus Box 428, Boulder, Colorado 80302.

Teufel, L. W., Sandia National Laboratories, Geomechanics Research 5532, P.O. Box 5800, Albuquerque, New Mexico 87185.

Ting, E., School of Civil Engineering, Purdue University, West Lafayette, Indiana 47907.

Ting, T. C. T., Department of Civil Engineering, Mechanics and Metallurgy University of Illinois at Chicago, Box 4348, Chicago, Illinois 60680.

Uittenbogaard, R. E., Department of Dredging Technology, Delft Hydraulics Lab. Rotterdamseweg 185, P.O. Box 177, 2600 MH Delft, The Netherlands.

Van Mier, J., University of Technology, Den Dolech 2, P.O. Box 513, 5600 MB Eindhoven, The Netherlands.

Vardoulakis, I., Department of Civil Engineering, University of Minnesota, 500 Pillsbury Drive, S. E., Minneapolis, Minnesota 55455.

Wang, Ming-Liang, # 56, Lee-Chan St., San-Min District, Kaoshiung, Taiwan, Rep. of China.

Wang, Tong-Sheng, The Huai River Commission of the Water Conservancy and Water Power Ministry, Bangbu, Anhui, China.

Warner, R. F., Department of Civil Engineering, University of Adelaide, Box 489 G. P. O. Adelaide 5001, South Australia.

Wawersik, W. R., Organization 5163, Sandia Laboratories, Albuquerque, New Mexico 87115.
Wecharatana, M., Department of Civil and Environmental Engineering, New Jersey Institute of Technology, 323 High Street, Newark, New Jersey 07102.
Weertman, J., Department of Material Science, Northwestern University, Evanston, Illinois 60201.
Willam, K., Department of Civil Engineering, University of Colorado, Campus Box 428, Boulder, Colorado 80309.
Wittmann, F. H., Ecole Polytéchnique Féderale de Lausanne, Départment des Matériaux, 32, ch. de. Bellerive, CH-1107 Lausanne, Switzerland.
Wnuk, M., Department of Civil Engineering, College of Engineering and Applied Science, The University of Wisconsin–Milwaukee, Milwaukee, Wisconsin 53201.
Wong, T.-F., Department of Earth and Space Sciences, State University of New York at Stony Brook, Long Island, New York 11794.
Wu, H. C., Division on Materials Engineering, The University of Iowa, Iowa City, Iowa 52242.
Ziegler, F., Institut Für Allgemeine Mechanik, Technische Universität Wien, A 1040 Wien, Karlsplatz 13, Austria.
Zienkiewicz, O. C., Department of Civil Engineering, University College of Swansea, Singleton Park, Swansea SA2 8PP, West Glamorgan, United Kingdom.

List of Contributed Papers Presented at the Symposium*

Constitutive Equations for Concrete Under Plane Stress Conditions	K. Maekawa and H. Okamura
Endochronic Description of Sand Behaviour	H. C. Wu
Remarks on Some Incrementally Non-linear Constitutive Relations	F. Darve
Constitutive Relation for Rock-like Material, Based on a Shear Crack Model	H. Lippmann
Deformation and Thermodynamic Response During Brittle Fracture	R. B. Stout
Computational Stability and Uniqueness of Strain-softening Materials	K. J. Willam
Analytical Models for Cracking of Concrete Subject to Shear	P. G. Gambarova
Crack Growth in Cement Based Composites	R. Ballarini, S. P. Shah, and L. M. Keer
Examples of Practical Results Achieved by Means of the Fictitious Crack Model	A. Hillerborg
Vibrothermography of Granular Soils	M. P. Luong
Gravity Flow of Bulk Solids	A. U. Jenike
Mechanical Modelling of the Shrinkage and Swelling of Porous Soils	S. C. Cowin
Stress–Strain Curves of Concrete Under Multiaxial Load Histories	K. H. Gerstle and H. Y. Ko
Plain Concrete in Uniaxial Post-peak Cyclic Tensile and Tensile Compressive Loading	H. W. Reinhardt

*Published in Preprints, W. Prager IUTAM Symposium on :'Mechanics of Geomaterials : Rocks, Concretes, Soils', ed. by Z. P. Bažant, Northwestern University, Evanston Illinois, September 1983 (available from NTIS, 5285 Port Royal Road, Springfield, Va. 22161, USA).

The Use of the Rough Crack Model of Walraven and the Fictitious Crack Model of Hillerborg in F. E. Analysis	J. Blaauwendraad
Plasticity-based Analysis of Reinforced Concrete Structures	E. C. Ting and M. Yener
A Generalized Basis for Modelling Plastic Behaviour of Materials	C. S. Desai
A Modelling of Nonlinear Mechanical Response and Failure of Solid	A. Sawczuk
Plasticity in Reinforced Concrete—Potential and Limitations	P. Marti
Plastic Analysis of Structural Concrete	M. W. Braestrup

Index

Activation energy 200, 435
Activation volume 437
Admixtures, air-entraining 362
Ageing 464
    coefficient 408
    effects 407
    effects in rocks 412
    model, linear 407
    viscoelastic solids 410
Aggregate 110
Anisotropy 68
Apparent volume fraction of voids 317
Arrhenius equation 372

Bangham equation 442
Benchmark problems 551
Biaxial conditions 72
Bifurcation 61, 102
    surface 547
Bilinear failure envelope 264
Bilinear strain path 84
Biot theory 316–321
Biot type formulation 472
Blunt crack band model 257–270
Boltzmann statistics 148
Bond slip length 274, 275, 298
Bond slip of reinforcement 274–275
Boundary conditions 171
Boundary element fracture analysis program 243–245, 309
Boundary values problems 79, 228, 551
Boussinesq approximation 414

Brazilian test 309
Brittle
    failure of concrete structures 278
    fracture 72, 308, 311
    material 71
    state 71
Bulk coefficient of expansion 414
Bulk modulus 318, 472, 504
    of the skeleton 473

Cap model 79, 513
Capillary
    condensation 443
    layer temperature 384
    pores 350
Cauchy stress 463
Cavity 62
    boundary 332
Cement paste 99
CENRBB 220, 221
Characteristic
    length 289, 290
    time 328
Clays, overconsolidated 335
Colloidal system 430
Compliance elastic matrix 264
Composite material 448, 462
Consistency conditions, Prager's 32, 34, 36
Constitutive
    modelling 66
    relations 21, 47, 53, 62, 64, 74, 169
Continuity condition 546

Continuous damage mechanics 264
Continuum modelling 157
Convergence with mesh refinement 262, 280
Convexity condition 68, 546
Coulomb–Mohr yield surface 42, 80
Coupling force 439
Crack
  band front width 265, 266
  band model 257–270
  branches 110
  concrete 272, 274
  curvilinear 228
  elastic brittle 214
  extension parameter 293
  formulation 72, 451
  inclined 116
  increment length 239–243
  iniation 94
  interfacial 121
  length 297
  microcrack 22
  Modes (I, II) 90, 98, 115, 117, 119, 197
  numerical modelling 227–243
  opening 104
  opening nodal displacements 234
  path 119
  pattern 121
  pre-existing 161
  propagation 106
  random 89
  rock 229–235
  shear 298
  spacing 298
  speed 330
  stability 292
  tip process zone 225
  zone 482
Cracked zone area 275
Cracking
  drying induced 370
  parameter 264
  pore pressure induced 306
  smeared 260
  tensile 480
Creep 75, 464
  basic 369
  concrete 369, 386
  corrosion 386
  effects in rocks 412
  function 407
  laws 371

  long time 435
  short time 434
Creep function, specific 455
Critical
  load 91
  surface 484
Cut-off criterion 72
Cyclic
  instability 560
  load 51, 69, 486
  mobility 486

D'Alembert principle 282
Damage 27, 36, 99
  parameters 158
Darcy's law 318
Dashpot 479
DC(T) 220
Degradation 35, 105
  progressive 40
Desorption isotherm 355
Deterioration 490
Deviatoric plane 68
  stress, strain 105
Diagonal shear failure of beams 279
Difference operator 7, 288
Diffusion
  coefficient 440, 449
  effects, coupled deformation 328, 335
  elements 372
  equation 320
  length 330
  pore fluid 307
  theory, linear 352
  theory, nonlinear 359, 362
  time 326
Dilatancy 102
  factor 336
Dilatant hardening 337, 338
  effects on shear fracture 341
  inclusion model 339
  stability 338
Dilation 56, 102
Dimensional analysis 292, 296
Dipole dislocation representation 308
Dirichlet series 408
Discrete equations 475
Discretization, spatial 368, 529
Disjoining pressure 430, 439, 443
Dislocation
  plane strain shear 322
  steadily moving 327

Displacement
  discontinuity 322, 325
  skeleton 473
Double-power law 438
Drained response 324, 325
Drive stress 413
Drucker–Prager model 79
Drying, accelerated 363
Drying of hardening cement paste 439
Ductile state 71
Dufour effect 364, 365
Dynamic loading 161, 501
Dynamics, saturated geomaterials 379, 383

Earthquake 78
Effective stress 311, 336, 472
Einstein equation 432
Elastic behaviour 41
Electromechanical effect 434
Elliptic equation 211
Elsasser equation 397
Endochronic 480
  theory 71
Energy
  balance 141
  criterion 72
  criterion, in LEFM 278
  dissipation 25, 34, 39
  interfacial 430, 442
  release rate 231, 265, 269, 276, 291, 293
  transport equation 414
Entropy production inequality 465
Equilibrium 49, 76
  water content 352
Equivalent hydration period 410
Eshelby relation 332, 340
Explicit method 287, 470

Failure
  criterion 94, 134
  damage, dynamic 134, 157
  energy 286
  length, finite 329, 331
  of concrete 89
  post 73
  process 115
  rule 34
  sequential 557
  single 557
  stress, nominal 276
  surface 68, 490
  surface of Drucker–Prager type 80

  tensile 480
  zone 211, 325
Faulting 211, 575
Fick's first law 353
Fick's second law 354
Finite difference equation 284
Finite element
  fracture analysis 260, 261
  mesh 274
  program 243, 245, 250, 270–275
Fire resistant 363
Fissures 104
Flow rule 34
  associated 36, 504
Fluid
  infiltrated geomaterial 581
  infiltrated porus solid 316
  mass dipole 332
  mass flux 316, 332
Fluidity parameter 494, 495
Fracture
  analysis 309–311
  Boltzmann limit 147
  concrete 259
  criteria, dynamic 129, 134
  data for concrete 266–70
  discussion 305–311
  ductile 146
  dynamic 129, 137, 228, 263
  elastic 131
  elastic plastic 311
  energy 577
  equivalent linear analysis 291
  hydraulic 242, 331
  insensitiveness 271
  loading function 105
  mechanics 105, 219, 221
  micro scale 306, 307, 308, 309
  mixed mode 235–239
  process zone 263, 266, 272
  program 243–252
  progressive 36
  propagation 577
  propagation modelling 243–244
  quasi-static analysis 237
  rock 219
  surface 37
  testing 219–227
  toughness 133, 139, 237
Fragmentation
  catastrophic 136, 137
  dynamic 157

Free energy function 464
Friction
  law 175
  static–dynamic 175
Frictional
  sliding 194, 339
  slip 211, 575
  surfaces 325, 335

Galerkin procedure 367
Gas constant, universal 410, 413
Geomaterials 385
Geomechanics 471
Gibbs free energy 354
Gibbs thermodynamic relation 381
Granular soil 49
Green's function 131
Griffith criterion 67, 74
Griffith theory 150

Hardening 559
  Drucker–Prager's 32, 33, 35
  function 80
  isotropic 36, 42
  kinematic 518
Heat and moisture transfer 362, 363, 364, 367
Heat capacity 414
Heaviside tensile stress 131
Helmholtz function 317
Hertzian spectroscopy 429
Heterogeneity 89
Hierarchic system 426
High rate stress 129
High strength concrete 119
Hindered absorbed layers 351
Homogeneous matrix 119
Homogenization 159, 160
Hooke's law 28, 33, 35, 75
Hopkinson bar test 143
Hydrostatic
  axis 68
  stress 336
Hygral length 454
Hygrothermal coefficient 358
Hyperelasticity 69
Hypoelasticity 30, 77

Il'iushin's postulate 35, 36, 38
Imbricate continuum 281, 285, 287
Imbricate elements 281
Imperfections 338

Implicit method 476
Inclusion 118, 431
  elliptic 335
Inelastic
  deformation 315
  hardening modulus 336
  properties 89
  reponse of rock 335
Instability
  local 211
  runaway 333
  rupture 305
Intercrystallite pore 428
Internal variable theory 323
Intracrystallite pore 428
Intrinsic flaws 307
Invariants, stress, strain 66
Isothermal
  drying 353, 356, 357, 362
  sorption 351, 352
Isotropic
  compression 156
  material 29
  materials, linear 309

J-integral 272, 311, 578
  critical values 223
Joints 482

Kelvin chain model 408
Kinematical condition 51
Kinetic energy 137, 138

Light weight concrete 119
Linear surface 547
Lithosphere structure 391
Loading
  function 30, 34, 40
  modes 53, 102
  quasi-static 342
Local elements 382
Localization 100, 102, 189, 190, 202, 211, 212

Macrolevel 433
Mass flux 353
Material
  constants 60
  response 305
Material engineering model 426
Material science model 426
Maturity 465

Maxwell
  chain model 408
  model 371
  relation 318
Mean strain 284
Mesh refinement 280–291
Mesolevel 431
Metamorphism 200, 204
Micro-cracks 72, 89, 98, 105, 263, 297
Microlevel 427
Microstructure 40
Mixture theory 316
Moduli for stress analysis 305
Mohr–Coulomb
  model 483
Moiré interferometry 263
Moisture
  concentration 364
  flux 364
  migration 427
Monotonic paths 53, 84
Monte Carlo method 106, 158, 161
Mössbauer effect 429
Multi-laminate model 480
Munich model 429, 431, 443

Newton–Raphson
  iteration 532
  method 62
Newtonian viscous 201
Nondimensional parameter for crack band failures 276
Nonlinear elasticity 480
Nonlocal continuum theory 281, 291
Notched annular core 311
No-tension material 38
Numerical modelling 471, 587
Nusselt number 418

Octahedral stresses 29, 68
Overlay model 480

Particle size 136
Permeability
  coefficient 474
  scalar 319
  tensor 318
Phase interaction force 380
Plane strain earth pressure 49
Plastic
  deformation 31, 43
  flow 75
  loading 211
Plasticity
  boundary surface plasticity 490
  dilatant 337, 386, 388
  endochronic 65
  generalized 490
  models 70
  non-associative 102, 484
  perfect 34, 42, 76
  potential 504
  reflective surface 490
  soil 75
  strain-hardening 79, 484
Poisson statistics 146
Poisson's ratio 319, 322
  model 509
Pore
  compaction 432
  fluid pressure 317, 320
  humidity 349, 356
  space 316
  water 349, 350, 352
Pores 431
Porosity 388
Porous materials, saturated 343
Post-failure range 66, 73
Potential
  energy function 27
  energy function, complementary 27
  energy release 275, 276
Powers model 427
Precursor time 326, 333, 335, 339, 340
Principal stress, strain 48
Probabilistic concept of crack 94
Proportional path 48, 53

Quarter-point singular elements 252
Quasi-brittle material 150
Quasi-static problems 158

Random distribution 160
Rankine criterion 67, 74
Rate
  effect 73, 490
  energy 32
  theory 435
Rate-type formulation 408
Rayleigh
  number 415
  number convection 412
  wave 163
Rayleigh–Bernard convection 413

R-curves 267, 270, 291, 293, 294, 295, 297
Reinforcement
  role in fracture 274
  effect of bond slip 275
Relaxation
  function 408
  method 477
  rate 547
  time 409
Remeshing, automatic 244
Representative volume 280, 289
Retardation time 408
Rheological, linear 467
Rock 65, 189
  behaviour 104
  rheology 160
  strength, dynamic 134

Saturated
  clay 49, 51
  soil 85
Scale effect 463
Secant modulus 6, 281
Second-order work 35
SECRBB 220, 221
Shear
  band 554
  compaction 335
  crack 329
  dislocation 326
  localization 169, 189, 211, 575
  modulus 105, 504
Short rod testing system 222–227
Shrinkage 441, 464
  concrete 369
  hardened cement paste 441
  unrestrained 443
Similitude for crack band failures 276
Size effect, structural 275, 280
Size effect law 276–277
Slider 479
Slip 172, 335
  planes 41
  steady 179, 183
  surface 49, 170, 172, 335
  surface, fluid infiltrated 326
Softening
  geometric 200
  isotropic 39
  modulus 264, 265, 271, 276
  strain 200

strain rate 200
structural 200
thermal 200
zone 481
Soil 65, 74
  mechanics 47, 75
Soret effect 364
Specific surface 429
Specimen behaviour, non-uniform 370
Spherical cavities 332
Spherical inclusion 340
Splitting force 224
Stability 7, 211
  dilatant hardening 338
Stabilization 325
Stiffness matrix 264
Strain
  elastic, plastic 30
  energy density 310
  gradient 288
  path 49, 53
Strain softening 263, 272, 284, 290
  modulus 270
Strength 120
  criterion 72
  ratio effect 233
  static 132
Stress
  broad-range 288
  deviator 504
  distribution 305
  path 51, 53, 55, 57
  rate 36
  total 289
Stress intensity factor 131, 330
  computation 234, 235
  critical value 222, 226
  effective 225, 242
  normalized values 222, 226
Sublayer model 480
Superposition, principle of 369
Surface
  energy 135
  energy density 138
  tension 429
Swelling factor, geomaterial 384

Tangent
  modulus 284
  stiffness 531
Taylor's model 429

Index

Tectonic deformations 189, 339
Temperature
  effects, modelling 410
  high 363
Tensile strength
  equivalent 270, 271, 275
  limit 271
Thermal
  conductivity 414
  diffusivity 415
  effects 405, 412
Thermodymanics 139
Time-dependent effect 75
TL-approach 426
Toughness, effective 225, 242
Transient loading 51, 486
Transition probability 94
Tresca criterion 67
Triaxial
  apparatus 52
  behaviour, nonlinear 569
  compression 29
  extension 52
  test 48, 66, 69
  test, undrained 56
Trimethylsilylation 428

Undrained condition 326, 337
  deformation 318
Uniaxial
  compression 56, 69, 107, 115
  tension 98, 109

test 48
Uniform stress–strain 159
Unloading 30

Viscoelastic response 342
Viscoelasticity 65, 75, 480
Viscoplasticity 480
Viscosity 50, 85
  effective 418
  kinematic 415
  soil 58
Volume strain 105
Von Mises criterion 504

Water
  absorbed 350
  capillary 350
  chemically bound 350
Wave
  propagation 320
  tectonic 396
Westergaad's solution 261
Work 25
Working–hardening model 70, 76

Xerogel 427

Yield surface 31, 32, 34, 41, 484, 503
Yielding criteria, Drucker–Prager 505

Zigzag crack band 273
Zigzagging history 534